Let Food Be Your Medicine, and Medicine Be Your Food

Scientific Insights into the Healing Power of Nutrition

nova science publishers

www.novapublishers.com

Food Science and Technology Series

Natural Pigments: Sources, Properties, and Health Benefits
Rashidi Othman, PhD,
Qurratu Aini Mat Ali, PhD,
Razanah Ramya, PhD,
Farah Ayuni Mohd Hatta, PhD,
Wan Syibrah Hanisah Wan Sulaiman, PhD,
and Nur Hanie Mohd Latiff, PhD
ISBN: 979-8-90134-055-4
eBook ISBN: 979-8-90134-119-3

Advances in Meat Alternatives: Principles, Engineering, and Safety Challenges
Amin Abbasi, PhD, Amin Paydareh,
Irandokht Vakili, Melika Dashab,
Mahdis Nadian, Fahimeh Ebrahimi, PhD
and Hedayat Hosseini, D.V.M., PhD
ISBN: 979-8-89530-714-4
eBook ISBN: 979-8-89530-801-1

Xanthan Gum: Science, Technology, and Applications
Hedayat Hosseini, Amin Abbasi, PhD,
Amin Paydareh, Negin Kamali,
Tina Jamshidi, Yasaman Kordmaslak,
Maryam Ebrahimi,
and Anahita Ghorbani Tajani, PhD
ISBN: 979-8-89530-628-4
eBook ISBN: 979-8-89530-651-2

Process Engineering in the Obtention and Preservation of Food Bioactive Ingredients
Pedro Aguilar-Zárate, PhD,
Romeo Rojas, PhD,
Mariela R. Michel, PhD,
and Guillermo Cristian G. Martínez-Ávila,
PhD (Editors)
ISBN: 979-8-89113-985-5
eBook ISBN: 979-8-89530-043-5

The Cassava Crop: Cultivation, Potential Uses, and Food Security
Pedro Manuel Villa, PhD, Rajesh Jha, PhD,
Ima Mulyama Zainuddin, PhD,
and Alexis Manuel Faneite Noguera
(Editors)
ISBN: 979-8-89113-844-5
eBook ISBN: 979-8-89113-931-2

Food Byproducts: Valorization Through Nutraceutical Production
Sandra Teresita Martín-del-Campo, PhD,
Anaberta Cardador-Martínez, PhD,
and Jessica del Pilar Ramírez-Anaya, PhD
(Editors)
ISBN: 979-8-88697-612-0
eBook ISBN: 979-8-88697-876-6

More information about this series can be found at

bitly

Gulcin Yildiz
Arzu Imece
Editors

Let Food Be Your Medicine, and Medicine Be Your Food

Scientific Insights into the Healing Power of Nutrition

nova
science publishers
www.novapublishers.com

DOI: https://doi.org/10.52305/YOGZ7456

Library of Congress Cataloging-in-Publication Data

ISBN: 979-8-90134-204-6 (Hardcover)
ISBN: 979-8-90134-261-9 (eBook)

Published by Nova Science Publishers, Inc. † New York

Contents

Preface

The profound connection between diet and health has been echoed for centuries, from Hippocrates' timeless wisdom, *"Let food be thy medicine and medicine be thy food,"* to the cutting-edge biomedical research of the 21st century. As we navigate an era where chronic diseases, environmental stressors, and lifestyle-induced disorders are on the rise, the role of nutrition as both a preventive and therapeutic tool has gained unprecedented importance.

Let Food Be Your Medicine, and Medicine Be Your Food: Scientific Insights into the Healing Power of Nutrition emerges from this global need to bridge traditional dietary knowledge with modern scientific inquiry. This volume gathers the contributions of leading scholars, researchers, and practitioners from various Turkish universities, each offering deep, evidence-based insight into the evolving landscape of food science, human health, and nutrition.

This book is organized into 13 comprehensive chapters, beginning with the foundational science behind nutrition and health and progressing toward more specific themes such as the influence of antioxidants, the gut-brain axis, cultural perspectives, functional foods, sustainability, and novel food innovations. It highlights not only nutrient composition and bioactivity but also the physiological, psychological, and socio-environmental impacts of dietary choices.

A central theme of this book is functional foods—natural or processed foods that provide health benefits beyond basic nutrition. Readers will find detailed examinations of plant-based antioxidants, nutraceuticals, phytochemicals, and their mechanisms in chronic disease modulation, immune support, and mental well-being. Importantly, several chapters explore the transformation of byproducts into health-promoting ingredients, aligning nutritional science with zero-waste and circular economy principles.

This book is intended to serve as a scholarly reference for a diverse audience: nutrition scientists, food technologists, public health professionals, medical practitioners, and graduate students. It is also a timely resource for policy makers, industry innovators, and educators seeking to integrate scientific findings into actionable food-based strategies for population health.

We are grateful to all contributing authors for their rigorous work, interdisciplinary collaboration, and shared vision. Their contributions have ensured that this book not only educates but also inspires further research and innovation in the dynamic field of food and health.

As editors, we hope this volume encourages deeper exploration into how the foods we consume—when chosen with intention and informed by science—can support lifelong wellness, prevent disease, and create more sustainable food systems for future generations.

Gulcin Yildiz and Arzu Imece

Chapter 1

The Connection Between Food and Health

Arzu Imece*
Department of Gastronomy and Culinary Arts, Igdir University, Iğdır, Turkiye

Abstract

This chapter explores the strong relationship between the evolutionary history of human nutrition and modern health issues from a multidimensional perspective. From the Paleolithic era to the present day, changes in dietary habits—without full adaptation at the genetic, metabolic, and microbial levels—have laid the foundation for many chronic diseases. Modern dietary patterns, especially those high in refined sugar andtrans fats but low in fiber, are closely associated with prevalent health problems such as obesity, diabetes, cardiovascular diseases, and inflammatory bowel disorders. This chapter also provides a detailed explanation of the biological functions of macro- and micronutrients, their optimal intake levels, and their impacts on health, while emphasizing the physiological importance of water. It highlights that healthy eating is not only a matter of individual choice but is also linked to socioeconomic access, cultural context, and health policies. Following Hippocrates' saying, "Let food be thy medicine," this chapter concludes that a balanced and natural diet is very important for both personal and public health.

Keywords: food and health relationships, macronutrients, micronutrients, food policies

Introduction

The saying by Hippocrates - one of the earliest authorities in medicine - "Let food be thy medicine and medicine be thy food," expresses an enduring universal truth: health and nutrition are inseparable. The concept that food provides not only the essential energy and structural components for life but also plays a direct role in maintaining health and preventing disease is well established. With the global rise in chronic diseases, the influence of nutrition on human health has become increasingly evident (Wang et al., 2023). Recent studies demonstrate that a

* Corresponding Author's Email: arzuodunkiran@gmail.com.

In: Let Food Be Your Medicine, and Medicine Be Your Food
Editors: Gulcin Yildiz and Arzu Imece
ISBN: 979-8-90134-204-6

substantial proportion of diseases worldwide is directly associated with unhealthy dietary patterns (Bonekamp et al., 2024; Pikula et al., 2024; Lu et al., 2024).

Afshin et al., (2019), in a large-scale international study, reported that in 2017, dietary risk factors were responsible for 11 million deaths and 255 million disability-adjusted life years. A substantial proportion of these deaths were attributed to cardiovascular diseases, cancer, and type 2 diabetes, highlighting the critical role of nutrition in public health.

The health effects of food are determined by its content of macro- and micronutrients as well as bioactive compounds. For example, fiber-rich diets support gastrointestinal health (Gill et al., 2021); omega-3 fatty acids reduce cardiovascular risk (Bowen et al., 2016); and antioxidants such as vitamins A, C, and E provide cellular protection, as demonstrated in numerous studies (Johnson et al., 2003). In contrast, diets high in saturated fats, trans fats, sugar, and sodium, but low in fruits and vegetables, are well-known to be associated with heart disease, hypertension, obesity, and diabetes (WHO, 2020).

Changes in dietary habits are influenced not only by individual choices but also by socioeconomic, cultural, and environmental factors. In the past 50 years, urbanization, globalization, and the expansion of the food industry have driven a phenomenon known as the "nutrition transition," in which traditional, plant-based, and low-calorie diets have been replaced by energy-dense, processed, and long-shelf-life foods. This shift has led to significant increases in obesity and diabetes, even in low- and middle-income countries. As a result, both undernutrition and overnutrition-related chronic diseases are now observed simultaneously (Popkin, 2006). Moreover, socioeconomic status is widely recognized in the literature as a major determinant of access to healthy food. Individuals with lower income levels often rely on inexpensive, energy-dense but nutrient-poor foods, which contributes to the higher prevalence of obesity and diet-related diseases in disadvantaged groups (Darmon & Drewnowski, 2008). This situation may cause mental health challenges.

In light of this evidence, it becomes clear that the relationship between food and health is not solely a matter of personal choice but also a structural and political issue. Therefore, health policies should not focus only on treatment but also promote environmental, economic, and cultural interventions that support healthy eating. Many countries today are striving to improve public dietary habits through strategies such as taxation on sugary beverages, mandatory labeling of processed foods, and regulation of school cafeterias (WHO, 2022).

This chapter will explore the impact of food on health from a multidimensional perspective, encompassing a broad spectrum of considerations—from individual biology to global health policy.

The Evolutionary Role of Food in Human Health

For millions of years, humans survived by hunting and gathering food from the environment. In fact, for over 99% of human history, diets were based on wild plants and game animals, while the transition to agriculture and animal husbandry occurred only about 10–12 thousand years ago (Alt et al., 2022). This dramatic shift brought profound changes in both dietary patterns and lifestyle. Today's dietary model—shaped by industrialization and globalization—is vastly different from ancestral diets in terms of both content and food processing methods. However, the human genetic makeup and physiology have not fully adapted to this rapid transformation, and this mismatch has contributed to a surge in obesity, diabetes, cardiovascular diseases, and other chronic disorders often referred to as "diseases of civilization" (Alt et al., 2022; Assaf et al., 2024).

An evolutionary perspective helps us understand the historical changes in human dietary habits and their impact on health. This section will explore the nutritional transitions from the Paleolithic era to the agricultural revolution, the industrial era, and the emergence of modern processed foods. It will emphasize the role of genetic, metabolic, and microbial adaptations throughout these phases. By doing so, the evolutionary role of food in human health will be examined in the context of disease prevention and health maintenance.

The Paleolithic Era: Hunter-Gatherer Nutrition

The Paleolithic era (the Stone Age) holds a defining place in the evolutionary history of human nutrition. Our ancestors sustained themselves through foraging edible plants—such as fruits, leaves, seeds, and roots—and hunting wild animals. Although the precise composition of Paleolithic diets varied by region and season, the consensus is that these diets were predominantly plant-based and supplemented with animal protein and fat when available. Studies of contemporary hunter-gatherer societies offer valuable insights into prehistoric diets. in many traditional communities, the majority of daily energy intake comes from a wide range of plant sources, while meat is consumed more sporadically and seasonally (Moles & Otaegui, 2020). During this time, food was minimally processed—nuts were cracked open, and leaves were chewed—but there were no industrial practices such as refining or milling.

During the Paleolithic era, important changes in the human body were closely linked to diet. Increases in brain size and body stature brought about greater energy and nutrient demands. To meet these needs, early humans were subject to evolutionary pressures that favored the acquisition of more calorie-dense and nutrient-rich foods. The use of stone tools allowed for more effective consumption of animal tissues such as marrow and meat, providing access to essential nutrients found in bones (Bragazzi et al., 2024). Likewise, the controlled use of fire—marking the beginning of cooking—was a revolutionary advancement in human nutrition. Cooking gelatinizes starches and denatures proteins enhancing digestibility, while also eliminating pathogens and improving food safety (Moles & Otaegui, 2020). Foods that were difficult to digest raw became more energy-accessible, enabling humans to evolve with a smaller digestive tract—conserving energy and supporting brain expansion (Alt et al., 2022).

The health profile of hunter-gatherer societies was markedly different from that of modern populations. People living in this era appeared largely free from chronic diseases prevalent today, such as obesity, type 2 diabetes, and coronary heart disease. Research indicates that these conditions are extremely rare among traditional-living populations; however, when these groups adopt a Westernized diet, they rapidly become susceptible to such illnesses (Cordain et al., 2005). Diet-related health issues at the time included tooth wear and, to a much lesser extent, dental caries. People had fewer cavities without sugar, but chewing tough foods could damage their teeth. Overall, the Paleolithic diet can be considered evolutionarily optimized, and the human body has fundamentally adapted to these ancestral nutritional conditions.

The Agricultural Revolution: A Radical Shift in Human Diet

One of the most transformative events in human history was the Neolithic (Agricultural) Revolution. Around 10,000 BCE, beginning in the Near East, humans began domesticating plants and animals, transitioning from a nomadic hunter-gatherer lifestyle to settled agricultural societies. This shift brought fundamental changes to dietary habits. Most notably, dietary diversity declined: while hunter-gatherers consumed hundreds of plant species according to seasonal availability, agriculture introduced a food production system centered on a limited number of staple grains and legumes. Another major consequence of the agricultural revolution was the advancement of food processing techniques. With the introduction of grinding stones, grains were processed into flour and bulgur, and new cooking methods—such as pottery and clay ovens for baking bread—were developed (Alt et al., 2022). These innovations allowed humans to consume hard grains in more digestible forms. Additionally, the discovery of fermentation enhanced food preservation and nutritional value. Early agricultural societies pioneered bread leavening, beer brewing, and the production of yogurt and cheese (Moles & Otaegui, 2020). Fermented foods not only provided calories but also supported gut health through probiotic effects.

The transition to farming had both positive and negative health impacts. Agricultural societies enjoyed more consistent food supplies, enabling higher reproduction rates and triggering a dramatic increase in global population. However, the human body did not initially adapt well to the new diet, leading to emerging health issues (Alt et al., 2022). Archaeological skeletal evidence from the post-agricultural era shows a decline in average height, deterioration in dental health, and signs of nutrient deficiencies. Dental caries (tooth decay) is considered one of the direct markers of agriculture: the increased consumption of carbohydrate-rich grains and later refined starches altered oral microbiota and significantly raised the incidence of tooth decay (Assaf et al., 2024). While hunter-gatherers rarely exhibited dental caries, Neolithic farmers displayed a noticeable rise in decay rates (Alt et al., 2022).

Micronutrient deficiencies also emerged in agricultural diets. A reliance on just a few staple foods, such as cereals, often ledto insufficient intake of minerals like iron and zinc, or certain vitamins. This is supported by archaeological findings such as bone lesions associated with iron deficiency (e.g., cribra orbitalia) and developmental defects in children's teeth (e.g., enamel hypoplasia), indicating periodic food scarcity. The immune system was also affected by the Agricultural Revolution: monotonous, nutrient-poor diets weakened immunity and increased vulnerability to infections. Furthermore, close proximity to newly domesticated animals created opportunities for zoonotic pathogens to enter human populations. In densely populated and unhygienic settlements, infectious diseases like smallpox, measles, and tuberculosis spread rapidly—many of which are believed to have originated post-agriculture (Assaf et al., 2024). In this sense, the Agricultural Revolution marks an epidemiological transition characterized by a rise in infectious diseases.

Human genetics are partially adapted to the innovations introduced by agriculture. Natural selection favored certain genetic variants that enhanced survival in agricultural societies. One prominent example is lactase persistence. In most mammals, the production of lactase—the enzyme that digests milk sugar (lactose)—declines after weaning. However, as Neolithic humans began consuming milk in adulthood (especially in pastoralist communities), individuals with mutations allowing lifelong lactase production gained a nutritional advantage

(Alt et al., 2022). This mutation underwent strong positive selection, particularly in Northern Europe and parts of Africa (Ségurel & Bon, 2017). Genetic and anthropological research suggests that this adaptation became widespread in Europe during the Bronze Age but was still rare in the early agricultural millennia (Evershed et al., 2022). With lactase persistence, milk became a major food source, providing calcium, protein, and fat that supported growth and reproductive success in adapted populations (Piperata & Dufour, 2021).

Another genetic adaptation linked to agriculture involves starch digestion. In societies that consumed more cereals and root vegetables, the number of gene copies coding for salivary amylase (AMY1) —the enzyme responsible for starch breakdown—increased. There is considerable variation in AMY1 gene copy number among humans; populations with historically high-starch diets tend to have more copies than those with low-starch diets (Perry et al., 2007). This genetic trait enhances the efficiency and speed of starch digestion, and studies suggest that increased AMY1 copy number may have undergone positive selection in farming societies (Perry et al., 2007; Elder et al., 2018). Thus, agriculture left clear adaptive marks on the human genome.

As the Neolithic era progressed, human societies grew larger, urbanization began, and socioeconomic stratification emerged. Dietary patterns began to reflect social status: elites and the wealthy had access to more animal products, sugars, and processed foods, while peasants and laborers relied primarily on staple grains. This stratification also led to disparities in health. For instance, studies of ancient Egyptian mummies revealed a surprisingly high prevalence of atherosclerosis among the nobility, likely linked to their rich diets—including honey, beer, red meat, and refined wheat bread (Allam et al., 2011).

In summary, the Agricultural Revolution marked a pivotal moment in human dietary history, bringing both innovation and new health challenges. It reshaped not only what people ate but also how they lived, adapted, and evolved.

The Industrial Revolution and the Rise of Processed Foods

The Industrial Revolution of the 18th and 19th centuries marked the next major transformation in human nutrition. Following the agricultural era, industrialization brought mechanization in food production, urbanization, and the expansion of global trade. It also introduced key innovations in food preservation and distribution. Technologies such as canning, pasteurization (developed by Louis Pasteur in the 1860s), refrigeration, and iceboxes made it possible to transport food over long distances and to consume it year-round. Out-of-season fruits and vegetables and foods from distant regions became accessible to broader populations. While these developments contributed positively to dietary diversity, the emerging food industry also introduced a range of negative consequences.

Driven by commercial interests, numerous additives began to be incorporated into foods—for instance, salt and nitrates for preservation or sugar for flavor enhancement. Artificial products like margarine were developed to replace traditional fats such as butter. Processed food products, designed for quick preparation and convenient consumption by urban workers, became widespread. The second half of the 20th century was defined by the global dominance of processed foods. Following World War II, advancements in food technology and the chemical industry accelerated; fast food, packaged snacks, and ready-made meals became integrated into daily life. This shift in dietary patterns gave rise to the frequently cited concept of the "Western diet."

The Western diet is characterized by high levels of saturated fat, refined carbohydrates (e.g., sugar and white flour), red meat, and salt, while being low in dietary fiber, vitamins, and antioxidants. Under the influence of globalization, this dietary pattern has spread widely across both developed and developing nations (Moles & Otaegui, 2020).

The health impacts of modern processed food consumption became increasingly evident by the late 20th century (Yildiz et al., 2020; Yildiz and Aadil, 2022; Yıldız et al., 2022). The obesity epidemic emerged as a global phenomenon in the 1980s, affecting large segments of the adult population. In parallel, the incidence of type 2 diabetes, hypertension, cardiovascular diseases, strokes, and certain types of cancer also began to rise. These chronic illnesses are largely associated with diet and lifestyle factors (Caballero, 2007). Contemporary dietary patterns place a substantial burden on the human immune system and metabolism, contributing to low-grade chronic inflammation, which underlies obesity, insulin resistance, and vascular disorders (Yildiz et al., 2019).

For example, the excessive intake of sugar-sweetened beverages promotes liver fat accumulation and impairs insulin regulation, while high salt consumption chronically elevates blood pressure. Insufficient fiber intake disrupts the balance of gut microbiota, leading to digestive disorders and systemic inflammation (Moles & Otaegui, 2020). Thus, the human body—never before exposed in its evolutionary history to such a high volume of ultra-processed, synthetic-laden foods—reacts negatively to this novel dietary environment. As a result, modern societies are increasingly confronted with a broad spectrum of diet-related chronic diseases.

Genetic, Metabolic, and Microbial Adaptations

In response to evolutionary changes in human nutrition, various adaptations have occurred over time at the genetic, metabolic, and microbial levels. While these adaptations have facilitated adjustment to new dietary environments, the rapid pace of modern dietary change has often outstripped the body's ability to keep up, contributing to various health problems.

Genetic Adaptations: The Agricultural Revolution and subsequent periods have triggered strong selective pressures on the human genome related to diet. One of the most well-known examples is lactase persistence - the ability to digest the milk sugar lactose into adulthood—which evolved in populations with a tradition of dairy consumption (Ségurel & Bon, 2017). This genetic adaptation enabled these populations to utilize milk as a vital nutritional resource rich in calcium and protein.

Another example is the increase in AMY1 gene copy number, associated with amylase production in saliva. Populations whose ancestors consumed high-starch diets tend to have more copies of this gene, which enhances starch digestion efficiency (Perry et al., 2007). Similarly, Arctic hunter-gatherers, such as the Inuit, developed metabolic gene variants that support high-fat diets (Fumagalli et al., 2015). These cases highlight how diet can act as an evolutionary selective force. However, genetic adaptation is typically a slow process, taking thousands of years. The abrupt dietary changes of the past few centuries have not allowed sufficient time for such adaptations, leaving modern humans genetically attuned to ancestral diets but physiologically ill-equipped to cope with current dietary excesses (Alt et al., 2022).

Metabolic Adaptations: The human body's energy production and storage mechanisms evolved under conditions of periodic food scarcity. These "thrifty" metabolic traits allowed early humans to store energy as fat during periods of abundance and use it during times of famine. In today's context—characterized by constant caloric availability and reduced physical activity—this same trait predisposes individuals to obesity.

Moreover, the human body is evolutionarily adapted to intermittent eating patterns. Prolonged satiety and excessive meal portions, common today, may disrupt metabolic regulation. For instance, insulin sensitivity—a mechanism honed for sporadic sugar exposure—is impaired under constant high-sugar intake. The pancreas is forced to secrete insulin continuously, leading to insulin resistance and chronically elevated glucose levels, which pave the way for type 2 diabetes (Inchauspé, 2022). Similarly, while salt was once scarce in the human diet, its modern overconsumption overwhelms the kidneys and blood pressure regulation systems, contributing to hypertension (Bragazzi et al., 2024).

Microbial Adaptations (Microbiota): The human digestive system is home to trillions of microbial cells that form a complex ecosystem known as the gut microbiota, which is highly responsive to diet. Over evolutionary time, human populations developed distinct microbiota profiles based on their traditional diets. For example, those consuming fiber-rich diets harbor bacteria capable of fermenting complex carbohydrates, while those with high protein and fat intake have greater levels of Bacteroides species.

The modern diet has severely disrupted this co-evolved microbial ecosystem. Declines in dietary fiber intake and increased consumption of high-fat, high-sugar, and ultra-processed foods have led to a significant reduction in beneficial microbial diversity. Western-style diets are associated with lower levels of short-chain fatty acid-producing bacteria such as *Bifidobacterium* and *Faecalibacterium* (Moles & Otaegui, 2020). This microbial imbalance, or dysbiosis, is linked to obesity, inflammatory bowel diseases, and even mental health disorders such as depression.

The rapid dietary changes of recent decades have created a mismatch between our evolved microbiota and current food environments. Whereas Paleolithic humans likely consumed over 100 grams of fiber daily, modern individuals often consume only 15–20 grams, essentially "starving" fiber-dependent bacteria (Krautkramer et al., 2021). This deficit deprives the intestinal epithelium of essential microbial metabolites, thins the protective mucus layer, and allows opportunistic pathogens to flourish. Furthermore, food additives commonly found in processed foods—such as emulsifiers and artificial sweeteners—have been shown to negatively impact microbial health (Zhou et al., 2023).

The disturbance of this microbial balance may contribute to a wide range of conditions, from immune dysfunction to metabolic syndrome. Interventions such as increasing the consumption of fermented foods and diverse prebiotic fiber sources from various traditional cuisines are promising strategies to help restore microbial diversity and support health (Assaf et al., 2024).

Evolutionary Perspectives on Nutrition and Health

The evolutionary journey of human nutrition sheds light on the origins of many contemporary health issues. The stark contrast between our ancestors' dietary and lifestyle conditions and those of the modern world plays a key role in understanding so-called "diseases of civilization." For example, insulin resistance and type 2 diabetes largely stem from our bodies not being evolutionarily adapted to frequent and excessive sugar intake—a reflection of sugar's historical rarity. Similarly, elevated LDL cholesterol levels, a major risk factor for cardiovascular disease, are associated with an unprecedented consumption of animal fats and trans fats. High blood pressure, too, is linked to the body's inherited sodium-retention mechanisms—developed in response to ancestral salt scarcity—being overwhelmed by today's excessive salt intake (Bragazzi et al., 2024). These examples illustrate how many modern chronic diseases arise from evolutionary mismatches.

An evolutionary perspective can also guide strategies for disease prevention through diet. The most physiologically suitable diet for humans is one that aligns with our evolutionary adaptations. This does not mean strictly replicating the Paleolithic diet but suggests that modern diets should emphasize more natural, whole, and diverse foods. For instance, a diet rich in unprocessed vegetables, fruits, nuts, seeds, and legumes—high in fiber and phytochemicals—provides a nutritional environment more consistent with our evolutionary heritage. Traditional dietary patterns such as the Mediterranean diet exemplify this approach. Featuring natural fats like olive oil, seafood, a wide variety of vegetables and fruits, and moderate amounts of animal products, the Mediterranean diet has been shown to reduce cardiovascular risk (Willett et al., 2019). Likewise, the high-fiber, plant-based diets found in rural African and Asian populations have been linked to lower incidences of colorectal cancer and inflammatory bowel diseases (O'Keefe et al., 2015). These findings suggest that diets closely resembling those of our ancestors may help prevent modern diseases.

Of course, today's living conditions differ greatly from those of the Paleolithic era. Therefore, evolutionarily informed nutrition must be adapted to the modern context using scientific evidence. The field of evolutionary medicine emphasizes that recreating aspects of ancestral environments—such as increased physical activity, reduced stress, and natural food intake—can offer powerful strategies for disease prevention (Stearns & Medzhitov, 2024). Indeed, in combating obesity and metabolic syndrome, teaching individuals to "eat and move like our ancestors" has been shown to facilitate behavioral change (Arlinghaus & Johnston, 2019). These interventions aim to help individuals develop healthy behaviors aligned with our evolutionary past. Examples include minimizing processed food consumption, choosing whole grains and natural protein sources, replacing sugary drinks with water or herbal teas, and eating slowly at set intervals rather than constantly snacking.

In conclusion, the evolutionary history of human nutrition offers valuable insights for protecting health and preventing disease. While it is not possible to fully return to the diets of our ancestors, we can develop more balanced and biologically appropriate eating strategies by acknowledging the legacy they left in our genes, metabolism, and microbiota. When making dietary choices, we are not just feeding our palates—we are also nourishing our genetic heritage, metabolic systems, and microbial companions. Embracing the principle of evolutionary alignment can help us avoid the pitfalls of modern eating patterns and restore the balance between humans and nature that has evolved over millennia.

Macronutrients and Micronutrients: The Cornerstones of Health

Nutrients are chemical components essential for growth, development, and the maintenance of health. They are generally categorized as macronutrients (proteins, carbohydrates, and fats) and micronutrients (vitamins and minerals). Macronutrients provide energy or serve as structural elements and are therefore required in larger amounts in the daily diet. Though not energy-providing, they are vital for catalyzing metabolic reactions and regulating bodily functions. Water is also considered an indispensable nutrient, essential for all life processes. Below, the physiological and biochemical roles, related metabolic functions, and optimal intake recommendations for these fundamental nutrients in healthy adults are explained (Taşğın, 2017).

Proteins

Proteins are composed of amino acids and serve as both structural units and functional molecules of the body. Their physiological roles are highly diverse: they are the main constituents of structural components such as muscles, connective tissues, and skin, and they also form vital functional molecules such as enzymes, hormones, transport proteins, and antibodies. For instance, hemoglobin is responsible for oxygen transport, insulin regulates blood glucose levels, and immunoglobulins are involved in immune responses. Through these various functions, proteins play a critical role in tissue repair, growth, and the maintenance of homeostasis (Mueller & Pereira, 2015; Lee et al., 2016; Yildiz et al., 2017; Yildiz et al., 2018; Jiang et al., 2019; Hudson et al., 2020).

The biochemical and metabolic functions of proteins largely depend on their amino acid composition. Dietary proteins are broken down into amino acids during digestion; these amino acids are then used for the synthesis of new proteins or, when necessary, metabolized for energy. Essential amino acids (e.g., leucine, valine, lysine) cannot be synthesized by the body and must be obtained from the diet. In the liver, amino acids can undergo deamination and contribute to gluconeogenesis or the citric acid cycle for energy production. However, their primary role is not energy provision but the synthesis of tissues and regulation of metabolic processes. For example, in the presence of vitamin B6, amino acids are used to synthesize neurotransmitters (like dopamine and serotonin), highlighting the impact of protein metabolism on the nervous system (Wolfe, 2017).

Optimal protein intake varies depending on age, body weight, and physiological status, but for healthy adults, the recommended daily intake is approximately 0.8 grams per kilogram of body weight (RDA: 0.8 g/kg/day). This level is generally sufficient for maintaining protein balance and supporting tissue renewal and metabolic needs (Hudson et al., 2020). In their meta-analysis, Hudson et al. (2020) reported that this amount was adequate for preserving lean body mass in sedentary adults. However, higher protein requirements may apply for older adults and individuals engaging in intense physical activity. Excessive protein intake (e.g., > 2.5–3 g/kg/day) does not offer additional benefits in healthy adults and may pose risks such as increased kidney workload. In general, ensuring both the quality (essential amino acid content) and quantity of protein is crucial for tissue repair, immune function, and overall health

Carbohydrates

Carbohydrates are not only the main source of energy in human nutrition but also exert multifaceted effects on health. Particularly, high intake of complex carbohydrates such as dietary fiber and whole grains has been associated with significantly reduced risks of cardiovascular diseases, type 2 diabetes, stroke, and certain types of cancer. These types of carbohydrates also positively influence body weight, blood pressure, and lipid profiles. On the other hand, excessive consumption of refined sugars and high-glycemic index foods has been linked to various metabolic disorders, notably obesity. In this context, dietary strategies focusing on carbohydrate quality offer more than energy balance—they also serve as preventive approaches against non-communicable diseases (Chambers et al., 2019).

The health effects of carbohydrates should be evaluated not only based on quantity but also on sources, structure, and glycemic response. As such, it is becoming increasingly clear that nutrition policies and individualized diet plans should be restructured with a focus on carbohydrate quality.

Beyond energy provision, carbohydrates also have important biochemical roles. For example, the backbone of DNA and RNA consists of five-carbon sugars called ribose and deoxyribose. In cell membranes, carbohydrates are components of glycoproteins and glycolipids, playing roles in cell recognition, receptor function, and signal transduction. A portion of dietary carbohydrates consists of indigestible dietary fiber. Although fiber does not provide energy directly, it undergoes fermentation in the colon, producing short-chain fatty acids (SCFAs) that are critical for gut microbiota health. Soluble fibers form gels in the gut, slowing nutrient absorption and improving blood cholesterol levels; insoluble fibers add bulk to stools and promote healthy bowel movements. Indeed, it has been shown that dietary fiber promotes microbiota diversity and the growth of beneficial bacteria. This supports the positive effects of whole grains, legumes, fruits, and vegetables on both digestive and metabolic health (Clemente-Suárez et al., 2022).

The health effects of carbohydrates depend largely on quality. Excessive intake of simple sugars and refined flour can cause blood sugar fluctuations and insulin resistance, increasing the risk of obesity and type 2 diabetes. In contrast, complex carbohydrate sources like whole grains and vegetables provide sustained satiety and support glycemic control due to their high fiber content. Clemente-Suárez et al. (2022) also note that high-sugar diets can negatively affect mood, whereas fiber-rich carbohydrates support mental well-being through the gut-brain axis.

While optimal carbohydrate intake depends on individual energy needs, general dietary guidelines recommend that 45–65% of daily energy should come from carbohydrates. For the average adult, at least ~130 grams of carbohydrates per day is needed to meet the brain's minimal glucose requirement. The World Health Organization (WHO, 2023) also recommends that no more than 30% of total energy should come from fat, and that most carbohydrates should be derived from quality sources such as whole grains, fruits, vegetables, and legumes. Limiting refined grains and added sugars while favoring natural fiber-rich carbohydrates is beneficial for long-term weight control and metabolic health. In summary, carbohydrates should form the foundation of the daily diet, but source and quantity must be carefully planned. For adults, most carbohydrates should come in complex forms, and a daily intake of around 300 grams - adjusted for physical activity levels—should meet energy requirements. Additionally, WHO (2023) recommends a minimum of 25 grams of dietary fiber per day, making sufficient carbohydrate intake an essential part of overall health.

Fats

Fats (lipids) are essential nutrients that, in addition to providing energy, play a wide variety of structural and functional roles in the body. Physiologically, adipose tissue is the body's main energy reservoir, supplying the highest energy density among macronutrients with 9 kcal per gram. The subcutaneous fat layer helps maintain body temperature, while fat cushions surrounding internal organs provide protection against mechanical shocks. At the cellular level, phospholipids and cholesterol are fundamental components of cell membranes, critical for maintaining membrane integrity and signal transmission. A large portion of brain tissue consists of fats; particularly, the lipids in the myelin sheath enable rapid conduction of nerve impulses. Biochemically, fats are involved in important processes such as the synthesis of steroid hormones (e.g., estrogen, testosterone) and bile acids. Dietary fat is also necessary for the absorption of fat-soluble vitamins (A, D, E, and K), as these vitamins require a lipid environment in the small intestine for proper absorption—insufficient fat intake can impair their absorption (Nelson et al., 2008; Gropper and Smith, 2013).

The type and balance of fatty acids are critically important for health. Dietary fats are primarily classified into saturated fatty acids (common in animal foods and tropical oils), monounsaturated fatty acids (found in olive oil, avocado, etc.), and polyunsaturated fatty acids (such as fish oil, walnuts, flaxseeds). Excessive intake of saturated fats may adversely affect blood cholesterol profiles and increase the risk of cardiovascular diseases. Therefore, WHO (2023) recommends that saturated fat intake should not exceed 10% of total energy, and trans fats should be less than 1%. In contrast, adequate consumption of monounsaturated and polyunsaturated fats is beneficial for cardiovascular health; for example, the Mediterranean diet enriched with monounsaturated fats like olive oil is associated with lower LDL cholesterol and reduced risk of heart disease.

Some polyunsaturated fatty acids, known as essential fatty acids, cannot be synthesized by the body and must be obtained from the diet. The primary ones are linoleic acid (LA) from the omega-6 family and alpha-linolenic acid (ALA) from the omega-3 family. These fatty acids integrate into cell membrane phospholipids to regulate membrane fluidity and serve as precursors for signaling molecules called eicosanoids (prostaglandins, leukotrienes, thromboxanes). For instance, the omega-6 fatty acid arachidonic acid is a substrate for prostaglandins involved in inflammation and blood clotting. Omega-3 fatty acids (EPA and DHA) are known for their anti-inflammatory effects and are important in protecting heart health as well as brain and vision functions. DHA is highly concentrated in the retina and cerebral cortex, contributing to optimal functioning of photoreceptor and nerve cells. Adequate intake of essential fatty acids is also critical for brain and nervous system development in children. Especially during pregnancy and early childhood, sufficient omega-3 intake is important for cognitive development and visual acuity (Calder, 2015).

Optimal fat intake is generally recommended to be around 20–35% of total energy. An average adult diet should include approximately 60–80 grams of fat daily, with the majority coming from healthy monounsaturated sources such as olive oil, nuts, and avocado, and omega-3-rich polyunsaturated sources such as fish, walnuts, and flaxseed. Saturated fat consumption should be minimized (preferably less than 20 grams per day), and trans fats should be avoided altogether. Adequate intake levels for essential fatty acids are approximately 1.1 grams of ALA and 12 grams of LA daily for women, and 1.6 grams of ALA and 17 grams of LA for men. For example, consuming at least two servings of fatty fish per week helps meet long-chain omega-

3 fatty acid (EPA, DHA) requirements (Phillips, 2021; WHO, 2023). In summary, both the amount and type of fat consumed should be balanced: sufficient fat intake is essential for bodily functions, but fat quality (favoring unsaturated over saturated fats) is a critical factor for long-term health.

Vitamins

Vitamins are organic compounds that the body cannot synthesize and must be obtained from diet in small amounts. There are 13 essential vitamins for humans: four fat-soluble vitamins (A, D, E, and K) that can be stored in tissues to some extent, and nine water-soluble vitamins (vitamin C and the B-complex vitamins: B1, B2, B3, B5, B6, B7, B9, B12) that are not stored significantly and are excreted in urine, thus requiring regular intake. Vitamins participate in vital biochemical reactions by acting as cofactors, antioxidants, or hormone-like substances (WHO, 2004).

Fat-soluble vitamins generally accumulate in body stores, and deficiency symptoms develop more slowly. In contrast, water-soluble vitamins must be consumed continuously because excess amounts are not retained. Biochemically, most B vitamins function as coenzymes in enzymatic reactions. For example, vitamins B1 (thiamine), B2 (riboflavin), and B3 (niacin) act as coenzymes in fundamental energy metabolism reactions such as glycolysis, the citric acid cycle, and the electron transport chain. Without these vitamins, the body cannot efficiently convert fuel sources into ATP energy sources. Vitamin B6 (pyridoxine) serves as a coenzyme in amino acid metabolism and neurotransmitter synthesis, while folic acid (B9) and vitamin B12 play critical roles in one-carbon transfers and DNA synthesis. Deficiencies in folate and B12 lead to megaloblastic anemia. Vitamin C (ascorbic acid) is a powerful antioxidant that neutralizes free radicals and is essential for collagen synthesis. Since collagen is the main protein in connective tissue, vitamin C deficiency causes scurvy, characterized by symptoms such as gum bleeding and poor wound healing (Barker, 2023).

Among fat-soluble vitamins, vitamin A (retinol) is necessary for the composition of visual pigments in the retina; deficiency causes night blindness and corneal damage. Vitamin A also plays a role in maintaining epithelial integrity and immune function (e.g., resistance to infections). Vitamin D (calciferol) acts as a prohormone that regulates calcium and phosphorus metabolism to maintain bone health. Synthesized in the skin under sunlight, vitamin D enhances calcium absorption from the intestines and is essential for bone mineralization. Deficiency leads to rickets in children and osteomalacia (soft bones) in adults and also modulates immune system function. Vitamin E (tocopherol) is the main antioxidant protecting polyunsaturated fatty acids in cell membranes from oxidative damage and is thought to protect against cardiovascular diseases by preventing oxidation of cell membranes and LDL cholesterol. Vitamin K is essential for the synthesis of clotting factors (particularly prothrombin and Factors VII, IX, X) in the liver; deficiency increases bleeding tendency. Vitamin K also activates osteocalcin, a bone protein, thereby influencing bone metabolism (Andrès et al., 2024).

Optimal vitamin intakes vary throughout life, and there are specific recommended daily allowances (RDAs) for each vitamin in healthy adults, approximately as follows:

- Vitamin A: ~700–900 µg
- Vitamin D: 15 µg (600 IU)
- Vitamin E: 15 mg
- Vitamin K: ~90–120 µg
- Vitamin C: ~75–90 mg
- Thiamine (B1): ~1.1–1.2 mg
- Riboflavin (B2): ~1.1–1.3 mg
- Niacin (B3): ~14–16 mg
- Vitamin B6: ~1.3–1.7 mg
- Folic acid: ~400 µg
- Vitamin B12: ~2.4 µg

These values may slightly differ between men and women and increase during pregnancy and lactation. Vitamin needs are generally met through a balanced diet that includes a variety of food groups. For example, orange vegetables like carrots and pumpkin, and liver are good sources of vitamin A; citrus fruits and leafy green vegetables provide vitamin C; B vitamins are abundant in whole grains, meat, dairy, and legumes (Yildiz, 2018; Yildiz, 2019; Yildiz and Feng, 2019; Yildiz and Aadil, 2020). While vitamin D is partly produced in the skin upon sun exposure, supplementation may be necessary during winter or in older adults with limited sun exposure. Vitamin deficiencies cause specific diseases (e.g., scurvy from vitamin C deficiency, beriberi from B1 deficiency, pellagra from niacin deficiency, anemia from B12 or folate deficiency, rickets from vitamin D deficiency). Conversely, excessive intake (hypervitaminosis), particularly fat-soluble vitamins (A and D), can be toxic (WHO, 2004). Therefore, vitamin supplements should be used cautiously, and ideally, vitamin requirements should be met through a balanced diet.

Minerals

Minerals are inorganic nutrients that play numerous structural and functional roles in the body. Macrominerals (such as calcium, phosphorus, magnesium, sodium, potassium, and chloride) are minerals required in larger amounts, whereas trace elements or microminerals (iron, zinc, copper, iodine, selenium, manganese, etc.) are needed in smaller quantities but are vital for health. Their physiological roles are as diverse as those of vitamins, spanning from bone formation and oxygen transport to enzyme function and fluid balance (WHO, 2004).

Among macrominerals, calcium (Ca) and phosphorus (P) are the primary building blocks of the skeletal system. Approximately 99% of the body'scalcium and the majority of phosphorus are found in bones and teeth; adequate intake of these minerals is essential for strong bone structure. Calcium is also a critical electrolyte for muscle contraction and nerve transmission. Temporary increases in calcium levels in muscle cells trigger contraction, while calcium flux in nerve cells is involved in neurotransmitter release. Calcium also acts as a cofactor in blood clotting mechanisms. Magnesium (Mg) serves as a cofactor for over 300 enzymes and is especially important in ATP stabilization and energy metabolism. Processes such as muscle relaxation, nerve signal transmission, and regulation of heart rhythm depend on magnesium. Sodium (Na), potassium (K), and chloride (Cl) are the main electrolytes in the body and regulate osmotic balance and fluid distribution. Sodium and potassium maintain the intracellular and extracellular ion balance, facilitating nerve impulse transmission; sodium is mostly found in extracellular fluid, while potassium is concentrated inside cells. The balance of these electrolytes also plays a crucial role in blood pressure regulation. For example, diets rich in potassium and low in sodium are well known to reduce hypertension risk. Chloride, part of stomach acid (HCl), contributes to digestion and balances osmotic pressure together with sodium (WHO, 2004; Farag et al., 2023).

Trace elements, though required in smaller amounts, are integral components of many enzymes or play roles in hormone functions. Iron (Fe) is indispensable as the central atom of hemoglobin, essential for oxygen transport capacity. The hemoglobin molecule in red blood cells carries oxygen from lungs to tissues via Fe^{2+}/Fe^{3+} valence changes. Iron is also found in myoglobin, cytochromes, and enzymes like catalase, making it critical for energy metabolism and cellular respiration. Iron deficiency causes iron-deficiency anemia, which results in insufficient oxygen transport to tissues, causing fatigue, weakness, and difficulty concentrating. Zinc (Zn) is a cofactor activating over 100 enzymes and is centrally important for immune function, wound healing, DNA synthesis, and growth processes. Zinc deficiency weakens immune response, increasing susceptibility to infections and causing growth retardation in children. Copper (Cu) participates in iron metabolism, energy production, and connective tissue synthesis; it is also a component of antioxidant enzymes such as superoxide dismutase (Prashanth et al., 2015). Iodine (I) is an indispensable part of thyroid hormones (T3 and T4). Insufficient iodine intake impairs thyroid hormone production and causes thyroid gland enlargement (goiter). Iodine deficiency during pregnancy and childhood can lead to intellectual disability and developmental disorders (Hatch-McChesney and Lieberman, 2022). Selenium (Se) is located at the active site of antioxidant enzymes such as glutathione peroxidase, protecting cells from oxidative damage. Selenium is also necessary for converting thyroid hormones to their active forms. Despite their small required amounts, these trace elements are indispensable for maintaining health (Prashanth et al., 2015; Bai et al., 2024).

For example, regarding metabolic roles, elements like zinc and copper support antioxidant defense systems, while iron enables oxygen transport and cellular energy production. Magnesium and calcium regulate muscle functions together: calcium induces muscle contraction, while magnesium aids relaxation. Selenium and iodine are essential for healthy endocrine system function. Deficiencies of many minerals such as zinc, copper, iron, magnesium, manganese, selenium, and calcium are associated with inflammation, oxidative stress, and impaired organ functions (WHO, 2004). For instance, Fekete et al. (2024) reviewed that zinc and copper modulate immune responses and reduce inflammation, iron is vital for oxygen transport, and magnesium, manganese, selenium, and calcium are critical for muscle function, respiratory health, and antioxidant defense. These findings indicate that these minerals serve interconnected functions across multiple systems rather than a single role.

Recommended daily mineral intakes vary over a wider range than vitamins, with specific recommendations for each mineral. For adults, recommended daily intakes are approximately: 1000 mg calcium (slightly higher for youth and elderly), 4700 mg potassium, 400 mg magnesium for men and 310 mg for women, 8 mg iron for adult men and 18 mg for women (higher in menstruating women). Zinc recommendations are about 8–11 mg, iodine 150 µg, selenium 55 µg. The upper limit for sodium intake is 2300 mg per day (approximately 5–6 grams of salt);however, for optimal blood pressure control, a lower intake of about 1500 mg may be targeted. These values are based on Dietary Reference Intakes (DRI) or similar guidelines established by national and international nutrition bodies. For example, the World Health Organization recommends iodizing table salt and a daily iodine intake of 150 µg to prevent iodine deficiency disorders globally (NIH, 2022; WHO, 2023).

In a balanced diet, consuming a variety of mineral sources is important for bioavailability. For example, vitamin C enhances the absorption of plant-based iron, while compounds like phytates can inhibit zinc and iron absorption. Therefore, dietary diversity is key to adequate mineral intake. Important sources include dairy products and leafy green vegetables for calcium; red meat, organ meats, and legumes for iron; meat, seafood, and nuts for zinc; leafy greens and legumes for magnesium; seafood and iodized salt for iodine; and seafood, red meat, and Brazil nuts for selenium. Mineral deficiencies can cause specific clinical conditions (e.g., anemia due to iron deficiency, goiter from iodine deficiency, osteoporosis from calcium and vitamin D deficiency), while excessive intake can lead to toxicity (especially iron, zinc, and selenium overloads) (Wazir and Ghobrial, 2017). Therefore, maintaining mineral balance with neither deficiency nor excess is essential for healthy living.

Water

Water is the most fundamental nutrient for sustaining life and participates in nearly all physiological processes in the body. It constitutes approximately 50–60% of an adult's body weight, and maintaining this level is critical for health (Popkin et al., 2010). Physiologically, water performs numerous functions ranging from digestion, absorption, and transport of nutrients to removal of metabolic wastes and regulation of body temperature (Jéquier and Constant, 2010). Water makes up the majority of blood volume, enabling oxygen and nutrients to be delivered to cells. As the main component of intercellular fluid, it provides a medium for cells to receive substances they need and remove metabolic wastes (Izli et al., 2018; Yildiz and Izli, 2019; Yildiz, 2021; İzli et al., 2022; Yildiz, 2022). The kidneys rely on water to excrete harmful metabolites through urine; sufficient water intake increases urine output, reducing risks of bladder and kidney infections (Valtin, 2002). Water also aids in digestion and nutrient absorption, facilitates intestinal transit, and helps prevent constipation. Body temperature regulation is achieved through sweating: evaporating water from sweat glands removes heat from the body, cooling it down (Jéquier and Constant, 2010). Special fluids such as synovial fluid in joints and cerebrospinal fluid are largely water; water reduces friction in joints, protecting cartilage and enabling pain-free movement. Fluids surrounding the brain, eyes, and spinal cord cushion these organs against shocks. Moreover, water is crucial in regulating blood pressure and maintaining electrolyte balance. The distribution of intracellular and extracellular fluids and proper electrolyte distribution (e.g., sodium, potassium) depend on adequate water availability (Popkin et al., 2010). In short, water is a vital medium and solvent that contributes to the proper functioning of every body system. Its main roles include nutrient and oxygen transport, waste removal, blood pressure regulation, lubrication of joints, protection of organs and tissues, and maintaining body temperature.

The body's water balance is maintained by a delicate equilibrium between intake and loss. We continuously lose water through respiration, sweat, urine, and feces. Recommendations for optimal water intake have been developed to replace these losses. Although individual water needs vary depending on factors such as climate, physical activity, and metabolic rate, average daily total water intake is about 3.7 liters for men and 2.7 liters for women; this includes water from both beverages and food. This roughly corresponds to about 13 cups per day for men and 9 cups for women (based on 8-ounce cup size) (Moyer, 2024). According to the Harvard Health Report (2023), most people can maintain basic hydration by drinking 4–6 cups of water daily, but fluid from food and individual needs can alter this amount. For example, individuals who exercise intensely, sweat a lot, or live in hot climates may require significantly more water than standard recommendations. Thirst is generally a reliable indicator of hydration needs; however, especially in older adults, the thirst mechanism may be impaired, so it is advisable to drink water regularly throughout the day.

Adequate water intake has many positive health effects: optimal hydration supports physical performance and cognitive functions, helps maintain body temperature, and ensures cardiovascular efficiency. Even mild dehydration (around 2%) can impair cognitive abilities such as attention, memory, and reaction time (Lieberman, 2007). Severe dehydration can be life-threatening, causing drops in blood pressure, electrolyte imbalances, and heat stroke. Therefore, spreading water consumption evenly throughout the day without waiting to feel thirsty is the best practice. A general rule is that pale yellow urine indicates adequate hydration; dark yellow urine often signals the need for more water. In conclusion, water is an indispensable nutrient that provides an essential environment for all macro- and micronutrients to function properly. Drinking sufficient clean water daily is a crucial part of a healthy diet.

The Relationship Between Nutrition and the Immune System

The immune system is a complex defense mechanism that protects the body against infectious agents and foreign substances. Its effective functioning is closely related to nutritional status. Inadequate and unbalanced nutrition weakens immune responses, increasing susceptibility to infections. Indeed, in a state of malnutrition characterized by deficiencies in macro- and micronutrients, immune system effectiveness decreases; host defense is compromised, and the proper elimination of pathogens becomes difficult. For example, energy-protein deficiency impairs the function of immune cells, especially T cells, leading to an increase in opportunistic infections and related complications (Daly et al., 1990).

Moreover, in developed countries, diets rich in calories but poor in nutrient quality also have adverse effects on immunity. Consumption of highly processed foods and imbalanced fat intake can trigger chronic low-grade inflammation in the body, disrupting immune homeostasis. For instance, the increased intake of omega-6 fatty acids in modern diets is excessive relative to omega-3 fatty acids; this imbalance promotes pro-inflammatory pathways. As a result, chronic inflammation underlies many chronic diseases and immune dysfunctions (Poggioli et al., 2023).

On the other hand, a nutrient-rich, balanced diet is one of the most important factors supporting the immune system. Consuming a variety of foods rich in vitamins, minerals, proteins, healthy fats, and phytochemicals ensures the optimal functioning of both innate and adaptive immune responses. Numerous recent studies in the field of nutritional immunology have revealed that certain dietary components can have direct immunomodulatory and anti-inflammatory effects (Tourkochristou et al., 2021; Munteanu and Schwartz, 2022).

Immunomodulatory and Anti-inflammatory Mechanisms of Nutrition

To understand the effects of nutrients on immunity, it is necessary to examine how they function at cellular and molecular levels. Macronutrients (carbohydrates, proteins, fats) and micronutrients (vitamins, minerals, and other bioactive compounds) obtained through diet play roles not only in the body's energy production, tissue repair, and metabolism but also in immune responses. Particularly, the gut mucosa is a critical immune site constantly interacting with dietary antigens and microbiota products. Antigens from the diet are presented to the immune system in the gut via GALT (gut-associated lymphoid tissue), where a delicate balance between tolerance and immune activation is maintained (Munteanu and Schwartz, 2022).

Many nutrient components regulate inflammatory responses by affecting processes such as the production of Toll-like receptors (TLRs), which are critical for innate immunity, and cytokines. For example, certain fatty acids and amino acids are part of the membrane structure of innate immune cells like macrophages and natural killer cells, modifying their antigen recognition and signal transduction capabilities. Some molecules derived from nutrient metabolism can act as cofactors or inhibitors in enzyme activity, influencing pathogen-killing mechanisms and oxidative stress levels. Compounds with antioxidant properties, such as vitamins and polyphenols, help control inflammation by preventing excessive free radical accumulation in cells (Tourkochristou et al., 2021).

On the other hand, nutrients also affect adaptive immune processes. The development, proliferation, and activation of B and T lymphocytes depend on sufficient levels of vitamins and minerals obtained from the diet. For example, vitamins such as folic acid and B12 are necessary for DNA synthesis, and their deficiencies can impair lymphocyte proliferation and immune memory formation. Moreover, certain nutrients can modulate chronic inflammation and autoimmunity by suppressing pro-inflammatory responses and promoting anti-inflammatory responses (Tourkochristou et al., 2021). Indeed, signaling molecules derived from nutrients can either activate or inhibit key pathways regulating inflammatory gene expression, such as NF-κB (Munteanu and Schwartz, 2022).

The example of omega-3 and omega-6 fatty acids clearly illustrates the effect of diet on inflammation balance. Eicosanoids derived from omega-6 fatty acids (e.g., linoleic acid) such as certain prostaglandins and leukotrienes generally have pro-inflammatory effects, whereas lipid mediators derived from omega-3 fatty acids (EPA and DHA), such as resolvins and protectins, exert inflammation-resolving and regulatory effects. The high omega-6/omega-3 ratio in Western diets is a factor promoting chronic inflammation. Conversely, adequate omega-3 intake provides benefits by reducing the production of inflammatory cytokines like TNF-α and IL-6 and by dampening excessive immune cell activity when necessary (e.g., in autoimmune processes) (Poggioli et al., 2023).

In summary, nutritional components have multifaceted regulatory effects on both innate and adaptive immunity. The mechanisms underlying these effects include modifications of gene expression, alteration of receptor signal transduction, reprogramming of intracellular metabolic pathways, and epigenetic regulation (Ferreira et al., 2024). The following sections will examine in detail the scientifically established effects of specific nutrient groups (vitamins, minerals, macronutrients, phytochemicals, and probiotics) on immune functions.

Effects of Vitamins on the Immune System

Vitamins are indispensable micronutrients for the development and function of immune cells. Deficiencies in many vitamins lead to signs of immunodeficiency, and severe vitamin shortages are associated with increased frequency and severity of infections. On the other hand, optimal vitamin intake plays a critical role in maintaining the integrity of protective barriers (e.g., skin and mucous membranes) and ensuring proper immune cell responses (Ahvanooei et al., 2021).

Vitamin A (retinoids) is one of the essential vitamins for a healthy immune system. Beyond its role in vision and the retina, vitamin A is necessary for maintaining epithelial tissue integrity and for the development and differentiation of innate immune cells such as macrophages and neutrophils. Active derivatives of vitamin A, such as retinoic acid, regulate the maturation and function of macrophages and dendritic cells, which play key roles in the first line of defense against pathogens (Huang et al., 2018). Vitamin A deficiency causes significant immune impairments. Even mild deficiency increases susceptibility to respiratory and gastrointestinal infections. In children with severe deficiencies, disruption of epithelial barriers makes mucosal surfaces more vulnerable to pathogen entry. Vitamin A deficiency also suppresses neutrophil and macrophage function and can impair the maturation and proliferation of T and B lymphocytes. Consequently, resistance to infections decreases and diseases tend to be more severe. According to WHO reports, vitamin A deficiency is a major factor in morbidity and mortality from infectious diseases in children in developing countries (Munteanu and Schwartz, 2022). Indeed, vitamin A supplementation has been shown to reduce complications and mortality from infections like measles in deficient children (WHO, 2011).

B vitamins are water-soluble vitamins that act as coenzymes in metabolism. Their effects on the immune system are largely indirect, through their roles in cellular energy production and new cell synthesis. Rapidly dividing and activated immune cells require B vitamins; deficiencies can adversely affect the production of lymphoid tissues and white blood cells (Gay and Meydani, 2001). For example, folic acid (B9) deficiency impairs thymus and spleen function and reduces T lymphocyte counts. Vitamin B12 deficiency lowers the phagocytic capacity of neutrophils. Deficiencies in B12 and folate impair DNA synthesis, slowing immune cell proliferation and restricting immune cell production particularly in gastrointestinal mucosa and bone marrow. Vitamin B6 (pyridoxine) serves as a cofactor for enzymes involved in many immune response stages. Insufficient B6 intake significantly reduces lymphocyte proliferation and cytokine production. Controlled human studies have reported that low B6 intake decreases mitogenic responses of T lymphocytes and IL-2 production, weakening antibody formation. Animal studies also showed B6 deficiency slows lymphocyte growth, impairs T cell differentiation, and shifts the Th1/Th2 balance (lower IL-2 and higher IL-4 levels) (Qian et al., 2017). Other B vitamins like B1 (thiamine), B2 (riboflavin), and B3 (niacin) also have indirect roles in immune function. Since riboflavin and niacin are key in energy metabolism, they are critical to meeting immune cells' energy demands. Low niacin levels affect NAD/NADP ratios, impairing macrophage and other immune cell functions (Munteanu and Schwartz, 2022).

Vitamin C is a well-known potent antioxidant that supports the immune system in multiple ways. Because humans cannot synthesize vitamin C, regular dietary intake is essential. Its effects are observed in both innate and adaptive immunity. First, vitamin C is necessary for the integrity of physical barriers and connective tissue by supporting collagen synthesis, helping maintain healthy skin, blood vessels, and mucous membranes, thereby hindering pathogen entry. Second, it enhances the function of phagocytic cells; neutrophils and macrophages rely on vitamin C to capture and kill microbes. Neutrophils accumulate high levels of vitamin C upon migrating to infection sites; this vitamin protects them from oxidative damage during the respiratory burst and aids in effective pathogen killing. Furthermore, vitamin C facilitates neutrophil apoptosis and clearance of debris by macrophages after phagocytosis, limiting tissue damage (Carr and Maggini, 2017).

In adaptive immunity, vitamin C supports the maturation and proliferation of B and T lymphocytes. Adequate vitamin C accelerates T cell division and differentiation, increases antibody-producing B cell numbers, and enhances antibody secretion. Its antioxidant effects reduce oxidative stress in inflammatory environments, protecting immune cell functions. Vitamin C also helps balance cytokine production: it can normalize excessive pro-inflammatory cytokine release (e.g., IL-6, TNF-α) during infection while supporting production of essential defense molecules such as interferons (Jang et al., 2014). High doses of vitamin C have also been reported to reduce histamine levels, alleviating inflammatory symptoms (Jarisch et al., 2014).

Historically, vitamin C deficiency (scurvy) was linked to impaired immune function and fatal infections. Although severe deficiencies are rare today, marginal deficiencies occur, especially among smokers, elderly, and poorly nourished individuals. Clinical studies show adequate daily vitamin C intake can shorten respiratory infection duration and mildly reduce symptom severity (Hemilä, 2004). Regular supplementation notably reduces common cold incidence in people under heavy physical stress (e.g., marathon runners, soldiers), likely due to protection against oxidative stress and enhanced first-line defense (Constantini et al., 2011). In summary, vitamin C is an indispensable immune supporter via maintaining epithelial integrity, improving phagocyte function, supporting lymphocyte proliferation, and controlling inflammation. Daily intake of fresh fruits and vegetables rich in vitamin C (citrus fruits, rosehip, kiwi, leafy greens, peppers) is important for maintaining immunity, especially during infection seasons.

Vitamin D, classically known for regulating calcium-phosphorus metabolism and bone health, has recently been intensely studied for its hormone-like effects on the immune system. The active form, 1,25-dihydroxyvitamin D (calcitriol), binds to vitamin D receptors expressed in many immune cells including monocytes, macrophages, dendritic cells, and T and B lymphocytes, modulating gene expression (Ghaseminejad-Raeini et al., 2023). Therefore, vitamin D is regarded as a broad immunoregulatory agent capable of modulating innate immune responses (Aranow, 2011).

A key effect of vitamin D on immunity is strengthening innate defense. In the presence of vitamin D, macrophages and monocytes produce antimicrobial peptides (e.g., cathelicidin, β-defensin), natural antibiotic-like molecules effective against bacteria, viruses, and fungi. For example, respiratory epithelial cells from individuals with sufficient vitamin D secrete more cathelicidin upon pathogen exposure, neutralizing infectious agents (Bilezikian et al., 2020). Vitamin D also prevents overactivation of macrophages and dendritic cells, reducing excessive inflammatory responses, protecting tissues, and preventing cytokine storms. Calcitriol suppresses pro-inflammatory cytokines such as TNF-α and IL-6 in macrophages (Skrobot et al., 2018).

In adaptive immunity, vitamin D primarily acts as a regulatory and tolerance-promoting factor. It induces a tolerogenic state by guiding dendritic cells toward a more quiescent phenotype and enhancing their capacity to induce regulatory T cells. This mechanism is critical in preventing excessive or inappropriate immune responses (e.g., autoimmune reactions).

Indeed, vitamin D deficiency is linked to increased prevalence of autoimmune diseases such as lupus, rheumatoid arthritis, multiple sclerosis, and type 1 diabetes, underscoring its role in immune tolerance. Vitamin D also modulates the Th1/Th2 balance by suppressing excessive Th1 (cellular) responses and supporting Th2 (humoral) responses as needed, balancing immune reactions. Adequate vitamin D maintains antibody responses while reducing tissue damage caused by overactive cytotoxic T cells (Ghaseminejad-Raeini et al., 2023).

Vitamin D deficiency causes significant immune alterations. Epidemiological studies report higher respiratory infection rates among individuals with low serum vitamin D levels (Smetana et al., 2021). Severe deficiency is known to worsen infectious disease courses like tuberculosis. This may be due to reduced microbicidal activity of macrophages and impaired functions of neutrophils and NK cells. Moreover, vitamin D deficiency may increase susceptibility to chronic inflammatory diseases such as inflammatory bowel disease and asthma. Therefore, maintaining optimal vitamin D levels through adequate sunlight exposure and consumption of vitamin D-rich foods (fatty fish, fish oil, fortified milk and cereals) is important for infection protection and inflammatory disease prevention (Ghaseminejad-Raeini et al., 2023).

Vitamin E is a fat-soluble vitamin with antioxidant functions in cell membranes. Immune system cells require vitamin E to protect the polyunsaturated fatty acid-rich membranes from oxidative damage. Vitamin E neutralizes free radicals, preserving membrane integrity and ensuring proper immune cell function. It supports both humoral immunity (antibody production) and cellular immunity (T cell responses) (Lee and Han, 2018).

Adequate vitamin E intake increases plasma cell differentiation from B cells and enhances T lymphocyte proliferation and IL-2 secretion. It has been especially important in revitalizing immune responses in elderly individuals. Natural killer (NK) cell activity, which declines with age, has been partially restored with high-dose vitamin E supplementation. Some studies report improved antibody responses to influenza and other respiratory infections in the elderly following vitamin E supplementation. Another critical effect of vitamin E is its ability to limit inflammation by reducing production of pro-inflammatory cytokines (e.g., IL-1, IL-6, TNF-α). Thus, vitamin E helps balance inflammatory mediators from macrophages and other immune cells, contributing to the balance between infection control and tissue damage (Venkatraman and Chu, 1999).

Vitamin E deficiency is rare but can occur especially in individuals with fat malabsorption and significantly affects immunity. Deficiency weakens antibody-mediated and cellular immune responses and reduces NK cell activity. Vitamin E-deficient experimental animals show increased susceptibility to infections and structural damage to immune organs (thymus, spleen). In humans, severe deficiency mainly presents with neurological symptoms, but suppressed immune responses have also been observed (Mainardi et al., 2024).

Optimal vitamin E intake in the general population is provided by consumption of healthy fat sources such as nuts, almonds, sunflower seeds, avocado, and vegetable oils (Pignitter, 2019). Although some studies have investigated the potential benefits of vitamin E supplementation on immune function in elderly people (with doses ranging 200–800 IU daily), results are inconsistent and high-dose vitamin E does not always provide additional benefits (Meydani et al., 2004; Niki, 2015). Nevertheless, maintaining sufficient vitamin E intake is necessary for sustaining healthy immune responses.

Effects of Minerals on the Immune System

Like vitamins, minerals are essential for a healthy immune response. Zinc, selenium, iron, copper, magnesium, and other trace elements are necessary for immune cell development, enzymatic reactions, and antioxidant defense systems. Deficiencies of these minerals increase susceptibility to infections, while excess levels can sometimes be harmful.

Zinc (Zn) is one of the most critical minerals for the immune system. Present in the structure of over 1,000 enzymes and transcription factors, zinc is especially required for the development and communication of immune cells. Zinc deficiency is a widespread micronutrient deficiency worldwide that severely impairs immune functions (Bonaventura et al., 2015). In zinc deficiency, the formation, activation, and maturation of lymphocytes (T and B cells) are disrupted; the thymus shrinks and peripheral T cell counts decrease. Additionally, the ratio of helper T cells ($CD4^+$) to suppressor/cytotoxic T cells ($CD8^+$) falls, IL-2 production declines, and the activity of natural killer (NK) cells and cytotoxic T cells weakens (Keusch, 2020). This multifaceted effect stems from zinc's fundamental role in both innate and adaptive immunity.

It is unsurprising that individuals with zinc deficiencies experience frequent and severe infections. For example, zinc deficiency markedly increases the incidence of diarrhea and pneumonia in children, and weakened immunity causes these infections to be more severe. The World Health Organization has emphasized the effect of zinc supplementation in reducing childhood diarrhea and respiratory infections in developing countries (Mandal and Lu, 2017). Zinc also plays a role in antiviral immunity; low zinc levels have been reported as factors worsening disease progression in infections such as HIV and Hepatitis C (Sumaily, 2022). Studies in the elderly have shown that zinc deficiency increases the risk of pneumonia and worsens prognosis (Barnett et al., 2010).

Zinc supports the immune system partly through its presence in antioxidant defense enzymes (e.g., superoxide dismutase), thus reducing oxidative stress. At the same time, zinc regulates inflammatory responses and prevents excessive inflammation from damaging tissues. For instance, zinc controls the NF-κB pathway and expression of related cytokine genes, allowing it to downregulate inflammatory reactions when necessary (Vasto et al., 2007). Therefore, zinc is indispensable for barrier immunity (integrity of skin and mucosa) as well as cellular and humoral immunity. It is important that dietary zinc sources such as seafood, meat, legumes, seeds, and nuts are included. In cases of zinc deficiency (especially in children, the elderly, and vegetarians), physician-guided zinc supplementation can effectively improve immunity. However, the benefits of zinc supplementation are mainly evident in those with deficiency; routine high-dose zinc intake by healthy individuals with adequate levels has limited evidence for additional benefits.

Selenium (Se) is a trace element found in essential antioxidant enzymes like glutathione peroxidase and is important for immune functions. Selenium deficiency can affect immune responses in various ways. First, selenium deficiency reduces the body's antioxidant capacity, which can trigger chronic inflammation and impair immune cell functions. Second, selenium deficiency has been shown to increase the virulence of viral pathogens. For example, Keshan disease, a cardiomyopathy caused by Coxsackie virus seen in selenium-poor soils in China, is associated with the virus becoming more aggressive under selenium deficiency (Lockitch, 1989). This is a striking example of selenium's effect on the course of viral infections.

At the immune cell level, selenium plays a role in both innate and adaptive immunity. The activity of enzymes involved in microbe killing by macrophages and neutrophils (e.g., myeloperoxidase system) depends on selenium. Similarly, T cell proliferation and B cell antibody production are sensitive to selenium availability. Selenium deficiency has been reported to weaken adaptive responses (e.g., reduced IgG production) and disrupt cytokine balance (Huang et al., 2012).

Selenium deficiency is not as widespread as zinc deficiency globally and is more closely related to regional dietary differences, such as the selenium content of soil and water. Selenium sources include seafood, offal (e.g., kidney), Brazil nuts, and whole grains. The recommended selenium intake (~55 μg/day for adults) is sufficient for maintaining immune health. Some studies have shown that selenium supplementation can enhance antibody responses to influenza vaccines or improve immune parameters in HIV patients (Broome et al., 2004), although there is no clear consensus on routine supplementation. Nevertheless, considering the effects of selenium deficiency on immune responses and viral mutations, ensuring adequate selenium intake remains an important public health goal.

Iron (Fe) is critical not only for blood cell production and oxygen transport but also for the immune system. Iron deficiency impairs energy production and DNA synthesis required for immune cell division and differentiation. Rapidly proliferating lymphocytes reduce production when iron is insufficient, weakening cellular immunity. Neutrophil and macrophage microbicidal mechanisms (myeloperoxidase system, NADPH oxidase complex) require iron as a cofactor; deficiency impairs pathogen elimination by these cells (Walker and Walker, 2000). However, findings on infection courses in iron-deficient individuals are mixed; while iron deficiency weakens defense against some microbes, limiting iron can also restrict growth of others (Abuga et al., 2023). Overall, iron-deficient persons' immune cells are less resilient to stress and infection, and recovery is prolonged.

Excess iron, conversely, creates an environment favorable for bacterial growth and increases infection risk. In conditions like hemochromatosis, where iron cannot be eliminated effectively, some pathogenic bacteria (e.g., Yersinia, Listeria) cause severe infections (Ganz, 2018). Therefore, during infection, the body increases hepcidin hormone to restrict iron absorption and release, limiting pathogen access to iron.

Preventing iron deficiency is necessary to maintain strong immunity, but iron supplementation should be controlled when infection is present. A balanced diet including iron sources such as red meat, poultry, fish, legumes, and green vegetables is recommended (Suardi et al., 2021).

Copper (Cu) acts as a cofactor for various enzymes (lysyl oxidase, cytochrome c oxidase, superoxide dismutase, etc.) involved in immunity (Angelova et al., 2011). Copper deficiency impairs neutrophil and macrophage microbe-killing functions and increases infection risk, especially in newborns and the elderly. It can also reduce antibody responses and lymphocyte counts (Prohaska et al., 1990). Excess copper can generate free radicals that damage cells, so the body tightly regulates its levels (Uriu-Adams and Keen, 2005).

Magnesium (Mg) is needed for numerous immune cell processes ranging from energy production to signal transduction. Low magnesium levels are linked to decreased immune cell activity, increased oxidative stress, and elevated inflammation. In particular, levels of inflammatory cytokines such as IL-6 can rise with magnesium deficiency, raising chronic inflammation risk. Magnesium is also important for antibody production and lymphocyte microenvironment function. Good dietary sources include leafy green vegetables, nuts, and whole grains. Supplementation has been reported in some studies to reduce inflammatory markers in deficient individuals (Ashique et al., 2023).

Iodine, while primarily required for thyroid hormone synthesis, is indirectly related to the immune system since thyroid hormones regulate immune function. Severe iodine deficiency can reduce resistance to infections. The relationship between immunity and iodine in autoimmune thyroid diseases remains an active research area (Sorrenti et al., 2021).

The European Society for Clinical Nutrition and Metabolism (ESPEN) reports that low levels of vitamins A, E, B6, B12, zinc, and selenium are associated with worse clinical outcomes in patients with viral infections (Pecora et al., 2020). Therefore, adequate dietary intake of each mineral at recommended levels is necessary for the holistic health of the immune system.

Effects of Macronutrients on Immunity

Macronutrients—proteins, fats, and carbohydrates—not only serve as energy sources but also fulfill the structural and functional requirements of the immune system. Their inadequate or excessive intake can disrupt immune response balance.

Proteins are the building blocks of the immune system: immune cells themselves, antibodies (immunoglobulins), complement system proteins, cytokines, and receptors are all proteins. Therefore, protein deficiency severely weakens immune functions. Protein-energy malnutrition (PEM) can cause immune system collapse, especially in children and the elderly. In PEM, thymus atrophy, decreased T lymphocyte counts, and striking reductions in cellular immunity occur, which can lead to severe infections. For example, children with severe acute malnutrition (kwashiorkor, marasmus) have very high mortality rates from infections such as measles, diarrhea, and pneumonia due to collapsed host immunity (Cruzat et al., 2014).

Adequate protein nutrition is a prerequisite for a strong immune response against infections. Certain amino acids have special roles in modulating immune responses. For instance, L-arginine is critical for T cell functions. Under injury and stress, arginine becomes "conditionally essential," meaning the body cannot meet demand and must obtain it from the diet. Research shows arginine supplementation strengthens T cell responses, increases lymphocyte numbers, and boosts helper T cell activity. Especially under surgical stress or severe infections, arginine support improves wound healing and immune responses (Daly et al., 1990). Similarly, glutamine is a preferred fuel for intestinal epithelial and immune cells; glutamine supplementation in intensive care patients helps maintain the gut barrier and reduce infection rates. Branched-chain amino acids (BCAAs; leucine, isoleucine, valine) support protein synthesis in muscles and immune cells, aiding recovery after trauma or infection (Xu et al., 2025).

Protein quality is also important. Animal proteins provide a complete amino acid profile, whereas vegetarians may need to combine various plant protein sources to obtain all essential amino acids. Well-planned plant-based diets can support immunity; for example, proteins from legumes complemented with cereals provide balanced amino acids (Mariotti and Gardner, 2019). In conclusion, adequate and balanced protein intake is indispensable for producing the immune system's "soldiers" (cells and antibodies). Protein deficiency must be corrected, and medical nutrition support should be provided if necessary to prevent immunodeficiency. On the other hand, there is no evidence that excessive protein intake directly enhances immunity; the key point is meeting needs and, especially in critical conditions, providing specific amino acid supplementation.

Fats serve as energy sources and cell membrane components in the immune system (Yildiz et al., 2016). However, the amount and especially the type of fatty acids consumed can profoundly affect immune responses. Dietary fatty acid composition is a major determinant of the balance of inflammatory mediators in the body. Excess saturated fatty acids and omega-6 polyunsaturated fatty acids (PUFAs) can activate pro-inflammatory pathways. Particularly, some eicosanoids derived from arachidonic acid (e.g., PGE2, LTB4) promote inflammation and immune cell migration. In obese individuals consuming excessive fat and calories, adipose tissue produces continuous low-level inflammatory signals via adipokines; this condition is known as chronic inflammation or "metaflammation." This chronic inflammatory state associated with obesity adversely affects immune cell function, disrupts responses to infections, and predisposes to autoinflammatory diseases. For example, more severe courses of influenza and pneumonia in obese individuals are partly linked to this chronic inflammation and immune dysfunction (Rohm et al., 2022).

Conversely, omega-3 fatty acids (eicosatetraenoic acid—EPA and docosahexaenoic acid—DHA) have potent anti-inflammatory effects. Adequate omega-3 intake alters the phospholipid composition of cell membranes, modifying the structure and effects of eicosanoids produced during inflammation. Lipid mediators derived from EPA and DHA, such as resolvins, protectins, and maresins, accelerate resolution of inflammation, shorten neutrophil tissue residency, and promote macrophage tissue repair. Omega-3s also directly modulate T and B cell functions, dampening excessive immune responses when needed. Therefore, omega-3 supplementation has been reported to alleviate symptoms and reduce inflammatory markers in autoimmune diseases such as rheumatoid arthritis and lupus. There is also preliminary evidence that omega-3 supplementation may help control cytokine storms in viral infections (e.g., COVID-19). Adequate omega-3 intake may slow "inflammaging" (age-related chronic inflammation) and delay immunosenescence in older adults (Poggioli et al., 2023).

Other fats: Medium-chain triglycerides (MCTs) provide rapid energy but have limited immunomodulatory effects. Monounsaturated fatty acids (e.g., olive oil fatty acids) are considered neutral or mildly anti-inflammatory and contribute to benefits of the Mediterranean diet. Trans fatty acids are known for pro-inflammatory effects and are not recommended for immune health (Poggioli et al., 2023). Both the quantity and quality of fat intake matter. Total fat intake should be kept to about 30% of energy, saturated and trans fats limited, and omega-3-rich and antioxidant-rich seeds/seafood preferred. This ensures structural components needed by the immune system while reducing chronic inflammation risk.

Carbohydrates provide rapid energy for immune cells. During infection, immune cells proliferate and activate using glucose. Therefore, diets that completely restrict carbohydrates can weaken immune responses over time. However, the type of carbohydrates matters for immune health. Excess refined sugars and high glycemic index foods increase obesity and inflammation risk. Persistent postprandial hyperglycemia can trigger low-level inflammatory signals and temporarily suppress neutrophil functions (Marik, 2009).

Dietary fiber, the indigestible portion of carbohydrates, has indirect but significant effects on immunity. Soluble fibers and resistant starch are fermented by gut microbiota to produce short-chain fatty acids (SCFAs). SCFAs such as acetate, propionate, and butyrate serve not only as nutrient sources for intestinal epithelial cells but also as powerful immunomodulatory molecules. Butyrate especially maintains colonic epithelial integrity and strengthens the gut barrier. A robust gut barrier prevents pathogen and toxin translocation into systemic circulation, reducing chronic inflammation. Additionally, SCFAs regulate Treg production via G-protein coupled receptors (GPR41, GPR43) in the gut lumen and suppress excessive inflammatory T cell responses (Parada Venegas et al., 2019). Butyrate exerts anti-inflammatory effects in inflammatory bowel disease and reduces gut inflammation in experimental models. SCFAs also affect immune cell production in bone marrow and spleen, shaping systemic immunity (Kim, 2023).

A prebiotic, fiber-rich diet supports beneficial microbial populations (e.g., *Bifidobacterium* and *Lactobacillus* spp.) and suppresses potential pathogens. This competitive inhibition reduces infection risk and promotes healthy microbiota metabolites, such as vitamins and SCFAs. For example, oligosaccharides in breast milk nourish beneficial bifidobacteria in infants' guts, contributing to newborn protection from infections (Stuivenberg et al., 2022). Carbohydrates thus have dual effects on immunity: a diet rich in sufficient and complex carbohydrates (whole grains, legumes, vegetables, fruits) meets energy needs and positively influences the gut–immune axis, whereas excessive simple sugar intake promotes obesity and inflammation, weakening immunity. Maintaining this balance is essential for robust infection responses and reduced chronic inflammation. Phytochemicals in plant foods, though not "essential nutrients" in the classical sense, exhibit beneficial biological activities. Hundreds of phytochemicals—polyphenols, flavonoids, carotenoids, and organosulfur compounds—are regarded as supporters of the immune system due to their antioxidant and anti-inflammatory effects. Increasing evidence supports their immunomodulatory roles. Polyphenols, abundant in fruits, vegetables, spices, tea, coffee, and cocoa, include flavonoids (quercetin, kaempferol), stilbenes (resveratrol), phenolic acids (ferulic and chlorogenic acid), and curcuminoids (curcumin in turmeric) (Yildiz et al., 2021; Cemal and Yildiz, 2022; Barel and Yildiz, 2024; Yildiz et al., 2025). Their immune effects mainly occur through modulation of inflammatory signaling pathways. These compounds bind to various receptors and enzymes in immune cells, altering cell functions. For example, some polyphenols inhibit nuclear factor kappa B (NF-κB), reducing the expression of pro-inflammatory genes, including cytokines and adhesion molecules. Higher flavonoid intake has been associated with decreased inflammatory markers, such as C-reactive protein (CRP) (Munteanu and Schwartz, 2022).

Polyphenols also affect intracellular signaling proteins like kinases and phosphatases. Research shows that they activate antioxidant defense mechanisms (e.g., Nrf2 pathway) while preventing overactivation of inflammatory complexes such as the NLRP3 inflammatory complexes. Resveratrol has been found to regulate mTOR and AMPK pathways, steering immune cell metabolism toward a less inflammatory phenotype. These effects indicate polyphenols are not mere antioxidants but also regulators of cellular signaling and gene expression (Ferreira et al., 2024).

Polyphenols influence adaptive immunity by modulating T cell polarization and suppressing autoimmunity. For instance, curcumin promotes tolerogenic dendritic cells, enhancing Treg and Th2 cells while inhibiting pathogenic Th17 cells (Hewlings and Kalman, 2017). Flavonoids like quercetin also possess antiviral effects, inhibiting replication of influenza and other viruses in cell culture. Regular consumption of polyphenol-rich foods is thus recommended to prevent infections and reduce chronic inflammation (Mehrbod et al., 2020).

Synergy of phytochemicals: Polyphenols naturally occur as complex mixtures, often enhancing each other's effects. For example, a cup of green tea contains catechins and other polyphenols, and different colors of fruits and vegetables represent various phytochemicals (Doughari, 2012). This diversity supports the immune system through multiple targets. Their antioxidant capacity protects immune cells from oxidative damage and preservs their functions. They also positively affect microbiota, as many polyphenols are metabolized in the gut and nourish beneficial bacteria, providing indirect immune benefits (Yahfoufi et al., 2018).

Therefore, it is unsurprising that Mediterranean-style diets rich in vegetables, fruits, legumes, olive oil, and nuts are associated with lower inflammation and better immune balance compared to Western diets. Clinical studies report reductions in inflammatory markers like CRP and IL-6 and improved immune cell function after adopting the Mediterranean diet (Estruch et al., 2018). This effect is largely attributed to the diet's rich phytochemical content.

Other Bioactive Compounds

Beyond polyphenols, plant-based foods contain other compounds that influence immunity:

- *Carotenoids (beta-carotene, lycopene, lutein):* These pigments can have provitamin A activity and also act directly as antioxidants protecting cells. Beta-carotene can convert to vitamin A when needed, supporting immune responses, while carotenoids such as lycopene reduce LDL oxidation and inflammation primarily through their antioxidant capacity. A diet rich in carotenoids (carrots, tomatoes, sweet potatoes, dark green leafy vegetables) may help build a body more resistant to infections (Dewanjee et al., 2021).
- *Allium compounds (components of garlic and onion):* Allicin and its derivatives found in garlic are known for their antimicrobial properties. There is evidence that these sulfur-containing compounds can enhance macrophage activity and natural killer (NK) cell function. Small studies have suggested that garlic consumption may shorten the duration of the common cold (El-Saadony et al., 2024).
- *Gingerol and capsaicin:* Gingerol compounds in ginger and capsaicin in chili peppers exhibit anti-inflammatory effects by suppressing pain and inflammation mediators. Laboratory studies have observed that these compounds can modulate cytokine secretion from immune cells (Surh, 1998).
- *Probiotic metabolites:* Organic acids, bacteriocins, and peptides found in fermented foods can have immune-supportive effects (Avci et al., 2024).

All these bioactive compounds collectively form a network that supports immunity when consumed together. Therefore, a diet including a variety of plant-based foods is generally more effective than taking individual compounds as supplements. Although absorption rates and bioavailability of phytochemicals vary, consistent intake can maintain protective levels in the body.

Probiotics, Prebiotics, and Gut Microbiota

Trillions of microorganisms living in the human gut (microbiota) act like a "forgotten organ," playing a vital role in the development and regulation of the immune system. Probiotics are live microorganisms that, when consumed in adequate amounts, confer health benefits to the host (typically lactic acid bacteria and bifidobacteria). Prebiotics are indigestible food components (e.g., inulin, fructooligosaccharides) that support the growth of these beneficial microorganisms. Together, probiotics and prebiotics are called symbiotic and influence the immune system through gut health.

Probiotic bacteria modulate immune responses particularly via GALT (gut-associated lymphoid tissue). Gut epithelial cells and M cells capture microbial antigens in the presence of probiotics and present them to underlying immune cells (dendritic cells, macrophages, B and T cells). Probiotics have been shown to alter cytokine and chemokine profiles secreted by the gut epithelium, shaping the immune microenvironment in the lamina propria. For example, some *Lactobacillus* species promote the secretion of anti-inflammatory cytokines like IL-10 from epithelial cells while suppressing pro-inflammatory chemokines like IL-8 induced by pathogens. This helps establish a balanced immune response (immune homeostasis) in the gut (Mazziotta et al., 2023).

One proven effect of probiotics is the increased production of IgA antibodies. Well-studied mechanisms show that probiotics stimulate intestinal B cells to differentiate into plasma cells that secrete IgA antibodies. This secretory IgA binds pathogens in the gut lumen, preventing their adhesion and invasion of epithelium. Consequently, individuals consuming probiotics may have higher IgA levels in the gut mucosa and secretions like saliva, strengthening the first line of defense against infection (Mazziotta et al., 2023).

Additionally, probiotics have been found to enhance natural killer (NK) cell activity and macrophage phagocytosis capacity. Some studies with healthy volunteers reported increased NK cell activity in peripheral blood mononuclear cells after several weeks of consuming specific probiotic strains (e.g., *Lactobacillus casei* Shirota) (Dong et al., 2013). This suggests probiotics can influence not only local but also systemic immune parameters. Likely, some compounds produced by probiotics or cytokines secreted by dendritic cells interacting with probiotics enter systemic circulation and affect distant immune cells.

Clinically, protective effects of probiotics have been investigated in various infections. Meta-analyses show probiotic use shortens the duration of acute diarrhea (e.g., rotavirus diarrhea) and reduces the risk of antibiotic-associated diarrhea. Data also supports benefits in respiratory tract infections: for example, studies in school-age children reported reduced frequency and duration of common colds in groups receiving probiotics (Luoto et al., 2014). The likely role of probiotics here is enhancing general immune resistance and strengthening mucosal immunity. During the COVID-19 pandemic, probiotics and prebiotics were proposed as adjunctive measures since gut microbiota may influence lung immunity (gut-lung axis) and thus respiratory infection outcomes (Dhar and Mohanty, 2020).

It should be noted that the effects of probiotics vary by strain. The *Lactobacillus* and *Bifidobacterium* genera are the most used probiotics, generally regarded as safe and known for their immune-supporting properties. However, differences exist even between species and strains. For instance, *L. rhamnosus* GG is effective in diarrheal patients, while *L. reuteri* is prominent infant colic treatment. Therefore, probiotic selection should align with the intended purpose. Prebiotic foods (e.g., inulin, oligofructose, galacto-oligosaccharides) ferment in the colon, promoting growth of beneficial bacteria. Thus, even without probiotic supplementation, activity of resident beneficial microbiota can increase, strengthening the microbiota's positive immune effects. Some studies showed that prebiotic intake reduces respiratory infection risk in infants and children and prevents development of atopy and eczema (Osborn and Sinn, 2013).

Regarding the overall health effects of microbiota composition, highly hygienic and fiber-poor diets reduce microbial diversity, a factor hypothesized to be linked to increased autoimmune and allergic diseases in modern societies (Hygiene hypothesis and microbiota). A rich and balanced microbiota shaped by sufficient fiber intake, fermented foods (e.g., homemade pickles, kefir, yogurt), and probiotics, when necessary, acts like an immune education center. Tolerance develops to harmless antigens while effective responses to pathogens are maintained (Kucuksezer et al., 2023).

In summary, probiotics and prebiotics are important tools that strengthen the immune system through the gut microbiota. Probiotic bacteria support immunological balance in the gut, enhance antibody production and phagocytosis, and regulate inflammation. Prebiotics maintain these effects by nourishing beneficial microbes. Including probiotic foods such as yogurt, kefir, fermented soy (miso, tempeh), sauerkraut, alongside prebiotic-rich vegetables, fruits, whole grains, and legumes in the diet, supports immune health in daily life.

Conclusion

Nutrition is a multidimensional process that directly influences not only the physical aspects of human health but also its evolutionary, metabolic, sociocultural, and political dynamics. Historically, the transition from the diverse and natural dietary patterns of the Paleolithic era to the widespread consumption of processed and refined foods in modern times has brought numerous health challenges. The insufficient adaptation of our bodies at genetic, metabolic, and microbial levels to this evolutionary shift has contributed to the increasing prevalence of chronic diseases such as obesity, type 2 diabetes, cardiovascular diseases, inflammatory bowel diseases, and mental disorders in contemporary societies.

While a balanced intake of macronutrients (carbohydrates, fats, proteins) and micronutrients (vitamins, minerals) forms the cornerstone of a healthy life, the quality and source of these components are as crucial as their quantity. Diets high in refined sugars, trans fats, and low in fiber negatively affect not only energy balance but also the gut microbiota, immune system, and hormonal regulation. In contrast, dietary patterns rich in unprocessed vegetables and fruits, whole grains, legumes, healthy fats, and fermented foods tend to align better with our genetic makeup and exhibit protective effects on metabolic and microbial health. From this perspective, the evolutionary medicine approach emphasizes that nutrition should be considered beyond calorie intake or food choices, integrating genetic, environmental, and behavioral factors. The naturalness, diversity, and low level of processing inherent in Paleolithic and traditional diets appear more compatible with health compared to modern dietary patterns. However, this approach does not advocate a return to the past but aims to develop sustainable nutritional models scientifically adapted to present-day conditions by learning from ancestral diets.

Moreover, access to healthy nutrition cannot be achieved solely through individual knowledge and awareness; it is directly linked to economic, social, and political factors. Food insecurity, income inequality, and disparities in education levels impede large segments of the population from maintaining healthy eating habits. Therefore, health policies must adopt a holistic approach that encompasses not only individuals but also the food system as a whole, prioritizing regulations that protect public health, ensure equitable access to nutritious food, and embrace environmentally friendly practices.

In conclusion, Hippocrates' famous dictum, "Let food be your medicine," has evolved into not only a philosophical maxim but a scientifically grounded public health strategy today. Human physiology is designed for a natural and balanced diet aligned with our evolutionary history. Hence, it is equally important for individuals to make informed dietary choices and for societies to take measures ensuring equitable and sustainable access to food. Recognizing food as a tool for health represents the most valuable investment in cultivating healthier generations for the future.

References

Abuga, K. M., Nairz, M., MacLennan, C. A., & Atkinson, S. H. (2023). Severe anemia, iron deficiency, and susceptibility to invasive bacterial infections. *Welcome Open Research*, *8*, 48.

Afroza, U., Abrar, A. K., Nowar, A., Sobhan, S. M. M., Ide, N., & Choudhury, S. R. (2024). Global overview of government-endorsed nutrition labeling policies of packaged foods: a document review. *Frontiers in Public Health*, *12*, 1426639.

Afshin, A., Sur, P. J., Fay, K. A., Cornaby, L., Ferrara, G., Salama, J. S., ... & Murray, C. J. (2019). Health effects of dietary risks in 195 countries, 1990–2017: a systematic analysis for the Global Burden of Disease Study 2017. *The lancet*, *393*(10184), 1958-1972.

Ahvanooei, M. R., Norouzian, M. A., & Vahmani, P. (2021). Beneficial effects of vitamins, minerals, and bioactive peptides on strengthening the immune system against COVID-19 and the role of cow's milk in the supply of these nutrients. *Biological trace element research*, 1-14.

Allam, A. H., Thompson, R. C., Wann, L. S., Miyamoto, M. I., Nur el-Din, A. E. H., el-Maksoud, G. A., ... & Thomas, G. S. (2011). Atherosclerosis in ancient Egyptian mummies: the Horus study. *JACC: Cardiovascular Imaging*, *4*(4), 315-327.

Alt, K. W., Al-Ahmad, A., & Woelber, J. P. (2022). Nutrition and health in human evolution–past to present. *Nutrients*, *14*(17), 3594.

Andrès, E., Lorenzo-Villalba, N., Terrade, J. E., & Méndez-Bailon, M. (2024). Fat-soluble vitamins A, D, E, and K: review of the literature and points of interest for the clinician. *Journal of Clinical Medicine*, *13*(13), 3641.

Angelova, M., Asenova, S., Nedkova, V., & Koleva-Kolarova, R. (2011). Copper in the human organism. *Trakia journal of sciences*, *9*(1), 88-98.

Antonelli, M., & Donelli, D. (2023, October). Precision Nutrition and Artificial Intelligence Mobile Apps: A Narrative Review. In *Biology and Life Sciences Forum*, 29 (1), 25. MDPI.

Aranow, C. (2011). Vitamin D and the immune system. *Journal of investigative medicine*, *59*(6), 881-886.

Arlinghaus, K. R., & Johnston, C. A. (2019). The importance of creating habits and routine. *American journal of lifestyle medicine*, *13*(2), 142-144.

Ashique, S., Kumar, S., Hussain, A., Mishra, N., Garg, A., Gowda, B. J., ... & Taghizadeh-Hesary, F. (2023). A narrative review on the role of magnesium in immune regulation, inflammation, infectious diseases, and cancer. *Journal of Health, Population and Nutrition*, *42*(1), 74.

Assaf, S., Park, J., Chowdhry, N., Ganapuram, M., Mattathil, S., Alakeel, R., & Kelly, O. J. (2024). Unraveling the Evolutionary Diet Mismatch and Its Contribution to the Deterioration of Body Composition. *Metabolites*, *14*(7), 379.

Avci, G. A., Yilmaz, Ü. İ., & Avci, E. (2024). Efficacy of probiotics, paraprobiotics, and postbiotics in colorectal cancer cell line and their role in immune response. *Revista da Associação Médica Brasileira*, *70*(6), e20240226.

Bai, S., Zhang, M., Tang, S., Li, M., Wu, R., Wan, S., ... & Feng, S. (2024). Effects and Impact of Selenium on Human Health, A Review. *Molecules*, *30*(1), 50.

Barel, O., and Yildiz, G. (2024). The effect of ultrasound pre-treatment on the color, antioxidant capacity and total phenolic content of freeze-dried banana slices, Latin American Applied Research Journal, 54(3): 329-335.

Barker, T. (2023). Vitamins and human health: Systematic reviews and original research. *Nutrients, 15*(13), 2888.

Barnett, J. B., Hamer, D. H., & Meydani, S. N. (2010). Low zinc status: a new risk factor for pneumonia in the elderly?. *Nutrition reviews, 68*(1), 30-37.

Bilezikian, J. P., Bikle, D., Hewison, M., Lazaretti-Castro, M., Formenti, A. M., Gupta, A., ... & Giustina, A. (2020). Mechanisms in endocrinology: vitamin D and COVID-19. *European journal of endocrinology, 183*(5), R133-R147.

Bonaventura, P., Benedetti, G., Albarède, F., & Miossec, P. (2015). Zinc and its role in immunity and inflammation. *Autoimmunity reviews, 14*(4), 277-285.

Bonekamp, N. E., Visseren, F. L., Cramer, M. J., Dorresteijn, J. A., van der Meer, M. G., Ruigrok, Y. M., ... & Koopal, C. (2024). Long-term lifestyles change and risk of mortality and type 2 diabetes in patients with cardiovascular disease. *European journal of preventive cardiology, 31*(2), 205-213.

Bowen, K. J., Harris, W. S., & Kris-Etherton, P. M. (2016). Omega-3 fatty acids and cardiovascular disease: are there benefits?. *Current treatment options in cardiovascular medicine, 18*, 1-16. DOI

Bragazzi, N. L., Del Rio, D., Mayer, E. A., & Mena, P. (2024). We are what, when, and how we eat: the evolutionary impact of dietary shifts on physical and cognitive development, health, and disease. *Advances in Nutrition*, 100280.

Broome, C. S., McArdle, F., Kyle, J. A., Andrews, F., Lowe, N. M., Hart, C. A., ... & Jackson, M. J. (2004). An increase in selenium intake improves immune function and poliovirus handling in adults with marginal selenium status. *The American journal of clinical nutrition, 80*(1), 154-162.

Caballero, B. (2007). The global epidemic of obesity: an overview. *Epidemiologic reviews, 29*(1), 1-5.

Calder, P. C. (2015). Functional roles of fatty acids and their effects on human health. *Journal of parenteral and enteral nutrition, 39*, 18S-32S.

Carr, A. C., & Maggini, S. (2017). Vitamin C and immune function. *Nutrients, 9*(11), 1211.

Cemal, D., and Yildiz, G. (2022). The effect of drying temperature, cutting types, storage degree and period on physical, chemical and physicochemical properties of red capia pepper, Latin American Applied Research Journal, 52 (1):67-72.

Chambers, E. S., Byrne, C. S., & Frost, G. (2019). Carbohydrate and human health: is it all about quality?. *The Lancet, 393*(10170), 384-386.

Clemente-Suárez, V. J., Mielgo-Ayuso, J., Martín-Rodríguez, A., Ramos-Campo, D. J., Redondo-Flórez, L., & Tornero-Aguilera, J. F. (2022). The burden of carbohydrates in health and disease. *Nutrients, 14*(18), 3809.

Constantini, N. W., Dubnov-Raz, G., Eyal, B. B., Berry, E. M., Cohen, A. H., & Hemilä, H. (2011). The effect of vitamin C on upper respiratory infections in adolescent swimmers: a randomized trial. *European journal of pediatrics, 170*, 59-63.

Cordain, L., Eaton, S. B., Sebastian, A., Mann, N., Lindeberg, S., Watkins, B. A., ... & Brand-Miller, J. (2005). Origins and evolution of the Western diet: health implications for the 21st century1, 2. *The American journal of clinical nutrition, 81*(2), 341-354.

Cruzat, V. F., Krause, M., & Newsholme, P. (2014). Amino acid supplementation and impact on immune function in the context of exercise. *Journal of the international Society of Sports Nutrition, 11*(1), 61.

Daly, J. M., Reynolds, J., Sigal, R. K., Shou, J. I. A. N., & Liberman, M. D. (1990). Effect of dietary protein and amino acids on immune function. *Critical care medicine, 18*(2), S94.

Daly, J. M., Reynolds, J., Sigal, R. K., Shou, J. I. A. N., & Liberman, M. D. (1990). Effect of dietary protein and amino acids on immune function. *Critical care medicine, 18*(2), S94.

Darmon, N., & Drewnowski, A. (2008). Does social class predict diet quality?. *The American journal of clinical nutrition, 87*(5), 1107-1117.

Dewanjee, S., Bhattacharjee, N., Chakraborty, P., & Bhattacharjee, S. (2021). Carotenoids as antioxidants. *Carotenoids: structure and function in the human body*, 447-473.

Dhar, D., & Mohanty, A. (2020). Gut microbiota and Covid-19-possible link and implications. *Virus research, 285*, 198018. https://doi.org/10.1016/j.virusres.2020.198018.

Dong, H., Rowland, I., Thomas, L. V., & Yaqoob, P. (2013). Immunomodulatory effects of a probiotic drink containing Lactobacillus casei Shirota in healthy older volunteers. *European journal of nutrition, 52*, 1853-1863.

Doughari, J. H. (2012). *Phytochemicals: extraction methods, basic structures and mode of action as potential chemotherapeutic agents* (pp. 1-33). Rijeka, Croatia: INTECH Open Access Publisher.

Elder, P. J., Ramsden, D. B., Burnett, D., Weickert, M. O., & Barber, T. M. (2018). Human amylase gene copy number variation as a determinant of metabolic state. *Expert Review of Endocrinology & Metabolism*, *13*(4), 193-205.

El-Saadony, M. T., Saad, A. M., Korma, S. A., Salem, H. M., Abd El-Mageed, T. A., Alkafaas, S. S., ... & Ibrahim, S. A. (2024). Garlic bioactive substances and their therapeutic applications for improving human health: A comprehensive review. *Frontiers in immunology*, *15*, 1277074.

Estruch, R., Ros, E., Salas-Salvadó, J., Covas, M. I., Corella, D., Arós, F., ... & Martínez-González, M. A. (2018). Retraction and republication: primary prevention of cardiovascular disease with a Mediterranean diet. N Engl J Med 2013; 368: 1279-90.

Evershed, R. P., Davey Smith, G., Roffet-Salque, M., Timpson, A., Diekmann, Y., Lyon, M. S., ... & Thomas, M. G. (2022). Dairying, diseases and the evolution of lactase persistence in Europe. *Nature*, *608*(7922), 336-345.

Fanzo, J., Bellows, A. L., Spiker, M. L., Thorne-Lyman, A. L., & Bloem, M. W. (2021). The importance of food systems and the environment for nutrition. *The American Journal of Clinical Nutrition*, *113*(1), 7-16.

Farag, M. A., Abib, B., Qin, Z., Ze, X., & Ali, S. E. (2023). Dietary macrominerals: Updated review of their role and orchestration in human nutrition throughout the life cycle with sex differences. *Current Research in Food Science*, *6*, 100450.

Fekete, M., Lehoczki, A., Csípő, T., Fazekas-Pongor, V., Szappanos, Á., Major, D., ... & Varga, J. T. (2024). The Role of Trace Elements in COPD: Pathogenetic Mechanisms and Therapeutic Potential of Zinc, Iron, Magnesium, Selenium, Manganese, Copper, and Calcium. *Nutrients*, *16*(23), 4118.

Ferreira, C., Vieira, P., Sá, H., Malva, J., Castelo-Branco, M., Reis, F., & Viana, S. (2024). Polyphenols: Immunonutrients tipping the balance of immunometabolism in chronic diseases. *Frontiers in Immunology*, *15*, 1360065.

Ferreira, C., Vieira, P., Sá, H., Malva, J., Castelo-Branco, M., Reis, F., & Viana, S. (2024). Polyphenols: Immunonutrients tipping the balance of immunometabolism in chronic diseases. *Frontiers in Immunology*, *15*, 1360065.

Food and Agriculture Organization (FAO) of the United Nations, International Fund for Agricultural Development (IFAD), United Nations Children's Fund (UNCEF), World Food Programme (WFP), & World Health Organization (WHO). (2020). The State of Food Security and Nutrition in the World 2020: Transforming food systems for affordable healthy diets. Rome: FAO.

Food and Agriculture Organization of the United Nations (FAO). (2023). Sustainable Food and Agriculture. Rome: FAO.

Fumagalli, M., Moltke, I., Grarup, N., Racimo, F., Bjerregaard, P., Jørgensen, M. E., ... & Nielsen, R. (2015). Greenlandic Inuit show genetic signatures of diet and climate adaptation. *Science*, *349*(6254), 1343-1347.

Ganz, T. (2018). Iron and infection. *International journal of hematology*, *107*(1), 7-15.

Gay, R., & Meydani, S. N. (2001). The effects of vitamin E, vitamin B6, and vitamin B12 on immune function. *Nutrition in clinical care*, *4*(4), 188-198.

Ghaseminejad-Raeini, A., Ghaderi, A., Sharafi, A., Nematollahi-Sani, B., Moossavi, M., Derakhshani, A., & Sarab, G. A. (2023). Immunomodulatory actions of vitamin D in various immune-related disorders: A comprehensive review. *Frontiers in Immunology*, *14*, 950465.

Gill, S. K., Rossi, M., Bajka, B., & Whelan, K. (2021). Dietary fiber in gastrointestinal health and disease. *Nature Reviews Gastroenterology & Hepatology*, *18*(2), 101-116.

Gropper, S. S., & Smith, J. L. (2013). *Advanced nutrition and human metabolism*. Cengage Learning.

Harvard Health Publishing. (2023, May 22). How much water should you drink? Harvard Medical School.

Hatch-McChesney, A., & Lieberman, H. R. (2022). Iodine and iodine deficiency: a comprehensive review of a re-emerging issue. *Nutrients*, *14*(17), 3474.

Hemilä, H. (2004). Vitamin C supplementation and respiratory infections: a systematic review. *Military medicine*, *169*(11), 920-925.

Hewlings, S. J., & Kalman, D. S. (2017). Curcumin: A review of its effects on human health. *Foods*, *6*(10), 92.

Huang, Z., Liu, Y., Qi, G., Brand, D., & Zheng, S. G. (2018). Role of vitamin A in the immune system. *Journal of clinical medicine*, *7*(9), 258.

Huang, Z., Rose, A. H., & Hoffmann, P. R. (2012). The role of selenium in inflammation and immunity: from molecular mechanisms to therapeutic opportunities. *Antioxidants & redox signaling*, *16*(7), 705-743.

Hudson, J. L., Wang, Y., Bergia III, R. E., & Campbell, W. W. (2020). Protein intake greater than the RDA differentially influences whole-body lean mass responses to purposeful catabolic and anabolic stressors: a systematic review and meta-analysis. *Advances in Nutrition*, *11*(3), 548-558.

Inchauspé, J. (2022). *Glucose Revolution: The life-changing power of balancing your blood sugar*. Simon and Schuster.

IPCC. (2018). Global warming? Nature.

Izli, G., Izli, N., Taskin, O., and Yildiz, G. (2018). Convective drying of kumquat slices: Comparison of different drying temperatures on drying kinetics, colour, total phenolic content and antioxidant capacity. *Latin American Applied Research Journal*, 48:37-42.

İzli, G., Yildiz, G., and Berk, S. E. (2022). Quality retention in pumpkin powder dried by combined microwave-convective drying, *Journal of Food Science and Technology*, 59 (4): 1558-1569.

Jang, I. S., Ko, Y. H., Moon, Y. S., & Sohn, S. H. (2014). Effects of vitamin C or E on the pro-inflammatory cytokines, heat shock protein 70 and antioxidant status in broiler chicks under summer conditions. *Asian-Australasian journal of animal sciences*, *27*(5), 749.

Jarisch, R., Weyer, D., Ehlert, E., Koch, C. H., Pinkowski, E., Jung, P., ... & Koch, A. (2014). Impact of oral vitamin C on histamine levels and seasickness. *Journal of Vestibular Research*, *24*(4), 281-288.

Jéquier, E., & Constant, F. (2010). Water as an essential nutrient: the physiological basis of hydration. *European journal of clinical nutrition*, *64*(2), 115-123.

Jiang, S., Yildiz, G., Ding., J., Andrade, J., Rababah, T. M., Almajwalc, A., Abulmeatyc, M. M., Feng, H. (2019). Pea Protein Nanoemulsion and Nanocomplex as Carriers for Protection of Cholecalciferol (vitamin D3). *Food and Bioprocess Technology*, 12 (6): 1031-1040.

Johnson, L. J., Meacham, S. L., & Kruskall, L. J. (2003). The antioxidants-vitamin C, vitamin E, selenium, and carotenoids. *Journal of agromedicine*, *9*(1), 65-82.

Keusch, G. T. (2020). Malnutrition and the thymus gland. In *Nutrient modulation of the immune response* (pp. 283-300). CRC Press.

Kim, C. H. (2023). Complex regulatory effects of gut microbial short-chain fatty acids on immune tolerance and autoimmunity. *Cellular & molecular immunology*, *20*(4), 341-350.

Krautkramer, K. A., Fan, J., & Bäckhed, F. (2021). Gut microbial metabolites as multi-kingdom intermediates. *Nature Reviews Microbiology*, *19*(2), 77-94.

Kucuksezer, U. C., Ozdemir, C., Yazici, D., Pat, Y., Mitamura, Y., Li, M., ... & Akdis, C. A. (2023). The epithelial barrier theory: development and exacerbation of allergic and other chronic inflammatory diseases. *Asia Pacific Allergy*, *13*(1), 28-39.

Lee, H., Yildiz, G., Dos Santos, L. C., Jiang, S., Andrade, J., Engeseth, N. C., and Feng, H. (2016). Soy protein nano-aggregates with improved functional properties prepared by sequential pH treatment and ultrasonication. *Food Hydrocolloids*, 55: 200–209.

Lee, G. Y., & Han, S. N. (2018). The role of vitamin E in immunity. *Nutrients*, *10*(11), 1614.

Lieberman, H. R. (2007). Hydration and cognition: a critical review and recommendations for future research. *Journal of the American College of Nutrition*, *26*(sup5), 555S-561S.

Lockitch, G. (1989). Selenium: clinical significance and analytical concepts. *Critical reviews in clinical laboratory sciences*, *27*(6), 483-541.

Lu, L., Jing, W., Qian, W., Fan, L., & Cheng, J. (2024). Association between dietary patterns and cardiovascular diseases: A review. *Current Problems in Cardiology*, *49*(3), 102412.

Luoto, R., Ruuskanen, O., Waris, M., Kalliomäki, M., Salminen, S., & Isolauri, E. (2014). Prebiotic and probiotic supplementation prevents rhinovirus infections in preterm infants: a randomized, placebo-controlled trial. *Journal of Allergy and Clinical Immunology*, *133*(2), 405-413.

Mainardi, E., Corino, C., & Rossi, R. (2024). The effect of vitamins on the immune systems of pigs. *Animals*, *14*(14), 2126.

Mandal, K., & Lu, H. (2017). Zinc deficiency in children. *IJSIT*, *6*(1), 9-019.

Marik, P. E. (2009). Glycemic control in critically ill patients: What to do post NICE-SUGAR?. *World Journal of Gastrointestinal Surgery*, *1*(1), 3.

Mariotti, F., & Gardner, C. D. (2019). Dietary protein and amino acids in vegetarian diets—a review. *Nutrients*, *11*(11), 2661.

Mazziotta, C., Tognon, M., Martini, F., Torreggiani, E., & Rotondo, J. C. (2023). Probiotics mechanism of action on immune cells and beneficial effects on human health. *Cells*, *12*(1), 184.

Mehrbod, P., Hudy, D., Shyntum, D., Markowski, J., Łos, M. J., & Ghavami, S. (2020). Quercetin as a natural therapeutic candidate for the treatment of influenza virus. *Biomolecules*, *11*(1), 10.

Meydani, S. N., Leka, L. S., Fine, B. C., Dallal, G. E., Keusch, G. T., Singh, M. F., & Hamer, D. H. (2004). Vitamin E and respiratory tract infections in elderly nursing home residents: a randomized controlled trial. *Jama*, *292*(7), 828-836.

Moles, L., & Otaegui, D. (2020). The impact of diet on microbiota evolution and human health. Is diet an adequate tool for microbiota modulation?. *Nutrients*, *12*(6), 1654.

Moyer, M. W. (2024). How Much Water Do I Need to Drink?. *The New York Times*, D6-D6.

Mueller, N. T., & Pereira, M. A. (2015). Leg length and type 2 diabetes: what's the link?. *Current Opinion in Clinical Nutrition & Metabolic Care*, *18*(5), 452-456.

Munteanu, C., & Schwartz, B. (2022). The relationship between nutrition and the immune system. *Frontiers in nutrition*, *9*, 1082500.

National Institutes of Health (NIH). (2022). *Dietary supplement fact sheet*.

Nelson, D. L., Lehninger, A. L., & Cox, M. M. (2008). *Lehninger principles of biochemistry*. Macmillan.

Niki, E. (2015). Evidence for beneficial effects of vitamin E. *The Korean journal of internal medicine*, *30*(5), 571.

Nkonya, E. M. (2019). Climate Change and Land: an IPCC special report on climate change, desertification, land degradation, sustainable land management, food security, and greenhouse gas fluxes in terrestrial ecosystems.

O'Keefe, S. J., Li, J. V., Lahti, L., Ou, J., Carbonero, F., Mohammed, K., ... & Zoetendal, E. G. (2015). Fat, fiber and cancer risk in African Americans and rural Africans. *Nature communications*, *6*(1), 6342.

Osborn, D. A., & Sinn, J. K. (2013). Prebiotics in infants for prevention of allergy. *Cochrane Database of Systematic Reviews*, (3).

Owino, V., Kumwenda, C., Ekesa, B., Parker, M. E., Ewoldt, L., Roos, N., ... & Tome, D. (2022). The impact of climate change on food systems, diet quality, nutrition, and health outcomes: A narrative review. *Frontiers in Climate*, *4*, 941842.

Parada Venegas, D., De la Fuente, M. K., Landskron, G., González, M. J., Quera, R., Dijkstra, G., ... & Hermoso, M. A. (2019). Short chain fatty acids (SCFAs)-mediated gut epithelial and immune regulation and its relevance for inflammatory bowel diseases. *Frontiers in immunology*, *10*, 277.

Pecora, F., Persico, F., Argentiero, A., Neglia, C., & Esposito, S. (2020). The role of micronutrients in support of the immune response against viral infections. *Nutrients*, *12*(10), 3198.

Perry, G. H., Dominy, N. J., Claw, K. G., Lee, A. S., Fiegler, H., Redon, R., ... & Stone, A. C. (2007). Diet and the evolution of human amylase gene copy number variation. *Nature genetics*, *39*(10), 1256-1260.

Phillips, J. A. (2021). Dietary guidelines for Americans, 2020–2025. *Workplace health & safety*, *69*(8), 395-395.

Pignitter, M., Grosshagauer, S., & Somoza, V. (2019). Stability of Vitamin E in foods. *Vitamin E in human health*, 215-232.

Pikula, A., Gulati, M., Bonnet, J. P., Ibrahim, S., Chamoun, S., Freeman, A. M., & Reddy, K. (2024). Promise of lifestyle medicine for heart disease, diabetes mellitus, and cerebrovascular diseases. *Mayo Clinic Proceedings: Innovations, Quality & Outcomes*, *8*(2), 151-165.

Piperata, B. A., & Dufour, D. L. (2021). Food insecurity, nutritional inequality, and maternal–child health: A role for biocultural scholarship in filling knowledge gaps. *Annual Review of Anthropology*, *50*(1), 75-92.

Poggioli, R., Hirani, K., Jogani, V. G., & Ricordi, C. (2023). Modulation of inflammation and immunity by omega-3 fatty acids: a possible role for prevention and to halt disease progression in autoimmune, viral, and age-related disorders. *European Review for Medical & Pharmacological Sciences*, *27*(15).

Popkin, B. M. (2006). Global nutrition dynamics: the world is shifting rapidly toward a diet linked with noncommunicable diseases. *The American journal of clinical nutrition*, *84*(2), 289-298.

Popkin, B. M., D'Anci, K. E., & Rosenberg, I. H. (2010). Water, hydration, and health. *Nutrition reviews*, *68*(8), 439-458.

Prashanth, L., Kattapagari, K. K., Chitturi, R. T., Baddam, V. R. R., & Prasad, L. K. (2015). A review on role of essential trace elements in health and disease. *Journal of Dr. YSR University of Health Sciences*, *4*(2), 75-85.

Prohaska, J. R., & Lukasewycz, O. A. (1990). Effects of copper deficiency on the immune system. *Antioxidant Nutrients and Immune Functions*, 123-143.

Qian, B., Shen, S., Zhang, J., & Jing, P. (2017). Effects of vitamin B6 deficiency on the composition and functional potential of T cell populations. *Journal of immunology research*, *2017*(1), 2197975.

Reddy, K. R., Aggarwal, M., & Freeman, A. M. (2024). Food Is Medicine: The Time Is Now. *The American journal of medicine*.

Rohm, T. V., Meier, D. T., Olefsky, J. M., & Donath, M. Y. (2022). Inflammation in obesity, diabetes, and related disorders. *Immunity*, *55*(1), 31-55.

Ségurel, L., & Bon, C. (2017). On the evolution of lactase persistence in humans. *Annual review of genomics and human genetics*, *18*(1), 297-319.

Seid, A., Fufa, D. D., & Bitew, Z. W. (2024). The use of internet-based smartphone apps consistently improved consumers' healthy eating behaviors: a systematic review of randomized controlled trials. *Frontiers in Digital Health*, *6*, 1282570.

Skrobot, A., Demkow, U., & Wachowska, M. (2018). Immunomodulatory role of vitamin D: a review. *Current trends in immunity and respiratory infections*, 13-23.

Sorrenti, S., Baldini, E., Pironi, D., Lauro, A., D'Orazi, V., Tartaglia, F., ... & Ulisse, S. (2021). Iodine: Its role in thyroid hormone biosynthesis and beyond. *Nutrients*, *13*(12), 4469.

Springmann, M., Clark, M., Mason-D'Croz, D., Wiebe, K., Bodirsky, B. L., Lassaletta, L., ... & Willett, W. (2018). Options for keeping the food system within environmental limits. *Nature*, *562*(7728), 519-525.

Stearns, S. C., & Medzhitov, R. (2024). *Evolutionary medicine*. Oxford University Press.

Stuivenberg, G. A., Burton, J. P., Bron, P. A., & Reid, G. (2022). Why are bifidobacteria important for infants?. *Microorganisms*, *10*(2), 278.

Suardi, C., Cazzaniga, E., Graci, S., Dongo, D., & Palestini, P. (2021). Link between viral infections, immune system, inflammation and diet. *International journal of environmental research and public health*, *18*(5), 2455.

Sumaily, K. M. (2022). The roles and pathogenesis mechanisms of a number of micronutrients in the prevention and/or treatment of chronic hepatitis, COVID-19 and type-2 diabetes mellitus. *Nutrients*, *14*(13), 2632.

Surh, Y. J. (1998). Chemopreventive Properties of Hot Chili Pepper and Ginger. *Cancer and Nutrition*, 193.

Taşğın, E. (2017). Macronutrients and micronutrients in nutrition. *International Journal of Innovative Research and Reviews*, *1*(1), 10-15.

Tourkochristou, E., Triantos, C., & Mouzaki, A. (2021). The influence of nutritional factors on immunological outcomes. *Frontiers in immunology*, *12*, 665968.

Uriu-Adams, J. Y., & Keen, C. L. (2005). Copper, oxidative stress, and human health. *Molecular aspects of medicine*, *26*(4-5), 268-298.

Valtin, H. (2002). "Drink at least eight glasses of water a day." Really? Is there scientific evidence for "8× 8"?. *American Journal of Physiology-Regulatory, Integrative and Comparative Physiology*.

Vasto, S., Mocchegiani, E., Malavolta, M., Cuppari, I., Listì, F., Nuzzo, D., ... & Caruso, C. (2007). Zinc and inflammatory/immune response in aging. *Annals of the New York Academy of Sciences*, *1100*(1), 111-122.

Venkatraman, J. T., & Chu, W. C. (1999). Effects of dietary ω-3 and ω-6 lipids and vitamin E on serum cytokines, lipid mediators and anti-DNA antibodies in a mouse model for rheumatoid arthritis. *Journal of the American College of Nutrition*, *18*(6), 602-613.

Walker, E. M., & Walker, S. M. (2000). Effects of iron overload on the immune system. *Annals of Clinical & Laboratory Science*, *30*(4), 354-365.

Wang, P., Song, M., Eliassen, A. H., Wang, M., Fung, T. T., Clinton, S. K., ... & Giovannucci, E. L. (2023). Optimal dietary patterns for prevention of chronic disease. *Nature medicine*, *29*(3), 719-728.

Wazir, S. M., & Ghobrial, I. (2017). Copper deficiency, a new triad: anemia, leucopenia, and myeloneuropathy. *Journal of community hospital internal medicine perspectives*, *7*(4), 265-268.

Willett, W., Rockström, J., Loken, B., Springmann, M., Lang, T., Vermeulen, S., ... & Murray, C. J. (2019). Food in the Anthropocene: the EAT–Lancet Commission on healthy diets from sustainable food systems. *The lancet, 393*(10170), 447-492.

Wolfe, R. R. (2017). Branched-chain amino acids and muscle protein synthesis in humans: myth or reality?. *Journal of the International Society of Sports Nutrition, 14*(1), 30.

World Health Organization (WHO). (2004). *Vitamin and mineral requirements in human nutrition*. World Health Organization.

World Health Organization (WHO). (2023). updates guidelines on fats and carbohydrates. *World Health Organization [Internet]*.

World Health Organization (WHO). (2023). *Global report on the use of sugar-sweetened beverage taxes, 2023*. World Health Organization.

World Health Organization (WHO). (2024). *Global status report on drowning prevention 2024*. World Health Organization.

World Health Organization (WHO). (2020). Healthy Diet: Fact sheet.

World Health Organization (WHO). (2022). *Fiscal policies to promote healthy diets: policy brief*. World Health Organization.

World Health Organization (WHO). (2023). *WHO global report on sodium intake reduction*. World Health Organization.

World Health Organization (WHO). (2011). vitamin A supplementation in infants and children 6–59 months of age. *Geneva: World Health Organization, 269*, 16.

Xu, M., Hu, D., Liu, X., Li, Z., & Lu, L. (2025). Branched-Chain Amino Acids and Inflammation Management in Endurance Sports: Molecular Mechanisms and Practical Implications. *Nutrients, 17*(8), 1335.

Yacoub Bach, L., Jana, B. E., Adaeze Egwatu, C. F., Orndorff, C. J., Alanakrih, R., Okoro, J., & Gahl, M. K. (2023). A sustainability analysis of environmental impact, nutritional quality, and price among six popular diets. *Frontiers in Sustainable Food Systems, 7*, 1021906.

Yahfoufi, N., Alsadi, N., Jambi, M., & Matar, C. (2018). The immunomodulatory and anti-inflammatory role of polyphenols. *Nutrients, 10*(11), 1618.

Yıldız, G., Yildiz, G., Khan, M. R., and Aadil, R. M. (2022). High-intensity ultrasound treatment to produce and preserve the quality of fresh-cut kiwifruit, *Journal of Food Processing and Preservation*, 46 (5): e16542.

Yildiz, G., Andrade, J., Engeseth, N. C., and Feng, H. (2017). Functionalizing soy protein nano-aggregates with pH-shifting and mano-thermo-sonication. *Journal of Colloid and Interface Science*, 505: 836-846.

Yildiz, G. (2018). The Effect of Different Chemical Agents on the Prevention of Enzymatic Browning in Banana. *Journal of Food Science and Engineering*, 7: 91-96.

Yildiz, G., Ding, J., Gaur, S., Andrade, J., Engeseth, N. J., and Feng, H. (2018). Microencapsulation of docosahexaenoic acid (DHA) with four wall materials including pea protein-modified starch complex. *International Journal of Biological Macromolecules*, 114: 935–941.

Yildiz, G. (2019). Application of ultrasound and high pressure homogenization against high temperature-short time in peach juice. *Journal of Food Process Engineering*, 42 (3): e12997.

Yildiz, G., and Feng, H. (2019). Sonication of Cherry Juice: Comparison of Different Sonication Times on Color, Antioxidant Activity, Total Phenolic and Ascorbic Acid Content, *Latin American Applied Research Journal,* 49 (4): 255-260.

Yildiz, G, Palma, S., and Feng, H. (2019). Ultrasonic cutting as a new method to produce fresh-cut red delicious and golden delicious apples, *Journal of Food Science,* 84 (12): 3391-3398.

Yildiz, G., and Izli, G. (2019). The effect of ultrasound pretreatment on quality attributes of freeze-dried quince slices: Physical properties and bioactive compounds. *Journal of Food Process Engineering*, 42 (5): e13223.

Yildiz, G., Izli, G., and Aadil, R. M. (2020). Comparison of chemical, physical, and ultrasound treatments on the shelf life of fresh-cut quince fruit (Cydonia oblonga Mill.), *Journal of Food Processing and Preservation,* 44 (3): e14366.

Yildiz, G., and Aadil, R. M. (2020). Comparison of high temperature-short time and sonication on selected parameters of strawberry juice during room temperature storage, *Journal of Food Science and Technology,* 57 (4): 1462-1468.

Yildiz, G. (2021). The Effect of High Intensity Ultrasound Pre-treatment on the Functional Properties of Microwave-dried Pears (*Pyrus communis*), *Latin American Applied Research Journal*, 51(2):133-137.

Yildiz, G., Cox, G., and Moran, L. (2021). Identification of Aroma Compounds of Dried and Fresh Thyme (*Thymus Vulgaris L.)* by Gas Chromatography-Olfactometry and Gas Chromatography-Mass spectroscopy Analysis, *Latin American Applied Research Journal*, 51 (1): 51-55.

Yildiz, G. (2022). Color, microstructure, physicochemical, textural and sensory properties with the retention of secondary metabolites in convective-, microwave- and freeze-dried carrot (Daucus carota) slices, *British Food Journal*, 124 (11): 3922-3935.

Yildiz, G., and Aadil, R. M. (2022). Comparative Analysis of Antibrowning Agents, Hot Water and High-intensity Ultrasound Treatments to Maintain the Quality of Fresh-cut Mangoes, *Journal of Food Science and Technology*, 59(1): 202-211.

Yildiz, G., Gao, Y., Ding, J., Zhu, S., Chen, G., and Feng, H. (2025). Enhancing physicochemical, bioactive, and nutritional properties of sweet potatoes: Ultrasonic contact drying with slot jet nozzles compared to hot-air drying and freeze drying, *Ultrasonics Sonochemistry, 112*: 107216.

Zhou, X., Qiao, K., Wu, H., & Zhang, Y. (2023). The impact of food additives on the abundance and composition of gut microbiota. *Molecules*, *28*(2), 631.

Ziska, L. H. (2022). Rising carbon dioxide and global nutrition: evidence and action needed. *Plants*, *11*(7), 1000.

Chapter 2

Functional Foods and Their Effects on Health

Dilara Sokmen
and Gulcin Yildiz*
Department of Food Engineering, Igdir University, Iğdır, Turkiye

Abstract

Functional foods have emerged as a pivotal category within modern nutritional science, characterized by their ability to provide health benefits beyond fundamental nutritional requirements. These foods, which include both natural and formulated products, contain physiologically active components such as dietary fiber, phytochemicals, prebiotics, probiotics, and omega-3 fatty acids, which exert specific biochemical and molecular effects in the human body. Scientific evidence from in vitro studies, in vivo models, epidemiological surveys, and clinical trials has demonstrated that functional foods play a significant role in modulating immune responses, reducing systemic inflammation, enhancing gut microbiota composition, and regulating oxidative stress and lipid metabolism. Their regular consumption has been associated with the prevention and management of non-communicable chronic diseases, including cardiovascular disorders, type 2 diabetes mellitus, certain types of cancer, and neurodegenerative conditions. This chapter provides a comprehensive overview of the classification, mechanisms of action, and clinical relevance of various functional foods. Special attention is given to the bioavailability and synergistic effects of bioactive compounds, the challenges of substantiating health claims, and the evolving landscape of regulatory frameworks governing functional food development. Additionally, the chapter discusses the future of personalized nutrition and omics-based approaches in optimizing the therapeutic potential of functional foods.

Keywords: bioactive compounds, functional ingredients, health benefits

* Corresponding Author's Email: gulcn86@gmail.com.

In: Let Food Be Your Medicine, and Medicine Be Your Food
Editors: Gulcin Yildiz and Arzu Imece
ISBN: 979-8-90134-204-6

Introduction

The understanding of nutrition has evolved significantly over the past century, transitioning from merely preventing nutrient deficiencies to promoting overall health and preventing chronic diseases. This shift is particularly crucial in addressing non-communicable diseases (NCDs) such as cardiovascular diseases, diabetes, obesity, and certain cancers, which now represent the leading causes of morbidity and mortality worldwide (Han et al., 2024). These diseases are increasingly linked to poor dietary habits, sedentary lifestyles, and environmental factors, posing a major challenge to global public health. In this context, functional foods have emerged as vital components of both preventive and therapeutic strategies, offering health benefits beyond basic nutrition. Functional foods are defined as foods that, when consumed regularly, provide additional physiological benefits that reduce the risk of chronic diseases (Doyon & Labrecque, 2008). Given the rising prevalence of NCDs, functional foods present a promising strategy to mitigate this global health crisis (Shahnaz et al., 2024), as they may not only prevent deficiencies but also manage or reverse certain health conditions (Vignesh et al., 2024).

Functional foods differ from traditional foods by their ability to offer health benefits beyond basic nutrition. They include bioactive compounds—naturally occurring chemical substances that exert physiological effects when consumed in adequate amounts. These compounds, such as phytochemicals (e.g., polyphenols, flavonoids, carotenoids), omega-3 fatty acids, dietary fibers, probiotics, and prebiotics, work synergistically to promote health and combat diseases (Panou & Karabagias, 2025). Phytochemicals, including flavonoids and polyphenols found in fruits, vegetables, and whole grains, have demonstrated antioxidant, anti-inflammatory, and anti-carcinogenic properties (Ullah et al., 2020). Omega-3 fatty acids, predominantly found in fatty fish and some plant oils, are well-established for their cardiovascular benefits, such as reducing blood triglycerides and lowering heart disease risk (Kaur et al., 2024). Additionally, dietary fibers, particularly soluble fiber, regulate blood sugar levels and improve gut health, while prebiotics and probiotics support a balanced gut microbiota, which plays a crucial role in immune function and metabolic health (Fu et al., 2022). As research continues to uncover the full range of health benefits, functional foods are increasingly recognized as a critical element in public health promotion.

The growing awareness of diet's role in health has propelled functional foods into the spotlight as preventive tools. The global rise in lifestyle-related diseases underscores the need for a broader understanding of food, which now encompasses its role in disease prevention and health maintenance (Neuhouser, 2019). As consumers become more proactive in managing their health, they seek dietary solutions that not only prevent deficiencies but also actively prevent disease and maintain health. This shift towards preventative healthcare is driving the demand for functional foods, which can alleviate the burden of chronic diseases, particularly as aging populations and rising healthcare costs strain healthcare systems worldwide. Moreover, the emerging field of nutrigenomics, which examines how diet influences gene expression and disease risk, highlights the potential of functional foods to modulate biological pathways involved in chronic diseases (Tkaczenko & Kurhaluk, 2025). Nutrigenomics offers insights into the molecular mechanisms by which functional foods affect health, enabling personalized nutrition strategies that optimize health outcomes based on individual genetic profiles (Lagoumintzis & Patrinos, 2023).

Table 1. Validation of health claims for functional foods

Regulatory Body	Process for Validating Health Claims	Requirements	References
European Food Safety Authority (EFSA)	Scientific evidence-based reviews of health claims	Requires rigorous scientific evidence, including randomized controlled trials (RCTs), epidemiological studies, and clinical investigations. The studies must demonstrate the claimed health benefits beyond general nutrition.	Domínguez Díaz et al., 2020; Banerjee et al., 2023; Frumuzachi et al., 2025
U.S. Food and Drug Administration (FDA)	Clinical evidence and substantial documentation supporting health claims	The FDA requires substantial evidence to support the health claims, including clinical studies that validate the functional foods' impact on health outcomes. Documentation must be thorough to substantiate claims made by manufacturers.	
General Criteria	Proven physiological benefits and risk reduction	Health claims must be substantiated by scientific studies, including clinical trials and laboratory studies, that demonstrate the physiological benefits of the food. These benefits should ideally show clear evidence of risk reduction for diseases or enhancement of health conditions.	

Table 2. Functional foods classification

Category	Examples	Description	References
Naturally Functional Foods	Fruits, vegetables, whole grains	These foods naturally contain bioactive compounds such as antioxidants, fiber and essential fatty acids, which contribute to overall health by preventing disease and enhancing bodily functions. These bioactive compounds support various physiological processes, including reducing oxidative stress, improving immune function, and lowering inflammation.	Ma et al., 2025
Fortified/Enriched Foods	Breakfast cereals (fortified with vitamins), dairy (fortified with probiotics or calcium)	These are foods that have been modified by the addition of extra nutrients or bioactive components to enhance their health benefits. Examples include breakfast cereals that are fortified with vitamins (e.g., folic acid) or dairy products enriched with probiotics or calcium. These added nutrients help address potential dietary gaps and further promote health, especially in populations at risk of deficiencies.	Obayomi et al., 2024

Functional foods are increasingly integrated into public health strategies to combat the global rise in NCDs. Regulatory bodies, such as the European Food Safety Authority (EFSA) and the U.S. Food and Drug Administration (FDA), have established frameworks for validating the health claims of functional foods (Domínguez Díaz et al., 2020) (Table 1). These frameworks rely on rigorous scientific evidence, including randomized controlled trials (RCTs), epidemiological studies, and clinical investigations, ensuring that functional foods can be reliably marketed and recommended as part of preventive healthcare (Frumuzachi et al., 2025). However, further research is needed to clarify the precise mechanisms, optimal dosages, and long-term effects of functional foods (Banerjee et al., 2023). While the potential benefits of functional foods are widely recognized, more studies are required to confirm their safety and efficacy across diverse populations. This ongoing research will provide the foundation for evidence-based dietary guidelines and therapeutic interventions involving functional foods.

Functional foods can be classified into categories based on their source and bioactive components, including naturally functional foods (e.g., fruits, vegetables, and whole grains) and fortified or enriched foods (e.g., dairy products with probiotics or breakfast cereals enriched with vitamins and minerals) (Galanakis, 2021) (Table 2). Naturally functional foods are integral to a health-promoting diet, offering a range of bioactive compounds such as antioxidants, fiber, and essential fatty acids (Ma et al., 2025). Fortified foods, on the other hand, are modified to include nutrients or bioactive compounds that may be deficient in a typical diet, such as cereals enriched with folic acid or dairy products fortified with calcium and vitamin D (Obayomi et al., 2024). Additionally, the emerging field of personalized nutrition is investigating how functional foods can be tailored to individual genetic, metabolic, and microbiome profiles, offering a more precise approach to health promotion and disease prevention (Lee et al., 2022). This personalized approach holds the potential to further enhance the effectiveness of functional foods by aligning dietary recommendations with an individual's unique genetic and environmental factors.

This chapter provides a comprehensive exploration of functional foods, their bioactive components, and their potential mechanisms of action. We will review current literature on how functional foods influence health through immune modulation, reduction of oxidative stress and inflammation, and modulation of the gut microbiota. The modulation of the immune system is a key area of research, as it is closely linked to chronic disease prevention and management. Functional foods that reduce oxidative stress and inflammation have shown promise in mitigating the progression of cardiovascular diseases, diabetes, and certain cancers (Chanda et al., 2023). Moreover, the gut microbiome plays a pivotal role in overall health, influencing digestion, immune function, and metabolism. Emerging research suggests that functional foods can positively impact the gut microbiota, promoting a balanced microbial ecosystem that supports health (Zhang, 2022). Furthermore, we will examine the growing body of evidence linking functional foods to the prevention and management of various chronic diseases, including cardiovascular disease, metabolic syndrome, and neurodegenerative conditions. As functional foods gain prominence in clinical nutrition, their role in managing chronic diseases becomes increasingly vital, especially as healthcare systems shift towards preventive care.

The integration of functional foods into both clinical nutrition and public health practices underscores their potential to play a pivotal role in modern healthcare. By addressing the molecular and physiological effects of functional foods, this chapter highlights their therapeutic potential and provides a framework for their effective use in preventing and managing diseases that significantly impact global health. As scientific research continues to uncover the full range of bioactive compounds in functional foods, the potential for these foods to improve health outcomes becomes more evident. Functional foods are not only a dietary choice but also a dynamic and evolving field of research with the capacity to transform global healthcare.

Definition and Classification of Functional Foods

Functional foods are defined as foods that offer health benefits beyond basic nutritional value. They are consumed not only to meet daily energy and nutrient requirements but also to support optimal health and help prevent chronic diseases. Unlike conventional foods, functional foods contain bioactive components—such as phytochemicals, antioxidants, dietary fibers, probiotics, prebiotics, vitamins, and minerals—that interact with the body to improve physiological functions. These interactions can strengthen the immune system, enhance digestive health, regulate metabolism, and reduce oxidative stress and inflammation. As a result, functional foods have become increasingly popular for their potential to improve quality of life and overall well-being (Hasler, 2002).

The concept of functional foods first emerged in Japan during the 1980s under the term "Foods for Specified Health Use" (FOSHU). These were foods with scientifically proven health benefits that targeted specific health issues. Since then, the concept has expanded globally, driven by rising consumer awareness, increased healthcare costs, and a shift in public health approaches from treatment to prevention. Functional foods are now considered a central component of dietary strategies aimed at reducing the burden of lifestyle-related diseases (Ma et al., 2025).

Functional foods can be classified into several major categories based on their composition, preparation, and health effects. The following categories illustrate the diverse nature of functional foods in today's food systems:

Naturally Functional Foods

Naturally functional foods inherently contain bioactive compounds that promote health without the need for modification or enhancement. These foods offer a wide range of health benefits due to the presence of specific nutrients and phytochemicals that support the body's physiological functions. They are an essential part of a balanced diet, providing nutrients that work synergistically to improve overall health and prevent chronic diseases (Frumuzachi et al., 2025).

Fruits and vegetables are prime examples of naturally functional foods. They are rich in flavonoids, antioxidants, vitamins, and minerals (İzli and Yildiz, 2021). For example, berries such as blueberries, strawberries, and raspberries are high in anthocyanins and other antioxidants, which help neutralize harmful free radicals and reduce oxidative stress, a key factor in aging and the development of diseases such as cancer. Citrus fruits like oranges, lemons, and grapefruits are excellent sources of vitamin C, which supports immune function, skin health, and collagen production. Leafy greens like spinach, kale, and swiss chard are abundant in vitamins A, C, K, and folate, along with minerals like calcium and magnesium, which contribute to bone health and anti-inflammatory effects. The diverse array of phytonutrients in these fruits and vegetables has been shown to help lower the risk of chronic diseases, such as cardiovascular disease, diabetes, and certain cancers (Yildiz and Izli, 2019).

Whole grains such as oats, brown rice, quinoa, and barley are rich in dietary fiber, which plays a crucial role in digestive health by promoting regular bowel movements and preventing constipation. Fiber also helps regulate blood sugar levels and lower cholesterol, contributing to cardiovascular health (Yildiz et al., 2025). In addition to fiber, whole grains contain lignans, which are plant compounds with antioxidant properties that have been linked to a reduced risk of breast cancer. The consumption of whole grains is associated with a lower incidence of heart disease, type 2 diabetes, and certain cancers (Xie et al., 2019). Quinoa, a highly nutritious pseudocereal, is also a complete protein, making it an excellent choice for plant-based diets (Kumar et al., 2025).

Legumes such as lentils, chickpeas, and beans, as well as nuts like almonds, walnuts, and pistachios, are rich in a combination of fiber, plant-based proteins, and healthy fats. These foods are known to promote heart health by lowering cholesterol levels, improving blood vessel function, and reducing inflammation (Hernández-López et al., 2022). For example, almonds are high in monounsaturated fats, which have been shown to reduce the risk of heart disease by improving cholesterol levels (Kalita et al., 2018). Legumes are also an excellent source of plant-based proteins, making them a valuable addition to vegetarian and vegan diets. They provide essential amino acids that support muscle maintenance and repair while being low in saturated fat (Langyan et al., 2022).

Fatty fishlike salmon, mackerel, sardines, and herring are naturally rich in omega-3 fatty acids, which have been extensively studied for their benefits on heart and brain health (Tsoupras et al., 2022; Rodrigues, et al., 2024). Omega-3s, particularly EPA and DHA, help reduce inflammation, lower blood pressure, and decrease triglyceride levels, all of which contribute to a reduced risk of heart disease. These fatty acids also support cognitive function and mental health by promoting brain health and reducing the risk of age-related cognitive decline. The inclusion of fatty fish in the diet is associated with improved cardiovascular outcomes and a lower risk of stroke (Yildiz et al., 2018).

Many herbs and spices, such as turmeric, ginger, garlic, and cinnamon, are considered naturally functional foods due to their bioactive compounds that offer anti-inflammatory, antioxidant, and antimicrobial properties (Ajanaku et al., 2022). Turmeric contains curcumin, which has been shown to have potent anti-inflammatory effects and may play a role in preventing chronic diseases such as arthritis, diabetes, and certain cancers (Peng et al., 2021). Garlic is known for its ability to support heart health by reducing cholesterol and blood pressure (Ried, 2020), while ginger is often used to relieve nausea and promote digestion (Aregawi et al., 2024). Cinnamon contains polyphenols that have been found to improve insulin sensitivity and regulate blood sugar levels, making it beneficial for individuals with type 2 diabetes (Anderson et al., 2016).

Fermented foods, such as kimchi, sauerkraut, kefir, and yogurt, are naturally functional due to their probiotic content. These foods contain live microorganisms that contribute to gut health by improving the balance of beneficial bacteria in the digestive system (Leeuwendaal et al., 2022). A healthy gut microbiome is essential for digestion, immune function, and even mental health, as it has been linked to the production of neurotransmitters like serotonin (Mhanna et al., 2024). Fermented foods are also rich in bioactive peptides, which may have antioxidant and antihypertensive effects, supporting cardiovascular health (Fabbri et al., 2024).

These naturally functional foods are central to health-promoting dietary patterns like the Mediterranean diet, which emphasizes the consumption of fruits, vegetables, whole grains, legumes, nuts, seeds, and fatty fish. This diet has been associated with a reduced risk of cardiovascular and metabolic diseases, as well as improved longevity. Incorporating these foods into daily meals can help promote overall well-being, enhance immune function, and prevent the onset of chronic diseases.

Fortified Functional Foods

Fortified foods are conventional foods that have been enhanced with additional nutrients or bioactive substances to improve their health benefits. This process, known as fortification, is commonly used as a public health strategy to address micronutrient deficiencies in populations. Fortified foods help bridge the nutritional gap where certain essential vitamins and minerals might be lacking due to dietary patterns, lifestyle factors, or limited access to a varied diet. Fortification can significantly reduce the prevalence of nutrient deficiencies, thereby promoting overall health and preventing related health conditions (Olson et al., 2021).

One of the most well-known examples of fortified functional foods is milk fortified with vitamin D (Zahedirad et al., 2019). Vitamin D is crucial for calcium absorption, bone health, and immune function. Fortifying milk with this vitamin has helped combat vitamin D deficiencies, particularly in regions with limited sunlight exposure, where natural synthesis of the vitamin through skin is insufficient. This fortification strategy has been instrumental in reducing the prevalence of rickets and osteomalacia, conditions caused by a lack of vitamin D, especially in children and elderly populations (Aiello et al., 2024).

Salt enriched with iodine is another classic example of fortification aimed at preventing iodine deficiency, which can lead to thyroid disorders such as goiter (Santos et al., 2019). Iodine deficiency is a global issue, particularly in areas where iodine-rich foods like seafood and dairy are not widely consumed. The addition of iodine to salt, a staple in most diets, has been a highly effective measure to prevent iodine deficiency and related conditions, such as cognitive impairments and developmental issues in children (Hatch-McChesney and Lieberman, 2022).

Orange juice fortified with calcium is another common fortified food, offering a convenient way to support bone and dental health (Tangpricha et al., 2003). Calcium is essential for the development and maintenance of strong bones and teeth, and fortifying orange juice with this mineral helps people meet their daily calcium needs, especially for individuals who are lactose intolerant or do not consume dairy products regularly. This fortification is beneficial in preventing osteoporosis and other bone-related conditions, particularly among older adults and individuals who may have a limited intake of calcium-rich foods (Beto, 2015).

Breakfast cereals are often fortified with B vitamins and iron, making them a convenient and nutrient-dense option for many individuals (Williams, 2014). Iron is critical for the formation of red blood cells and the prevention of anemia, while B vitamins play a key role in energy metabolism, nerve function, and the production of DNA. Many cereals, especially those consumed by children and adolescents, are fortified to help meet the nutritional needs of growing individuals, particularly when their diet may not be well-balanced. These fortified cereals serve as a quick and easily accessible source of important nutrients (Donker et al., 2021).

Beyond these common examples, fortification can extend to a wide variety of other foods and beverages, including yogurt fortified with probiotics, water enriched with vitamins or minerals, and bakery products enhanced with additional fiber or antioxidants (Ahmad et al., 2022). Fortified foods are particularly beneficial in addressing the needs of vulnerable populations, such as pregnant women, infants, the elderly, and individuals with restricted diets. For instance, folic acid is often added to wheat flour to reduce the risk of neural tube defects in pregnant women, while vitamin B12 fortification in plant-based products helps prevent deficiency in those following vegan diets (Ismail et al., 2023).

Fortification plays a crucial role in public health efforts to prevent chronic diseases related to micronutrient deficiencies, such as cardiovascular disease, bone fractures, and cognitive impairments. It is especially vital in developing regions, where access to a diverse range of nutrient-rich foods can be limited due to economic, geographical, or cultural barriers. Additionally, fortification programs are often more cost-effective than other public health interventions, making them a viable solution to improve nutrition at a population level (Abdul Majeed et al., 2025). While fortified foods provide important health benefits, it is also essential to ensure that fortification is done in a balanced manner. Over-fortification of certain nutrients can lead to adverse effects, such as toxicity or imbalances in nutrient intake. Therefore, fortification efforts must be carefully regulated to meet the nutritional needs of the population without causing harm.

In summary, fortified functional foods are an important tool in the fight against micronutrient deficiencies, offering a practical and efficient means of improving public health. By enriching everyday foods with essential nutrients, fortification helps ensure that individuals have access to the vitamins and minerals necessary for optimal health, reducing the burden of preventable diseases and improving quality of life across populations.

Enriched or Enhanced Functional Foods

Enriched or enhanced functional foods are those that have been modified to increase the levels of health-promoting components that are already naturally present in the food. This modification enhances the nutritional value of the food without fundamentally altering its structure or sensory qualities. These foods serve as an effective way to boost the intake of beneficial nutrients, especially for individuals seeking to optimize their health through their diet (Tafese Awulachew, 2024).

One well-known example of an enriched functional food is omega-3 enriched eggs. These eggs are produced by feeding hens a diet that is high in omega-3 fatty acids, typically from sources like flaxseed or fish oil. As a result, the eggs contain higher levels of omega-3s, which are essential for heart and brain health (Antony et al., 2024). Omega-3 fatty acids have been shown to reduce inflammation, lower blood pressure, and support cognitive function (Dighriri et al., 2022). By enhancing the natural omega-3 content of eggs, consumers can enjoy a convenient source of these beneficial fats without having to rely on supplements or more specialized food products.

Similarly, high-fiber bread is an example of an enriched functional food (Feili et al., 2013). Bread, a staple in many diets, is often enriched by adding ingredients like wheat bran or inulin, a type of soluble fiber. These added fibers support digestive health, help regulate blood sugar levels, and contribute to a feeling of fullness, which can aid in weight management (Naveed et al., 2024). By enriching bread with additional fiber, manufacturers create a more nutritious version of a commonly consumed product, offering a simple way for individuals to increase their fiber intake without changing their dietary habits (Brandl et al., 2022).

Tomatoes are another example of a food that can be selectively enhanced for functional benefits. Through selective breeding or genetic modification, tomatoes can be developed to have higher levels of lycopene, a powerful antioxidant that has been linked to a reduced risk of certain cancers and cardiovascular diseases (Collins et al., 2022). Lycopene is naturally present in tomatoes, but enhancing its concentration can further boost the health benefits of this widely consumed fruit. By focusing on increasing the nutrient density of common foods like tomatoes, food producers can offer more nutritionally potent options without compromising on taste or texture.

The main goal of food enrichment is to amplify the food's natural nutritional properties while maintaining its familiar taste, texture, and overall appeal. This makes enriched foods an attractive option for consumers who are looking to improve their diets without significantly changing their food preferences or preparation methods. Enrichment is especially useful for addressing specific health concerns, such as improving heart health, digestive function, or antioxidant intake. In addition to enhancing individual foods, this concept can extend to entire food systems. For instance, crops may be biofortified to increase the levels of essential nutrients, such as vitamins and minerals, in the plants themselves. This approach can play a significant role in combating nutrient deficiencies in regions where certain vitamins and minerals are lacking from the diet. While enriched functional foods offer clear health benefits, it is important to ensure that the enrichment process is done thoughtfully. Excessive enrichment can lead to imbalances in nutrient intake, and the potential for overconsumption of certain nutrients must be considered. As with all dietary interventions, moderation and balance are key.

Overall, enriched or enhanced functional foods provide an effective strategy to promote better health, offering a simple yet powerful way to boost nutrient intake without drastically altering traditional eating habits.

Altered or Modified Functional Foods

Altered or modified functional foods are those that have been engineered to remove or replace harmful components with healthier alternatives. This modification aims to improve the nutritional profile of the food, making it a better option for individuals seeking to manage or reduce the risk of chronic diseases. By addressing specific health concerns through the removal of undesirable ingredients and the inclusion of beneficial alternatives, modified foods can play an important role in disease prevention and management.

One common example of a modified functional food is margarine that is formulated with plant sterols (Ntanios, 2001). Plant sterols are naturally occurring substances found in plants that have been shown to reduce cholesterol levels by inhibiting the absorption of cholesterol in the digestive tract (Barkas et al., 2023). By adding plant sterols to margarine, the product can help individuals lower their cholesterol levels, which is particularly beneficial for those with high cholesterol or at risk for cardiovascular diseases. This modification provides a healthier alternative to traditional margarine, which may contain high levels of trans fats that can contribute to heart disease (Kozłowska-Wojciechowska et al., 2003).

Low-sodium soups are another example of modified functional foods. These soups are specifically designed to reduce sodium content, which is a key factor in managing blood pressure (Zandstra et al., 2018). High sodium intake is linked to an increased risk of hypertension, a condition that can lead to heart disease and stroke (Grillo et al., 2019). By lowering the sodium content in soups, these modified foods cater to individuals with hypertension or those looking to prevent the condition, offering a more heart-healthy option without compromising taste or convenience (Jachimowicz-Rogowska and Winiarska-Mieczan, 2023).

Sugar-free products, such as beverages, snacks, and baked goods, are also considered modified functional foods. These products are often sweetened with natural alternatives like stevia or monk fruit, which provide sweetness without the calorie content or the negative health effects associated with refined sugars (Saraiva et al., 2020). For individuals with diabetes, obesity, or those seeking to reduce their sugar intake, these sugar-free products offer a healthier choice. They allow consumers to enjoy sweet flavors while managing their blood sugar levels and reducing the risk of conditions related to excessive sugar consumption, such as obesity, type 2 diabetes, and metabolic syndrome.

The primary benefit of altered or modified functional foods is their ability to cater to individuals with specific health needs, such as those managing chronic conditions like cardiovascular disease, diabetes, or obesity (Essa et al., 2023). By making these foods healthier without sacrificing taste or convenience, modified functional foods provide consumers with practical, everyday options that support long-term health (Baker et al., 2022). Additionally, these foods can contribute to a broader public health strategy focused on prevention, helping to reduce the incidence of chronic diseases and the burden on healthcare systems. However, it is important to note that the modification of foods should be done thoughtfully, as over-processing or excessive alteration can sometimes lead to unintended consequences. For example, altering the natural composition of a food too much may affect its nutritional value or introduce new risks, such as the overconsumption of certain additives or substitutes. Therefore, a balanced approach to food modification is necessary to ensure that the end product remains beneficial and sustainable for consumers.

In conclusion, altered or modified functional foods play a vital role in supporting health by providing healthier alternatives to traditionally high-risk foods. They help individuals make better dietary choices, particularly for those managing or preventing chronic health conditions. Through the thoughtful engineering of these foods, consumers can enjoy a more health-conscious diet without giving up familiar tastes and textures.

Functional Beverages

Functional beverages are a rapidly growing category of drinks designed to offer specific health benefits through the inclusion of bioactive ingredients, vitamins, minerals, or herbal extracts. These beverages are created to go beyond basic hydration, providing targeted benefits to consumers seeking ways to enhance their health and well-being. They cater to a wide range of health-conscious individuals, offering convenience and a variety of options that can address specific dietary and lifestyle needs.

Examples of functional beverages include vitamin-enriched waters, which are fortified with essential vitamins like B vitamins, vitamin C, or electrolytes to support energy levels and hydration (Panou and Karabagias, 2025). Antioxidant-infused teas, such as green tea, matcha, and hibiscus tea, are popular for their ability to neutralize free radicals, which can help reduce oxidative stress and promote overall health (Jakubczyk et al., 2020). Protein shakes are another example, designed to support muscle maintenance and recovery, making them especially popular among athletes, bodybuilders, and individuals looking to maintain a healthy weight. These shakes often contain proteins derived from whey, soy, or pea, along with added vitamins and minerals to enhance their nutritional profile (Patel et al., 2023).

Beverages containing adaptogenic herbs like ginseng, ashwagandha, and rhodiola are also gaining popularity, particularly for their stress-reducing and mood-balancing effects (Tóth-Mészáros et al., 2023). Adaptogens are natural substances that help the body adapt to stress and restore balance in the body's systems. These herbs are thought to support the body's ability to cope with physical and mental stress, improve energy levels, and enhance resilience (Panossian and Wikman, 2010). Functional beverages containing adaptogens are often marketed as solutions for stress management, anxiety reduction, and overall mental well-being (Gulati et al., 2016).

The functional beverage market is expanding rapidly, driven by innovation in product development and growing consumer interest in preventative health. Companies are continually introducing new beverages that combine unique ingredients, such as probiotics, prebiotics, and herbal extracts, to target a wide range of health concerns. These drinks are not only marketed for their functional properties but also for their appeal as convenient and enjoyable wellness products. As the demand for functional foods continues to rise, functional beverages are poised to become a central part of health-focused diets, offering a practical and effective way to support various aspects of physical and mental health.

In conclusion, functional beverages represent an exciting and dynamic segment of the food and beverage industry. By incorporating bioactive ingredients, vitamins, minerals, and herbal extracts, these drinks provide an innovative way to support overall health and wellness. Whether for hydration, stress management, immune support, or recovery, functional beverages are increasingly seen as a convenient and appealing option for those looking to enhance their well-being in a busy world.

Bioactive Components in Functional Foods

Functional foods are distinguished by their rich content of bioactive compounds, which are naturally occurring substances that have a significant positive impact on human health. These compounds are not only essential for providing vital nutrients but also contribute to physiological effects that can improve health, prevent diseases, and support overall bodily functions. The bioactive components in functional foods are a diverse group that includes phytochemicals, dietary fiber, probiotics and prebiotics, omega-3 fatty acids, as well as vitamins and minerals. Each of these components plays a critical role in maintaining health, preventing chronic diseases, and promoting well-being. Phytochemicals, for example, are powerful plant-based compounds with antioxidant, anti-inflammatory, and anticancer properties that help combat oxidative stress and inflammation, which are linked to many chronic diseases (Muscolo et al., 2024). Dietary fiber supports digestive health, regulates blood sugar, and lowers cholesterol levels, while probiotics and prebiotics enhance gut health by promoting a healthy balance of beneficial bacteria in the gastrointestinal tract (Holscher, 2017). Omega-3 fatty acids are essential fats that contribute to heart health, reduce inflammation, and are critical for cognitive function (Banaszak et al., 2024). Finally, essential vitamins and minerals not only help maintain basic bodily functions but also have therapeutic effects in preventing or managing various health conditions such as bone loss, immune deficiencies, and heart disease (Tardy et al., 2020).

Phytochemicals

Phytochemicals are bioactive compounds found in plants that have been shown to offer numerous health benefits. These naturally occurring substances are largely responsible for the protective properties of fruits, vegetables, whole grains, legumes, and herbs (Hossain et al., 2025). Phytochemicals have a wide range of health-promoting effects, including antioxidant, anti-inflammatory, anticancer, and immune-boosting properties (Paudel et al., 2023). There are many types of phytochemicals, each with distinct mechanisms of action, and they are classified based on their chemical structure. Among the most studied groups of phytochemicals are flavonoids, carotenoids, and polyphenols (Kumar et al., 2023).

Flavonoids are polyphenolic compounds found in a variety of foods such as fruits, vegetables, and beverages like tea. Known for their potent antioxidant activity, flavonoids help protect cells and tissues from damage caused by free radicals and oxidative stress. This antioxidant effect reduces inflammation and supports cardiovascular health by improving blood circulation, lowering blood pressure, and preventing the oxidation of low-density lipoprotein (LDL) cholesterol, which can lead to plaque buildup in the arteries (Ullah et al., 2020). Well-known flavonoids such as quercetin, catechins, and anthocyanins are found in apples, onions, green tea, berries, and red cabbage. These compounds have been shown to reduce the risk of chronic diseases like heart disease, stroke, and even certain cancers (Jomova et al., 2025).

Carotenoids are pigments that give color to fruits and vegetables, particularly those that are orange, yellow, and red. These include well-known carotenoids such as beta-carotene, lutein, and zeaxanthin (Maoka, 2020). Beta-carotene, for instance, is a precursor to vitamin A, which supports vision health and immune function. It also has antioxidant properties that help protect against the harmful effects of UV radiation and other environmental stressors (Farhana et al., 2025). Lutein and zeaxanthin, found predominantly in green leafy vegetables like spinach and kale, play an essential role in protecting the eyes from age-related macular degeneration (AMD), a leading cause of vision loss in older adults (Mrowicka et al., 2022).

Polyphenols are another important group of phytochemicals with strong antioxidant properties that have been linked to a variety of health benefits, including protection against chronic diseases such as cardiovascular disease, cancer, and neurodegenerative disorders. Polyphenols like resveratrol, found in red wine, grapes, and berries, and curcumin, found in turmeric, are particularly well-studied for their anti-inflammatory and anticancer effects (Thirumalaisamy et al., 2022; Farhan and Rizvi, 2023; Eren et al., 2024) Research has shown that polyphenols can reduce oxidative damage, lower blood pressure, and even improve the health of the gut microbiome, further contributing to overall health and disease prevention (Wu et al., 2020; Wang et al., 2022; Cano et al., 2024; Zambrano et al., 2024).

Dietary Fiber

Dietary fiber, found in plant-based foods such as fruits, vegetables, whole grains, and legumes, is a crucial component of a healthy diet. Unlike other nutrients that are digested and absorbed by the body, fiber remains largely undigested and passes through the digestive tract, offering numerous health benefits (Ioniță-Mîndrican et al., 2022). Fiber is classified into two main types—soluble and insoluble—each of which plays a distinct and vital role in maintaining health. Both types are important for maintaining optimal digestive health, regulating blood sugar, and preventing various chronic diseases such as obesity, diabetes, and colorectal cancer (Alahmari, 2024).

Soluble fiber dissolves in water to form a gel-like substance, which slows down digestion and absorption. This helps to regulate blood sugar levels by preventing rapid spikes in glucose following meals, making it especially beneficial for individuals with diabetes. In addition to its role in blood sugar regulation, soluble fiber has the ability to lower blood cholesterol levels by binding to cholesterol molecules in the digestive system and preventing their absorption into the bloodstream. Foods rich in soluble fiber include oats, barley, legumes, and fruits such as apples, citrus fruits, and pears. The consumption of soluble fiber is associated with a reduced risk of cardiovascular diseases, as it helps lower LDL (bad) cholesterol levels and improve overall heart health (Abutair et al., 2016).

Insoluble fiber, on the other hand, does not dissolve in water and instead adds bulk to stool, which helps to move waste through the digestive system more efficiently. This type of fiber is essential for promoting regular bowel movements, preventing constipation, and supporting overall digestive health. Insoluble fiber is found in whole grains, vegetables, and nuts. It is particularly important for maintaining the health of the colon by facilitating the elimination of waste and toxins. In addition, insoluble fiber has been linked to a reduced risk of colorectal cancer, as it helps to speed up the transit of food and waste through the intestines, thus reducing the time that harmful substances are in contact with the intestinal walls (Khalid et al., 2022).

Probiotics and Prebiotics

Probiotics and prebiotics are two components that play an essential role in maintaining the health of the gut microbiome, which is increasingly recognized for its impact on overall health (Bistas and Tabet, 2023). The gut microbiome is made up of trillions of microorganisms, including bacteria, viruses, and fungi, that live in the digestive tract and play a central role in digestion, immune function, and even mental health. Probiotics and prebiotics are important for supporting the growth and activity of beneficial gut bacteria, which help to maintain a balanced microbiome and promote overall health (Chaudhry et al., 2023; Chandrasekaran et al., 2024).

Probiotics are live microorganisms that, when consumed in adequate amounts, confer health benefits, particularly for the digestive system (Ashaolu, 2020). Probiotics are found in fermented foods such as yogurt, kefir, kimchi, and sauerkraut (Maftei et al., 2024). These beneficial bacteria help maintain the balance of the gut microbiome by inhibiting the growth of harmful pathogens and promoting the growth of beneficial bacteria (Mafe et al., 2025). Research has shown that the consumption of probiotics can improve digestion, enhance immune function, and reduce the risk of gastrointestinal disorders such as irritable bowel syndrome (IBS) and inflammatory bowel disease (IBD) (Markowiak and Śliżewska, 2017). Additionally, probiotics have been linked to improved mental health and may help reduce symptoms of anxiety and depression (Aziz et al., 2024).

Prebiotics, on the other hand, are non-digestible food components, often in the form of fiber, that stimulate the growth or activity of beneficial bacteria in the gut (Davani-Davari et al., 2019). Prebiotics are found in a wide variety of plant-based foods, including garlic, onions, bananas, asparagus, and leeks. By nourishing the beneficial microbes in the gut, prebiotics help to enhance the production of short-chain fatty acids (SCFAs), which have anti-inflammatory properties and support the integrity of the gut lining (Kaur et al., 2021). Prebiotics also play a role in improving mineral absorption, regulating blood sugar levels, and promoting satiety, making them beneficial for weight management (Basnet et al., 2024). The combined consumption of both probiotics and prebiotics can significantly improve gut health, leading to a more robust immune system, better nutrient absorption, and improved overall well-being (Pandey et al., 2015; Fazilah et al., 2018; Al-Habsi et al., 2024).

Omega-3 Fatty Acids

Omega-3 fatty acids are a group of essential polyunsaturated fats that are crucial for human health. These fats are called essential because the body cannot synthesize them on its own, meaning they must be obtained through the diet. Omega-3 fatty acids are primarily found in fatty fish, such as salmon, mackerel, and sardines, as well as in plant-based sources like flaxseeds, chia seeds, and walnuts. Omega-3s are known for their potent anti-inflammatory properties and their role in supporting heart health (Rodrigues et al., 2024).

There are three main types of omega-3 fatty acids: alpha-linolenic acid (ALA), eicosapentaenoic acid (EPA), and docosahexaenoic acid (DHA). ALA is primarily found in plant-based sources such as flaxseeds, chia seeds, and walnuts. While ALA is a precursor to the other two omega-3 fatty acids, it must be converted into EPA and DHA in the body, though this conversion process is not very efficient (Burns-Whitmore et al., 2019). EPA, which is found in fatty fish, has been shown to reduce inflammation, lower blood triglyceride levels, and improve cardiovascular health by enhancing blood vessel function (Khan et al., 2021). DHA is particularly important for brain health, as it constitutes a large portion of the brain and retina. It plays a critical role in cognitive function, visual health, and fetal development during pregnancy. Adequate DHA intake has been associated with a reduced risk of neurodegenerative diseases such as Alzheimer's and dementia (Weiser et al., 2016).

Omega-3 fatty acids are not only beneficial for heart and brain health but also have significant effects on mental health. Research has shown that omega-3s can help reduce symptoms of depression, anxiety, and bipolar disorder (Wani et al., 2015; Psara et al., 2025). Furthermore, their anti-inflammatory effects can assist in managing chronic conditions such as rheumatoid arthritis and inflammatory bowel disease (Grosso et al., 2014; Lange, 2020).

Vitamins and Minerals with Therapeutic Effects

Vitamins and minerals are essential micronutrients that play a variety of critical roles in maintaining overall health. In addition to supporting basic bodily functions, many vitamins and minerals also possess therapeutic properties that can help prevent or manage specific health conditions. These micronutrients are involved in numerous biochemical processes and are vital for everything from energy production to immune function, bone health, and even mood regulation.

Vitamin D is crucial for bone health as it facilitates the absorption of calcium in the intestines, helping to maintain strong bones and teeth. Beyond its role in bone health, vitamin D also supports immune function and has been linked to a reduced risk of autoimmune diseases such as multiple sclerosis, as well as certain cancers. The main sources of vitamin D include fatty fish, fortified dairy products, and exposure to sunlight, which triggers the production of vitamin D in the skin (Burns-Whitmore et al., 2024).

Vitamin C, a powerful antioxidant, plays a key role in protecting the body from oxidative stress and supporting the immune system. It helps neutralize free radicals, which can damage cells and tissues, leading to premature aging and increased risk of chronic diseases. Vitamin C is found in high concentrations in citrus fruits, berries, and leafy green vegetables. It also supports the healing of wounds and has been shown to reduce the severity and duration of colds (Alberts et al., 2025).

Magnesium is involved in over 300 biochemical reactions in the body and is essential for proper muscle function, nerve transmission, and energy production. It also helps regulate blood pressure and supports bone health by assisting in calcium absorption. Magnesium is found in a variety of foods, including nuts, seeds, whole grains, and leafy green vegetables. Adequate magnesium intake has been associated with a reduced risk of cardiovascular disease, type 2 diabetes, and osteoporosis (Fatima et al., 2024).

Zinc is another essential mineral with antioxidant properties. It is involved in numerous biological processes, including immune function, wound healing, and DNA synthesis. Zinc deficiency can lead to impaired immune function, delayed wound healing, and increased susceptibility to infections. Good dietary sources of zinc include oysters, red meat, poultry, and legumes (Maywald and Rink, 2022).

Health Benefits

Functional foods provide a wide array of health benefits, making them an integral part of a well-rounded diet. These foods, which are often naturally rich in essential nutrients, bioactive compounds, and antioxidants, offer more than basic nourishment; they actively contribute to the prevention of chronic diseases and help maintain optimal health. By including functional foods in the daily diet, individuals can improve their overall well-being and reduce the risk of developing various health conditions.

One of the most significant benefits of functional foods is their ability to support cardiovascular health. Foods that are rich in omega-3 fatty acids, such as fatty fish, flaxseeds, and walnuts, play a crucial role in reducing inflammation and lowering levels of harmful cholesterol. These foods are essential for maintaining healthy blood vessels, promoting good circulation, and supporting overall heart function. Additionally, foods high in soluble fiber, like oats, beans, and fruits, help lower cholesterol levels by binding to bile acids and promoting their excretion. Regular consumption of such foods has been shown to help maintain healthy blood pressure and reduce the risk of heart disease, strokes, and other cardiovascular diseases.

Furthermore, plant sterols found in nuts, seeds, and vegetable oils work to lower bad cholesterol levels, contributing to heart health (Arumugam et al., 2024).

In the context of metabolic health, functional foods can be particularly beneficial for managing and preventing type 2 diabetes. Many functional foods, especially those that are high in fiber, help regulate blood sugar levels by slowing the absorption of glucose into the bloodstream. Whole grains like brown rice, quinoa, and oats are prime examples of foods that help stabilize blood sugar and prevent spikes. In addition to their fiber content, whole grains provide a sustained source of energy, making them an excellent choice for people seeking to manage their weight and reduce their risk of diabetes. Legumes, such as lentils, chickpeas, and beans, are also beneficial as they provide a combination of protein and fiber, both of which help to control blood sugar levels. By improving insulin sensitivity and promoting better blood sugar control, functional foods can help reduce the likelihood of developing type 2 diabetes and support those already managing the condition (Alahmari, 2024).

Another significant benefit of functional foods is their positive impact on gastrointestinal (GI) health. A healthy gut is critical for overall health, as the gut microbiome plays a crucial role in digestion, nutrient absorption, and even immune function. Functional foods that contain probiotics, such as yogurt, kefir, and fermented vegetables, are excellent for promoting a balanced gut microbiome. Probiotics introduce beneficial bacteria to the digestive system, helping to improve digestion and reduce symptoms of digestive disorders, such as irritable bowel syndrome (IBS) and inflammatory bowel disease (IBD). These foods also aid in the absorption of nutrients and support the immune system by helping to balance the gut flora. On the other hand, prebiotics, found in foods like garlic, onions, bananas, and asparagus, nourish the good bacteria in the gut, helping them to thrive and create short-chain fatty acids that support gut health. The combination of probiotics and prebiotics ensures a robust digestive system and may help prevent common GI issues, such as constipation and bloating, while promoting long-term gastrointestinal health (Obayomi et al., 2024).

Beyond digestive health, functional foods play an important role in supporting brain function and mental well-being. Foods that are rich in antioxidants, such as berries, leafy greens, and nuts, help protect brain cells from oxidative stress, which is a major contributor to cognitive decline and neurodegenerative diseases like Alzheimer's and Parkinson's. Berries, especially blueberries, are packed with anthocyanins and other flavonoids that have been shown to enhance cognitive function, improve memory, and reduce the effects of aging on the brain. These antioxidant-rich foods play a vital role in protecting the brain from oxidative damage and promoting long-term cognitive health. In addition, certain functional foods, such as turmeric, which contains curcumin, have been shown to possess potent anti-inflammatory properties. These properties can help reduce neuroinflammation, a common factor in cognitive decline and conditions like Alzheimer's disease. Regular inclusion of such foods in the diet can contribute to better cognitive function, memory retention, and potentially reduce the risk of neurodegenerative disorders as people age (Kalogerakou and Antoniadou, 2024).

Functional foods are also known for their ability to enhance immune function, helping to protect the body from infections and illnesses. Many foods, such as citrus fruits, garlic, ginger, and mushrooms, are rich in bioactive compounds that support the immune system. Citrus fruits are an excellent source of vitamin C, which is vital for immune health, as it helps stimulate the production of white blood cells that fight off pathogens. Garlic, on the other hand, contains sulfur compounds that have antimicrobial and immune-boosting effects, helping the body ward off infections. Ginger, with its anti-inflammatory properties, is often used to support the immune system and reduce symptoms of colds and flu. Functional foods that are rich in antioxidants, vitamins, and minerals strengthen the body's natural defense mechanisms, helping to fight off infections and diseases, and promoting overall vitality (Dixit et al., 2023).

Finally, functional foods contribute significantly to general well-being by improving skin health, supporting healthy aging, and maintaining a healthy weight. Foods rich in omega-3 fatty acids, such as fatty fish and flaxseeds, help maintain skin elasticity and hydration, preventing signs of premature aging. Additionally, the antioxidants found in fruits and vegetables, such as vitamin C in citrus fruits and lycopene in tomatoes, protect the skin from UV damage and oxidative stress, promoting healthy, glowing skin. Moreover, functional foods high in fiber help with weight management by promoting satiety, reducing overeating, and supporting a healthy metabolism (Singh et al., 2025).

All in all, functional foods are much more than just nourishing; they are a powerful tool for promoting long-term health and preventing a range of chronic diseases. From supporting heart and brain health to improving digestive function and boosting immunity, functional foods offer a variety of benefits that can enhance the quality of life. By incorporating these nutrient-dense foods into a daily diet, individuals can support their overall health and reduce the risk of developing chronic conditions, leading to a healthier and more vibrant life.

Emerging Trends and Challenges

The field of functional foods is rapidly evolving, driven by scientific advancements, changing consumer demands, and new opportunities for innovation. This dynamic landscape has given rise to several key trends and challenges that are shaping the future of the industry. As consumers seek natural health solutions and personalized nutrition, the demand for functional foods is increasing. However, the industry faces several obstacles, including regulatory challenges, the need for more clinical research, and the complexity of defining and marketing these products.

Growing Consumer Demand for Natural Health Solutions

Consumers are becoming increasingly interested in natural, holistic approaches to health and wellness. This shift away from synthetic medications and supplements has fueled a growing demand for functional foods that offer health benefits through naturally occurring bioactive compounds found in everyday foods. Consumers are actively seeking products that align with their values, preferring natural ingredients that are minimally processed and free from artificial additives, preservatives, and genetically modified organisms (GMOs). This trend toward natural health solutions is evident in the rising popularity of organic, plant-based, and non-GMO food options, which are perceived as safer and more beneficial for long-term health.

As health-conscious lifestyles continue to grow, functional foods are increasingly seen as essential tools for disease prevention and overall wellness. Many individuals are integrating functional foods into their daily diets as a way to support key areas of health, including immunity, digestive health, cardiovascular function, and mental well-being. The demand for functional foods that promote these benefits is expected to continue growing, as more people focus on maintaining their health and preventing chronic diseases through nutrition.

Regulatory Challenges in Defining and Marketing Functional Foods

One of the most significant challenges facing the functional foods industry is the lack of standardized definitions and clear regulations regarding what qualifies as a "functional food." In many countries, regulatory bodies impose strict requirements for health claims made by functional food manufacturers, and these claims must be substantiated with scientific evidence, often derived from clinical studies. This creates hurdles for manufacturers who wish to market their products effectively, especially when the regulations surrounding health claims are particularly stringent in certain regions.

For instance, in the European Union, the regulations governing health claims are among the most restrictive globally, making it challenging for manufacturers to market their products as beneficial for specific health conditions without undergoing a rigorous approval process. This situation is compounded by the lack of consistency in regulations across regions. While countries like Japan and South Korea have broadly accepted definitions for functional foods, regions like the United States and the European Union maintain more rigid definitions that focus on specific health benefits or disease prevention, adding complexity for companies attempting to enter global markets.

The Need for More Human-Based Clinical Studies

While research into the benefits of functional foods has demonstrated promising results in laboratory and animal models, there is a critical need for more human-based clinical trials to validate the health benefits of these foods. The results of animal studies do not always translate directly to human health outcomes, and human trials are essential for determining the bioavailability, proper dosage, and long-term effects of bioactive compounds in functional foods.

Many functional foods have been extensively studied in animal models, yet the lack of large-scale, well-designed clinical trials involving human participants limits the ability of manufacturers to make definitive health claims. Without robust clinical evidence, the health benefits of many functional foods remain uncertain, which creates challenges for both consumers and manufacturers in terms of product trust and marketability. More human-based studies are necessary to provide the scientific backing required to support the marketing of functional foods and to ensure that the products are both safe and effective for human consumption.

Personalized Nutrition and Functional Food Innovation

As we look to the future of functional foods, personalized nutrition is poised to play a central role in shaping the industry. Personalized nutrition involves tailoring dietary recommendations and functional food interventions to an individual's unique needs, based on genetic, lifestyle, and environmental factors. This approach could revolutionize how functional foods are consumed and how they contribute to health outcomes.

Advances in genomics and microbiome research are opening up exciting possibilities for customizing functional foods to meet individual health needs. By understanding the genetic variations and microbiome composition of individuals, functional foods can be optimized for specific health conditions or risk factors. For example, individuals with certain genetic predispositions may benefit more from foods rich in antioxidants, omega-3 fatty acids, or other bioactive compounds that target specific health concerns. Personalized nutrition could lead to more precise and effective use of functional foods in preventing and managing diseases, improving overall health, and enhancing individual well-being.

Furthermore, the growing demand for personalized nutrition is driving innovation in the development of new functional foods. Manufacturers are exploring ways to create products tailored to specific health conditions, such as diabetes, heart disease, or neurological disorders. These innovations are likely to result in a more targeted approach to functional food consumption, allowing consumers to select foods that align closely with their health needs and preferences. This shift toward personalized functional foods holds the potential to improve efficacy, increase consumer satisfaction, and open new avenues for market growth in the functional foods industry.

Conclusion

Functional foods have become an essential component in the evolving landscape of modern healthcare and nutrition. As global health challenges, especially chronic diseases such as cardiovascular disease, type 2 diabetes, cancer, and neurodegenerative disorders, continue to rise, functional foods offer a viable and sustainable strategy for both disease prevention and health promotion. These foods, which extend beyond basic nutrition, provide physiological benefits that help manage disease risks, improve overall health, and enhance the quality of life. By offering natural alternatives to pharmaceuticals, functional foods serve as powerful tools in the fight against chronic diseases, contributing significantly to the well-being of individuals and populations worldwide.

Over the past few decades, scientific research has increasingly revealed the profound impact that functional foods can have on human health. Bioactive compounds found in these foods—such as phytochemicals, omega-3 fatty acids, fiber, and probiotics—have been shown to modulate various metabolic pathways, reduce inflammation, and enhance immune function. For example, phytochemicals like flavonoids and carotenoids found in fruits and vegetables have been linked to reduced oxidative stress, a key factor in the development of chronic diseases. Similarly, omega-3 fatty acids, present in fatty fish and certain plant oils, are well-documented for their anti-inflammatory properties, supporting cardiovascular health and reducing the risk of heart disease. Probiotics and prebiotics also play a critical role in gut health, influencing digestion, immunity, and even mood regulation.

The health benefits of functional foods extend beyond disease prevention; they show significant promise in managing existing conditions, reducing the need for medical interventions, and improving the quality of life for patients. For instance, functional foods used in the management of metabolic syndrome and obesity have yielded encouraging results, with dietary interventions focusing on whole grains, legumes, and plant-based foods demonstrating improvements in insulin sensitivity, blood sugar regulation, and promoting healthy body weight.

Despite the growing body of evidence supporting the efficacy of functional foods, several challenges remain that hinder their widespread adoption and integration into mainstream healthcare practices. One of the primary barriers lies in the regulatory landscape. While the scientific community recognizes the potential of functional foods, regulatory bodies have yet to establish comprehensive guidelines for defining and approving health claims associated with these foods. This lack of standardization can lead to confusion and misinterpretation by consumers, who may be uncertain about which foods genuinely offer health benefits. Establishing clear, evidence-based regulatory frameworks is crucial to ensuring that functional foods can be safely marketed, allowing consumers to trust the health claims associated with these products.

Another significant challenge is the limited availability of large-scale, human-based clinical studies that provide robust evidence of the long-term health effects of functional foods. Although numerous animal studies and in vitro experiments have demonstrated the potential of bioactive compounds, more clinical trials are needed to confirm their benefits in humans and establish definitive guidelines for their use. The variability in individual responses to functional foods further underscores the need for personalized nutrition, where dietary recommendations are tailored to an individual's genetic profile, health status, and lifestyle. Advancements in personalized nutrition may lead to more targeted interventions that maximize the health benefits of functional foods for each person.

Additionally, the growing consumer demand for natural and holistic health solutions has driven the rapid expansion of the functional food market. As awareness of lifestyle-related diseases continues to rise, consumers are increasingly seeking foods that not only nourish the body but also offer additional health benefits. As this demand grows, the functional food industry is expected to innovate, developing new and improved products tailored to the specific needs of different populations, from those looking to enhance their immune systems to individuals with chronic health conditions requiring specialized diets.

The future of functional foods holds immense promise. Ongoing research into novel bioactive compounds, innovative food processing techniques, and new formulations that enhance the bioavailability and effectiveness of functional ingredients will be essential in advancing the field. Personalized nutrition, based on a deeper understanding of genetic, microbiome, and metabolic profiles, is poised to be a key driver in the next phase of functional food development. As our understanding of the human body and the complex interactions between diet and health continues to evolve, functional foods are likely to become even more integral to individualized health management, allowing people to make dietary choices that proactively support their long-term well-being.

In conclusion, functional foods represent a critical and transformative approach to nutrition, disease prevention, and health promotion. As our global society faces the dual challenges of increasing chronic disease prevalence and rising healthcare costs, the integration of functional foods into everyday diets offers a sustainable and effective solution to these pressing health concerns. The ability of functional foods to reduce the risk of chronic diseases, manage existing health conditions, and improve overall quality of life positions them as a cornerstone of modern nutrition and healthcare. By embracing the potential of functional foods, we can move toward a future where nutrition-based healthcare solutions are at the forefront of disease prevention, ultimately leading to healthier individuals and communities worldwide.

References

Abdul Majeed, S., Said, S., Hassan, D. A., Sadiq, F., Alhosani, M., Al-Jawaldeh, A., El-Obeid, T., & Tayyem, R. (2025). Evaluating the effectiveness and risks of bread fortification programs in the Middle Eastern region: A comprehensive review. *Frontiers in Public Health, 13*, 1530380.

Abutair, A. S., Naser, I. A., & Hamed, A. T. (2016). Soluble fibers from psyllium improve glycemic response and body weight among diabetes type 2 patients (randomized control trial). *Nutritional Journal, 15*(1), 86.

Ahmad, I., Hao, M., Li, Y., Zhang, J., Ding, Y., & Lyu, F. (2022). Fortification of yogurt with bioactive functional foods and ingredients and associated challenges – A review. *Trends in Food Science & Technology, 129*, 558–580.

Aiello, G., Lombardo, M., & Baldelli, S. (2024). Exploring Vitamin D Synthesis and Function in Cardiovascular Health: A Narrative Review. *Applied Sciences, 14*(11), 4339.

Ajanaku, C. O., Ademosun, O. T., Atohengbe, P. O., Ajayi, S. O., Obafemi, Y. D., Owolabi, O. A., Akinduti, P.A., & Ajanaku, K. O. (2022). Functional bioactive compounds in ginger, turmeric, and garlic. *Frontiers in Nutrition, 9*, 1012023.

Alahmari, L. A. (2024). Dietary fiber influence on overall health, with an emphasis on CVD, diabetes, obesity, colon cancer, and inflammation. *Frontiers in Nutrition, 11*, 1510564.

Alberts, A., Moldoveanu, E. T., Niculescu, A. G., & Grumezescu, A. M. (2025). Vitamin C: A comprehensive review of its role in health, disease prevention, and therapeutic potential. *Molecules*, 30(3), 748.

Al-Habsi, N., Al-Khalili, M., Haque, S. A., Elias, M., Olqi, N. A., & Al Uraimi, T. (2024). Health benefits of prebiotics, probiotics, synbiotics, and postbiotics. *Nutrients, 16*(22), 3955.

Anderson, R. A., Zhan, Z., Luo, R., Guo, X., Guo, Q., Zhou, J., Kong, J., Davis, P. A., & Stoecker, B. J. (2016). Cinnamon extract lowers glucose, insulin and cholesterol in people with elevated serum glucose. *Journal of Traditional and Complementary Medicine, 6*(4), 332-336.

Antony, B., Benny, M., Jose, S., Jacob, S., Nedumpilly, V., M S, A., & Abraham, G. (2024). Development of omega-3 enriched egg using fish-oil based fowl feed supplement. *Journal of Applied Poultry Research, 33*(3), 100429.

Aregawi, L. G., Gebremeskel, T. G., & Zoltan, C. (2024). Preventive and therapeutic effects of ginger on bowel disease: A review of clinical trials. *Pharmacological Research - Modern Chinese Medicine, 12*, 100457.

Arumugam, V., Cheeran Amal, T., Sarvalingam, A., & Vasanth, K. (2024). A review on the influence of nutraceuticals and functional foods on health. *Food Chemistry Advances, 5*, 100749.

Ashaolu, T. J. (2020). Immune boosting functional foods and their mechanisms: A critical evaluation of probiotics and prebiotics. *Biomedicine & Pharmacotherapy, 130*, 110625.

Aziz, T., Hussain, N., Hameed, Z., & Lin, L. (2024). Elucidating the role of diet in maintaining gut health to reduce the risk of obesity, cardiovascular and other age-related inflammatory diseases: recent challenges and future recommendations. *Gut Microbes, 16*(1), 2297864.

Baker, M. T., Lu, P., Parrella, J. A., & Leggette, H. R. (2022). Consumer acceptance toward functional foods: A scoping review. *International Journal of Environmental Research and Public Health, 19*(3), 1217.

Banaszak, M., Dobrzyńska, M., Kawka, A., Górna, I., Woźniak, D., Przysławski, J., & Drzymała-Czyż, S. (2024). Role of omega-3 fatty acids eicosapentaenoic (EPA) and docosahexaenoic (DHA) as modulatory and anti-inflammatory agents in noncommunicable diet-related diseases – Reports from the last 10 years. *Clinical Nutrition ESPEN, 63*, 240-258.

Banerjee, A., Somasundaram, I., Das, D., Jain, M. S., Banu, H., Mitta, S. P., Paul, S., Bisgin, A., Zhang, H., Sun, X. F., Duttaroy, A. K., & Pathak, S. (2023). Functional foods: A promising strategy for restoring gut microbiota diversity impacted by SARS-CoV-2 variants. *Nutrients, 15*(11), 2631.

Barkas, F., Bathrellou, E., Nomikos, T., Panagiotakos, D., Liberopoulos, E., & Kontogianni, M. D. (2023). Plant Sterols and Plant Stanols in Cholesterol Management and Cardiovascular Prevention. *Nutrients, 15*(13), 2845.

Basnet, J., Eissa, M. A., Yanes Cardozo, L. L., Romero, D. G., & Rezq, S. (2024). Impact of probiotics and prebiotics on gut microbiome and hormonal regulation. *Gastrointestinal Disorders, 6*(4), 801–815.

Beto, J. A. (2015). The role of calcium in human aging. *Clinical Nutrition Research, 4*(1), 1–8.

Bistas, K. G., & Tabet, J. P. (2023). The benefits of prebiotics and probiotics on mental health. *Cureus, 15*(8), e43217.

Brandl, B., Rennekamp, R., Reitmeier, S., Pietrynik, K., Dirndorfer, S., Haller, D., Hofmann, T., Skurk, T., & Hauner, H. (2022). Offering fiber-enriched foods increases fiber intake in adults with or without cardiometabolic risk: A randomized controlled trial. *Frontiers in Nutrition, 9*, 816299.

Burns-Whitmore, B., Froyen, E., Heskey, C., Parker, T., & San Pablo, G. (2019). Alpha-Linolenic and Linoleic Fatty Acids in the Vegan Diet: Do They Require Dietary Reference Intake/Adequate Intake Special Consideration? *Nutrients*, 11(10), 2365.

Burns-Whitmore, B., Froyen, E. B., & Isom, K. A. (2024). Vitamin D and calcium—An overview, review of metabolism, and the importance of co-supplementation. *Dietetics*, 3(4), 588-608.

Cano, R., Bermúdez, V., Galban, N., Garrido, B., Santeliz, R., Gotera, M. P., Duran, P., Boscan, A., Carbonell-Zabaleta, A.-K., Durán-Agüero, S., & et al. (2024). Dietary polyphenols and gut microbiota cross-talk: Molecular and therapeutic perspectives for cardiometabolic disease: A narrative review. *International Journal of Molecular Sciences, 25*, 9118.

Chanda, D., Ha, H., & Lee, I. K. (2023). Editorial: The role of oxidative stress and systemic inflammation in diabetes and chronic kidney disease. *Frontiers in Endocrinology (Lausanne), 14*, 1272525.

Chandrasekaran, P., Weiskirchen, S., & Weiskirchen, R. (2024). Effects of probiotics on gut microbiota: An overview. *International Journal of Molecular Sciences, 25*(11), 6022.

Chaudhry, T. S., Senapati, S. G., Gadam, S., Mannam, H. P. S. S., Voruganti, H. V., Abbasi, Z., Abhinav, T., Challa, A. B., Pallipamu, N., Bheemisetty, N., Arunachalam, S. P. (2023). The impact of microbiota on the gut-brain axis: Examining the complex interplay and implications. *J Clin Med, 12*(16), 5231.

Collins, E. J., Bowyer, C., Tsouza, A., & Chopra, M. (2022). Tomatoes: An extensive review of the associated health impacts of tomatoes and factors that can affect their cultivation. *Biology (Basel), 11*(2), 239.

Davani-Davari, D., Negahdaripour, M., Karimzadeh, I., Seifan, M., Mohkam, M., Masoumi, S. J., Berenjian, A., & Ghasemi, Y. (2019). Prebiotics: Definition, types, sources, mechanisms, and clinical applications. *Foods, 8*(3), 92.

Dighriri, I. M., Alsubaie, A. M., Hakami, F. M., Hamithi, D. M., Alshekh, M. M., Khobrani, F. A., Dalak, F. E., Hakami, A. A., Alsueaadi, E. H., Alsaawi, L. S., Alshammari, S. F., Alqahtani, A. S., Alawi, I. A., Aljuaid, A. A., & Tawhari, M. Q. (2022). Effects of Omega-3 Polyunsaturated Fatty Acids on Brain Functions: A Systematic Review. *Cureus, 14*(10), e30091.

Dixit, V., Kamal, J., William, S., Chole, B., Pranjali, D., Dayal, D., Chaubey, K. K., Pal, A. K., Xavier, J., Manjunath, B. T., & Bachheti, R. K. (2023). Functional foods: Exploring the health benefits of bioactive compounds from plant and animal sources. *Journal of Food Quality*, 2023, 5546753.

Domínguez Díaz, L., Fernández-Ruiz, V., & Cámara, M. (2020). An international regulatory review of food health-related claims in functional food products labeling. *Journal of Functional Foods, 68*, 103896.

Donker, A. E., van der Staaij, H., & Swinkels, D. W. (2021). The critical roles of iron during the journey from fetus to adolescent: Developmental aspects of iron homeostasis. *Blood Reviews, 50*, 100866.

Doyon, M., & Labrecque, J. (2008). Functional foods: A conceptual definition. *British Food Journal, 110*(11), 1133-1149.

Eren, E., Das, J., & Tollefsbol, T. O. (2024). Polyphenols as immunomodulators and epigenetic modulators: An analysis of their role in the treatment and prevention of breast cancer. *Nutrients, 16*, 4143.

Essa, M. M., Bishir, M., Bhat, A., Chidambaram, S. B., Al-Balushi, B., Hamdan, H., Govindarajan, N., Freidland, R. P., & Qoronfleh, M. W. (2023). Functional foods and their impact on health. *Journal of Food Science and Technology, 60*(3), 820-834.

Fabbri, L. P., Cavallero, A., Vidotto, F., & Gabriele, M. (2024). Bioactive peptides from fermented foods: Production approaches, sources, and potential health benefits. *Foods, 13*(21), 3369.

Farhan, M., & Rizvi, A. (2023). The pharmacological properties of red grape polyphenol resveratrol: Clinical trials and obstacles in drug development. *Nutrients, 15*(20), 4486.

Farhana, A., Saleem Khan, Y., Alsrhani, A., Manni, E., Alameen, A. M., Derafa, W., Alhathiaul, N., Atif, M., & Bashier Eltayeb, L. (2025). Antioxidant and prooxidant functions of carotenoids in human health: Trigger factors, mechanism and application. In *IntechOpen.*

Fatima, G., Dzupina, A., Alhmadi, H. B., Magomedova, A., Siddiqui, Z., Mehdi, A., & Hadi, N. (2024). Magnesium matters: A comprehensive review of its vital role in health and diseases. *Cureus*, 16(10), e71392.

Fazilah, N. F., Ariff, A. B., Khayat, M. E., Rios-Solis, L., & Halim, M. (2018). Influence of probiotics, prebiotics, synbiotics, and bioactive phytochemicals on the formulation of functional yogurt. *Journal of Functional Foods, 48*, 387-399.

Feili, R., WahiduZzaman, W. N. W. A., & Yang, T. A. (2013). Physical and Sensory Analysis of High Fiber Bread Incorporated with Jackfruit Rind Flour. *Food Science and Technology, 1*(2), 30-36.

Frumuzachi, O., Flanagan, A., Rohn, S., & Mocan, A. (2025). The dichotomy between functional and functionalized foods – A critical characterization of concepts. *Food Research International, 208*, 116173.

Fu, J., Zheng, Y., Gao, Y., & Xu, W. (2022). Dietary fiber intake and gut microbiota in human health. *Microorganisms, 10*(12), 2507.

Galanakis, C. M. (2021). Functionality of food components and emerging technologies. *Foods, 10*(1), 128.

Grillo, A., Salvi, L., Coruzzi, P., Salvi, P., & Parati, G. (2019). Sodium intake and hypertension. *Nutrients, 11*(9), 1970.

Grosso, G., Galvano, F., Marventano, S., Malaguarnera, M., Bucolo, C., Drago, F., & Caraci, F. (2014). Omega-3 fatty acids and depression: scientific evidence and biological mechanisms. *Oxidative Medicine and Cellular Longevity, 2014*, 313570.

Gulati, K., Anand, R., & Ray, A. (2016). Chapter 16 - Nutraceuticals as adaptogens: Their role in health and disease. In R. C. Gupta (Ed.), *Nutraceuticals* (pp. 193-205). Academic Press.

Han, T., Wei, W., Jiang, W., Geng, Y., Liu, Z., Yang, R., Jin, C., Lei, Y., Sun, X., Xu, J., Chen, J., & Sun, C. (2024). The future landscape and framework of precision nutrition. *Engineering, 42*, 15-25.

Hasler, C. M. (2002). Functional foods: Benefits, concerns, and challenges—A position paper from the American Council on Science and Health. *The Journal of Nutrition, 132*(12), 3772-3781.

Hatch-McChesney, A., & Lieberman, H. R. (2022). Iodine and iodine deficiency: A comprehensive review of a re-emerging issue. *Nutrients, 14*(17), 3474.

Hernández-López, I., Ortiz-Solà, J., Alamprese, C., Barros, L., Shelef, O., Basheer, L., Rivera, A., Abadias, M., & Aguiló-Aguayo, I. (2022). Valorization of local legumes and nuts as key components of the Mediterranean diet. *Foods, 11*(23), 3858.

Holscher, H. D. (2017). Dietary fiber and prebiotics and the gastrointestinal microbiota. *Gut Microbes, 8*(2), 172-184.

Hossain, M. S., Wazed, M. A., Asha, S., Amin, M. R., & Shimul, I. M. (2025). Dietary phytochemicals in health and disease: Mechanisms, clinical evidence, and applications – A comprehensive review. *Food Science & Nutrition, 13*(3), e70101.

Ioniță-Mîndrican, C. B., Ziani, K., Mititelu, M., Oprea, E., Neacșu, S. M., Moroșan, E., Dumitrescu, D. E., Roșca, A. C., Drăgănescu, D., & Negrei, C. (2022). Therapeutic benefits and dietary restrictions of fiber intake: A state of the art review. *Nutrients, 14*(13), 2641.

Ismail, S., Eljazzar, S., & Ganji, V. (2023). Intended and unintended benefits of folic acid fortification—A narrative review. *Foods, 12*(8), 1612.

İzli, G., & Yildiz, G. (2021). Evaluation of the high intensity ultrasound pre-treatment effects on the physical properties and bioactive compounds of convective dried quince samples. *International Journal of Fruit Science, 21*(1), 645-656.

Jachimowicz-Rogowska, K., & Winiarska-Mieczan, A. (2023). Initiatives to reduce the content of sodium in food products and meals and improve the population's health. *Nutrients, 15*(10), 2393.

Jakubczyk, K., Kochman, J., Kwiatkowska, A., Kałduńska, J., Dec, K., Kawczuga, D., & Janda, K. (2020). Antioxidant properties and nutritional composition of matcha green tea. *Foods, 9*(4), 483.

Jomova, K., Alomar, S. Y., Valko, R., Liska, J., Nepovimova, E., Kuca, K., & Valko, M. (2025). Flavonoids and their role in oxidative stress, inflammation, and human diseases. *Chemico-Biological Interactions, 413*, 111489.

Kalita, S., Khandelwal, S., Madan, J., Pandya, H., Sesikeran, B., & Krishnaswamy, K. (2018). Almonds and cardiovascular health: A review. *Nutrients, 10*(4), 468.

Kalogerakou, T., & Antoniadou, M. (2024). The role of dietary antioxidants, food supplements and functional foods for energy enhancement in healthcare professionals. *Antioxidants (Basel), 13*(12), 1508.

Kaur, A. P., Bhardwaj, S., Dhanjal, D. S., Nepovimova, E., Cruz-Martins, N., Kuča, K., Chopra, C., Singh, R., Kumar, H., Şen, F., Kumar, V., Verma, R., & Kumar, D. (2021). Plant prebiotics and their role in the amelioration of diseases. *Biomolecules, 11*(3), 440.

Kaur, G., Mason, R. P., Steg, P. G., & Bhatt, D. L. (2024). Omega-3 fatty acids for cardiovascular event lowering. *European Journal of Preventive Cardiology, 31*(8), 1005–1014.

Khalid, W., Arshad, M. S., Jabeen, A., Muhammad Anjum, F., Qaisrani, T. B., & Suleria, H. A. R. (2022). Fiber-enriched botanicals: A therapeutic tool against certain metabolic ailments. *Food Science & Nutrition, 10*(10), 3203–3218.

Khan, S. U., Lone, A. N., Khan, M. S., Virani, S. S., Blumenthal, R. S., Nasir, K., Miller, M., Michos, E. D., Ballantyne, C. M., Boden, W. E., & Bhatt, D. L. (2021). Effect of omega-3 fatty acids on cardiovascular outcomes: A systematic review and meta-analysis. *EClinicalMedicine*, 38, 100997.

Kozłowska-Wojciechowska, M., Jastrzębska, M., Naruszewicz, M., & Foltyńska, A. (2003). Impact of margarine enriched with plant sterols on blood lipids, platelet function, and fibrinogen level in young men. *Metabolism, 52*(11), 1373-1378.

Kumar, A., P N, Kumar, M., Jose, A., Tomer, V., Oz, E., Proestos, C., Zeng, M., Elobeid, T., K S, & Oz, F. (2023). Major phytochemicals: Recent advances in health benefits and extraction method. *Molecules, 28*(2), 887. https://doi.org/10.3390/molecules28020887.

Kumar, R., Bhardwaj, S., Sikarwar, M., Kumar, A., Singh, B. R., Gupta, M., & Shukla, R. (2025). Innovative food applications of quinoa: Exploring novel insights into its nutritional potential and biological activities. *Journal of Food Composition and Analysis, 139*, 107133.

Lagoumintzis, G., & Patrinos, G. P. (2023). Triangulating nutrigenomics, metabolomics, and microbiomics toward personalized nutrition and healthy living. *Human Genomics, 17*, 109.

Lange, K. W. (2020). Omega-3 fatty acids and mental health. *Global Health Journal*, 4(1), 18–30.

Langyan, S., Yadava, P., Khan, F. N., Dar, Z. A., Singh, R., & Kumar, A. (2022). Sustaining protein nutrition through plant-based foods. *Frontiers in Nutrition, 8*, 772573.

Lee, B. Y., Ordovás, J. M., Parks, E. J., Anderson, C. A. M., Barabási, A. L., Clinton, S. K., de la Haye, K., Duffy, V. B., Franks, P. W., Ginexi, E. M., Hammond, K. J., Hanlon, E. C., Hittle, M., Ho, E., Horn, A. L., Isaacson, R. S., Mabry, P. L., Malone, S., Martin, C. K., Mattei, J., Meydani, S. N., Nelson, L. M., Neuhouser, M. L., Parent, B., Pronk, N. P., Roche, H. M., Saria, S., Scheer, F. A. J. L., Segal, E., Sevick, M. A., Spector, T. D., Van Horn, L., Varady, K. A., Voruganti, V. S., & Martinez, M. F. (2022). Research gaps and opportunities in precision nutrition: An NIH workshop report. *The American Journal of Clinical Nutrition, 116*(6), 1877–1900.

Leeuwendaal, N. K., Stanton, C., O'Toole, P. W., & Beresford, T. P. (2022). Fermented foods, health and the gut microbiome. *Nutrients, 14*(7), 1527.

Ma, Z. F., Fu, C., & Lee, Y. Y. (2025). The modulatory role of bioactive compounds in functional foods on inflammation and metabolic pathways in chronic diseases. *Foods, 14*, 821.

Mafe, A. N., Edo, G. I., Majeed, O. S., Gaaz, T. S., Akpoghelie, P. O., Isoje, E. F., Igbuku, U. A., Owheruo, J. O., Opiti, R. A., Garba, Y., Essaghah, A. E. A., Ahmed, D. S., & Umar, H. (2025). A review on probiotics and dietary bioactives: Insights on metabolic well-being, gut microbiota, and inflammatory responses. *Food Chemistry Advances, 6*, 100919.

Maftei, N. M., Raileanu, C. R., Balta, A. A., Ambrose, L., Boev, M., Marin, D. B., & Lisa, E. L. (2024). The potential impact of probiotics on human health: An update on their health-promoting properties. *Microorganisms, 12*(2), 234.

Maoka, T. (2020). Carotenoids as natural functional pigments. *Journal of Natural Medicines, 74*(1), 1-16.

Markowiak, P., & Śliżewska, K. (2017). Effects of probiotics, prebiotics, and synbiotics on human health. *Nutrients, 9*(9), 1021.

Maywald, M., & Rink, L. (2022). Zinc in human health and infectious diseases. *Biomolecules*, 12(12), 1748.

Mhanna, A., Martini, N., Hmaydoosh, G., Hamwi, G., Jarjanazi, M., Zaifah, G., Kazzazo, R., Haji Mohamad, A., & Alshehabi, Z. (2024). The correlation between gut microbiota and both neurotransmitters and mental disorders: A narrative review. *Medicine (Baltimore), 103*(5), e37114.

Mrowicka, M., Mrowicki, J., Kucharska, E., & Majsterek, I. (2022). Lutein and zeaxanthin and their roles in age-related macular degeneration-neurodegenerative disease. *Nutrients, 14*(4), 827.

Muscolo, A., Mariateresa, O., Giulio, T., & Mariateresa, R. (2024). Oxidative stress: The role of antioxidant phytochemicals in the prevention and treatment of diseases. *International Journal of Molecular Sciences, 25*(6), 3264.

Naveed, H., Sultan, W., Awan, K. A., Imtiaz, A., Yaqoob, S., Al-Asmari, F., Faraz, A., Qian, J.-Y., Sharma, A., Mugabi, R., Alotaibi, S. S., & Nayik, G. A. (2024). Glycemic impact of cereal and legume-based bakery products: Implications for chronic disease management. *Food Chemistry: X, 24*, 101959.

Neuhouser, M. L. (2019). The importance of healthy dietary patterns in chronic disease prevention. *Nutritional Research, 70*, 3-6.

Ntanios, F. (2001). Plant sterol-ester-enriched spreads as an example of a new functional food. *European Journal of Lipid Science and Technology, 103*(2), 102-106.

Obayomi, O. V., Olaniran, A. F., & Owa, S. O. (2024). Unveiling the role of functional foods with emphasis on prebiotics and probiotics in human health: A review. *Journal of Functional Foods, 119*, 106337.

Olson, R., Gavin-Smith, B., Ferraboschi, C., & Kraemer, K. (2021). Food fortification: The advantages, disadvantages and lessons from Sight and Life programs. *Nutrients, 13*(4), 1118.

Pandey, K. R., Naik, S. R., & Vakil, B. V. (2015). Probiotics, prebiotics and synbiotics—a review. *Journal of Food Science and Technology, 52*(12), 7577–7587.

Panossian, A., & Wikman, G. (2010). Effects of adaptogens on the central nervous system and the molecular mechanisms associated with their stress-protective activity. *Pharmaceuticals (Basel), 3*(1), 188-224.

Panou, A., & Karabagias, I. K. (2025). Composition, properties, and beneficial effects of functional beverages on human health. *Beverages, 11*(2), 40.

Patel, V., Aggarwal, K., Dhawan, A., Singh, B., Shah, P., Sawhney, A., & Jain, R. (2023). Protein supplementation: the double-edged sword. *Proc (Bayl Univ Med Cent), 37*(1), 118-126.

Paudel, S., Mishra, N., & Agarwal, R. (2023). Phytochemicals as immunomodulatory molecules in cancer therapeutics. *Pharmaceuticals (Basel), 16*(12), 1652.

Peng, Y., Ao, M., Dong, B., Jiang, Y., Yu, L., Chen, Z., Hu, C., & Xu, R. (2021). Anti-inflammatory effects of curcumin in the inflammatory diseases: Status, limitations, and countermeasures. *Drug Design, Development and Therapy, 15*, 4503–4525.

Psara, E., Papadopoulou, S. K., Mentzelou, M., Voulgaridou, G., Vorvolakos, T., Apostolou, T., & Giaginis, C. (2025). Omega-3 fatty acids for the treatment of bipolar disorder symptoms: A narrative review of the current clinical evidence. *Marine Drugs*, 23(2), 84.

Ried, K. (2020). Garlic lowers blood pressure in hypertensive subjects, improves arterial stiffness and gut microbiota: A review and meta-analysis. *Experimental and Therapeutic Medicine, 19*(2), 1472–1478.

Rodrigues, M., Rosa, A., Almeida, A., Martins, R., Ribeiro, T., Pintado, M., Gonçalves, R. F. S., Pinheiro, A. C., Fonseca, A. J. M., Maia, M. R. G., Cabrita, A. R. J., Barros, L., & Caleja, C. (2024). Omega-3 fatty acids from fish by-products: Innovative extraction and application in food and feed. *Food and Bioproducts Processing, 145*, 32-41.

Santos, J. A. R., Christoforou, A., Trieu, K., McKenzie, B. L., Downs, S., Billot, L., Webster, J., & Li, M. (2019). Iodine fortification of foods and condiments, other than salt, for preventing iodine deficiency disorders. *Cochrane Database of Systematic Reviews, 2019*(2), CD010734.

Saraiva, A., Carrascosa, C., Raheem, D., Ramos, F., & Raposo, A. (2020). Natural sweeteners: The relevance of food naturalness for consumers, food security aspects, sustainability, and health impacts. *International Journal of Environmental Research and Public Health, 17*(17), 6285.

Shahnaz, T., Fawole, A. O., Adeyanju, A. A., & Onuh, J. O. (2024). Food proteins as functional ingredients in the management of chronic diseases: A concise review. *Nutrients, 16*(14), 2323.

Singh, H., Kamal, Y. T., Pandohee, J., Mishra, A. K., Biswas, A., Mohanto, S., Kumar, A., Nag, S., Mishra, A., Singh, M., Gupta, H., & Chopra, H. (2025). Dietary phytochemicals alleviate the premature skin aging: A comprehensive review. *Experimental Gerontology, 199*, 112660.

Tafese Awulachew, M. (2024). Functional foods: Functional ingredients, sources and classification, health claims, food intolerance, and allergy. *IntechOpen.*

Tangpricha, V., Koutkia, P., Rieke, S. M., Chen, T. C., Perez, A. A., & Holick, M. F. (2003). Fortification of orange juice with vitamin D: A novel approach for enhancing vitamin D nutritional health. *The American Journal of Clinical Nutrition, 77*(6), 1478–1483.

Tardy, A. L., Pouteau, E., Marquez, D., Yilmaz, C., & Scholey, A. (2020). Vitamins and minerals for energy, fatigue, and cognition: A narrative review of the biochemical and clinical evidence. *Nutrients, 12*(1), 228.

Thirumalaisamy, R., Bhuvaneswari, M., Haritha, S., Jeevarathna, S., Sai Janani, K. S., & Suresh, K. (2022). Curcumin, naringenin, and resveratrol from natural plant products hold promising solutions for modern world diseases – A recent review. *South African Journal of Botany, 151*(Part B), 567-580.

Tkaczenko, H., & Kurhaluk, N. (2025). Antioxidant-rich functional foods and exercise: Unlocking metabolic health through Nrf2 and related pathways. *International Journal of Molecular Sciences, 26*, 1098.

Tóth-Mészáros, A., Garmaa, G., Hegyi, P., Bánvölgyi, A., Fenyves, B., Fehérvári, P., Harnos, A., Gergő, D., Nguyen Do To, U., & Csupor, D. (2023). The effect of adaptogenic plants on stress: A systematic review and meta-analysis. *Journal of Functional Foods, 108*, 105695.

Tsoupras, A., Brummell, C., Kealy, C., Vitkaitis, K., Redfern, S., & Zabetakis, I. (2022). Cardio-protective properties and health benefits of fish lipid bioactives; The effects of thermal processing. *Marine Drugs, 20*(3), 187.

Ullah, A., Munir, S., Badshah, S. L., Khan, N., Ghani, L., Poulson, B. G., Emwas, A. H., & Jaremko, M. (2020). Important flavonoids and their role as a therapeutic agent. *Molecules, 25*(22), 5243.

Vignesh, A., Amal, T. C., Sarvalingam, A., & Vasanth, K. (2024). A review on the influence of nutraceuticals and functional foods on health. *Food Chemistry Advances, 5*, 100749.

Wang, X., Qi, Y., & Zheng, H. (2022). Dietary polyphenol, gut microbiota, and health benefits. *Antioxidants (Basel), 11*(6), 1212.

Wani, A. L., Bhat, S. A., & Ara, A. (2015). Omega-3 fatty acids and the treatment of depression: A review of scientific evidence. *Integrative Medicine Research*, 4(3), 132–141.

Weiser, M. J., Butt, C. M., & Mohajeri, M. H. (2016). Docosahexaenoic acid and cognition throughout the lifespan. *Nutrients*, 8(2), 99.

Williams, P. G. (2014). The benefits of breakfast cereal consumption: A systematic review of the evidence base. *Advances in Nutrition, 5*(5), 636S–673S.

Wu, M., Luo, Q., Nie, R., Yang, X., Tang, Z., & Chen, H. (2020). Potential implications of polyphenols on aging considering oxidative stress, inflammation, autophagy, and gut microbiota. *Critical Reviews in Food Science and Nutrition, 61*(13), 2175–2193.

Xie, M., Liu, J., Tsao, R., Wang, Z., Sun, B., & Wang, J. (2019). Whole grain consumption for the prevention and treatment of breast cancer. *Nutrients, 11*(8), 1769.

Yildiz, G., Ding, J., Gaur, S., Andrade, J., Engeseth, N. J., & Feng, H. (2018). Microencapsulation of docosahexaenoic acid (DHA) with four wall materials including pea protein-modified starch complex. *International Journal of Biological Macromolecules, 114*, 935–941.

Yildiz, G., & Izli, G. (2019). The effect of ultrasound pretreatment on quality attributes of freeze-dried quince slices: Physical properties and bioactive compounds. *Journal of Food Process Engineering, 42*(5), e13223.

Yildiz, G., Gao, Y., Ding, J., Zhu, S., Chen, G., & Feng, H. (2025). Enhancing physicochemical, bioactive, and nutritional properties of sweet potatoes: Ultrasonic contact drying with slot jet nozzles compared to hot-air drying and freeze drying. *Ultrasonics Sonochemistry, 112*, 107216.

Zahedirad, M., Asadzadeh, S., Nikooyeh, B., Neyestani, T. R., Khorshidian, N., Yousefi, M., & Mortazavian, A. M. (2019). Fortification aspects of vitamin D in dairy products: A review study. *International Dairy Journal, 94*, 53–64.

Zambrano, A. K., Cadena-Ullauri, S., Ruiz-Pozo, V. A., Tamayo-Trujillo, R., Paz-Cruz, E., Guevara-Ramírez, P., Frias-Toral, E., & Simancas-Racines, D. (2024). Impact of fundamental components of the Mediterranean diet on the microbiota composition in blood pressure regulation. *Journal of Translational Medicine, 22*(1), 417.

Zandstra, E. H., Willems, A. A., & Lion, R. (2018). Making salt-reduced products more appealing to consumers: impact of front-of-pack messages on liking and table salt use over time. *Public Health Nutrition, 21*(15), 2762-2772.

Zhang, P. (2022). Influence of foods and nutrition on the gut microbiome and implications for intestinal health. *International Journal of Molecular Sciences, 23*(17), 9588.

Chapter 3

Nutrition and Noncommunicable Diseases: Risk Factors, Prevention Strategies, and Policies

**Merve Aydın Mende*
and Derya Dikmen**
Department of Nutrition and Dietetics, Hacettepe University, Ankara, Turkiye

Abstract

Noncommunicable diseases (NCDs) represent one of the most pressing public health challenges at both global and national levels. Unhealthy dietary habits are a major underlying cause of many chronic conditions, including cardiovascular diseases, cancers, diabetes, and chronic respiratory diseases. Excessive intake of salt, free sugars, saturated and trans fats, along with insufficient consumption of fiber and protective nutrients, are among the primary dietary risk factors for NCDs. Western-type dietary patterns contribute to disease development through mechanisms such as inflammation, insulin resistance, oxidative stress, and unfavorable alterations in the gut microbiota, whereas Mediterranean, DASH, and plant-based dietary patterns exert protective effects due to their anti-inflammatory and antioxidant properties. Evidence from global and national studies indicates that nutrition-based interventions hold strong potential for reducing the burden of NCDs. Policies such as front-of-pack labeling, restrictions on food marketing, school food standards, product reformulation, and taxation serve as cost-effective tools to promote healthier eating behaviors. In Turkey, initiatives such as salt and sugar reduction programs, diabetes control strategies, and food labeling regulations have emerged as significant steps in this regard. Nevertheless, the dissemination of sustainable and culturally adapted dietary models remains essential for achieving long-term success in combating NCDs. This chapter explores the relationship between nutrition and NCDs from multiple perspectives, including biological mechanisms, protective dietary patterns, and policy approaches.

Keywords: noncommunicable diseases, nutrition, dietary patterns, nutrition policies, public health

---

* Corresponding Author's Email: merveaydin@hacettepe.edu.tr

In: Let Food Be Your Medicine, and Medicine Be Your Food
Editors: Gulcin Yildiz and Arzu Imece
ISBN: 979-8-90134-204-6

Noncommunicable Diseases as a Global Public Health Challenge

Noncommunicable diseases (NCDs) are long-term, slow-progressing, and multifactorial chronic conditions (World Health Organization, 2003). The four major NCD groups defined by the World Health Organization (WHO) are cardiovascular diseases, cancers, chronic respiratory diseases, and diabetes (Nishida et al., 2004). In addition, obesity, chronic kidney disease, neurological disorders, and mental health conditions are increasingly included within the broader NCD landscape (World Health Organization, 2024b). An unhealthy diet, physical inactivity, tobacco use, harmful alcohol consumption, and environmental factors (e.g., air pollution) are the key modifiable determinants of NCD onset and progression (World Health Organization, 2020).

Currently, NCDs account for approximately 74% of all global deaths, corresponding to more than 41 million deaths annually (World Health Organization, 2022). Of these, 17.9 million are caused by cardiovascular diseases, 9 million by cancers, nearly 4 million by chronic respiratory diseases, and 2 million by diabetes (Watkins et al., 2022). Premature deaths among those aged 30–70 years highlight the critical public health burden of NCDs worldwide (World Health Organization, 2023). According to the Global Burden of Disease (GBD) 2021 Study, the leading contributors to disability-adjusted life years (DALYs) attributable to NCDs include high blood pressure, unhealthy diet, elevated body mass index (BMI), and high fasting plasma glucose (Institute for Health Metrics and Evaluation, 2024).

The WHO's Sustainable Development Goal (SDG) 3.4 aims to reduce premature mortality from NCDs by one-third by 2030 (Watkins et al., 2022; World Health Organization, 2023). However, current trajectories indicate that achieving this target will be highly challenging in low- and middle-income countries (LMICs), where limited health system capacity, inequitable food environments, and socioeconomic barriers accelerate the rise in NCDs (Watkins et al., 2022; World Health Organization, 2025). Population aging, urbanization, and lifestyle changes are additional structural drivers likely to intensify the global NCD burden in the coming decades (World Health Organization, 2023).

In Turkey, the situation mirrors global patterns. More than one-third of all deaths are due to cardiovascular diseases, followed by cancers and diabetes (Republic of Turkey Ministry of Health, 2025a). According to the Turkey Nutrition and Health Survey (TNHS) 2017, the prevalence of obesity among adults (≥15 years) reached 31.5% (Republic of Turkey Ministry of Health, 2019). The Turkey Diabetes Program 2023–2027 estimates that nearly 9 million people are currently living with diabetes (Republic of Turkey Ministry of Health, 2023). Similarly, the Turkey Kidney Diseases Prevention and Control Program 2025–2030 reports that approximately one in six adults has kidney disease (Republic of Turkey Ministry of Health, 2025b). Chronic obstructive pulmonary disease (COPD) and asthma remain leading contributors to premature mortality and diminished quality of life (Republic of Turkey Ministry of Health, 2025c). Gender and age disparities are notable: obesity prevalence is higher among women, while the prevalence of diabetes and hypertension increases significantly with age (Republic of Turkey Ministry of Health, 2019, 2023, 2025a). Urban populations are further exposed to higher rates of unhealthy diets and physical inactivity, exacerbating the NCD burden (Jakab et al., 2014; Presidency of the Republic of Turkey, Strategy and Budget Presidency, 2023; Republic of Turkey Ministry of Health, 2017).

Beyond mortality, NCDs are associated with substantial morbidity and economic burden, undermining both quality of life and health system sustainability (World Health Organization, 2025). In the WHO European Region, economic losses linked to NCDs amount to hundreds of billions of U.S. dollars annually, primarily due to avoidable deaths and productivity losses (World Health Organization, Regional Office for Europe, 2025a). The 2025 WHO Europe report estimated that 1.8 million NCD deaths were preventable in that year alone, with an economic cost of USD 514 billion (World Health Organization, Regional Office for Europe, 2025b). In Turkey, health expenditures attributable to NCDs are estimated at 3.6% of gross domestic product (GDP) (Jakab et al., 2014). However, the economic implications extend beyond health systems: early retirement, reduced workforce productivity, and household financial strain contribute to worsening poverty, particularly among vulnerable populations. Thus, NCDs must be addressed not only as a medical challenge but also as a socio-economic development priority (Gupta et al., 2025; Presidency of the Republic of Turkey, Strategy and Budget Presidency, 2023; World Health Organization, 2025a).

The Role of Nutrition in Noncommunicable Diseases

Dietary habits are one of the key determinants in the prevention and control of noncommunicable diseases (NCDs). In Turkey, studies indicate that excessive salt intake is directly linked to hypertension and cardiovascular diseases, while high sugar consumption is associated with obesity and diabetes (Republic of Turkey Ministry of Health, 2016, 2021b). Excessive intake of saturated and trans fats represents a significant threat to cardiovascular health (Republic of Turkey Ministry of Health, 2022). Conversely, dietary patterns rich in vegetables, fruits, whole grains, and healthy fats are consistently associated with lower NCD risk (Presidency of the Republic of Turkey, Strategy and Budget Presidency, 2023).

According to the Global Burden of Disease (GBD) 2021 study, high sodium intake, together with low intake of whole grains, fruits, and dietary fiber, are the most critical nutrition-related risk factors for NCD-related mortality (Institute for Health Metrics and Evaluation, 2024). A policy analysis from Zambia and international data highlight that the widespread consumption of processed foods and sugar-sweetened beverages increases obesity and diabetes prevalence, creating a heavy economic burden on health systems (Mukanu et al., 2021). Similarly, Western-type diets, characterized by high energy density, added sugars, and saturated fats, contribute significantly to the risk of obesity, cardiovascular disease, diabetes, and cancer (Chen et al., 2023). The WHO's updated Best Buys report (2024) underscores that addressing unhealthy dietary patterns remains the most cost-effective intervention area for reducing the NCD burden, urging governments to integrate nutrition policies into health systems (World Health Organization, 2024a).

Unhealthy dietary patterns mediate NCD development through various biological mechanisms. Diets high in fat and refined carbohydrates promote systemic inflammation, impair intestinal permeability, and contribute to metabolic endotoxemia via lipopolysaccharides (LPS) (Canbolat & Cakıroğlu, 2023; Chen et al., 2023). Excess energy and saturated fat intake also increase oxidative stress, insulin resistance, and mitochondrial dysfunction (Canbolat & Cakıroğlu, 2023; Fričová, 2024). At the microbiota level, Western diets induce dysbiosis, reduce short-chain fatty acid (SCFA) production, and trigger inflammatory responses (Fragkou et al., 2021; Jovandaric et al., 2024). By contrast, plant-based diets and high fiber intake are linked to greater microbial diversity and anti-inflammatory effects (Fragkou et al., 2021). Furthermore, inadequate or unhealthy diets during early life can increase the risk of obesity and metabolic syndrome in later years via epigenetic mechanisms and immune programming (Alabduljabbar et al., 2021; Tsigalou et al., 2020).

In Turkey, average daily salt intake exceeds 10 grams, nearly double the WHO's 5 gram/day recommendation, while consumption of sugar and sugar-sweetened products is particularly high among children and adolescents (Republic of Turkey Ministry of Health, 2019). The Turkey Diabetes Program 2023–2027 emphasizes that increased sugar and refined carbohydrate intake worsens the national diabetes burden (Republic of Turkey Ministry of Health, 2023). The Turkey Salt Reduction Program (2017–2021) and the Sugar Reduction Guide (2021) clearly illustrate the direct links between these dietary factors and NCD risk (Republic of Turkey Ministry of Health, 2016, 2021b).

Protective dietary components include dietary fiber, omega-3 fatty acids, polyphenols, and antioxidants. The Mediterranean diet—rich in vegetables, fruits, whole grains, legumes, olive oil, and fish—has been linked to reduced inflammation, improved insulin sensitivity, and cardiovascular protection (Bravo Iniguez & Zhu, 2021; Oliveira et al., 2025). Polyphenols and other bioactive compounds (e.g., resveratrol, curcumin, xanthohumol) further exert beneficial effects on oxidative stress and inflammatory pathways (Diniz et al., 2023; Tekle et al., 2023). In contrast, excess intake of added sugars, saturated fats, trans fats, and sodium substantially increases NCD risk. High sodium intake is strongly linked with hypertension, cardiovascular disease, gastric cancer, and kidney disorders (World Health Organization, 2022). Sugar-sweetened beverages and refined carbohydrates contribute to obesity and diabetes, while high energy density is a key driver of overweight and metabolic disturbances (Institute for Health Metrics and Evaluation, 2024; World Health Organization, Regional Office for Europe, 2023). The updated 2023 WHO European Region nutrient profile model provides category-specific thresholds for these dietary risk factors and underscores the role of sugary drinks and snack foods as key determinants of childhood obesity (World Health Organization, Regional Office for Europe, 2023).

In summary, excessive intake of sodium, added sugars, and saturated/trans fats, coupled with low fiber and protective nutrient intake, significantly increases NCD risk. This relationship is supported by epidemiological evidence (GBD 2021), WHO synthesis reports, and nutrition policy guidelines (Institute for Health Metrics and Evaluation, 2024; World Health Organization, 2022, 2024a; World Health Organization, Regional Office for Europe, 2023).

Limiting Dietary Components and Their Associations with Noncommunicable Diseases

Dietary patterns play a central role in the development of NCDs. In particular, excess energy intake, saturated and trans fats, added sugars, and sodium are well-established nutrition-related risk factors, supported by evidence from numerous epidemiological and clinical studies.

Energy imbalance is a primary determinant of obesity and metabolic syndrome (Hong et al., 2023). Western-type diets, characterized by high energy density, promote abdominal adiposity, insulin resistance, and inflammatory processes, thereby elevating the risk of type 2 diabetes and cardiovascular disease (Chen et al., 2023). According to the Global Burden of Disease (GBD) 2021 study, high body mass index (BMI) ranks among the top five global risk factors for mortality, with the NCD burden continuing to rise (Institute for Health Metrics and Evaluation, 2024). In Turkey, obesity prevalence exceeding 30% highlights the national significance of energy-related health problems (Republic of Turkey Ministry of Health, 2019).

Excessive intake of saturated fatty acids raises serum LDL cholesterol, directly contributing to the development of atherosclerotic cardiovascular diseases (Republic of Turkey Ministry of Health, 2022). Trans fatty acids accelerate atherosclerosis by raising LDL and lowering HDL, thereby increasing myocardial infarction and stroke risk (World Health Organization, 2024a). The WHO classifies the elimination of industrial trans fats from the global food supply as one of the Best Buy interventions (Galea et al., 2025). The updated WHO Europe Nutrient Profile Model (2023) also prioritizes the removal of trans fats from the food chain as a top public health intervention (World Health Organization, Regional Office for Europe, 2023).

Added sugars—particularly through sugar-sweetened beverages are major risk factors for obesity and type 2 diabetes. Excess sugar intake promotes insulin resistance and hepatic lipogenesis, thereby increasing the risk of nonalcoholic fatty liver disease (NAFLD). Added sugars also have a long-established role in dental caries, one of the most prevalent diet-related health outcomes globally (Institute for Health Metrics and Evaluation, 2024; World Health Organization, 2024a; World Health Organization, Regional Office for Europe, 2022). WHO policy reports emphasize the taxation and labeling of sugar-sweetened beverages as priority strategies for reducing NCD burden, supported by policy analyses across different countries (Mukanu et al., 2021; World Health Organization, 2024a; World Health Organization, Regional Office for Europe, 2022). In Turkey, the Sugar Reduction Guide (2021) marks a significant step toward reducing sugar consumption at the national level (Republic of Turkey Ministry of Health, 2021b).

Excess sodium intake remains the strongest dietary determinant of hypertension (Tekle et al., 2023). WHO data indicate that high sodium intake contributes to approximately 1.8 million deaths annually, particularly by raising the risk of stroke and cardiovascular disease (World Health Organization, 2022). The WHO Europe Nutrient Profile Model (2023) sets sodium thresholds for key food categories such as bread, processed foods, and processed meats, providing a policy basis for national reformulation programs (World Health Organization, Regional Office for Europe, 2023). In Turkey, average salt intake remains well above recommended levels, contributing significantly to the national prevalence of hypertension (Republic of Turkey Ministry of Health, 2016).

The strength of evidence linking excess energy, saturated and trans fats, added sugars, and sodium to NCDs is consistently high. For this reason, both national and international policy documents place the restriction of these dietary components at the core of preventive strategies (World Health Organization, 2022, 2024a). At the same time, emerging evidence highlights the roles of gut microbiota alterations, food additives, and refined carbohydrates in NCD risk, suggesting that these may become stronger categories of concern in the near future (Fričová, 2024; Park et al., 2024). The WHO's 2022 and 2023 reports underscore that nutrient profile models are critical tools for translating scientific evidence into effective nutrition policies (World Health Organization, Regional Office for Europe, 2022; 2023).

In summary, reducing excess energy intake, saturated and trans fats, added sugars, and sodium is a top priority in NCD prevention. However, addressing NCDs cannot rely solely on restriction. Strengthening the intake of protective nutrients and promoting comprehensive dietary patterns remains equally crucial, which will be discussed in the following section on Nutrition-Based Protective Strategies.

Nutrition-Based Protective Strategies

Nutrition-based protective strategies represent a cornerstone of public health approaches in the prevention and management of NCDs. Rather than focusing on single nutrients, dietary models emphasizing whole dietary patterns have demonstrated protective effects at both clinical and epidemiological levels (Dominguez et al., 2021; Tsigalou et al., 2020). The WHO, in its 2022 report, emphasized that these positive dietary patterns should be promoted at the policy level and that nutrient profile models should be used not only as restrictive tools but also as mechanisms to support healthy diets (World Health Organization, Regional Office for Europe, 2022).

The Mediterranean diet is characterized by high intakes of vegetables, fruits, legumes, whole grains, olive oil, and fish While some definitions include moderate wine or alcohol consumption, this component remains controversial in terms of health benefits, and increasing alcohol intake is not recommended. The Mediterranean diet, due to its richness in antioxidant and anti-inflammatory components, is associated with reduced risk of cardiovascular diseases, type 2 diabetes, cancer, and neurodegenerative disorders (Oliveira et al., 2025; Xiao et al., 2024). Improvements in advanced glycation end-products (AGEs) and endothelial function have also been reported (Oliveira et al., 2025). The WHO recommends the Mediterranean diet as one of the most effective dietary models for NCD prevention (World Health Organization, 2022, 2024a; World Health Organization, Regional Office for Europe, 2022). Furthermore, the updated WHO Europe Nutrient Profile Model (2023) highlights the preventive potential of food groups such as fish, olive oil, and whole grains, serving as a reference point for policymakers (World Health Organization, Regional Office for Europe, 2023).

The Dietary Approaches to Stop Hypertension (DASH) diet is characterized by low sodium intake and high consumption of vegetables, fruits, whole grains, and low-fat dairy products. It has been shown to reduce blood pressure, decrease cardiovascular mortality, and improve metabolic syndrome components (World Health Organization, 2022, 2024a; World Health Organization, Regional Office for Europe, 2022). For this reason, it is recommended as a cost-effective strategy in populations with high hypertension prevalence (World Health Organization, 2024a; World Health Organization, Regional Office for Europe, 2022).

Plant-based diets (e.g., vegetarian and vegan) are rich in fiber, polyphenols, and phytochemicals, contributing to reduced inflammation and favorable gut microbiota modulation (Jovandaric et al., 2024). Epidemiological evidence supports associations between plant-based diets and reduced risk of type 2 diabetes, obesity, and cardiovascular diseases (Peña-Jorquera et al., 2023). Additionally, these diets contribute to sustainability by reducing greenhouse gas emissions (Hong et al., 2023; Tapkigen et al., 2024).

Table 1. Positive dietary patterns and their health effects on noncommunicable diseases

Dietary Pattern	Key Components	Potential Health Effects (Associations with NCDs)
Mediterranean Diet (Oliveira et al., 2025; Tsigalou et al., 2020; Xiao et al., 2024)	Olive oil (monounsaturated fat), fish (omega-3), vegetables, fruits, whole grains, legumes (some definitions include moderate alcohol; not recommended to increase consumption)	Reduced inflammation and oxidative stress; lower risk of cardiovascular disease, type 2 diabetes, certain cancers, and neurodegenerative disorders; improvements in glycation markers and endothelial function
DASH Diet (World Health Organization, 2022, 2024a; World Health Organization, Regional Office for Europe, 2022)	Vegetables, fruits, whole grains, low-fat dairy, low sodium	Blood pressure reduction and hypertension control; decreased cardiovascular and stroke risk; improvements in metabolic syndrome components; cost-effective in high-hypertension populations
Plant-Based Diets (Hong et al., 2023; Jovandaric et al., 2024; Peña Jorquera et al., 2023; Tapkigen et al., 2024)	Fiber, polyphenols, phytochemicals, low animal fat	Increased gut microbiota diversity; reduced inflammatory markers; lower risks of obesity, type 2 diabetes, and cardiovascular disease; additional environmental/sustainability benefits
Asian Dietary Patterns (Khemka et al., 2023; Tariq et al., 2023)	High vegetables and fish intake; low red meat	Reduced risk of obesity and type 2 diabetes; decreased mortality risk; suppression of inflammation and support for metabolic health via omega-3 and antioxidant compounds
General Healthy (Positive) Patterns (Canbolat & Cakıroğlu, 2023; Diniz et al., 2023)	Fiber, omega-3 fatty acids, antioxidants, low saturated fats and added sugars	Improved insulin sensitivity; better metabolic syndrome outcomes; reduced NCD risk

Asian-origin dietary patterns, characterized by high vegetable and fish intake and low red meat consumption, have been shown to reduce obesity and type 2 diabetes risk while lowering overall mortality (Tariq et al., 2023). Omega-3 fatty acids and antioxidant components derived from fish and plant-based foods may contribute to reduced inflammation and improved metabolic health (Khemka et al., 2023).

Functional foods and bioactive compounds are also gaining attention in protective strategies. Probiotic- and prebiotic-rich foods have been shown to positively influence gut microbiota and reduce inflammatory processes, while polyphenols, resveratrol, and curcumin are reported to suppress oxidative stress (Bravo Iniguez & Zhu, 2021; Diniz et al., 2023; Park et al., 2024).

The protective effects of holistic dietary patterns stem not only from individual components but also from the synergistic interactions among nutrients and foods. Therefore, the leading features and documented health benefits of different dietary models are often examined comparatively. Table 1 summarizes the main components and health effects of the Mediterranean, DASH, plant-based, and Asian-origin diets in relation to NCD prevention.

In summary, healthy (positive) dietary patterns are associated with reduced NCD risk not only through limiting harmful components but also by increasing protective nutrients (e.g., fiber, omega-3 fatty acids, antioxidants). Alongside the Mediterranean and DASH diets, Asian-origin and culturally specific dietary models—such as traditional Anatolian cuisine rich in legumes, yogurt, and fermented foods—may also contribute to reducing obesity and metabolic syndrome risk (Özvarış & Aktuna, 2020). Overall, healthy dietary patterns are associated with reduced NCD risk not only through the limitation of harmful components but also by increasing protective nutrients such as fiber, omega-3 fatty acids, and antioxidants. Alongside the Mediterranean and DASH diets, Asian-origin dietary patterns have also been reported to lower morbidity and mortality risks (Khemka et al., 2023; Tariq et al., 2023). These diets have demonstrated potential to suppress inflammatory responses, reduce oxidative stress, enhance insulin sensitivity, and support cardiovascular health (Canbolat & Cakıroğlu, 2023; Diniz et al., 2023).

The WHO Europe Nutrient Profile Model (2023) translates these findings into a policy perspective, emphasizing the integration of healthy dietary patterns into national strategies as a priority for public health (World Health Organization, Regional Office for Europe, 2023). Thus, strengthening diverse dietary models rooted in cultural traditions within a universal public health framework represents an effective and sustainable approach to preventing NCDs.

Public Health Interventions and Policy Approaches

The prevention of NCDs extends beyond individual lifestyle modifications; the implementation of comprehensive and sustainable population-level policies is essential. The World Health Organization (WHO) emphasizes that cost-effective public health interventions, known as "Best Buys," represent some of the most effective tools for reducing both premature mortality and health expenditures (Galea et al., 2025; World Health Organization, 2024a). The 2025 Quick Buys report highlights the importance of prioritizing short-term, highly cost-effective actions within national policy agendas (Galea et al., 2025).

Key policy approaches include front-of-pack labeling, advertising restrictions, school food standards, reformulation, taxation, and national strategic plans. In addition, community awareness campaigns, nutrition education, and collaboration with the food industry are identified as supportive measures in NCD prevention (Gupta et al., 2025).

Front-of-Pack Labeling (FOPL) Systems

Front-of-pack labeling (FOPL) systems enable consumers to evaluate products quickly and clearly, thereby promoting healthier choices. Examples include Nutri-Score, the UK traffic light system, and Chile's black warning labels, all of which have been shown to positively influence consumer behavior (World Health Organization, Regional Office for Europe, 2022). The WHO's 2022 report highlights FOPL systems as powerful tools in reducing unhealthy food consumption (World Health Organization, Regional Office for Europe, 2022). Furthermore, the 2023 WHO Europe Nutrient Profile Model reinforced the scientific foundation of these systems by introducing product-specific thresholds (World Health Organization, Regional Office for Europe, 2023). In Turkey, the Turkish Food Codex Food Labeling and Consumer Information Regulation (2024–2025) introduced new measures to enhance consumer information and strengthen alignment with international standards (Republic of Turkey Ministry of Agriculture and Forestry, 2024).

Advertising Restrictions

Unhealthy food marketing directed at children is a significant driver of obesity and poor dietary behaviors. The WHO designates protecting children from such advertising as a priority policy domain (World Health Organization, Regional Office for Europe, 2022). In Turkey, the Multisectoral Action Plan for NCDs (2017–2025) introduced measures to restrict unhealthy food advertising to children across television and digital media (Republic of Turkey Ministry of Health, 2017). However, strengthening independent monitoring and enforcement mechanisms remains essential for ensuring effectiveness (World Health Organization, 2023).

School Food Standards

School-aged children are a critical target group for establishing lifelong healthy eating habits. International evidence shows that restricting high-sugar, high-salt, and high-fat products in school canteens while promoting healthier options reduces obesity and diabetes risks (World Health Organization, Regional Office for Europe, 2022). School-based nutrition policies have been shown to shape long-term dietary behaviors and reduce NCD risk (Gupta et al., 2025). In Turkey, criteria and standards restricting unhealthy foods in school canteens and promoting healthier alternatives have already been introduced (Özvarış & Aktuna, 2020).

Table 2. Public health nutrition policies: global and turkish practices

Policy Domain	Global Practices	Practices in Turkey
Front-of-Pack Labeling (FOPL) Systems	Nutri-Score (France), Traffic Light System (UK), Warning Labels (Chile); FOPL recognized as a key policy tool; WHO Europe model provides scientific basis (World Health Organization, Regional Office for Europe, 2022, 2023)	Turkish Food Codex Food Labeling and Consumer Information Regulation (2024–2025) (Republic of Turkey Ministry of Agriculture and Forestry, 2024; World Health Organization, Regional Office for Europe, 2023)
Advertising Restrictions	Bans on unhealthy food marketing to children (Chile, UK) (World Health Organization, Regional Office for Europe, 2022)	National Multisectoral Action Plan for NCDs (2017–2025): restrictions on advertising unhealthy foods to children on TV and digital media (Republic of Turkey Ministry of Health, 2017; World Health Organization, 2023)
School Food Standards	Restricting unhealthy foods; promoting healthy options (USA, EU) (Gupta et al., 2025; World Health Organization, Regional Office for Europe, 2022)	Development of criteria and implementation of school food standards (Özvarış & Aktuna, 2020)
Reformulation	Salt reduction (UK); elimination of industrial trans fats (Denmark) (World Health Organization, Regional Office for Europe, 2022)	Salt reduction in bread (2011–2021), sugar reduction program (2021), fat reformulation initiatives (Republic of Turkey Ministry of Health, 2016, 2021a, 2022)
Taxation Policies	SSB taxes reduce consumption and promote reformulation (World Health Organization, 2024a; World Health Organization, Regional Office for Europe, 2022)	Comprehensive SSB tax not yet implemented; 12th Development Plan (2024–2028) under review (Presidency of the Republic of Turkey, 2023)
National Strategies and Plans	WHO Best Buys; WHO PEN for LMICs; success stories include Chile (labels), UK (salt), Mexico (SSB tax) (World Health Organization, 2024a; World Health Organization, Regional Office for Europe, 2022)	Turkey Diabetes Program (2023–2027); Salt & Sugar Reduction Programs; Turkish Food Codex Regulation; 12th Development Plan (2024–2028) (Presidency of the Republic of Turkey, 2023; Republic of Turkey Ministry of Health, 2016, 2021b, 2023)

Reformulation Policies

Reformulation strategies aim to reduce salt, sugar, and saturated fat levels in processed foods. The WHO's 2022 report highlights nutrient profile models as critical scientific tools and reference frameworks for reformulation (World Health Organization, Regional Office for Europe, 2022). In Turkey, initiatives include salt reduction in bread (since 2011), the 2021 sugar reduction program, and measures to limit saturated and trans fats (Republic of Turkey Ministry of Health, 2016, 2021a, 2022). The 2023 WHO Europe Nutrient Profile Model recommends expanding reformulation beyond salt and sugar reduction through category-specific thresholds to ensure a more comprehensive application (World Health Organization, Regional Office for Europe, 2023).

Taxation Policies

Fiscal measures are a powerful policy tool to reduce unhealthy food and beverage consumption. WHO policy documents consistently highlight sugar-sweetened beverage (SSB) taxation as a cost-effective measure to reduce intake and incentivize reformulation (World Health Organization, 2024a; World Health Organization, Regional Office for Europe, 2022). In Turkey, a comprehensive SSB tax has not yet been implemented; however, the 12th Development Plan (2024–2028) explicitly notes consideration of such measures (Presidency of the Republic of Turkey, 2023).

National Strategies and Plans

Globally, Chile's warning labels, the UK's salt reformulation program, and Mexico's SSB tax represent successful examples of public health interventions targeting NCDs (World Health Organization, 2024a; World Health Organization, Regional Office for Europe, 2022). In Turkey, strategies such as the Turkey Diabetes Program (2023–2027), the Salt Consumption Reduction Program, the Sugar Reduction Guide, and the Turkish Food Codex Labeling Regulation support multisectoral approaches to NCD prevention (Republic of Turkey Ministry of Agriculture and Forestry, 2024; Republic of Turkey Ministry of Health, 2021b, 2023). In addition, the WHO Package of Essential NCD Interventions (PEN) provides a cost-effective set of interventions for primary health care in low- and middle-income countries, serving as a guiding framework for national policy development (World Health Organization, 2020, 2024a). Both the WHO Europe Nutrient Profile Model (2023) and the WHO's 2022 report recommend stronger alignment of national policies with international standards (World Health Organization, Regional Office for Europe, 2022, 2023).

Public health interventions for NCDs extend beyond individual-level approaches by aiming to transform the broader food environment. Policies such as front-of-pack labeling, advertising restrictions, school food standards, reformulation, and taxation represent cost-effective strategies supported by robust scientific evidence, and they are further reinforced through WHO's updated nutrient profile models. Although Turkey has made notable progress in these areas in recent years, closer alignment of national policies with international standards remains essential. Such integration would not only strengthen public health outcomes but also reduce the economic burden on the health system. Table 2 summarizes public health nutrition policies across policy domains, comparing global practices with their implementation in Turkey.

Conclusion and Recommendations

Noncommunicable diseases (NCDs) remain among the most pressing public health challenges globally and nationally. A substantial proportion of the mortality and morbidity burden is directly attributable to modifiable lifestyle factors, particularly unhealthy dietary patterns (Institute for Health Metrics and Evaluation, 2024; World Health Organization, 2022). Nutrition therefore plays a central role in NCD prevention and should be considered a core focus of food and health policies (World Health Organization, 2003, 2024a).

For nutrition-based strategies to be effective, individual-level behavior change alone is insufficient; the broader food environment must be transformed to facilitate healthy choices (World Health Organization, 2024a). Policy tools such as front-of-pack labeling, advertising restrictions, school food standards, reformulation, and taxation are among the most evidence-based and effective measures to achieve this transformation (World Health Organization, Regional Office for Europe, 2022, 2023). In Turkey, regulatory frameworks under the Turkish Food Codex, initiatives promoting healthier school canteen standards, and national salt and sugar reduction programs represent important steps toward creating healthier food environments (Gupta et al., 2025; Republic of Turkey Ministry of Agriculture and Forestry, 2024; Republic of Turkey Ministry of Health, 2021a).

Success in tackling NCDs depends not only on scientific evidence but also on strong political commitment and active community engagement. At the political level, multisectoral strategies must be implemented, while at the community level, awareness campaigns and education initiatives should be expanded (World Health Organization, 2023). Furthermore, aligning scientific research more closely with policy processes will accelerate the effective translation of evidence into practice (Watkins et al., 2022).

Looking ahead, the policy agenda should prioritize the wider adoption of nutrient profile models, stronger incentives for industry reformulation, fiscal measures such as sugar-sweetened beverage taxes, and enhanced school-based nutrition programs (Gupta et al., 2025; Presidency of the Republic of Turkey, 2023; World Health Organization, Regional Office for Europe, 2025b). In addition, personalized nutrition approaches (Alabduljabbar et al., 2021), healthy nutrition in early life (Fragkou et al., 2021), and plant-based dietary patterns (Peña Jorquera et al., 2023) are expected to gain increasing relevance. Supporting healthy diets at the individual level while implementing comprehensive policy measures at the population level will be critical. Moreover, integrating nutrition policies with climate change and sustainability goals can provide a long-term vision for reducing the NCD burden (Hong et al., 2023; Khemka et al., 2023).

In conclusion, promoting healthy diets, transforming food environments, strengthening political and community engagement, and scaling up evidence-based policy measures are essential to prevent NCDs. Such actions will enhance public health gains while alleviating the economic burden on health systems (Galea et al., 2025; Republic of Turkey Ministry of Agriculture and Forestry, 2024; World Health Organization, Regional Office for Europe,

2025b). Integrating scientific evidence into policymaking for NCD prevention is not only critical for health systems but also indispensable for achieving the broader goals of sustainable development.

References

Alabduljabbar, S., Al Zaidan, S., Lakshmanan, A. P., & Terranegra, A. (2021). Personalized nutrition approach in pregnancy and early life to tackle childhood and adult non-communicable diseases. *Life (Basel), 11*(6), 467.

Bravo Iniguez, A., & Zhu, M.-J. (2021). Hop bioactive compounds in prevention of nutrition-related noncommunicable diseases. *Critical Reviews in Food Science and Nutrition, 61*(11), 1900–1913.

Canbolat, E., & Cakıroğlu, F. P. (2023). The importance of AMPK in obesity and chronic diseases and the relationship of AMPK with nutrition: A literature review. *Critical Reviews in Food Science and Nutrition, 63*(4), 449–456.

Chen, J., Xiao, Y., Li, D., Zhang, S., Wu, Y., Zhang, Q., & Bai, W. (2023). New insights into the mechanisms of high-fat diet mediated gut microbiota in chronic diseases. *iMeta, 2*(1), e69.

Diniz, M. S., Magalhães, C. C., Tocantins, C., Grilo, L. F., Teixeira, J., & Pereira, S. P. (2023). Nurturing through nutrition: Exploring the role of antioxidants in maternal diet during pregnancy to mitigate developmental programming of chronic diseases. *Nutrients, 15*(21), 4623.

Dominguez, L. J., Di Bella, G., Veronese, N., & Barbagallo, M. (2021, June 12). Impact of Mediterranean diet on chronic non-communicable diseases and longevity. *Nutrients, 13*(6), 2028.

Fragkou, P. C., Karaviti, L., Zemlin, A., & Skevaki, C. (2021). Impact of early life nutrition on children's immune system and noncommunicable diseases through its effects on the bacterial microbiome, virome and mycobiome. *Frontiers in Immunology, 12,* Article 644269.

Fričová, A. (2024). Role of diet in development of non-communicable diseases: Focus on gut microbiome. *Central European Journal of Public Health, 32*(3), 200–204.

Galea, G., Ekberg, A., Ciobanu, A., Corbex, M., Farrington, J., Ferreira Bores, C., Kokole, D., Lasierra Losada, M., Neufeld, M., Rakovac, I., Tsoy, E., Wickramasinghe, K., Williams, J., McKee, M., & Stuckler, D. (2025). Quick buys for prevention and control of noncommunicable diseases. *The Lancet Regional Health – Europe, 24,* 101281.

Gupta, S. K., Rawa, T., & Arora, M. (2025). Healthy choices, healthy futures: Promoting healthy nutrition to curb non-communicable diseases. *Trends in Endocrinology & Metabolism, 36*(6), 497–501.

Hong, J. Y., Kim, Y. J., Bae, S., & Kim, M. K. (2023). Associations of daily diet-related greenhouse gas emissions with the incidence and mortality of chronic diseases: A systematic review and meta-analysis of epidemiological studies. *Epidemiology and Health, 45,* e2023011.

Institute for Health Metrics and Evaluation. (2024, May 16). *Global Burden of Disease 2021: Findings from the GBD 2021 Study.* Institute for Health Metrics and Evaluation.

Jakab, M., Hawkins, L., Loring, B., Tello, J., Ergüder, T., & Kontaş, M. (2014). *Bulaşıcı olmayan hastalıklarda daha iyi sonuçlar: Sağlık sistemi için zorluklar ve fırsatlar – Turkey ülke değerlendirmesi* [Better outcomes in noncommunicable diseases: Challenges and opportunities for health systems – Turkey country assessment]. World Health Organization, Regional Office for Europe.

Jovandaric, M. Z., Jovanović, K., Raus, M., Babic, S., Igic, T., Kotlica, B., & Milicevic, S. (2024, November 29). The significance of plant nutrition in the creation of the intestinal microbiota prevention of chronic diseases: A narrative review. *Medicina (Kaunas), 60*(12), Article 1969.

Khemka, S., Reddy, A., Garcia, R. I., Jacobs, M., Reddy, R. P., Roghani, A. K., Pattoor, V., Basu, T., Sehar, U., & Reddy, P. H. (2023, October 11). Role of diet and exercise in aging, Alzheimer's disease, and other chronic diseases. *Ageing Research Reviews, 91,* 102091. \

Mukanu, M. M., Abdool Karim, S., Hofman, K., Erzse, A., & Thow, A. M. (2021). Nutrition-related non-communicable diseases and sugar-sweetened beverage policies: A landscape analysis in Zambia. *Global Health Action, 14*(1), Article 1872172.

Nishida, C., Uauy, R., Kumanyika, S., & Shetty, P. (2004). The joint WHO/FAO expert consultation on diet, nutrition and the prevention of chronic diseases: Process, product and policy implications. *Public Health Nutrition, 7*(1A), 245–250.

Oliveira, J. S., da Silva, J. A., de Freitas, B. V. M., Alfenas, R. C. G., & Bressan, J. (2025). A Mediterranean diet improves glycation markers in healthy people and in those with chronic diseases: A systematic review of clinical trials. *Nutrition Reviews, 83*(2), e317–e331.

Özvarış, Ş. B., & Aktuna, G. (Eds.). (2020). *Bulaşıcı olmayan hastalıklardan korunma* [Prevention of non-communicable diseases]. Hacettepe University, Institute of Public Health, Department of Noncommunicable Diseases.

Park, J. E., Park, H. Y., Kim, Y. S., & Park, M. (2024, December 13). The role of diet, additives, and antibiotics in metabolic endotoxemia and chronic diseases. *Metabolites, 14*(12), 704.

Peña Jorquera, H., Cid Jofré, V., Landaeta Díaz, L., Petermann Rocha, F., Martorell, M., Zbinden Foncea, H., Ferrari, G., Jorquera Aguilera, C., & Cristi Montero, C. (2023, July 1). Plant-based nutrition: Exploring health benefits for atherosclerosis, chronic diseases, and metabolic syndrome — A comprehensive review. *Nutrients, 15*(14), Article 3244.

Presidency of the Republic of Turkey, Strategy and Budget Presidency. (2023). *On ikinci kalkınma planı (2024–2028)* [Twelfth development plan (2024–2028)]. Presidency of the Republic of Turkey, Strategy and Budget Presidency.

Republic of Turkey Ministry of Agriculture and Forestry, General Directorate of Food and Control. (2024). *Türk gıda kodeksi gıda etiketleme ve tüketicileri bilgilendirme yönetmeliği kılavuzu (İlk yayım: 26 Kasım 2024; Rev. 001, 21 Haziran 2025)* [Turkish Food Codex food labeling and consumer information regulation guide (First published: November 26, 2024; Rev. 001, June 21, 2025)]. Republic of Turkey Ministry of Agriculture and Forestry, Department of Food Enterprises and Codex.

Republic of Turkey Ministry of Health. (2017, August 16). *Multisectoral action plan of Turkey for noncommunicable diseases 2017–2025 (English and Turkish)* [Technical document]. Republic of Turkey Ministry of Health.

Republic of Turkey Ministry of Health, General Directorate of Health Information Systems. (2025a). *Health statistics yearbook 2023.* Republic of Turkey Ministry of Health.

Republic of Turkey Ministry of Health, General Directorate of Public Health, Department of Chronic Diseases and Elderly Health. (2025b). *Turkey böbrek hastalıkları önleme ve kontrol programı (2025–2030)* [Turkey kidney diseases prevention and control program (2025–2030)]. Republic of Turkey Ministry of Health.

Republic of Turkey Ministry of Health, General Directorate of Public Health, Department of Chronic Diseases and Elderly Health. (2025c). *Turkey kronik hava yolu hastalıkları önleme ve kontrol programı (2025–2030)* [Turkey chronic respiratory diseases prevention and control program (2025–2030)]. Republic of Turkey Ministry of Health.

Republic of Turkey Ministry of Health, General Directorate of Public Health, Department of Healthy Nutrition and Active Life. (2019). *Turkey beslenme ve sağlık araştırması (TBSA) 2017* [Turkey nutrition and health survey (TNHS) 2017]. Republic of Turkey Ministry of Health.

Republic of Turkey Ministry of Health, General Directorate of Public Health, Department of Healthy Nutrition and Active Life. (2021a). *Gıda ve içecek sektörü için tuz azaltma protokolü uygulama rehberi (Yayın No: 1202, 1. baskı)* [Salt reduction protocol implementation guide for the food and beverage sector (Publication No. 1202, 1st ed.)]. Republic of Turkey Ministry of Health.

Republic of Turkey Ministry of Health, General Directorate of Public Health, Department of Healthy Nutrition and Active Life. (2021b). *Turkey şeker tüketimini/kullanımını azaltma rehberi (Yayın No: 1201)* [Turkey sugar consumption reduction guide (Publication No. 1201)]. Republic of Turkey Ministry of Health.

Republic of Turkey Ministry of Health, General Directorate of Public Health, Department of Healthy Nutrition and Active Life. (2022). *Ulusal beslenme konseyi yağ bilim komisyonu raporu (1. baskı, Yayın No: 1225)* [National Nutrition Council Fat Science Commission report (1st ed., Publication No. 1225)]. Republic of Turkey Ministry of Health.

Republic of Turkey Ministry of Health, General Directorate of Public Health, Department of Healthy Nutrition and Active Life. (2023). *Turkey diyabet programı 2023–2027* [Turkey diabetes program 2023–2027]. Republic of Turkey Ministry of Health.

Republic of Turkey Ministry of Health, Public Health Institution of Turkey, Department of Obesity, Diabetes and Metabolic Diseases. (2016). *Turkey aşırı tuz tüketiminin azaltılması programı 2017–2021 (2. baskı)* [Turkey salt consumption reduction program 2017–2021 (2nd ed.)]. Republic of Turkey Ministry of Health.

Tapkigen, J., Harding, S., Pulkki, J., Atkins, S., & Koivusalo, M. (2024, June 18). Climate change induced shifts in the food systems and diet-related non-communicable diseases in sub-Saharan Africa: A scoping review and a conceptual framework. *BMJ Open, 14*(6), e080241.

Tariq, M. N. M., Stojanovska, L., Al Dhaheri, A. S., Cheikh Ismail, L., Apostolopoulos, V., & Ali, H. I. (2023). Lifestyle interventions for prevention and management of diet-linked non-communicable diseases among adults in Arab countries. *Healthcare, 11*(1), 45.

Tekle, D. Y., Rosewarne, E., Santos, J. A., Trieu, K., Buse, K., Palu, A., Thow, A. M., Jan, S., & Webster, J. (2023). Do food and nutrition policies in Ethiopia support the prevention of non-communicable diseases through population-level salt reduction measures? A policy content analysis. *Nutrients, 15*(7), 1745.

Tsigalou, C., Konstantinidis, T., Paraschaki, A., Stavropoulou, E., Voidarou, C., & Bezirtzoglou, E. (2020). Mediterranean diet as a tool to combat inflammation and chronic diseases: An overview. *Biomedicines, 8*(7), Article 201.

Watkins, D. A., Msemburi, W. T., Pickersgill, S. J., Kawakatsu, Y., Gheorghe, A., Dain, K., Johansson, K. A., Said, S., Renshaw, N., Tolla, M. T., Twea, P. D., Varghese, C., Chalkidou, K., Ezzati, M., & Norheim, O. F.; NCD Countdown 2030 collaborators. (2022). NCD Countdown 2030: Efficient pathways and strategic investments to accelerate progress towards the Sustainable Development Goal target 3.4 in low income and middle income countries. *The Lancet, 399*(10331), 1266–1278.

World Health Organization. (2003). *Diet, nutrition and the prevention of chronic diseases: Report of a joint WHO/FAO expert consultation, Geneva, 28 January–1 February 2002 (WHO Technical Report Series No. 916).* World Health Organization.

World Health Organization. (2020). *WHO package of essential noncommunicable (PEN) disease interventions for primary health care.* World Health Organization.

World Health Organization. (2022, September 15). *Invisible numbers: The true extent of noncommunicable diseases and what to do about them* [Report]. World Health Organization.

World Health Organization. (2023, July 5). *Advancing the global agenda on prevention and control of noncommunicable diseases 2000 to 2020: Looking forwards to 2030.* World Health Organization.

World Health Organization. (2024a, April 30). *Tackling NCDs: Best buys and other recommended interventions for the prevention and control of noncommunicable diseases* (2nd ed.). World Health Organization.

World Health Organization. (2024b, December 23). *Noncommunicable diseases* [Fact sheet]. World Health Organization.

World Health Organization. (2025, May 12). *Assessing national capacity for the prevention and control of noncommunicable diseases: Report on 2023 global survey.* World Health Organization.

World Health Organization, Regional Office for Europe. (2022, October 6). *Use of nutrient profile models for nutrition and health policies: Meeting report on the use of nutrient profile models in the WHO European Region* (Technical document; WHO/EURO:2022-6201-45966-66383). World Health Organization, Regional Office for Europe.

World Health Organization, Regional Office for Europe. (2023, March 6). *WHO Regional Office for Europe nutrient profile model: Second edition* (Technical document; WHO Reference No. WHO/EURO:2023-6894-46660-68492). World Health Organization, Regional Office for Europe.

World Health Organization, Regional Office for Europe. (2025a). *Avoidable mortality, risk factors and policies for tackling noncommunicable diseases – Leveraging data for impact: Monitoring commitments in the WHO European Region ahead of the Fourth United Nations High Level Meeting* (Document No. WHO/EURO:2025-12445-52219-80271). World Health Organization, Regional Office for Europe.

World Health Organization, Regional Office for Europe. (2025b, June 27). *New data: Noncommunicable diseases cause 1.8 million avoidable deaths and cost US $ 514 billion every year, reveals new WHO/Europe report* [Media release]. World Health Organization.

Xiao, Y., Xiao, X., Zhang, X., Yi, D., Li, T., Hao, Q., Zhang, F., Li, X., & Wang, N. (2024, April 16). Mediterranean diet in the targeted prevention and personalized treatment of chronic diseases: Evidence, potential mechanisms, and prospects. *EPMA Journal, 15*(2), 207–220.

Chapter 4

Antioxidants and Their Role in Health

Umut Uyar[1,*]
and Bedriye Ural[2]
[1]Department of Nutrition and Dietetics, Igdir University, Iğdır, Turkiye
[2]Department of Nutrition and Dietetics, University of Health Sciences, Istanbul, Turkiye

Abstract

This chapter provides a comprehensive overview of the role of oxidative stress and antioxidants in human health. The discussion begins with the definition and generation of reactive oxygen species (ROS) and free radicals, which are produced both endogenously during normal metabolic processes and exogenously through factors such as ultraviolet radiation, environmental pollution, smoking, alcohol consumption, and stress. Excessive ROS production leads to oxidative stress, causing cellular damage to DNA, proteins, and lipids, and is implicated in the pathogenesis of chronic diseases including cancer, cardiovascular disorders, neurodegeneration, and metabolic dysfunctions. The chapter emphasizes the mechanisms of antioxidants, distinguishing between endogenous enzymatic antioxidants (such as superoxide dismutase, catalase, and glutathione peroxidase) and non-enzymatic antioxidants (bilirubin, albumin, glutathione, etc.), as well as exogenous antioxidants derived from diet. Dietary antioxidants, including vitamins A, C, and E, selenium, zinc, carotenoids, flavonoids, and polyphenols, are highlighted for their ability to neutralize ROS, reduce oxidative damage, and support immune and metabolic health. Special attention is given to bioactive compounds such as quercetin, curcumin, resveratrol, beta-carotene, and lycopene, which have demonstrated strong antioxidant, anti-inflammatory, anticarcinogenic, and cardioprotective properties in both experimental and clinical studies. Their synergistic effects, such as the combined activity of vitamin C and flavonoids, are shown to enhance bioavailability and biological effectiveness. Clinical evidence is reviewed, including trials linking antioxidant intake with reduced risks of cardiovascular disease, cognitive decline, and oxidative stress-related complications. The chapter also examines the paradox of antioxidant supplementation, noting that while diets rich in fruits and vegetables consistently reduce disease risk, isolated antioxidant supplements often yield inconsistent or limited results in large-scale trials. This underscores the importance of whole dietary patterns rather than single-compound supplementation.

---

* Corresponding Author's Email: umut.uyar@igdir.edu.tr

In: Let Food Be Your Medicine, and Medicine Be Your Food
Editors: Gulcin Yildiz and Arzu Imece
ISBN: 979-8-90134-204-6

Keywords: oxidative stress, reactive oxygen species (ROS), antioxidants, chronic diseases, dietary bioactive compounds

Introduction

Research on free radical processes and the effects of antioxidants on health began in the mid-20th century and gained momentum with the awarding of the Nobel Prize to Hinshelwood and Semenov (Koltover, 2010). Advances in technology, increasing environmental pollution, the use of agricultural chemicals, smoking and alcohol consumption, and exposure to ultraviolet (UV) radiation are among the many factors that expose living organisms, particularly humans, to various harmful agents. In addition, working and living conditions contribute to higher stress levels. As a result of these processes, environmental and psychological effects in humans can lead to the formation of free radicals. The increase in radical production in turn contributes to the development of various diseases, negatively affecting human health. Addressing these detrimental factors and diseases can primarily be achieved by eliminating the causes and preventing the onset of illnesses. In this regard, the consumption of natural foods and dietary sources, rather than the use of pharmaceuticals, has gained importance (Güleşçi & Aygül, 2016).

Reactive Oxygen Sources (ROS)

Oxygen, essential for the survival of living organisms, is a vital molecule (Güleşçi & Aygül). However, oxygen is highly reactive and oxidative, readily forming oxides with many elements and compounds. While oxygen in the atmosphere is stable in its ground state, it can undergo stepwise reduction. Living organisms utilize atmospheric oxygen in the oxidation process to produce 36–38 molecules of adenosine triphosphate (ATP) (Gulcin, 2020).

Reactive oxygen species (ROS) are generally defined as oxygen-containing molecules that are more reactive than oxygen itself. Some ROS are free radicals, containing one or more unpaired electrons, while others are non-radical species with all electrons paired. Certain ROS are produced accidentally, whereas others are generated to support processes such as pathogen defense and signal transduction (Halliwell, 2024). Under physiological conditions, excessive ROS are neutralized by antioxidants that mitigate their harmful effects (Parcheta et al., 2021).

Free radicals are atoms or molecules that contain unpaired electrons in their outer orbitals, making them highly reactive and unstable. They vary widely in shape, size, and chemical configuration. Reactive oxygen species (ROS) and non-oxygen radicals are the two main categories of free radicals. Radical ROS include superoxide anion (O2•–), hydroxyl radical (OH•), peroxyl radical (ROO•), nitric oxide (NO•), and organic peroxyl radicals (RCOO•), whereas non-radical ROS comprise hydrogen peroxide (H2O2), lipid hydroperoxides (LOOH), singlet oxygen (1O2), ozone (O3), nitrogen dioxide (NO2), and hypochlorous acid (HOCl). In the human body, approximately 1–3% of oxygen is converted into ROS under normal physiological conditions. Among these, three major ROS—superoxide radical, hydrogen peroxide, and hydroxyl radical—are continuously generated as metabolic by-products by mitochondria in actively growing cells (Meral et al., 2012).

The body generates free radicals as by-products while converting food into energy. Sources of radical formation include UV radiation, ionizing radiation, sunlight, combustion products of fossil fuels, viruses, cigarette smoke, stress, infection, inflammation, toxic products of fat metabolism, destructive chemicals, incomplete reduction of oxygen in the mitochondrial electron transport chain, pesticide residues, and metal-catalyzed reactions involving copper and iron, as well as organ damage observed after surgery and reperfusion of ischemic tissues. Radicals can react with all cellular macromolecules, causing cellular damage such as lipid peroxidation and DNA damage (Aslan et al., 1995; Meral et al., 2012; Güleşçi & Aygül, 2016).

Excessive production of ROS and free radicals in cells is termed "oxidative stress," which negatively affects cellular components. Hydroxyl radicals (OH•) and other free radicals can induce modifications of nucleic acids and bases, cause DNA strand breaks, and initiate processes leading to cancer, cellular aging, and cell death (Güleşçi & Aygül, 2016).

If the kinetics of free radical inactivation are sufficiently rapid, antioxidants can effectively suppress radical processes. These reaction kinetics are studied using fast time-resolved techniques such as pulse radiolysis and laser flash photolysis. Free radical neutralization and repair of biological radicals generally occur on microsecond (10–6 s) to millisecond (10–3 s) timescales (Jonavic & Simic, 2000).

Free radicals also form following exposure to cigarette smoke, air pollution, and sunlight. They can damage DNA sequences, alter cell membranes, and promote the accumulation and oxidation of circulating low-density lipoproteins (LDL) within blood vessels. Persistent high levels of free radicals in the body lead to oxidative stress, which can damage cells and contribute to chronic diseases. A balance exists between free radicals produced under pathological or physiological conditions and the antioxidant defense system that neutralizes them. A shift of this balance in favor of free radicals indicates oxidative stress (Lu et al., 2021). Factors causing this imbalance include xenobiotics, reduced antioxidant availability, or increased ROS/RNS production. Oxidative stress can be a contributing factor to cancer and cardiovascular diseases (Parcheta et al., 2021).

Oxidative Stress

Oxidative stress is a general term describing the imbalance between the production of free radicals and the body's ability to eliminate these reactive species through endogenous and exogenous antioxidants. The use of oxygen by cells for energy production can lead to the formation of partially oxidized intermediates with high reactivity (Yildiz et al., 2018). These compounds, known as reactive oxygen species (ROS), act as free radicals or precursors to radicals. In stable neutral molecules, electrons are paired in their molecular orbitals, which provide maximum natural stability. If an orbital contains an unpaired electron, the molecule becomes highly reactive, tending to capture an electron from another molecule to compensate for the electron deficiency (Santos-Sánchez et al., 2019).

During metabolic processes, a variety of reactions can produce ROS such as hydrogen peroxide and superoxide anion radical. Free radicals can also be generated by external factors like ultraviolet (UV) radiation from sunlight. Excessive ROS in the human body can lead to various pathological issues, ranging from cardiovascular diseases to cancer. Free radicals can also arise during disease conditions; for example, during a heart attack, the limited oxygen and glucose supply to heart muscle generates large amounts of free radicals. Another significant external source of free radicals is chemical poisoning, where the body produces free radicals as it detoxifies harmful substances. The toxicity of many drugs is also linked to their conversion into free radicals or their effects on radical formation (Santos-Sánchez et al., 2019).

Biological systems possess antioxidant mechanisms to control the damage caused by ROS through enzymatic and non-enzymatic structures. Endogenous antioxidants include enzymes such as superoxide dismutase (SOD), catalase (CAT), and glutathione peroxidase (GPx), as well as non-enzymatic compounds like bilirubin and albumin. When organisms are exposed to high levels of ROS, the endogenous antioxidant system can be overwhelmed. In such cases, exogenous antioxidants—obtained through diet or supplements—can help restore antioxidant balance. Key exogenous antioxidants include phenolic compounds, carotenoids, vitamin C, selenium, and zinc (Santos-Sánchez et al., 2019).

Lipids, a major energy source for humans, are highly susceptible to oxidation. Lipid-derived oxidative stress is categorized into lipid accumulation and lipid peroxidation. Lipid units are the primary drivers of lipid peroxidation. Increased dietary fat intake, particularly from Western-type diets, contributes to nutrition-related oxidative stress and impaired stimulus control. Changes in the level, location, or properties of phospholipids in mitochondrial membranes can increase ROS production. Maintaining ROS levels within a physiological range requires keeping intracellular lipid content low (Guo et al., 2020).

From the perspective of glucose metabolism, high blood glucose levels trigger oxidative stress by stimulating ROS production in various cell types. Excessive ROS generation alters mitochondrial structure, leading to mitochondrial dysfunction under high-glucose conditions. Antioxidant enzymes such as GPx, SOD, and CAT lose their activity and antigenicity under chronic high-glucose exposure. Chronic oxidative stress induced by hyperglycemia is closely linked to beta-cell dysfunction in the pancreas and can contribute to insulin resistance (Guo et al., 2020).

In overweight individuals, elevated levels of free fatty acids secreted by adipose tissue promote ROS production and reduce the activity of antioxidant enzymes, including SOD, GPx, and CAT. ROS in adipose tissue can also alter the synthesis and secretion of adipokines, contributing to an inflammatory response (Guo et al., 2020).

Antioxidant Mechanism

Antioxidants were clearly defined through a consensus reached in 2022, resulting in two distinct definitions:

1. "Any substance that delays, prevents, or removes oxidative damage to a target molecule."
2. "A substance that reacts with an oxidant to regulate reactions, thereby influencing redox-dependent biological signaling pathways and/or oxidative damage" (Halliwell, 2024).

The first definition includes mechanisms that remove oxidatively damaged molecules, whereas the second focuses on the physiological roles of ROS affected by antioxidants. Although the second definition is more commonly used, repair of oxidative damage remains important for maintaining health (Halliwell, 2024).

The primary effect of antioxidants on human health is their ability to scavenge free radicals (Yıldız, 2022; Yıldız and Yildiz, 2025; Yildiz et al., 2025). During metabolic activities, oxygen can be converted into reactive oxygen species (ROS) such as superoxide, hydrogen peroxide, singlet oxygen, and hydroxyl radicals, making oxygen potentially toxic to biological systems. In the presence of antioxidants, oxidative stress-induced damage is significantly reduced. Antioxidants act as hydrogen atom donors (Meral et al., 2012). Their effectiveness depends on their oxidation potential, defined as the single-electron reduction potential of the radical species. This potential can be determined from the equilibrium kinetics of electron transfer reactions studied via pulse radiolysis or laser flash photolysis. Although reduction potential is a primary factor, overall antioxidant activity is also influenced by other physicochemical properties (Jovanovic & Simic, 2000).

Aerobic metabolism, in which glucose is metabolized in the presence of oxygen via glycolysis, the tricarboxylic acid (TCA) cycle, and the electron transport chain to produce ATP and CO2, is the most efficient mechanism for converting food energy into usable energy (Burnside & Snowden, 2014). ATP serves as the primary energy source during rest and low-intensity exercise (Petterson et al., 2006). Mitochondria carry out this aerobic process through oxidation-reduction reactions linked to the TCA cycle and electron transport chain. However, these reactions also generate ROS, which are unstable due to unpaired electrons and can disrupt cellular homeostasis, potentially leading to cell death. Organisms are further affected by oxidative imbalances caused not only by metabolic reactions but also by external factors such as microbial infections, xenobiotics, dietary toxins, radiation, and environmental pollution (Aguilar et al., 2016).

Oxygen-utilizing systems have both physiological and biochemical defense mechanisms. Physiologically, a microvascular system helps maintain tissue oxygen levels, while biochemically, antioxidant defenses include enzymatic, non-enzymatic, and repair systems (Santos-Sánchez et al., 2019).

The primary enzymatic defense system in aerobic organisms includes superoxide dismutase (SOD), catalase (CAT), glutathione peroxidase (GPx), and DT-diaphorase. SOD converts superoxide radicals (O2•−) into hydrogen peroxide (H2O2), which CAT or GPx then breaks down into water and oxygen. As a metalloenzyme, SOD requires a metal cofactor that defines its isoforms. CAT, with iron or manganese cofactors, decomposes hydrogen peroxide, while GPx, mainly in mitochondria and sometimes in the cytosol, reduces H2O2 and lipid hydroperoxides using selenium. Humans express at least eight GPx isoforms (GPx1–GPx8), with GPx1 being the most abundant, protecting cells from oxidative stress by preventing lipid peroxidation (Santos-Sánchez et al., 2019).

The non-enzymatic system consists of antioxidants that scavenge free radicals by either neutralizing them or donating electrons, transforming themselves into less reactive radicals that are subsequently neutralized by other antioxidants. Cells utilize vitamin E, vitamin C, carotenoids, ferritin, ceruloplasmin, selenium, reduced glutathione, manganese, ubiquinone, zinc, flavonoids, coenzyme Q, melatonin, bilirubin, taurine, and cysteine to remove free radicals (Santos-Sánchez et al., 2019).

Repair systems include enzymes that repair or remove biomolecules damaged by ROS, such as lipids, proteins, and DNA. Examples include DNA repair and proteolytic enzymes found in both the cytosol and mitochondria of mammalian cells. Specific enzymes like GPx, glutathione reductase, and methionine sulfoxide reductase mediate the repair of oxidative damage caused by excessive ROS attacks (Santos-Sánchez et al., 2019).

Dietary Antioxidant Compounds in Foods

To control oxidative processes and reduce or prevent their harmful effects, various antioxidants are consumed through the diet. The average human lifespan exceeds 82 years in countries like Iceland and Japan, while in some regions it is less than 50 years. Nutrition is considered a significant factor influencing human longevity (Jovanovic & Simic, 2000).

Regular consumption of fruits and vegetables is believed to minimize the risk of chronic diseases due to their rich content of bioactive compounds, including vitamins, minerals, phenolics, and terpenoids (Gostner et al., 2015 Yildiz, 2025a).

A key characteristic of antioxidants is their polarity, which affects their distribution between polar and non-polar environments in the body. Vitamin E, for example, is located in non-polar cell membranes and provides minimal protection to extracellular components. Another important factor is the bioavailability and pharmacokinetics of different antioxidants (Yildiz, 2024). Many antioxidants are readily absorbed in the intestines and circulate in the blood either unchanged or as sulfated, methylated, conjugated, or metabolized forms. However, antioxidants with lower absorption, such as quercetin, may still play a critical role in chemoprevention of colorectal carcinogenesis (Jovanovic & Simic, 2000).

Many foods naturally contain antioxidant compounds capable of neutralizing ROS threats. The inclusion of compounds such as vitamin C, vitamin E (tocopherols), vitamin A (carotenoids), and polyphenols in the diet has been proposed to support the body's protective systems against ROS-induced damage and reduce disease risks (Finley et al., 2011; Basturk et al., 2023; Yildiz, 2025b).

Hundreds of substances can act as antioxidants. Prominent examples include vitamin C, vitamin E, beta-carotene, selenium, and zinc. Other notable antioxidants include glutathione, coenzyme Q10, lipoic acid, flavonoids, polyphenols, and phytoestrogens. Most of these antioxidants occur naturally in fruits and vegetables, serving either to prevent oxidation in foods or as a natural defense mechanism. Clinical studies have begun testing the effect of single antioxidant supplements on chronic diseases. For example, the Physicians' Health Study reported that beta-carotene supplementation over an 18-year follow-up was associated with a modest reduction in cognitive decline (WEB SITE 1).

Phenolic Compounds

Phenolic compounds are substances containing one or more hydroxyl groups attached to an aromatic ring, including their functional derivatives. They are among the most active natural antioxidants, exerting their effects by scavenging free radicals, forming metal chelates, and inhibiting lipoxygenase enzymes. Polyphenols, a subgroup of phenolic compounds, act as antioxidants by neutralizing radicals such as ROO- that break lipid and ROS bonds, similarly to the way metal ions form chelates. To be classified as an antioxidant, a polyphenol must possess two key properties: it should delay or prevent oxidation even at low concentrations, and it must remain structurally stable after converting into a free radical (Güleşçi & Aygül, 2016; Uylaşer & İnce, 2008). Many plant-based foods contain phenolic phytochemicals, which not only protect food from oxidation but also contribute antioxidants to the human body upon consumption. Phenolic compounds in plants are subdivided into lignans, phenolic acids, stilbenes, and flavonoids, with phenolic acids and flavonoids being particularly important for their antioxidant activity. Flavonoids, due to their antioxidant properties, are considered one of the most important anticancer agents in the diet.

While vitamin C is commonly used in dietary supplements for its health-promoting properties, phenolic compounds, especially flavonoids, are suggested to provide synergistic antioxidant and cardiovascular benefits. Chrysikopoulou et al. (2025) investigated a high-dose (1000 mg) vitamin C–low-dose (50 mg) bioflavonoid supplement to evaluate its antioxidant potential as well as anti-inflammatory and antiplatelet effects in human platelets using both in vitro and in vivo approaches. Among all phenolic standards, flavonoids—particularly catechin + quercetin mixtures—demonstrated the highest in vitro antioxidant capacity, anti-inflammatory, and antiplatelet activities. The vitamin C–flavonoid combination showed stronger antioxidant capacity and superior anti-inflammatory and antiplatelet activities compared to either vitamin C or the flavonoid mixture alone, indicating higher synergy and bioefficacy when used together.

Numerous studies have reported that carotenoids, flavonoids, and anthocyanidins possess a wide range of biological and pharmacological activities, including antioxidant, anticancer, antiviral, antibacterial, anti-inflammatory, antiallergic, antithrombotic, cardioprotective, hepatoprotective, and neuroprotective effects (Yıldız, 2018; Yıldız, 2019; Bayat and Yildiz, 2019; Yildiz and Yıldız, 2023). Carotenoids and their metabolites are thought to positively affect ROS-mediated conditions such as cardiovascular disease, osteoporosis, cancer, and myocardial infarction, particularly in smokers. Epidemiological studies suggest that carotenoid-rich foods and supplements may partially prevent prostate cancer (Lu et al., 2021).

ROS generated during oxidation reactions in organisms can damage DNA, RNA, lipids, and proteins, contributing to degenerative diseases like multiple sclerosis, cancer, and Alzheimer's disease. The conjugated polyunsaturated double-bond structure characteristic of carotenoids allows them to quench free radicals effectively. This structure also imparts a lipophilic character, enabling carotenoids to protect lipoproteins and cell membranes from free radical attack. Flavonoids, including anthocyanidins, can directly scavenge free radicals by donating hydrogen atoms or indirectly inhibit radicals by chelating free metal ions (Lu et al., 2021).

Key dietary antioxidants include vitamins A, C, and E, minerals such as selenium, chromium, and zinc, and bioactive compounds like carotenoids, flavonoids, polyphenols, and lignans. These compounds provide protection against free radicals and help maintain cognitive functions such as memory, thinking, and concentration. For example, a study found that daily supplementation with 75 mg of resveratrol significantly improved overall cognitive performance in middle-aged women (Fekete et al., 2023).

Vitamin E

Vitamin E was discovered in 1922 by Evans and Bishop as an essential nutrient for reproduction in rats (Traber & Atkinson, 2007). Vitamin E, along with carotenoids and the reduced form of coenzyme Q known as ubiquinol, is recognized as a natural, lipophilic radical-scavenging antioxidant (Niki & Noguchi, 2004). The term "vitamin E" is commonly used for plant-synthesized molecules that exhibit the antioxidant activity of α-tocopherol and encompasses eight naturally occurring compounds (Blaner et al., 2021). These eight related homologs are classified as R, α, γ, and δ forms, depending on the number and position of methyl substitutions on the chromanol ring and the corresponding tocopherols and tocotrienols (Niki & Noguchi, 2004). Tocopherols possess a saturated phytyl side chain, whereas tocotrienols have an unsaturated side chain. Among these forms, α-tocopherol is the most biologically active and antioxidant-rich form relevant to human nutrition, fulfilling dietary vitamin E requirements, as other forms cannot be converted to α-tocopherol in non-plant organisms (Blaner et al., 2021).

Numerous studies suggest that vitamin E acts as a ROS scavenger, particularly inhibiting the propagation of lipid peroxidation chain reactions. The exact mechanisms of α-tocopherol in reproduction and fetal development, as well as its potential effects in adults beyond its antioxidant function, have been investigated for decades. The literature indicates that α-tocopherol may have additional molecular roles beyond antioxidant activity, although these are often observed only in cell culture studies lacking co-antioxidants such as selenium (Blaner et al., 2021).

Vitamin E is known to protect polyunsaturated fatty acids, especially highly oxidizable docosahexaenoic acid (DHA), from lipid peroxidation. Vitamins C and E, along with endogenous reducing agents like GSH and NADPH, form an integrated antioxidant network. In this cycle, the lipophilic peroxyl radical is neutralized by vitamin E, forming a one-electron tocopheroxyl radical. This radical is then regenerated back to α-tocopherol by vitamin C (Blaner et al., 2021).

Being fat-soluble, vitamin E is found in plant oils, nuts, seeds, fruits, and vegetables. (Caglar et al., 2023a). The content and type of tocopherols vary across plant parts; leaves contain the highest α-tocopherol, while seeds are richer in β-, γ-, and δ-tocopherols (Caglar et al., 2023b). For humans, α-tocopherol—possessing the strongest antioxidant activity—is most abundant in wheat germ oil. Considering total tocopherol content, soybean oil, corn oil, and sunflower oil are particularly rich sources. The amount of vitamin E extracted from a plant can also vary depending on oil extraction and cooking methods (Ungurianu et al., 2021).

Insufficient intake of vitamin E is associated with a higher risk of developing various diseases linked to low-grade inflammation. Low-grade inflammation is characterized by a moderate increase in pro-inflammatory molecules, such as C-reactive protein (CRP), without typical signs of inflammation. Initially, this represents a prolonged state of inflammation aimed at restoring tissue homeostasis, but persistent low-grade inflammation can lead to alterations or loss of tissue function and is associated with metabolic syndrome, cardiovascular diseases, type 2 diabetes mellitus, non-alcoholic fatty liver disease, and neurodegenerative disorders (Ungurianu et al., 2021).

Beta-Carotene and Lycopene

Vitamin A is a fat-soluble group of retinoids, primarily including retinol and retinyl esters. Vitamin A plays a crucial role in immune function, cellular communication, growth and development, and the reproductive system. It supports cell growth and differentiation and is critical for the normal formation and maintenance of the heart, lungs, eyes, and other organs. Additionally, it is an essential component of rhodopsin, a light-sensitive protein in the retina that responds to incoming light, thereby supporting visual function, corneal health, and integrity (WEB 2).

The human diet contains two types of vitamin A: preformed vitamin A (retinol and retinyl esters) from animal sources and provitamin A carotenoids from plant sources. Preformed vitamin A is found in animal-derived foods, including milk and dairy products, eggs, fish, and organ meats. Provitamin A carotenoids, including beta-carotene, alpha-carotene, and beta-cryptoxanthin, are plant pigments mainly present in orange-red or dark green leafy vegetables such as pumpkin, carrot, apricot, spinach, and kale (Abrego-Guandique et al., 2023).

Beta-carotene is not only a precursor of vitamin A but also a powerful antioxidant increasingly investigated for its role in maintaining mental health and cognitive performance. However, its efficacy remains somewhat uncertain. A systematic review of 10 epidemiological studies found that in seven studies, dietary intake or serum levels of beta-carotene were associated with improved cognitive function. Intervention studies showed that beta-carotene alone did not significantly enhance cognitive performance in the short term, but long-term studies (up to 18 years) indicated beneficial effects. Studies also suggest that daily supplementation of beta-carotene (6–50 mg) together with multivitamins containing vitamin E, vitamin C, zinc, or selenium can be advantageous. Current evidence indicates a potential link between beta-carotene intake (from food or supplements) and the maintenance of cognitive function, likely acting synergistically with other micronutrients rather than alone (Abrego-Guandique et al., 2023).

The importance of natural antioxidants for human health has increased with scientific advances in antioxidant research. Across the world, societies have used natural antioxidants to treat various diseases and improve health. Poor nutrition is recognized as a primary contributor to numerous pathological conditions, including cardiovascular diseases. Cardiovascular diseases are characterized by proinflammatory conditions, excessive production of reactive oxygen species (ROS), and dyslipidemia. Antioxidants are among the most critical molecules for maintaining cardiovascular health and preventing disease. Plants prevent various oxidative stresses from ROS, and plant-based extracts are used for medicinal purposes to prevent disease. Moreover, plants protect their cells and genetic material via antioxidant enzymes (superoxide dismutase, catalase, and peroxidase) and radical-scavenging compounds such as ascorbic acid and carotenoids. These natural compounds are typically reducing agents, including polyphenols, thiols, carotenoids, and ascorbic acid (Khan et al., 2021).

Lycopene is a lipophilic carotenoid hydrocarbon pigment found in red, pink, and orange fruits and vegetables, such as tomatoes, apricots, melons, papaya, grapes, peaches, watermelon, and cranberries. Natural antioxidants like lycopene play an essential role in preventing lipid and food oxidation (Khan et al., 2021). Lycopene, synthesized by plants and some microorganisms, imparts red-orange color to fruits and vegetables. It has the ability to neutralize reactive species such as hydrogen peroxide, hydroxyl radicals, and nitrogen dioxide, thereby exhibiting antioxidant activity (Khan et al., 2021).

Lycopene has also been reported as a key antioxidant that enhances both enzymatic (superoxide dismutase, catalase, peroxidase) and non-enzymatic antioxidants (vitamins E and C) in cells. By protecting cellular biomolecules such as lipoproteins, lipids, DNA, and proteins, lycopene may help prevent carcinogenesis and atherogenesis. Epidemiological studies indicate that lycopene intake can be effective in protecting against prostate and colorectal cancers. In some cell culture studies, lycopene inhibited cell cycle progression and induced apoptosis in cancer cells, suppressing tumor growth. Lycopene has also been identified as an important molecule for inhibiting breast cancer cell proliferation by downregulating the insulin-like growth factor 1 receptor (IGF-1R) pathway (Khan et al., 2021).

Chronic diseases, including cancer and cardiovascular diseases, remain leading causes of death in Western countries. Oxidative stress caused by reactive oxygen species is a major focus of recent research related to these diseases. ROS are highly reactive oxidant molecules generated endogenously through metabolic processes, lifestyle factors (e.g., smoking, alcohol use, poor diet, and exercise), air pollution, UV radiation, and diet. They react with cellular components, causing oxidative damage to lipids, proteins, and DNA. Strong evidence suggests that this damage plays a significant role in the development of various chronic diseases (Agarwal & Rao, 2000).

Antioxidants are protective agents that neutralize reactive oxygen species, thereby delaying or preventing oxidative damage. Among dietary carotenoids, beta-carotene and lycopene are some of the most potent antioxidants. Serum and tissue lycopene levels have been inversely associated with the incidence of various cancers, including breast and prostate cancers (Agarwal & Rao, 2000).

Although the anti-prostate cancer activity of tomatoes and lycopene has been demonstrated in rodent models, significant knowledge gaps remain in understanding dose-response relationships and molecular mechanisms of action (Moran et al., 2022). The effects of carotenoid-rich vegetables (particularly lycopene and lutein) on metabolic syndrome have also been studied in obese Japanese men. Middle-aged men ($40 \leq$ age < 65) with high body mass index (BMI ≥ 25) were randomized into four groups: high lycopene + high lutein (HLyHLu), high lycopene + low lutein (HLyLLu), low lycopene + high lutein (LLyHLu), and low lycopene + low lutein (LLyLLu). Daily consumption increased plasma carotenoid levels without adverse effects, and visceral fat levels decreased in all groups, while waist circumference decreased significantly only in the HLyLLu group (Takagi et al., 2020).

In a randomized trial involving 5,956 men over 65 years, daily supplementation with 50 mg of beta-carotene or placebo for at least 15 years provided long-term cognitive benefits (Grodstein et al., 2007). UV radiation is a major cause of skin damage and photoaging. Beta-carotene, as a strong antioxidant, can reduce UV-induced oxidative stress and enhance skin photoprotection. Putthong et al. (2024) developed a chewable tablet containing hemp seed oil and banana (Musa AA) delivering 8 ± 2 mg of beta-carotene and 400 mg PUFA per day, administered five times daily for 16 weeks. After four weeks, significant reduction in UV-induced skin effects and increased serum beta-carotene levels were observed.

Regarding potential side effects, lycopenemia is a clinical condition characterized by yellow-orange skin pigmentation, resulting from excessive dietary intake of lycopene sources (Khan et al., 2021).

Quercetin

Polyphenols are a large family of natural compounds characterized by the presence of phenolic hydroxyl groups (aromatic rings) and include four major classes: phenolic acids, flavonoids, stilbenes, and lignans. Their antioxidant effects are based on the ability of aromatic hydroxyl groups to donate H^+ to free radicals such as hydroxyl and peroxyl radicals. In quercetin, these free radicals (FRs) are stabilized by forming a relatively stable flavonoid radical, illustrating the biochemical mechanism of FR scavenging (Andrés et al., 2024).

Quercetin possesses antioxidant, anti-inflammatory, anti-proliferative, anticarcinogenic, anti-diabetic, and antiviral properties. It is a lipophilic molecule that can easily cross the blood-brain barrier, making it potentially effective in protecting against neurodegenerative diseases. Various in vivo and in vitro studies have examined quercetin as a therapeutic agent in neurodegeneration, diabetes, cancer, and inflammation, highlighting its role in preventing age-related diseases (Deepika & Maurya, 2022).

The antioxidant properties of quercetin are associated with the prevention and treatment of cancer and cardiovascular diseases. Quercetin is a strong scavenger of free radicals among flavonoids. The hydroxyl groups in its structure act as scavengers by donating active hydrogen to neutralize free radicals. This process oxidizes the radicals, rendering them relatively stable, and prevents the oxidation of polyunsaturated fatty acids (Deepika & Maurya, 2022).

Regarding anticancer effects, quercetin can induce apoptosis and autophagy via both intrinsic and extrinsic pathways and can arrest the cell cycle (Wang et al., 2021).

In one study, quercetin was evaluated for its ability to prevent progression to the critical phase of COVID-19 and to reduce inflammatory markers associated with SARS-CoV-2 pathogenesis. Sixty severe cases were randomly assigned to control and intervention groups. Over seven days, the control group received antiviral treatment alone, while the intervention group received antivirals plus 1000 mg quercetin daily. Quercetin intake was significantly associated with partial early discharge and reductions in serum levels of q-CRP, LDH (lactate dehydrogenase), and ALP (alkaline phosphatase). Although the values remained within normal ranges, statistical analysis showed significant increases in hemoglobin levels and respiratory rates in patients receiving quercetin. Based on these observations, quercetin appears safe and effective in reducing serum levels of key markers involved in COVID-19 severity (Shohan et al., 2022).

Fanconi anemia (FA) is a rare hereditary disorder characterized by progressive bone marrow failure and increased susceptibility to malignancy. Systemic reactive oxygen species (ROS) and the heightened ROS sensitivity of FA hematopoietic progenitors play a critical role in the pathogenesis of bone marrow failure. Antioxidant therapy has been shown to improve hematopoietic function in FA mice and has also been investigated in humans. Patients tolerated quercetin at all administered dose levels, with a maximum dose of 4000 mg/day given for six weeks. In one patient cohort, reductions in ROS levels were observed in both peripheral blood and the bone marrow stem cell compartment. Patients receiving the recommended quercetin dose achieved optimal ROS responses compared to baseline, and hematologic improvements were observed in 8 out of 15 patients (53%) with bone marrow failure following quercetin therapy (Mehta et al., 2025).

Curcumin

Curcumin is a polyphenolic compound isolated from Curcuma longa (turmeric). Curcumin has strong antioxidant effects and has been shown to possess potent biochemical and biological activities, including anti-inflammatory, antiviral, antibacterial, antioxidant, and anticancer effects. In addition, it protects cells against protein carbonylation, lipid peroxidation, and mitochondrial permeability transition. Various studies in model organisms have demonstrated that curcumin treatment can extend lifespan, which is associated with increased superoxide dismutase (SOD) activity, and reduced levels of malondialdehyde (MDA) and lipofuscin (Zia et al., 2021).

In traditional herbal medicine, curcumin is used to enhance the immune system and to treat various respiratory disorders, such as allergies and asthma. Curcumin has also been traditionally applied in the treatment of numerous conditions, including ulcers, dysentery, jaundice, nausea, bloating, sprains, arthritis, acne, wounds, and eye and skin infections (Zia et al., 2021).

Vitamin E is a fat-soluble vitamin that is essential for human health. Vitamin E refers to eight fat-soluble compounds (α-, β-, γ-, δ-tocopherols and α-, β-, γ-, δ-tocotrienols), but only α-tocopherol meets the body's vitamin E requirement. Vitamin E is commonly found in foods and fruits, particularly in fats, nuts, and seeds. It functions as an antioxidant in the body. Its roles include protecting polyunsaturated fatty acids in cell membranes from oxidation, regulating the production of reactive oxygen species (ROS) and reactive nitrogen species (RNS), and modulating related signaling pathways.

Liu et al. (2023) combined previous meta-analyses to assess the relationship between vitamin E intake or circulating tocopherol levels and health outcomes. They concluded with recommendation-level findings regarding endothelial function (supplemental vitamin E). According to their results, serum C-reactive protein (CRP) concentrations (supplemental vitamin E), cervical cancer (dietary vitamin E), esophageal cancer (dietary vitamin E), cervical intraepithelial neoplasia (dietary vitamin E), pancreatic cancer (total vitamin E intake), and colorectal cancer (circulating α-tocopherol levels) showed protective effects consistent with the source of vitamin E (Liu et al., 2023).

In a study by Szewczyk et al. (2025), the concentrations of vitamin E isoforms (α- and γ-tocopherols, α- and γ-tocotrienols) were evaluated in 127 participants. Vitamin E isoform levels were associated with CRP concentrations. Similarly, higher concentrations of vitamin E isoforms were linked to lower body fat content, suggesting reduced inflammation and improved metabolic health.

In the Women's Health Study, 39,876 initially healthy women received 600 IU of natural-source vitamin E or placebo every other day for 10 years. At the end of the study, rates of major cardiovascular disease and cancer were not lower among vitamin E users compared to placebo. However, a significant 24% reduction in total cardiovascular mortality was observed (Lee et al., 2005).

Large vitamin E trials in people with coronary disease or at high risk generally did not show clear benefits. In the HOPE study, major cardiovascular event rates were similar between the vitamin E (21.5%) and placebo (20.6%) groups, but those taking vitamin E had higher risks of heart failure and related hospitalizations. The GISSI study also found no preventive effect in over 11,000 patients who had experienced a myocardial infarction after more than three years of vitamin E treatment. However, some studies indicate possible benefits in specific subgroups; for instance, a recent trial in Israel reported a significant reduction in coronary heart disease among type 2 diabetes patients with a genetic susceptibility to oxidative stress (Milman et al., 2008).

Some evidence indicates that oxidative stress plays an important role in cognitive decline and dementia, especially in neurodegenerative processes accompanying Alzheimer's disease. The brain consumes approximately 20% of the body's total oxygen, making it particularly vulnerable to reactive oxygen species (ROS). Beydoun et al. (2015) conducted a cross-sectional study of 1,274 adults aged 30–64 (541 men and 733 women), evaluating memory, language/verbal, attention, spatial, psychomotor speed, executive function, and overall cognitive performance. Depressive symptoms were assessed using the 20-item Center for

Epidemiologic Studies Depression Scale (CES-D). Dietary intake was estimated from two 24-hour recalls for total carotenoids and vitamins A, C, and E per 1,000 kcal. Key findings showed that higher vitamin E intake was associated with higher scores in verbal memory, immediate recall, and language/verbal fluency, particularly in younger participants. Women with higher vitamin E intake also performed better on psychomotor speed tests (Beydoun et al., 2015).

In the Prevention of Alzheimer's Disease by Vitamin E and Selenium (PREADViSE) study, over 3,700 men aged 60 and above were followed for six years. Vitamin E or selenium, alone or in combination, did not show protective effects against dementia compared to placebo (Kryscio et al., 2017).

Wang et al. (2024) used data from 90,167 UK Biobank participants to develop prediction models for cardiovascular disease (CVD) risk based on micronutrient intake. The results showed that intake of vitamin E, calcium, vitamin C, and potassium was associated with reduced CVD risk (Wang et al., 2024).

Shokri and Sadeghi (2020) conducted a comprehensive meta-analysis summarizing current evidence on the effects of vitamin E supplementation on inflammatory biomarkers in adults. Overall, 33 studies with a total of 2,102 participants aged 20–70 were analyzed. Data from 26 RCTs examining serum CRP concentrations indicated a significant reduction after vitamin E supplementation. Although the overall effect on serum interleukin-6 (IL-6) concentrations was not significant, studies using α-tocopherol and including participants with insulin resistance showed significant reductions. Additionally, high-dose vitamin E supplementation (≥700 mg/day) significantly reduced tumor necrosis factor-α (TNF-α) concentrations.

Vitamin C

Vitamin C, also known as ascorbic acid, is an essential water-soluble nutrient found in most fruits and vegetables. Many of its biological functions are linked to its role as an antioxidant and free radical scavenger. Case-control and cohort studies have suggested that a high intake of vitamin C may be associated with a reduced risk of cancers of the digestive system, including colorectal, stomach, and pancreatic cancers (Larsson, 2022). Unlike some other vitamins, vitamin C is present in substantial amounts in many foods. Globally, a significant portion of dietary vitamin C comes from fruits, vegetables, potatoes, and fruit juices. Animal-based foods generally contain low levels of vitamin C. In Europe and Asia, rose hips and sea buckthorn are particularly rich sources. Citrus fruits provide adequate amounts, while cruciferous vegetables such as broccoli, kale, and peppers also contribute significantly to dietary vitamin C intake (Doseděl et al., 2021).

Vitamin C (l-ascorbic acid or ascorbate) is an essential nutrient in humans, functioning as an indispensable electron donor and cofactor in various biological reactions, including collagen hydroxylation, carnitine biosynthesis, and tyrosine metabolism. Interestingly, vitamin C shows its highest concentrations in the brain. Animal models and in vitro studies have reported that vitamin C plays critical roles in brain function. It protects neurons from oxidative stress, supports neuronal differentiation and maturation, and regulates the synthesis or release of neuromodulators such as serotonin, catecholamines, and glutamate. These findings suggest that vitamin C is important for maintaining normal mental health. Low vitamin C levels in young adults have been linked to external factors such as smoking, excessive alcohol consumption, and unhealthy dietary habits lacking sufficient vitamin C (Sim et al., 2022).

Vitamin C is an antioxidant with a potential role in preventing digestive system cancers; however, there is not yet a consensus on whether it plays a causal role in these cancers. A meta-analysis indicated that genetically predicted higher circulating vitamin C is associated with a lower risk of small intestine and colorectal cancers. In contrast, no association was found with liver, esophageal, stomach, or pancreatic cancers (Larsson et al., 2022).

In a randomized controlled study, serum vitamin C concentrations were evaluated in relation to vitality (fatigue and attention) and mood (stress, depression, positive and negative affect). Participants received 500 mg of vitamin C twice daily or a placebo for four weeks. Compared with the placebo group, vitamin C supplementation was associated with improved attention, enhanced task performance, reduced fatigue, and greater overall work engagement. Vitamin C deficiency was linked to lower mental vitality, while supplementation effectively improved work motivation and attentional focus, contributing to better performance in cognitive tasks requiring sustained attention (Sim et al., 2022).

Strawberries contain numerous bioactive compounds, including vitamin C, folate, and polyphenols. Kishimoto et al. investigated the effects of acute strawberry consumption on serum vitamin C, folate levels, and low-density lipoprotein (LDL) antioxidant potential. In a randomized, placebo-controlled, double-blind, crossover study, twenty-three healthy female volunteers (mean age 22.5 ± 1.4 years) consumed 500 g of strawberry puree beverage or a sugar-matched placebo. Serum vitamin C and folate concentrations significantly increased 0.5–4 hours after consuming the strawberry beverage. Notably, one hour after intake, LDL oxidation lag time was significantly prolonged, indicating enhanced LDL antioxidant potential (Kishimoto et al., 2023).

In 1,923 postmenopausal women with diabetes, the relationship between vitamin C intake and mortality due to total cardiovascular disease, coronary artery disease, and stroke was examined. When dietary and supplemental vitamin C were analyzed separately, only supplemental vitamin C showed a positive association with mortality. High-dose vitamin C intake was linked to increased cardiovascular disease mortality in postmenopausal women with diabetes (Lee et al., 2004).

Uz et al. (2025) retrospectively examined the effects of low-dose (2 g/day) and high-dose (25 g/day) intravenous vitamin C (IV-VC) in hospitalized patients with moderate or severe COVID-19. Patients treated with high-dose IV-VC had significantly longer median hospital stays compared with low-dose IV-VC and control groups, while the need for ICU transfer was significantly lower in the low-dose IV-VC group (Uz et al., 2025).

In a study involving 60 reproductive-age women (15–45 years) with pelvic pain, participants were divided into two groups. One group received vitamin C (1000 mg/day, two 500 mg tablets) and vitamin E (800 IU/day, two 400 IU tablets) for 8 weeks, while the other group received placebo. Following supplementation with vitamin C and E, there was a significant reduction in malondialdehyde (MDA) and reactive oxygen species (ROS) compared with the placebo group. Pelvic pain, dysmenorrhea, and dyspareunia severity were significantly reduced in the treatment group after 8 weeks. These findings support the potential role of antioxidants in treating endometriosis. Vitamin C and E supplementation effectively reduced dysmenorrhea and improved dyspareunia and pelvic pain severity (Amini et al., 2021).

Postmenopausal women experience increased oxidative stress due to endocrine changes, which can lead to cardiovascular and metabolic alterations. A randomized, double-blind clinical study evaluated the effects of resveratrol and vitamin C on oxidative stress in postmenopausal women. Forty-six postmenopausal women with insulin resistance (HOMA-IR

> 2.5) were included and divided into three treatment groups: group A received resveratrol, group B received resveratrol + vitamin C, and group C received vitamin C. After supplementation, group B showed a 33% reduction in lipohydroperoxides, while reductions in MDA were 26%, 32%, and 38% in groups A, B, and C, respectively. Total antioxidant capacity increased by 30% in group B and 28% in group C. The combination of antioxidants effectively reduced oxidative stress markers in postmenopausal women (Montoya-Estrada et al., 2024).

A study investigated the effects of vitamin C on blood pressure, oxidative stress, and nitric oxide release after low intensity exercise in patients. Twenty-four patients with type 2 diabetes mellitus were followed for six weeks, receiving either placebo or 1000 mg of vitamin C. Compared with placebo, vitamin C supplementation significantly lowered blood pressure and increased plasma ascorbate and nitric oxide levels. The study suggested that 6 weeks of vitamin C supplementation may help improve oxidative stress and reduce pre- and post-exercise blood pressure due to increased nitric oxide release (Boonthongkaew et al., 2021).

Vitamin C is an electron donor and a potent water-soluble antioxidant that humans must obtain from dietary sources. Its antioxidant properties have been demonstrated in numerous vitro studies. In diseases such as atherosclerosis and cancer, oxidative damage in tissues may play a critical role. Oxidation of lipids, proteins, and DNA produces specific, measurable oxidation products. Although these oxidation biomarkers can be measured in humans, the analyses are not yet fully validated or standardized, and their relationship to disease states remains unclear. Epidemiological studies suggest that diets rich in fruits and vegetables are associated with a lower risk of cardiovascular disease, stroke, and cancer, as well as increased longevity. To attribute these protective effects specifically to vitamin C, the dosage, population (healthy or patient), and plasma and tissue vitamin C concentrations must be clearly defined (Padayatty et al., 2003).

Selenium

Aging is characterized by an imbalance between damage caused by reactive oxygen species (ROS) and the body's antioxidant defenses. Selenium, an essential trace element, can reshape the gradual and spontaneous physiological changes caused by oxidative stress and potentially lead to disease prevention and healthy aging. Selenium plays a role in enhancing antioxidant defense, immune functions, and metabolic homeostasis (Bjørklund et al., 2022).

Dietary selenium is obtained from plant-based products containing selenomethionine and methylselenocysteine (grains, cereals, fruits, Brazil nuts, broccoli, garlic, onions, and cabbage). Selenium from animal sources is obtained from meat, seafood, eggs, and dairy products in the form of selenocysteine (Andrés et al., 2024).

Amino acids, peptides, and enzymes are considered the main biologically important selenium-containing organic compounds. Selenocysteine, a substituted variant of the amino acid cysteine, is a fundamental component of selenoproteins. Therefore, selenium's physiological effects primarily arise from its incorporation into selenoproteins. To date, 25 selenoproteins have been discovered in humans. Selenoproteins have been shown to play roles in numerous metabolic and functional pathways, including aging, cancer, and infection (Bjørklund et al., 2022).

Selenium-containing glutathione peroxidase (GPx) enzymes help reduce H_2O_2 to H_2O and organic hydroperoxides to alcohols, thereby lowering free radical reactions to tolerable levels.

Selenium-dependent glutathione peroxidases and thioredoxin reductases directly suppress oxidative stress in the body. Glutathione peroxidase is a unique mammalian selenoenzyme that detoxifies ROS and maintains redox balance (Bjørklund et al., 2022).

Selenium performs its primary biological functions through selenoamino acids. These functions include defense against oxidative stress, maintenance of cellular redox status and signaling, lipid metabolism, and thyroid hormone metabolism (Bjørklund et al., 2022). A study by Avery and Hoffmann (2018) reported that selenium deficiency may lead to impaired immunity in humans, significantly increasing susceptibility to pathogens and cancer.

Selenium is an essential trace element required for thyroid hormone synthesis and exhibits antioxidant effects. Therefore, it may be important to manage Hashimoto's thyroiditis. A meta-analysis including 687 records and 35 specific studies reported that selenium was effective and safe in reducing TSH, thyroid peroxidase antibodies, and malondialdehyde (MDA) in hypertensive individuals (Huwiler et al., 2024).

Selenium supplementation has frequently been investigated as a therapeutic approach for preventing metabolic diseases, delaying their progression, and mitigating their outcomes. Recent evidence supports that selenium plays an important role in the prevention, early stages, and clinical outcomes of metabolic diseases such as cardiovascular disease, hypertension, non-alcoholic fatty liver disease, and type 2 diabetes (Huang et al., 2022).

Excessive selenium intake can be hazardous to health. According to the recommended daily intake, selenium intake ranges between 55 and 70 µg per day, which is generally based on the selenium required for adequate GPx activity. In healthy individuals, plasma selenium levels are approximately 100 µg/L, and urinary selenium levels range between 10 and 85 µg/L in the general population. Plasma selenium levels between 300 µg/L and 30,000 µg/L have been associated with mortality compared to the normal reference level of 100 µg/L (Huang et al., 2022).

Zinc

Zinc is an essential trace element required for the structure and function of numerous macromolecules, including enzymes that regulate cellular processes and signaling pathways. This mineral modulates immune responses and exhibits antioxidant and anti-inflammatory activity. Zinc delays oxidative processes in the long term by inducing metallothionein expression. Additionally, zinc enhances the activity of enzymes with known antioxidant properties, such as glutathione and catalase (Jarosz et al., 2017). Disrupted zinc homeostasis has been implicated in the pathogenesis of various chronic human diseases, including cancer, diabetes, depression, Wilson's disease, Alzheimer's disease, and other age-related conditions (Costa et al., 2023).

Antioxidants are generally defined as substances that react with and inhibit any type of oxidant. Zinc, however, does not fit this definition, as it has not been shown to directly interact with oxidant species; instead, it exerts its effects indirectly. One of the earliest examples of zinc's antioxidant role comes from metallurgy. For over 100 years, zinc has been used to galvanize iron or steel nails. In this process, nails are coated with zinc—either electrochemically or by hot dipping—which delays air oxidation and prevents rust formation (Powell, 2000).

Zinc's antioxidant effects in living organisms are classified into acute and chronic mechanisms. Chronic effects involve prolonged exposure to zinc, leading to the induction of another substance that acts as a final antioxidant. Zinc deficiency increases the body's susceptibility to oxidative stress (Powell, 2000).

Zinc is the second most abundant mineral in the human body after iron. An adult human contains approximately 2–3 grams of zinc, distributed across various tissues, with the highest concentrations in skeletal muscle and bone (approximately 90% of the total). Zinc content in the human body varies according to age, pathological condition, and other factors. Foods rich in zinc include meat, eggs, oysters, crabs, peanuts, almonds, and potatoes (Duan et al., 2021).

In a study investigating the role of zinc in oxidative stress and disease, 10 healthy volunteers received 45 mg of zinc in gluconate form for 8 weeks, while another 10 healthy volunteers received a placebo for 45 days. In the zinc group, plasma levels of lipid peroxidation products and DNA damage decreased, whereas no changes were observed in the placebo group. Additionally, messenger RNA (mRNA) levels of TNF-α and IL-1β were reduced in the zinc group compared to the placebo group. Studies in elderly populations have shown that zinc supplementation reduces infection incidence, oxidative stress, and inflammatory cytokine production. In patients with age-related macular degeneration, zinc supplementation decreased the incidence of blindness and enhanced atheroprotective effects (Prasad, 2014).

Studies in elderly populations have shown that zinc supplementation reduces infection incidence, oxidative stress, and inflammatory cytokine production. In patients with age-related macular degeneration, zinc supplementation decreased the incidence of blindness and enhanced atheroprotective effects (Prasad, 2014).

Green Tea

Green tea is a type of tea produced from fresh leaves of the plant *Camellia sinensis* without undergoing fermentation. Green tea was discovered in China around 3000 BCE and has been shown to have medicinal effects (Sarıca & Diktaş, 2008).

Green tea exhibits higher antioxidant activity compared to some other beverages rich in phenolic compounds. Green tea extracts, which have greater antioxidant activity than black tea, demonstrate a superior ability to break free radical chains and eliminate reactive oxygen. The polyphenolic compounds in tea scavenge reactive oxygen and nitrogen species, enhance the activity of intracellular enzymes (glutathione reductase, glutathione peroxidase, glutathione-S-reductase, catalase, and quinone reductase), and inhibit hydrolytic and oxidative enzymes (phospholipase A2, cytochrome, oxygenase, lipoxygenase) (Sarıca & Diktaş, 2008).

In adults, green tea has been shown to affect endothelial function, blood pressure, insulin sensitivity, cardiovascular diseases, various types of cancer, viral infections, and inflammatory conditions. Green tea contains polyphenolic compounds called catechins, which include epigallocatechin gallate (EGCG), epicatechin gallate (ECG), epigallocatechin (EGC), and epicatechin (EC). Catechins, as polyphenolic phytochemicals, are widely found in foods and medicinal plants, present in all types of tea, and have various health benefits.

Catechins obtained from green tea are powerful antioxidants, balance vasoconstrictive agents, inhibit lipid peroxidation, and help prevent cholesterol accumulation in blood vessels. The number of hydroxyl groups in the molecule is positively associated with the antioxidant activity of phenolic compounds. The catechins' free radical scavenging capacity follows the order: EGCG > ECG > EGC > EC > C. In addition to flavonoids, green tea contains vitamins such as vitamin C, riboflavin, niacin, folic acid, pantothenic acid, and minerals such as magnesium, potassium, manganese, and fluoride. Numerous clinical and epidemiological studies have investigated catechins for their anti-tumorigenic, anti-carcinogenic, anti-mutagenic, anti-diabetic, anti-obesity, and cardiovascular effects (Chen et al., 2016; Jianhua et al., 2016; Huang et al., 2018).

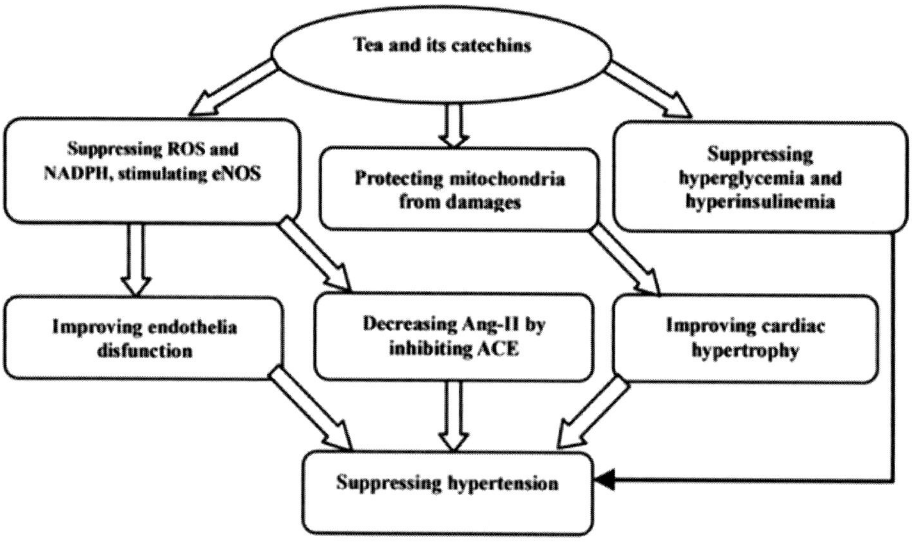

Figure 1. Actions of tea and catechins on the development of cardiovascular diseases and hypertension (Li et al., 2013).

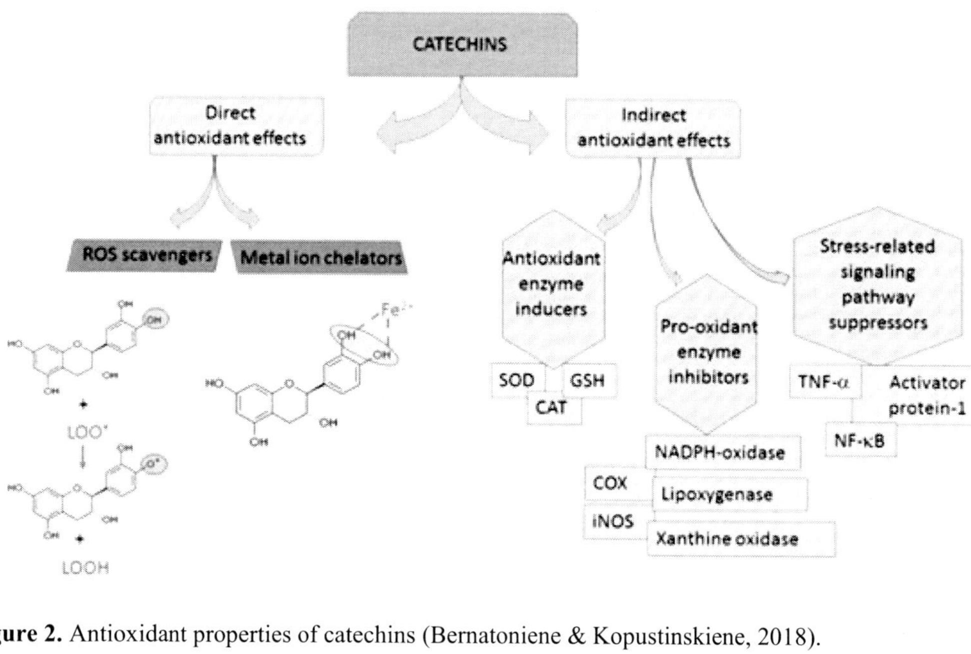

Figure 2. Antioxidant properties of catechins (Bernatoniene & Kopustinskiene, 2018).

Tea catechins contribute to the prevention of cardiovascular diseases through their beneficial effects in lowering plasma cholesterol levels, preventing hypertension, and improving endothelial function. The cholesterol-lowering effect is attributed to catechins' role in reducing cholesterol reabsorption and synthesis. An enhanced nitric oxide signaling system has been shown as an effective mechanism by which catechins in tea reduce blood pressure and the severity of myocardial infarction (Yang et al., 2018).

In a meta-analysis conducted by Pang et al. (2016), a total of 9 studies were evaluated. The results indicated that individuals who did not consume green tea had a higher risk of cardiovascular disease (CVD), intracerebral hemorrhage, and cerebral infarction compared to those consuming less than 1 cup per day. Additionally, drinking 1–3 cups of green tea daily was associated with a reduced risk of myocardial infarction and stroke compared to those drinking less than 1 cup per day (Pang et al., 2016).

In an animal study investigating the effects of green tea, particularly EGCG, on the progression of heart failure, mice were divided into three groups. The control group received drinking water, the second group received 10 mg/L EGCG, and the third group received 100 mg/L EGCG, with an eight-week follow-up. Compared to the control group, the EGCG-treated groups exhibited lower myocardial oxidative stress and free fatty acids, improved cardiac function, reduced inflammation, and decreased mortality (Oylama et al., 2017).

Due to the caffeine content in green tea, excessive consumption may cause insomnia, anxiety, tachycardia, and increased urinary output (diuretic effect). Therefore, it is not recommended for certain risk groups—including breastfeeding and pregnant women, adolescents, individuals with chronic conditions, and the elderly—to consume large amounts of green tea (Oylama et al., 2017).

Conclusion

Antioxidants play a critical role in maintaining human health by counteracting the damaging effects of reactive oxygen species (ROS) and free radicals. These reactive molecules, produced both endogenously during normal metabolism and exogenously through environmental and lifestyle factors, can cause oxidative stress when present in excess, leading to cellular damage and contributing to the development of chronic diseases such as cancer, cardiovascular disorders, neurodegeneration, and metabolic dysfunctions.

The human body relies on a complex antioxidant defense system that includes enzymatic antioxidants (e.g., superoxide dismutase, catalase, glutathione peroxidase) and non-enzymatic antioxidants (e.g., bilirubin, albumin, glutathione), while exogenous antioxidants obtained from diet—such as vitamins A, C, E, selenium, zinc, carotenoids, flavonoids, and polyphenols—further support oxidative balance. Bioactive compounds like quercetin, curcumin, resveratrol, beta-carotene, and lycopene exhibit strong antioxidant, anti-inflammatory, and disease-preventive effects, often demonstrating enhanced activity when combined synergistically.

Clinical and epidemiological evidence indicates that diets rich in natural antioxidants, particularly from fruits and vegetables, consistently reduce the risk of chronic diseases and support cognitive and cardiovascular health. However, isolated antioxidant supplementation often yields inconsistent results, highlighting the importance of whole dietary patterns over single-compound interventions. Overall, the findings underscore the significance of integrating dietary antioxidants into daily nutrition as a practical and effective strategy for promoting health, preventing disease, and mitigating oxidative stress-related damage throughout life.

References

Abrego-Guandique, D. M., Bonet, M. L., Caroleo, M. C., Cannataro, R., Tucci, P., Ribot, J., & Cione, E. (2023). The effect of beta-carotene on cognitive function: a systematic review. *Brain sciences,* 13(10), 1468.

Agarwal, S., & Rao, A. V. (2000). Tomato lycopene and its role in human health and chronic diseases. *Cmaj,* 163(6), 739-744.

Aguilar, T. A. F., Navarro, B. C. H., & Pérez, J. A. M. (2016). Endogenous antioxidants: a review of their role in oxidative stress. *IntechOpen.*

Amini, L., Chekini, R., Nateghi, M. R., Haghani, H., Jamialahmadi, T., Sathyapalan, T., & Sahebkar, A. (2021). The effect of combined vitamin C and vitamin E supplementation on oxidative stress markers in women with endometriosis: a randomized, triple-blind placebo-controlled clinical trial. *Pain Research and Management,* 2021(1), 5529741.

Andrés, C. M. C., Pérez de la Lastra, J. M., Juan, C. A., Plou, F. J., & Pérez-Lebeña, E. (2024). Antioxidant metabolism pathways in vitamins, polyphenols, and selenium: parallels and divergences. *International Journal of Molecular Sciences*, 25(5), 2600.

Aslan, R., Şekeroğlu, M. R., & Bayıroğlu, F. (1995). Serbest radikal türlerinin membran lipid peroksidasyonuna etkileri ve hücresel antioksidan savunma. *Van Sağlık Bilimleri Dergisi,* 1(2), 137-142.

Avery, J. C., & Hoffmann, P. R. (2018). Selenium, selenoproteins, and immunity. Nutrients, 10(9), 1203.

Bayat, G. and Yıldız, G. (2019). The special Fermented Turkish Drink: Boza. *Journal of Tourism and Gastronomy Studies,* 7(4): 2438 - 2446.

Bernatoniene, J., & Kopustinskiene, D. M. (2018). The role of catechins in cellular responses to oxidative stress. Molecules, 23(4), 965.

Beydoun, M. A., Fanelli-Kuczmarski, M. T., Kitner-Triolo, M. H., Beydoun, H. A., Kaufman, J. S., Mason, M. A., Evans, M. K., & Zonderman, A. B. (2015). Dietary antioxidant intake and its association with cognitive function in an ethnically diverse sample of US adults. *Psychosomatic medicine,* 77(1), 68–82.

Blaner, W. S., Shmarakov, I. O., & Traber, M. G. (2021). Vitamin A and vitamin E: will the real antioxidant please stand up?. *Annual review of nutrition,* 41(1), 105-131.

Boonthongkaew, C., Tong-Un, T., Kanpetta, Y., Chaungchot, N., Leelayuwat, C., & Leelayuwat, N. (2021). Vitamin C supplementation improves blood pressure and oxidative stress after acute exercise in patients with poorly controlled type 2 diabetes mellitus: A randomized, placebo-controlled, cross-over study. *Journal of Physiological Investigation,* 64(1), 16-23.

Burnside, W. S., & Snowden, C. (2014). Physiological basis of preoperative cardiopulmonary exercise testing. *Surgery (Oxford),* 32(2), 59-62.

Cao, J., Han, J., Xiao, H., Qiao, J., & Han, M. (2016). Effect of tea polyphenol compounds on anticancer drugs in terms of anti-tumor activity, toxicology, and pharmacokinetics. *Nutrients,* 8(12), 762.

Chen, X. Q., Hu, T., Han, Y., Huang, W., Yuan, H. B., Zhang, Y. T.,... & Jiang, Y. W. (2016). Preventive effects of catechins on cardiovascular disease. *Molecules*, 21(12), 1759.

Chrysikopoulou, V., Rampaouni, A., Koutsia, E., Ofrydopoulou, A., Mittas, N., & Tsoupras, A. (2025). Anti-Inflammatory, Antithrombotic and Antioxidant Efficacy and Synergy of a High-Dose Vitamin C Supplement Enriched with a Low Dose of Bioflavonoids; In Vitro Assessment and In Vivo Evaluation Through a Clinical Study in Healthy Subjects. *Nutrients,* 17(16), 2643.

Costa, M. I., Sarmento-Ribeiro, A. B., & Gonçalves, A. C. (2023). Zinc: from biological functions to therapeutic potential. *International Journal of Molecular Sciences,* 24(5), 4822.

Doseděl, M., Jirkovský, E., Macáková, K., Krčmová, L. K., Javorská, L., Pourová, J., ... & Oemonom. (2021). Vitamin C—sources, physiological role, kinetics, deficiency, use, toxicity, and determination. *Nutrients, 13*(2), 615.

Duan, M., Li, T., Liu, B., Yin, S., Zang, J., Lv, C.,... & Zhang, T. (2023). Zinc nutrition and dietary zinc supplements. *Critical Reviews in Food Science and Nutrition,* 63(9), 1277-1292.

Fekete, M., Lehoczki, A., Tarantini, S., Fazekas-Pongor, V., Csípő, T., Csizmadia, Z., & Varga, J. T. (2023). Improving cognitive function with nutritional supplements in aging: a comprehensive narrative review of clinical studies investigating the effects of vitamins, minerals, antioxidants, and other dietary supplements. *Nutrients,* 15(24), 5116.

Finley, J. W., Kong, A. N., Hintze, K. J., Jeffery, E. H., Ji, L. L., & Lei, X. G. (2011). Antioxidants in foods: state of the science important to the food industry. *Journal of agricultural and food chemistry,* 59(13), 6837-6846.

Gostner, J. M., Becker, K., Ueberall, F., & Fuchs, D. (2015). The good and bad of antioxidant foods: an immunological perspective. *Food and Chemical Toxicology,* 80, 72-79.

Grodstein, F., Kang, J. H., Glynn, R. J., Cook, N. R., & Gaziano, J. M. (2007). A randomized trial of beta carotene supplementation and cognitive function in men: the Physicians' Health Study *II. Archives of internal medicine,* 167(20), 2184-2190.

Gulcin, İ. (2020). Antioxidants and antioxidant methods: An updated overview. *Archives of toxicology,* 94(3), 651-715.

Guo, Q., Li, F., Duan, Y., Wen, C., Wang, W., Zhang, L.,... & Yin, Y. (2020). Oxidative stress, nutritional antioxidants and beyond. *Science China Life Sciences,* 63(6), 866-874.

Güleşci, N., & Aygül, İ. (2016). Beslenmede yer alan antioksidan ve fenolik madde içerikli çerezler. *Gümüşhane Üniversitesi Sağlık Bilimleri Dergisi,* 5(1), 109-129.

Halliwell, B. (2024). Understanding mechanisms of antioxidant action in health and disease. *Nature Reviews Molecular Cell Biology,* 25(1), 13-33.

Huang, J., Xie, L., Song, A., & Zhang, C. (2022). Selenium status and its antioxidant role in metabolic diseases. *Oxidative medicine and cellular longevity,* 2022(1), 7009863.

Huang, S., Li, J., Wu, Y., Ranjbar, S., Xing, A., Zhao, H.,... & Gao, X. (2018). Tea consumption and longitudinal change in high-density lipoprotein cholesterol concentration in Chinese adults. *Journal of the American Heart Association,* 7(13), e008814.

Huwiler, V. V., Maissen-Abgottspon, S., Stanga, Z., Mühlebach, S., Trepp, R., Bally, L., & Bano, A. (2024). Selenium supplementation in patients with Hashimoto thyroiditis: a systematic review and meta-analysis of randomized clinical trials. *Thyroid,* 34(3), 295-313.

Jovanovic, S. V., & Simic, M. G. (2000). Antioxidants in nutrition. *Annals of the New York Academy of Sciences*, 899(1), 326-334.

Khan, U. M., Sevindik, M., Zarrabi, A., Nami, M., Ozdemir, B., Kaplan, D. N.,... & Sharifi-Rad, J. (2021). Lycopene: food sources, biological activities, and human health benefits. *Oxidative medicine and cellular longevity*, 2021(1), 2713511.

Kishimoto, Y., Taguchi, C., Iwashima, T., Kobayashi, T., Kikoku, Y., Nishiyama, H.,... & Kondo, K. (2023). Effects of acute strawberry consumption on serum levels of vitamin C and folic acid, the antioxidant potential of LDL and blood glucose response: a randomised cross-over controlled trial. *Journal of Nutritional Science,* 12, e39.

Koltover, V. K. (2010). Antioxidant biomedicine: from free radical chemistry to systems biology mechanisms. *Russian Chemical Bulletin,* 59(1), 37-42.

Kryscio, R. J., Abner, E. L., Caban-Holt, A., Lovell, M., Goodman, P., Darke, A. K., Yee, M., Crowley, J., & Schmitt, F. A. (2017). Association of Antioxidant Supplement Use and Dementia in the Prevention of Alzheimer's Disease by Vitamin E and Selenium Trial (PREADViSE). *JAMA neurology,* 74(5), 567–573.

Larsson, S. C., Mason, A. M., Vithayathil, M., Carter, P., Kar, S., Zheng, J. S., & Burgess, S. (2022). Circulating vitamin C and digestive system cancers: Mendelian randomization study. *Clinical nutrition* (Edinburgh, Scotland), 41(9), 2031–2035.

Lee, D. H., Folsom, A. R., Harnack, L., Halliwell, B., & Jacobs Jr, D. R. (2004). Does supplemental vitamin C increase cardiovascular disease risk in women with diabetes?. *The American journal of clinical nutrition,* 80(5), 1194-1200.

Lee, I. M., Cook, N. R., Gaziano, J. M., Gordon, D., Ridker, P. M., Manson, J. E., Hennekens, C. H., & Buring, J. E. (2005). Vitamin E in the primary prevention of cardiovascular disease and cancer: the Women's Health Study: a randomized controlled trial. *JAMA,* 294(1), 56–65.

Li, N., Zhao, Y., & Liang, Y. (2013). Cardioprotective effects of tea and its catechins. *Health,* 5(4), 23-30.

Lu, W., Shi, Y., Wang, R., Su, D., Tang, M., Liu, Y., & Li, Z. (2021). Antioxidant activity and healthy benefits of natural pigments in fruits: A review. International journal of molecular sciences, 22(9), 4945.

Maurya, P. K. (2022). Health benefits of quercetin in age-related diseases. *Molecules,* 27(8), 2498.

Mehta, P. A., Nelson, A., Loveless, S., Lane, A., Fukuda, T., Teusink-Cross, A.,... & Davies, S. M. (2025). Phase 1 study of quercetin, a natural antioxidant for children and young adults with Fanconi anemia. *Blood Advances,* 9(8), 1927-1939.

Meral, R., Doğan, İ. S., & Kanberoğlu, G. S. (2012). Fonksiyonel gıda bileşeni olarak antioksidanlar. *Journal of the Institute of Science and Technology,* 2(2), 45-50.

Milman, U., Blum, S., Shapira, C., Aronson, D., Miller-Lotan, R., Anbinder, Y., Alshiek, J., Bennett, L., Kostenko, M., Landau, M., Keidar, S., Levy, Y., Khemlin, A., Radan, A., & Levy, A. P. (2008). Vitamin E supplementation reduces cardiovascular events in a subgroup of middle-aged individuals with both type 2 diabetes mellitus and the haptoglobin 2-2 genotype: a prospective double-blinded clinical trial. *Arteriosclerosis, thrombosis, and vascular biology*, 28(2), 341–347.

Montoya-Estrada, A., García-Cortés, A. Y., Romo-Yañez, J., Ortiz-Luna, G. F., Arellano-Eguiluz, A., Belmont-Gómez, A.,... & Reyes-Muñoz, E. (2024). The Administration of Resveratrol and Vitamin C Reduces Oxidative Stress in Postmenopausal Women—A Pilot Randomized Clinical Trial. *Nutrients,* 16(21), 3775.

Moran, N. E., Thomas-Ahner, J. M., Wan, L., Zuniga, K. E., Erdman, J. W., & Clinton, S. K. (2022). Tomatoes, Lycopene, and Prostate Cancer: What Have We Learned from Experimental Models?. *The Journal of nutrition*, 152(6), 1381–1403.

Niki, E., & Noguchi, N. (2004). Dynamics of antioxidant action of vitamin E. *Accounts of chemical research,* 37(1), 45-51.

Oyama, J. I., Shiraki, A., Nishikido, T., Maeda, T., Komoda, H., Shimizu, T.,... & Node, K. (2017). EGCG, a green tea catechin, attenuates the progression of heart failure induced by the heart/muscle-specific deletion of MnSOD in mice. *Journal of Cardiology,* 69(2), 417-427.

Padayatty, S. J., Katz, A., Wang, Y., Eck, P., Kwon, O., Lee, J. H.,... & Levine, M. (2003). Vitamin C as an antioxidant: evaluation of its role in disease prevention. *Journal of the American college of Nutrition,* 22(1), 18-35.

Pang, J., Zhang, Z., Zheng, T. Z., Bassig, B. A., Mao, C., Liu, X. & Peng, Y. (2016). Green tea consumption and risk of cardiovascular and ischemic related diseases: A meta-analysis. *International journal of cardiology*, 202, 967-974.

Parcheta, M., Świsłocka, R., Orzechowska, S., Akimowicz, M., Choińska, R., & Lewandowski, W. (2021). Recent developments in effective antioxidants: The structure and antioxidant properties. *Materials,* 14(8), 1984.

Petterson, S., Kuchta, C., & Snyder-Mackler, L. (2006). Aerobic metabolism during exercise. *Sports-Specific Rehabilitation,* 65.

Powell, S. R. (2000). The antioxidant properties of zinc. *The Journal of nutrition,* 130(5), 1447-1454.

Prasad, A. S. (2014). Zinc: an antioxidant and anti-inflammatory agent: role of zinc in degenerative disorders of aging. *Journal of Trace Elements in Medicine and Biology,* 28(4), 364-371.

Prasad, A. S., Bao, B., Beck, F. W., Kucuk, O., & Sarkar, F. H. (2004). Antioxidant effect of zinc in humans. *Free Radical Biology and Medicine,* 37(8), 1182-1190.

Putthong, C., Panmanee, T., Charoensit, P., Ross, S., Tongpoolsomjit, K., & Viyoch, J. (2024). Efficacy of Natural β-Carotene Chewable Tablets Derived from Banana (Musa AA) Pulp in Reducing UV-Induced Skin Erythema. *Nutrients,* 17(1), 65.

Santos-Sánchez, N. F., Salas-Coronado, R., Villanueva-Cañongo, C., & Hernández-Carlos, B. (2019). Antioxidant compounds and their antioxidant mechanism. *IntechOpen.*

Sarıca, Ş., & Diktaş, Ü. K. M. (2008). Çay (Camellia sinensis); içeriği, metabolizma ve sağlık üzerine etkileri, antioksidan aktivitesi ve etlik piliç karma yemlerinde kullanımı. *Journal of Agricultural Faculty of Gaziosmanpaşa University (JAFAG),* 2008(2), 79-85.

Shohan, M., Nashibi, R., Mahmoudian-Sani, M. R., Abolnezhadian, F., Ghafourian, M., Alavi, S. M., Sharhani, A., & Khodadadi, A. (2022). The therapeutic efficacy of quercetin in combination with antiviral drugs in hospitalized COVID-19 patients: A randomized controlled trial. *European journal of pharmacology*, 914, 174615.

Sim, M., Hong, S., Jung, S., Kim, J. S., Goo, Y. T., Chun, W. Y., & Shin, D. M. (2022). Vitamin C supplementation promotes mental vitality in healthy young adults: Results from a cross-sectional analysis and a randomized, double-blind, placebo-controlled trial. *European journal of nutrition,* 61(1), 447-459.

Szewczyk, K., Bryś, J., Brzezińska, R., & Górnicka, M. (2025). Nutritional Status of Vitamin E and Its Association with Metabolic Health in Adults. *Nutrients,* 17(3), 408.

Takagi, T., Hayashi, R., Nakai, Y., Okada, S., Miyashita, R., Yamada, M.,... & Itoh, Y. (2020). Dietary intake of carotenoid-rich vegetables reduces visceral adiposity in obese Japanese men—A randomized, double-blind trial. *Nutrients,* 12(8), 2342.

Traber, M. G., & Atkinson, J. (2007). Vitamin E, antioxidant and nothing more. *Free radical biology and medicine,* 43(1), 4-15.

Ungurianu, A., Zanfirescu, A., Nițulescu, G., & Margină, D. (2021). Vitamin E beyond its antioxidant label. *Antioxidants,* 10(5), 634.

Uz, B., İnce, Ö., Gümüş, C., Gökosmanoğlu, F., Özgür, E. G., & Bekiroğlu, G. N. (2025). Role of intravenous vitamin C on outcomes in hospitalized patients with moderate or severe COVID-19: a real life data of Turkish patients. *Inflammopharmacology,* 33(2), 833-843.

Wang, Y., Han, L., Ling, S., Sha, Y., & Sun, H. (2024). Dietary intake of potassium, vitamin E, and vitamin C emerges as the most significant predictors of cardiovascular disease risk in adults. *Medicine,* 103(32), e39180.

Wang, Z. X., Ma, J., Li, X. Y., Wu, Y., Shi, H., Chen, Y., Lu, G., Shen, H. M., Lu, G. D., & Zhou, J. (2021). Quercetin induces p53-independent cancer cell death through lysosome activation by the transcription factor EB and Reactive Oxygen Species-dependent ferroptosis. *British journal of pharmacology,* 178(5), 1133–1148.

Yang, C. S., Wang, H., & Sheridan, Z. P. (2018). Studies on prevention of obesity, metabolic syndrome, diabetes, cardiovascular diseases and cancer by tea. *journal of food and drug analysis,* 26(1), 1-13.

Yıldız, G. (2018). Physicochemical properties of soy protein concentrate treated with ultrasound at various amplitudes. *Journal of the Institute of Science and Technology,* 8(4): 133-139.

Yıldız, G. (2019). Control of enzymatic browning in potato with calcium chloride and ascorbic acid coatings. *Food and Health*, 5(2): 121-127.

Yıldız, G. (2022). Changes in Emulsifying Properties, Droplet Size, Turbidity and Lipid Oxidation of Pea Protein Nanoemulsions Exposed to High-Intensity Ultrasound and High-Pressure Homogenization during Storage. *Akademik Gıda,* 20 (4), 336-342.

Yıldız, G., & Yildiz, G. (2025). Insights into ultrasound-assisted postharvest preservation: Metabolic, structural, and quality changes in strawberries during storage. *Food and Bioprocess Technology,* 18: 7899–7914.

Yildiz, G., Ding, J., Andrade, J., Engeseth, N.J., and Feng, H. (2018). Effect of plant protein-polysaccharide complexes produced by mano-thermo-sonication and pH-shifting on the structure and stability of oil-in-water emulsions. *Innovative Food Science and Emerging Technologies,* 47: 317-325.

Yildiz, G., and Yıldız, G. (2023). A new approach to enhance quinoa protein nanoaggregates: Combined pH shifting – high pressure homogenization, *Food Chemistry*, 415: 135800.

Yildiz, G. (2024). The efficacy of power ultrasound treatment in preserving quality of fresh-cut cucumbers, *Latin American Applied Research Journal,* 54(4): 485–490.

Yildiz, G., Gao, Y., Ding, J., Zhu, S., Chen, G., and Feng, H. (2025). Enhancing physicochemical, bioactive, and nutritional properties of sweet potatoes: Ultrasonic contact drying with slot jet nozzles compared to hot-air drying and freeze drying, *Ultrasonics Sonochemistry, 112*: 107216.

Yildiz, G. (2025). Effects of high-intensity ultrasound, high-pressure processing, and their combination with pH-shifting on the techno-functionality and digestibility of melon seed protein isolate. *Food Research International, 208*: 116219.

Yildiz, G. (2025). From seed to solution: Enhancing techno-functionality and digestibility of pumpkin seed protein isolate through high-intensity ultrasound, high-pressure processing, and pH-shifting, *Food Chemistry, 474*: 143222.

Zia, A., Farkhondeh, T., Pourbagher-Shahri, A. M., & Samarghandian, S. (2021). The role of curcumin in aging and senescence: Molecular mechanisms. *Biomedicine & pharmacotherapy = Biomedecine & pharmacotherapie*, 134, 111119.

Website 1: https://nutritionsource.hsph.harvard.edu/antioxidants/ Access Date: 10.04.2025.

Website 2: https://ods.od.nih.gov/factsheets/VitaminA-HealthProfessional/ Access Date: 11.04.2025.

Chapter 5

The Healing Power of Food

Elif Feyza Topdas*
Department of Food Engineering, Faculty of Agriculture, Ataturk University, Erzurum, Turkiye

Abstract

Today, nutrition research seeks answers not only to the question of "What should we eat?" but also to multidimensional questions such as "How should public health be supported?" and "How should quality of life be increased?" through the foods we consume. Natural compounds derived from plants are at the root of the pharmacological knowledge that underpins modern medicine, and thanks to their bioactive properties, they offer significant opportunities to prevent chronic diseases and maintain health. Plant-based medicines have been used to treat diseases in various cultures for hundreds of years, and their health effects are supported by modern scientific studies. Spices have become an important component of functional foods, exhibiting various effects such as anticancer, antimicrobial, antioxidant, hepatoprotective, antidiabetic, and antidepressant, beyond their use as flavor enhancers from past to present, thanks to their high levels of bioactive components such as flavonoids, phenolic compounds, sulfur-containing compounds, tannins, alkaloids, phenolic diterpenes, and vitamins. Fermented foods such as yogurt, kefir, kombucha, cheese, sauerkraut, and kimchi, which are among the trending foods of recent times and are estimated to be more than 5000 in number, support digestive system health and contribute to the strengthening of the immune system through the probiotic microorganisms and bioavailable metabolites they contain. This section discusses the effects of plant-based ingredients, spices, fermented foods, and traditional medicines on health in light of scientific findings.

Keywords: food, bioactive compounds, fermented foods, health, traditional medicine

Introduction

Food is a physiological need and a multifaceted, vital element that shapes an individual's health, identity, cultural affiliation, social relations, and values. Feeding, sheltering, and maintaining a healthy life for the world's population, which is expected to reach 9.7 billion by

* Corresponding Author's Email: efeyza.topdas@atauni.edu.tr

In: Let Food Be Your Medicine, and Medicine Be Your Food
Editors: Gulcin Yildiz and Arzu Imece
ISBN: 979-8-90134-204-6

2050, and taking precautions against various diseases affecting the masses, not only on a human level but also on the necessity of sustainable food production and environmental protection. Human, natural, and animal health are inseparable. To ensure this integrity, interdisciplinary teams come together across subjects such as food engineering (especially food chemistry and microbiology), environmental engineering, ecology, natural resource management, climate sciences, agricultural sciences, public health, epidemiology, toxicology, and environment and animal health.

The primary purpose of nutrition research is to generate scientific data to maintain public health, improve quality of life, and prevent or delay the onset of disease. These studies reveal the role of nutrition in the development of widespread, chronic health problems and provide a scientifically based assessment of the effects of nutritional habits on disease risk factors. When considered comprehensively at the societal level, it becomes clear that healthy living practices developed at the individual level play an important role in shaping community-based preventive health policies.

Food has a very comprehensive role in human life. Today, the stress-loaded lifestyle brought about by modernization and inadequate and unbalanced eating habits, unfortunately, has adverse effects on human health and causes chronic diseases (such as obesity, type 2 diabetes, hypertension, cardiovascular diseases, osteoporosis, fatty liver, and cancer), which can result in death. In recent years, an increasing number of consumers have considered health-related factors as the primary factor in their purchasing decisions to prevent lifestyle-related diseases and improve health (Oliviero et al., 2018). This situation particularly affects manufacturers, who try to increase sales by emphasizing the health-related features of their products through product labels.

Biologically active components from natural sources, such as plants, are increasingly important in protecting human health and preventing diseases. Plant-based components offer potential contributions to the prevention and treatment of many chronic diseases thanks to their antioxidant, anti-inflammatory, immune system-regulating, neuroprotective, and antimicrobial properties. In particular, beyond their use as flavor enhancers from past to present, spices offer various health benefits thanks to their bioactive components, such as polyphenols, alkaloids, and volatile oils, and, as a result, become valuable candidates for functional foods and pharmaceutical applications. In addition, consuming fermented foods plays an important role in supporting digestive health and balancing the intestinal microbiota thanks to the probiotic microorganisms and bioavailable metabolites formed during fermentation. The pharmacological effects of some plants and bioactive compounds derived from them, which have been traditionally used in various cultures for centuries, have been confirmed by modern science, and various in vitro, animal, and clinical studies are ongoing on their contributions to health. This section discusses the potential effects of plant-based ingredients, spices, fermented foods, and traditional medicines on human health, based on scientific data.

Medicinal Use of Plant-Based Compounds

Natural compounds obtained from plants are at the root of the pharmacological knowledge that forms the basis of modern medicine. The adventure of developing drugs from plant sources, which began with the extraction of morphine from opium by the pharmacist Sertürner (1817)

(Chast, 2008), has enabled the identification of the active compounds of many powerful drugs used in the treatment of various diseases to this day. Over time, ongoing studies have contributed to the field of medicinal chemistry, sometimes with coincidental discoveries, sometimes with wrong hypotheses, or under the influence of impurities. Thus, between 1820 and 1850, the purification of alkaloids such as emetine, quinine, colchicine, sparteine, caffeine, atropine, codeine, and papaverine was achieved (Sneader, 1997). Purifying active ingredients from raw materials (although the chemical structures of these ingredients remain unclear) paved the way for the development of the ''single target-single drug-single disease'' approach, the basis for modern drug development. The use of fungi for penicillin production during World War II further encouraged the development of drug production processes (Sneader, 1997). Until the mid-1960s, only a few large classes of molecules (e.g., sulfonamides, benzodiazepines, and phenothiazines) were developed without any relation to natural products. However, studies have shown that sulfonamides are biologically similar to *p*-aminobenzoic acid, which plays a role in folic acid metabolism (Sneader, 1997). Benzodiazepines share a similar chemical structure (Klotz, 1991). The successes achieved in this field have led to the emergence of the concept of rational drug design in the following periods. Thus, drug development processes have been further accelerated by advances in compound synthesis (combinatorial chemistry), high-throughput screening, and target identification (genomics). While all these developments have occurred, natural products have remained outside these systematic approaches due to their limited resources, complex purification processes, and often having multiple biological effects. However, many plant-derived drugs or lead compounds, and their derivatives, continue to be developed. In modern medicine, striking examples of drugs developed to benefit from the therapeutic effects of natural compounds of plant origin are paclitaxel and its derivatives obtained from the yew tree (*Taxus* spp.), vincristine and vinblastine isolated from the Madagascar periwinkle [*Catharanthus roseus* (L.) G. Don] plant, and camptothecin and similar compounds obtained from the Chinese tree *Camptotheca acuminata* Decne (Cragg & Newman, 2013; Kinghorn et al., 2011).

Effective bioactive compounds, defined as secondary metabolites, can be produced by plants, animals, and microorganisms. It is easier to isolate secondary metabolites from plants because animals produce these compounds only when needed and store them in particular organs. On the contrary, microorganisms secrete their products into the external environment. Plants produce secondary metabolites as a defense strategy against damage caused by living organisms such as bacteria, fungi, herbivores, insects, nematodes, oomycetes, viruses, and weeds, in addition to conditions such as drought, salinity, flooding, heavy metals, and extreme temperatures (Fich et al., 2016; Verma et al., 2013; Umar et al., 2021). In short, they are metabolites not required for the plant's natural life cycle and are produced to adapt to competitive environmental conditions (Teoh & Teoh, 2016). They usually store these compounds in the roots, leaves, and fruits. Only 10% of the approximately 250 thousand plant species worldwide have been identified and chemically studied (Hawksworth, 2004). Therefore, it would not be wrong to describe plants as the primary sources of secondary metabolites.

To date, over two million secondary metabolites produced by plants have been reported. Secondary metabolites can be classified into three main groups: terpenoids, phenolics, and nitrogen-containing compounds. Different biosynthetic pathways produce other secondary metabolites and contribute to plant defense by inhibiting herbivore development (Chen, 2008).

Terpenoids

Terpenoids have the broadest variety among plant secondary metabolites and are found in almost all plant species. More than 22,000 compounds in this group have been identified (Tiku, 2018). In the synthesis of terpenes, isopentane units containing five carbon atoms serve as basic building blocks. These molecules are divided into various subclasses, such as monoterpenes, diterpenes, triterpenes, and sesquiterpenes, according to the number of isoprene units they contain (Tiku, 2018; Garas & Waiss, 1986; Kortbeek et al., 2019). Their biosynthesis occurs in plastids and the cytosol via the pyruvate-based 2-C-methyl-D-erythritol-4-phosphate pathway and the acetyl-CoA-based mevalonate pathway (Divekar et al., 2022). Some terpenes, such as menthol, camphor, pyrethrins, artemisinins, and farnesol, have protective properties against protozoa, bacteria, and fungi (Tiku, 2018). Some terpene derivatives, such as sterols, provide structural support by interacting with phospholipids to maintain the integrity of the cell membrane (Price et al., 1989). Carotenoids in the tetraterpenes group are precursor compounds of the abscisic acid hormone that regulate stress responses and developmental processes. In addition to serving as auxiliary pigments in photosynthesis, they protect photosynthetic tissues against photo-oxidation.

Sulfur-Containing Secondary Metabolites

Sulfur is a vital element for plants and all living things, and it is the fourth basic nutrient after nitrogen, potassium, and phosphorus. It plays an important role in the biosynthesis of amino acids, coenzymes, antioxidants, and secondary metabolites (Fike, 2015; Abadie & Tcherkez, 2019; Prasad & Shivay, 2018). Sulfate ion (SO_4^{2-}), which plants take from the soil through their roots, reacts with ATP and turns into adenosine 5′-phosphosulfate and then participates in the synthesis of important metabolites and sulfur-containing amino acids through sulfation or reduction (Na & Salt, 2011; Prasad & Shivay, 2018; Kopriva et al., 2019). This process plays a role in plants' redox balance and metabolic regulation by forming molecules such as glutathione and S-adenosylmethionine. Sulfur-containing secondary metabolites from plants include low-molecular-weight compounds with antimicrobial activity, such as camalexin, rapalexin A, glucosinolates, isothiocyanates, and allicin. Their effectiveness has been demonstrated in both laboratory and living systems (Künstler et al., 2020). Among the sulfolipids, especially sulfoquinovosyl diacylglycerol, is one of the main glycerolipids in the structure of chloroplast thylakoid membranes and has a critical biosynthetic role in the synthesis of the cell membrane (Shimojima, 2011). Some vegetable families, especially *Brassicaceae*, *Liliaceae*, *Capparaceae*, and *Caricaceae*, are rich in sulfur-containing compounds. Species such as broccoli, cabbage, and cauliflower attract attention for their immune-boosting glucosinolates. *Allium* species, such as garlic and onion, attract attention for their organosulfur compounds, including allicin, which may affect cardiovascular disease and cancer treatment (Aarabi et al., 2020; Miekus et al., 2020). The biological effects of these compounds result from the disulfide bonds in their structures. These bonds increase the antimicrobial and anticancer properties of thiosulfinides while also helping maintain intracellular redox balance (Abdalla & Mühling, 2019).

Phenolic Compounds

Phenolic compounds are a large group of at least 10,000 different compounds containing one or more aromatic hydrocarbon rings and one or more hydroxyl groups (González-Vallinas et al., 2013; Li et al., 2014). They are generally divided into two broad subgroups: flavonoids and phenolic acids. Until recently, phenolic compounds, which were considered unnecessary and lacked nutritional value, have been effective in the treatment of various diseases, including cardiovascular, neurodegenerative, and cancer, through their antioxidant, anti-inflammatory, and antimicrobial activities (Brglez Mojzer et al., 2016).

The effects of phenolic compounds on various diseases are mainly due to their antioxidant activity (Wu et al., 2016). Polyphenols act as antioxidants by neutralizing free radicals by donating hydrogen ions or electrons, thereby preventing the oxidation of compounds such as fatty acids. At the same time, phenolic compounds alter the redox potential of the environment by chelating transition metals, such as Fe^{2+} and Cu^{+} (Balasundram et al., 2006). It is reported that, thanks to their antioxidant activities, polyphenols not only prevent various diseases but also suppress their progression and even contribute to healing (Ramos 2008; González-Vallinas et al., 2013). In addition, some polyphenols have been reported to have hormonal effects and help prevent osteoporosis (Hendric, 2006). Phenolic compounds can donate a hydrogen atom to a free radical through their phenolic hydroxyl groups. Thus, the unpaired free radical electron can be delocalised by the conjugated aromatic systems of phenolic compounds. As a result of donating a hydrogen atom to the free radical by phenolic compounds, phenoxyl radical intermediates are formed. Thanks to the high stability of these intermediates, the free radical chain reaction can be suppressed. According to Dai and Mumper (2010), phenoxyl radical intermediates react with other free radicals that have formed, thereby halting their propagation.

Phenolic acids, which represent one-third of phenolic compounds (Yang et al., 2001), are an important class of polyphenols, mostly hydroxycinnamic acids, but also hydroxybenzoic acids (Cadenas & Packer, 2002). This group, called phenylpropanoids, is generally found bound to proteins or cell wall polymers as glycosides or organic acid esters (Yang et al., 2001). Phenolic acids are important compounds that play roles in vital processes such as growth, development, and pigment formation, as well as in defense mechanisms against stress conditions (Lattanzio et al., 2009; 2012). They contribute to the sensory properties of plant foods, such as taste and flavor, and attract attention through various biological effects that support human health. As chelators, phenolic acids are important free radical scavengers that act on hydroxyl and peroxyl radicals, superoxide anions, and peroxynitrites (Lall et al., 2015). Their antioxidant activities vary depending on the number of hydroxyl groups of the aromatic rings they contain and the binding sites and positions of these groups (Sroka & Cisowski, 2003). Scientific studies have reported that some compounds in this group undertake functions such as reducing oxidative stress, suppressing inflammation, providing protection against microbial agents, regulating blood sugar, and preventing cancer development (Kiokias & Oreopoulou, 2021; Kumar & Goel, 2019; Ruwizhi & Aderibigbe, 2020; Sova & Saso, 2020). Caffeic, sinapic, syringic, protocatechuic, ferulic, and 3,4-dihydroxyphenyl acetic acids found in olive oil reduce the proliferation of breast and prostate cancer cells, depending on time and dose (Kampa et al., 2007). It has been determined that dietary levels of 0.02% to 0.05% of caffeic acid, ellagic acid, chlorogenic acid, and ferulic acid prevent cancer formation in rats (Yang et al., 2001). On the other hand, it has been reported that gallic acid, tannic acid, and caffeic acid can induce oxidative DNA strand breaks in vitro (Dai & Mumper, 2010). In their studies,

Marinova and Yanishlieva (2003) determined that hydroxycinnamic acid derivatives showed higher antioxidant activity than hydroxybenzoic acid derivatives.

Hydroxybenzoic acids are compounds with potent antioxidative and anticarcinogenic activity in the C6-C1 phenylpropane structure (Cemeroğlu et al., 2011). This group of acids is a component of complex structures such as lignin (especially *p*-hydroxybenzoic, vanillic, syringic, and protocatechuic acids) and hydrolysable tannins (Brglez Mojzer et al., 2016). They can also be found as organic acids or sugar derivatives. Gallic acid esterifies with glucose and some other sugar alcohols to form various galloyl esters, precursors of hydrolysable tannins, especially gallotannins. One of the most basic examples of these structures is pentagalloylglucose. Gallotannins are transformed into more complex structures by attaching additional galloyl groups, while ellagitannins contain hexahydroxydiphenoyl units formed by oxidative C-C bonds between galloyl groups. Ellagic acid is formed by the hydrolysis of ellagitannins in acidic or basic environments, which explains the fundamental structural transformation that gives this group its name (Gross, 2008). Their limited distribution in foods results in limited information about the absorption and metabolism of hydroxybenzoic acids (Manach et al., 2005). The most important representatives of hydroxybenzoic acids are gallic and ellagic acids, abundant in beans and peanuts (Stoner & Mukhtar, 1995; Lall et al., 2015).

Flavonoids are polyphenols with a C6-C3-C6 skeleton, which is mainly composed of the combination of two phenyl rings with a propane chain and is divided into 12 subgroups, including flavanones, chalcones, stilbenes, aurones, isoflavones, dihydroflavonols, flavonols, phlobaphenes, leucoanthocyanidins, proanthocyanidins, and anthocyanins (Panche et al., 2016). Flavonoids provide plants with a wide range of colors and play an important role in their defense systems. More than 9000 different flavonoids have been identified in plants to date, and these compounds are among the primary pigments that give plants color. Anthocyanins provide shades ranging from orange to blue, while aurones and chalcones provide yellow tones. They function as protective phytoalexins against UV light, pathogen attack, low temperature, and insect damage. They can also protect cells from oxidative damage by scavenging reactive oxygen species (Liu et al., 2021). Flavonoids have various biochemical and antioxidant effects that are beneficial in treating various diseases such as cancer, Alzheimer's disease, and atherosclerosis (Castañeda-Ovando et al., 2009; Lee et al., 2009). Flavonoids have become indispensable components for a range of nutraceutical, pharmaceutical, medical, and cosmetic applications due to their health benefits. They are effective in health due to their antioxidant, anti-inflammatory, antimutagenic, and anticarcinogenic properties, as well as their capacity to regulate key cellular enzyme functions (Panche et al., 2016). Flavonoids are also potent inhibitors of many enzymes, such as xanthine oxidase, cyclooxygenase, lipoxygenase, and phosphoinositide 3-kinase (Metodiewa et al., 1997; Walker et al., 2000).

Alkaloids

Alkaloids are a group of compounds with nitrogen-containing ring structures, basic character, bitter taste, odorless, soluble in apolar solvents, mostly crystalline, with about 12 thousand members (Brugnerotto et al., 2021; Cinelli & Jones, 2021). Alkaloids are divided into 14 different subclasses; pyrrolizidine, pyrrole and pyrrolidine, quinoline, pyridine and piperidine, tropane, aporphine, isoquinoline, indolizidine, terpenoid, norlupinane, indole, purine, steroid and imidazole, pyrrole, tropane, indole, quinoline, steroid and imidazole according to whether they are amino acid derivatives or not and the rings they contain (Ahmed et al., 2017).

Alkaloids are primarily found in herbaceous dicotyledonous and monocotyledonous plants, but not in more primitive plants such as mosses and ferns. Alkaloids are found in various parts of plants, usually as salts of organic acids (e.g., oxalic acid, acetic acid, malic acid, lactic acid, tartaric acid, tannic acid, and aconitic acid) (Muthaura et al., 2007). Among these compounds, nicotine has been found in leaves, quinine and cinchonine in bark, nibidine and strychnine in seeds, rawelfininc and punarnavine in roots (Muthaura et al., 2007).

The solubility properties of alkaloids may differ from each other. For example, some alkaloids, pseudoalkaloids, and protoalkaloids are more easily soluble in water than classical alkaloids. These solubility differences between alkaloids and their salts play an important role in the formulation and substance recovery in drug design and manufacturing processes of therapeutic products (Das et al., 2020).

In modern and traditional medicine, quinine is considered antimalarial, ephedrine is antihistamine, vincristine is antitumor, vinblastine is antineoplastic and immunosuppressant, morphine is analgesic, theophylline is diuretic and antiasthmatic, aconitine is anaesthetic and anti-inflammatory, caffeine is analgesic and anorectans, piperine is a blood sugar regulator, and homoharringtonine is an adjunct in cancer treatment (Rios et al., 1989; Chan, 2009; Ahmed et al., 2017; Pereira et al., 2023). Current research suggests that some plant alkaloids have potential antiviral activity against SARS-CoV-2. For example, it has been suggested that thalictrine and sophoridine, isolated from *Thalictrum simplex* and *Sophora alopecuroides*, respectively, can inhibit the replication of this virus (Garg & Roy, 2020). Similarly, alkaloids such as cryptosminrin, cryptospirolepine, and biscriptolepine, obtained from the plant Cryptolepis sanguinolenta, can target SARS-CoV-2 proteins and exert inhibitory effects (Borquaye et al., 2020). On the other hand, while alkaloids have positive effects on human health, their toxic properties should not be ignored. Some compounds can cause harm, especially at high doses or when misused. It is known that psilocin has a hallucinogenic effect; cocaine, caffeine, and nicotine have stimulating effects on the central nervous system; and atropine and tubocurarine have the potential for toxic effects (Chiocchio et al., 2021; Griffiths et al., 2021). Such adverse effects are generally associated with the uncontrolled production of alkaloids, their sale, or their consumption by users without sufficient knowledge (Upadhyay et al., 2012).

Spices: More Than Flavor

Food taste is an important acceptance criterion for being preferred and consumed. Within the scope of the European Salt Network studies initiated by the World Health Organization in 2008, concrete steps have been taken to reduce salt intake at the community level. Foods need to be reformulated to reduce salt, added sugar, and saturated fat, which are included in food formulations and are effective in taste, and to eliminate the taste profile deficiencies caused by reduced ingredient levels. Adding spices to the product formulation as a substitute can offer a feasible approach to increasing taste. Spices are materials obtained from plant parts, including roots, leaves, and seeds, sometimes fresh but usually dried, and used to impart properties such as flavor, aroma, color, and odor to foods (Kaundal et al., 2022).

It would be a great injustice to consider spices as only flavoring components because many scientific studies have proven the multidimensional health benefits of spices, thanks to the various bioactive components they contain, such as flavonoids, phenolic compounds, sulfur-containing compounds, tannins, alkaloids, phenolic diterpenes, and vitamins (Yashin et al., 2017). When we consider that spices are among the 100 foods richest in polyphenols, as determined by the US Department of Agriculture (Pérez-Jiménez et al., 2010), we realize how important they are for our health. This section discusses the introduction and chemical composition of some spices commonly encountered and used today, along with their potential biological activities, as reported in current scientific research.

Saffron (*Crocus sativus* L.) is in the genus *Crocus* (family Iridaceae), which includes about 85 different plant species. It is a perennial herbaceous flowering plant. It is cultivated in Turkiye (Safranbolu), Greece, Pakistan, China, Japan, France, India, Spain, Switzerland, and Italy (Menia et al., 2018). It is also called "red gold" because it is the world's most expensive spice ($40-50 per gram). It is expensive because cultivation is time-consuming, harvesting takes only 2-3 weeks, flowers are hand-collected (Molina et al., 2005), and only 1 kg of dry saffron is obtained from 80 kg of flowers. Iran is the leading country in world saffron production, with an 80% share. Most of the saffron produced worldwide is used in the food industry as a flavoring and coloring agent. Its unique organoleptic properties and beneficial health effects have made saffron a valuable plant, particularly for the functional food industry (Melnyk et al., 2010). Although the chemical composition of saffron varies depending on growing conditions, a significant portion consists of non-nitrogenous extracts, crude protein, water, and minerals. It is a source of different minerals such as phosphorus, magnesium, calcium, iron, potassium, and sodium, and vitamins A, B1, B2, B6, and C (Hashemi & Erim, 2016; Karimi et al., 2010; Lim, 2012; D'Archivio & Maggi, 2017; Jadouali et al., 2019). It contains about 150 different volatile and non-volatile compounds. Terpenoids and terpene alcohol are the dominant volatile components. Safranal is the most prominent compound in this group and is used in therapeutic applications. Picrocrizin, crocetin, and crocin are non-volatile compounds of saffron (Pitsikas, 2016). The predominant fatty acids are palmitic, linoleic, and linolenic (USDA, 2013). The bioactive compounds in saffron have been shown to positively affect diseases such as cancer, Alzheimer's disease, male sexual dysfunction, diabetes, and cardiovascular diseases (Razak et al., 2017). Various animal studies have also shown saffron's protective effects on health, thanks to its antioxidant, antimicrobial, liver-protective, and cancer-fighting properties (Baba et al., 2015; Xu et al., 2009).

Curcuma longa L. (family Zingiberaceae), known as turmeric, is a valuable plant that humans have used for centuries as a spice in foods and for therapeutic purposes. The plant is a perennial with short stems, leaves with sharp tips, and yellow flowers. It is cultivated mainly in Asian countries such as India and China. It is known worldwide as curcuma, Indian saffron, haridra, Jianghuang, Kyoo, or Ukon (Goel et al., 2008). Turmeric, a spice obtained by drying and grinding the root of *C. longa*, is used as a flavoring and coloring agent, especially in Asian cuisine. Flavonoids are the active components of turmeric. The most dominant components in this group are curcumin (diferuloylmethane), monodexmethoxycurcumin, and bisdesmethoxycurcumin (Heath et al., 2004). Curcumin, the most active component of turmeric, accounts for around 90% of these components. In addition, turmeric contains turmerone, germacrone, atlantone, and zingiberene, as well as essential oils, sugars, proteins, and resins. Phenolic groups in the curcumin structure enable it to destroy free radicals such as OH, $^1O_2$, NO_2, and NO (Sreejayan et al., 1997). Various animal experiments and clinical studies have shown that turmeric and curcumin have anti-inflammatory (Shoskes et al., 2006), antioxidant (Gukovsky et al., 2003), hepatoprotective (Park et al., 2000), anticarcinogenic (Garg et al., 2008; Hong et al., 2004), antidiabetic (Wickenberg et al., 2010), antimicrobial (Khattak et al., 2005; Kundu et al., 2005), and antidepressant (Xu et al., 2006) properties. In addition, the beneficial effects of turmeric and curcumin on cardiovascular, neurological, and gastrointestinal disorders have been determined (Lee, 2006; Kim et al., 2005; Kulkarni & Dhir, 2010).

Ginger (*Zingiber officinale* Roscoe) is a plant in the Zingiberaceae family native mainly to East and South Asia (Shahrajabian et al., 2019). It is a perennial herb with bright green, grassy leaves. Ginger has been widely used worldwide as an important culinary spice and condiment for centuries (El Sayed & Moustafa, 2016; Azhari et al., 2017). The rhizome (the underground part) is the main part of the plant used for eating and medicine. This part generally consists of fixed fats (2-12%), carbohydrates (40-70% starch), protein (6-20%), fiber (3-8%), water (9-12%) and essential oil (1.5-3%), as well as coloring matter and minerals (Ali et al., 2008; Chan et al., 2009). USDA (2013) states that raw or dried ground ginger contains calcium, magnesium, iron, phosphorus, potassium, zinc, manganese, and sodium. Ginger is rich in compounds with antioxidant effects such as β-carotene, terpenoids, polyphenols, and rutin (Aruoma et al., 1997). It is reported that the antioxidant effect it exhibits due to its high phenolic and flavonoid content is positively correlated with its strong medicinal potential (Ghasemzadeh, 2010). The positive effects of ginger on health are associated with its bioactive compounds that exhibit biological and pharmaceutical effects. The compounds that give ginger its sharp odor are gingerol compounds (especially [6]-gingerol) found in its rhizome. These compounds are converted into shogaol structures by heat when ginger is cooked or dried (Semwal et al., 2015). In addition, the essential oil containing 1-3% mono and sesquiterpenoid derivatives (camphene, borneol, zingiberene, sesquiphellandrene, and bisabolene) has a significant effect on the odor (Langner et al., 1998; Ali et al., 2008; Chrubasik et al., 2005). There are many studies on the health benefits of ginger, its various extracts, and essential oils. In vitro and in vivo studies have shown that the plant has anti-pregnancy and chemotherapy nausea (Sharma et al., 1997; Chrubasik et al., 2005; Sonkusare, 2008; Ozgoli et al., 2009), anti-coagulant (Young et al., 2006), blood pressure lowering (Ali et al., 2008; Afzal et al., 2001; Ghayur et al., 2005), blood values regulator (blood glucose, cholesterol, triglycerides) (Bhandari & Pillai, 2005), antidiabetic (Islam et al., 2008), anticancer (Kim et al., 2009; Sung et al., 2009), antioxidant (Amin & Hamza, 2006; Ahmed et al., 2008; Mallikarjuna et al., 2008), anti-inflammatory (Ali et al., 2008; Chrubasik et al, 2005; Grzanna et al., 2005; Jung et al., 2009), painkiller (Grzanna et al., 2005; Cortright et al., 2007), and antimicrobial (Chen et al., 2007) effects.

Syzygium aromaticum, commonly known as clove, is a 10-20 m high tree species that remains green throughout the four seasons. The commercially valuable part of the tree is the flower buds. Clove flowers are rich in phenolic compounds, especially flavonoids, hydroxybenzoic acids, hydroxycinnamic acids, and hydroxyphenyl propanes (Shan et al., 2005; Jirovetz et al., 2006; Neveu et al., 2010). Clove flowers are used as a spice and flavor in various foods such as meat, rice, pickles, sauces, desserts, and tea. The most dominant component of the essential oil is eugenol, and it also contains eugenol acetate, humulene, β-caryophyllene, β-pinene, limonene, farnesol, benzaldehyde, 2-heptanone, and ethyl hexanoate (Jirovetz et al., 2006). Clove extracts and eugenol have been reported to show antioxidant activity (Gülçin et al., 2004; Gülçin, 2011). In addition, the broad-spectrum antimicrobial activity of clove against many bacterial and fungal strains has been investigated by various researchers (Dorman &

Deans, 2000; Ali et al., 2005; Sofia et al., 2007; Perez-Conesa et al., 2006; Rana et al., 2011). Kurokawa et al. (1998) reported that clove exhibits antiviral activity against herpesvirus strains. Clove has been known as a painkiller since the 13th century, and this effect has been proven by recent scientific studies (Li et al., 2008; Daniel et al., 2009).

Cinnamon (*Cinnamomum verum*) is a genus of the *Lauraceae* family's aromatic, fragrant, evergreen, 10-15 m high trees. Cinnamon is cultivated in various Asian countries, such as Sri Lanka and South India. Rolled sticks (quills) obtained by peeling the outer bark of the tree, removing the inner bark, and drying them after fermentation by keeping them in a humid environment for a while are used as flavorings, spices, and condiments in various foods (Pittman, 2011). Cinnamon bark essential oil contains cinnamaldehyde, cinnamyl acetate, eugenol (the most dominant compound), β-caryophyllene, linalool, 1.8-cineole, cinnamic acid, various procyanidins and phenolic acids, diterpenes, mucilage polysaccharides and sugars (Hansel et al., 1992; Cardoso-Ugarte et al., 2016). Recent scientific research has provided evidence of cinnamon's potent antioxidant, antimicrobial, antiviral, anti-inflammatory, antifungal, cognitive function supportive, immunomodulatory, antidiabetic, blood glucose regulating, gastric and heart protective, and anticarcinogenic effects (Gruenwald et al, 2010; Naveed et al., 2013; Faghani et al., 2014; Gopinath et al., 2014; Rahemi et al., 2015; Hong et al., 2016; Sharma et al., 2016; Fauvelle et al., 2017; Nematollahi et al., 2020). Li et al. (2013) reported the strong wound-healing effect of 3% ethanolic extract of cinnamon in mice.

Black pepper (*Piper nigrum*), known as the "king of spices" due to its global commercial and economic importance, is a perennial, woody, aromatic plant (Bui et al., 2017). Vietnam is the world's leading producer of black pepper (Hao et al., 2012; Ravindran & Kallupurackal, 2012). Black pepper is mainly preferred as a spice to add flavor to foods and benefit from its preservative effect (Ravindran & Kallupurackal, 2012). Piperine (1-peperonyl piperidine) is black pepper's most well-known chemical compound, and piperine and chavicine are responsible for its sharp taste. The amount of piperine in commercial white and black black pepper can vary between 5% and 10%. In addition, over 592 compounds, including valuable bioactive compounds such as polyphenols, terpenes, lignan derivatives, dihydropipereric acid, steroids, and piperamine, have been reported to be isolated from black pepper (Parmar et al., 1997). The extract of various parts of the plant such as seeds, leaves and fruits, its essential oil and various components have been reported to have antioxidant (Nahak & Sahu, 2011; Bagheri et al., 2014; Shanmugapriya et al., 2012), anticancer (Ee et al., 2009; Sriwiriyajan et al., 2014; Prashant et al., 2017), analgesic (Bukhari et al., 2013; Tasleem et al., 2014), neuroprotective (Hritcu et al., 2015; Yang et al., 2015), hypoglycemic (Onyesife et al., 2015), hypolipidemic (Vijayakumar & Nalini, 2006), and anti-inflammatory (Jeena et al., 2014) effects have been proven.

Fermented Foods and Digestive-Health Effects

Fermented foods, which are estimated to be more than 5000 in number, such as yogurt, kefir, kombucha, cheese, sourdough, sauerkraut, miso, tempeh, natto, kimchi, beer, and wine, which are among the trending foods of recent times (Tamang et al., 2010), have been a part of the human diet throughout history. Dietary patterns passed down from centuries ago to the present day, particularly in traditional Japanese and Mediterranean diets, have influenced the development of fermented foods (Selhub et al., 2014). Fermentation has been used as a low-cost, less energy-intensive way to prepare food for humans since approximately 10,000 BC.

Fermentation in foods occurs in two ways: aerobic and anaerobic. While fungal and alkaline processes dominate aerobic fermentation, alcoholic and lactic acid processes are present in anaerobic fermentation (Wilburn & Ryan, 2017). Marco et al. (2017) describe fermented products as foods and beverages obtained by converting food components into valuable products under controlled enzymatic conditions and microbial activity. The International Scientific Association for Probiotics and Prebiotics (ISAPP) defines fermented foods and beverages more broadly as "foods obtained by processing food components through desired microbial growth and enzymatic transformations".

Fermented foods are generally classified in two ways, depending on the method by which they are obtained: those obtained using starter cultures known as fermentation aids (controlled, modern fermentation or with active allochthonous microorganisms) and those resulting from natural fermentation (wild, uncontrolled, traditional, spontaneous or with active autochthonous microorganisms) that develops in a natural process spontaneously or without intervention (Marco et al., 2017). Fermentative microorganisms found in food or food processing environments are effective in natural fermentation (Dimidi et al., 2019). The aids required for the start of fermentation usually consist of bacteria or sometimes yeasts. Nowadays, fermentation using starter cultures under controlled conditions is generally preferred over fermentation using natural cultures. Thus, a fermentation environment free from pathogens and other unwanted microorganisms, as well as harmful by-products that may result from their activities or cause undesirable changes in the product's organoleptic properties (taste, odor, texture), can be created.

Most fermented foods contain live microorganisms, including probiotics that positively affect human health, prebiotics, and various bioactive compounds. Common fermentation microorganisms include lactic acid bacteria, acetic acid bacteria, *Bacillus*, *Staphylococcus*, *Brevibacterium*, and *Propionibacterium* species, yeasts, and moulds (Figure 1).

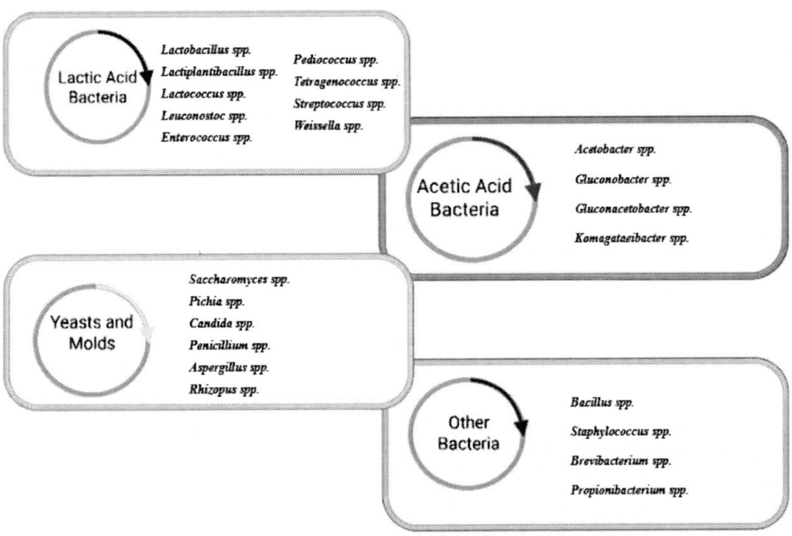

Figure 1. The most common microorganisms involved in the fermentation process (Batt, 2016; Marco et al., 2021; Šikić-Pogačar et al., 2022).

People prefer fermented products because of their health benefits, sensory properties, and ability to extend shelf life (Castellone et al., 2021). Nutritional studies on fermented foods support the importance of intestinal microbiota on human health. Many studies have shown that various bioactive components (e.g., peptides, organic acids, vitamins, short-chain fatty acids, and bacteriolytic enzymes) are produced by microorganisms during the formation process of fermented foods. Thus, the organoleptic properties of foods, nutritional value (e.g., easier digestibility, vitamin content), and positive effects on health (e.g., strengthening immunity, supporting digestion, antioxidative capacity, probiotic properties, antimicrobial and antitoxin effects) increase (Voidarou et al., 2020; Sharma et al., 2020). Probiotics in fermented products have been reported to reduce symptoms of ulcerative colitis and Crohn's disease, as well as to help manage inflammatory bowel diseases and irritable bowel syndrome, thanks to their antibacterial and anti-inflammatory properties (Estevinho et al., 2024; Gwee et al., 2025). Different fermented plant- and milk-based foods and their effects on health are mentioned below.

Fermented Plant-Based Products

Fermentation is one of the most widely used methods worldwide, especially in Asian societies (Tasdemir & Sanlier, 2020), for preserving vegetables that are rich in beneficial components but also perishable. The increase in the popularity of fermented vegetables in recent years is due to the beneficial effects of the bioactive components they contain on health, which have been learned (Siddiqui et al., 2023). Fermented vegetables are valuable sources of short-chain fatty acids, minerals, polyphenols, γ-aminobutyric acid (GABA), B vitamins, flavonoids, antibacterial peptides, and dietary fibers, which have beneficial effects on health (Behera et al., 2018; El Sheikha & Hu, 2020; Singhal et al., 2021; Major et al., 2022).

Although there are variations in the production processes of fermented vegetables, in general, they are obtained by pre-treatments that include cleaning and breaking down the raw material, salting, seasoning, and fermentation stages that can vary from a few days to a few months after the addition of starter cultures. Dry salting or mostly marinating are the most commonly applied techniques. The quality of a fermented product is closely related to the raw material used, the conditions and quality of the fermentation medium, and the population of microorganisms active in the fermentation process.

There are many fermented vegetable products, depending on the choice of raw material, formulation, and technology. In addition to fermented beverages such as fermented turnip juice (Kabak & Dobson, 2011) and hardaliye (Baschali et al., 2017), there are various homemade or commercially produced pickles such as broccoli (Aksay et al., 2022), cabbage (sauerkraut) (Fijan et al., 2024), paocai (Yang et al., 2022), zhacai (Zhang et al., 2021), kimchi (Lee et al., 2020; Fijan et al., 2024), hot pepper (Xiao et al., 2010), carrot (Llorca et al., 2001), ginger (Chen et al., 2023), cucumber (Wilson et al., 2015), eggplant (Ajayi & Awe, 2022), beetroot (Srivastava & Singh, 2016), garlic (Sadeghi, 2016), olive (Fayek et al., 2021), celery (Knez et al., 2024), onion (Bahram-Parvar & Lim, 2018), radish (Fijan et al., 2024). However, the number of studies examining the effects of pickles on the digestive system is quite limited. Therefore, sauerkraut and kimchi, which are known worldwide and have been studied most extensively, will be mentioned.

Sauerkraut, which is widely consumed in Europe and North America, is a pickle obtained by fermenting cabbage under certain conditions. During fermentation, the rapidly increasing number of heterofermentative lactic acid bacteria, such as *Leuconostoc mesenteroides*, leads to the production of lactic and acetic acids, increasing the medium's acidity and promoting anaerobic conditions by producing CO_2. Thus, favourable conditions are created for the growth of acid-tolerant homolactic LAB species, especially *Lactiplantibacillus plantarum*. It is reported that the bioavailability of vitamins, minerals, and phytonutrients in sauerkraut is increased compared to raw cabbage. In addition, the main feature that distinguishes sauerkraut from many other foods is its high probiotic content (Holzapfel, 2003).

Kimchi is a traditional Korean pickle, known worldwide. Its main ingredients are cabbage and radish, but it also includes auxiliary ingredients such as garlic, green onion, ginger, red pepper, mustard, parsley, jeotgal (fermented seafood), carrot, and salt (Jung et al., 2011). Many lactic acid bacteria, including *Lactobacillus plantarum* and *Leuconostoc citreum*, naturally participate in kimchi fermentation. Among these bacteria, *Lactobacillus kimchii* is a species specific to kimchi (Kim & Chun, 2005).

Two clinical studies conducted by Nielsen et al. (2018) and Kim et al. (2022) reported that consumption of sauerkraut and kimchi could alleviate symptoms of irritable bowel syndrome. They found that this condition was associated with decreased serum levels of inflammatory cytokines, decreased harmful fecal enzyme activity, and changes in gut microbiota composition. In addition, different studies have shown that kimchi consumption causes a decrease in body fat tissue, insulin insensitivity, systolic and diastolic blood pressure, fasting blood sugar, serum total cholesterol levels, and LDL/HDL cholesterol ratio (Baek et al., 2001; Choi et al., 2001; Kim et al., 2011; An et al., 2013; Choi et al., 2013; Kim & Park, 2018). The researchers stated that the decrease observed in the mentioned parameters can directly affect gene expression in metabolism and immune pathways in humans with kimchi consumption, and that changes in the intestinal microbiota can alter human metabolism. Simulated colonic fermentation of pig feces has shown that digested sauerkraut samples significantly alter the composition of the intestinal microbiota and that beneficial components, such as vitamins, short-chain fatty acids, and essential amino acids, are produced (Hashizume et al., 2003; Shetty et al., 2013).

It has been reported that various lactic acid bacteria species, such as *Lactobacillus plantarum*, Lactobacillus brevis, and Lactobacillus sakei, which are identified in sauerkraut and kimchi, provide immunomodulatory signals on the immune system, as well as assist in digestive processes, produce bioactive compounds such as isothiocyanates, and suppress pathogens through acid and bacteriocin production (Lee et al., 2011; Selhub et al., 2014; Touret et al., 2018; Shahbazi et al., 2021). Regarding irritable bowel syndrome, preventing the overgrowth of pathogenic microorganisms in the intestine is very important. This allows plant-based fermented foods to be considered a potential replacement for antibiotics in the fight against infections. In addition, when evaluated for dietary fiber intolerance, which is frequently seen in irritable bowel syndrome, it is critical that fermented vegetables can partially break down the fibers, making them more tolerable thanks to the microorganisms they contain (Chan et al., 2021).

In general, fermented vegetables play an important role in supporting intestinal health, thanks to the dietary fiber they contain. Dietary fibers, especially insoluble fiber fractions, stimulate intestinal peristalsis, increase stool volume, and promote regular bowel movements. In addition, the fermentable part of fibers is converted into short-chain fatty acids by the microbiota in the large intestine (Slavin, 2013). These metabolites (primarily acetate, propionate, and butyrate) serve as energy sources for colonic epithelial cells, maintain intestinal barrier integrity, and suppress inflammation (Koh et al., 2016). Thus, the fiber content of pickled vegetables supports mechanical digestive functions and positively affects intestinal metabolic and immunological health.

Kombucha is a non-alcoholic or low-alcohol fermented beverage that has gained popularity and has been commercially produced in the last 10 years. It is based on green or black tea (*Camellia sinensis*), sweetened with sugar, and prepared involving a combined microbial culture of bacteria and yeast (SCOBY) (Gaggia et al., 2018; Martinez Leal et al., 2018). Its microbial composition consists primarily of acetic acid bacteria, lactic acid bacteria, and yeasts belonging to the *Acetobacter* and *Gluconobacter* genera (Vīna et al., 2014). Thanks to the interactions between the bacteria and yeast species contained in kombucha, valuable metabolites such as organic acids (such as acetic, glucuronic, and gluconic acids), vitamins (vitamin C and B complex vitamins), enzymes, bioactive peptides, and phenolic compounds are formed (Diez-Ozaeta et al., 2022). In this regard, kombucha is of interest as a promising beverage regulating intestinal microbiota. In a randomized controlled clinical study conducted by Ecklu-Mensah et al. (2024), individuals who consumed kombucha for four weeks were compared with a control group, and the effects of kombucha consumption on intestinal microbiota and health parameters in healthy individuals fed a Western diet were investigated. The results showed that kombucha consumption increased intestinal colonization by probiotic bacteria, such as *Weizmannia coagulans, and* the number of bacteria that synthesize short-chain fatty acids, such as acetate and butyrate. The researchers reported improvements in insulin sensitivity, intestinal permeability, and parameters of low-grade inflammation. A different study investigated the effects of green tea-based kombucha consumption on overweight individuals' intestinal health, microbiota profile, and serum metabolome for 8 weeks (Fraiz et al., 2024). Researchers observed a significant increase in beneficial bacteria, such as *Bifidobacteria* and *Faecalibacterium prausnitzii*, in the intestinal flora of the kombucha group, along with decreased markers of intestinal permeability, reduced low-grade inflammation, and a more balanced microbiota profile. De Campos Costa et al. (2024) reported that, in a 6-week randomized controlled study, black tea-based kombucha consumption altered the intestinal microbiota in obese and healthy individuals; microbial diversity increased in both groups, and, notably, healthy microbiota members such as *Akkermansia muciniphila* and *Roseburia* increased in number. Positive improvements were observed in terms of gas, bloating, and stool quality in the kombucha group. Jung et al. (2019) aimed to determine the effects of kombucha tea on the intestinal microbiota in obese mice fed a methionine/choline-deficient diet and its potential protective role on non-alcoholic fatty liver disease. The researchers reported that in the group consuming kombucha and receiving a methionine/choline-deficient diet, the density of *Erysipelotrichia*, *Allobaculum*, *Turicibacter*, and *Clostridium* species decreased in the intestinal microbiota compared to the group receiving only a methionine/choline-deficient diet. In contrast, *Lactobacillus* species bacteria increased significantly, and Mucispirillum species was observed only in the kombucha-treated group. This suggests that kombucha has a probiotic effect, supporting the proliferation of beneficial bacteria while suppressing pathogens. It has been stated that this modulation of the microbiota may constitute a positive mechanism for alleviating fatty liver and improving overall gastrointestinal health. In addition, various animal model studies have shown that kombucha consumption may provide anti-hyperglycemic (Moreira et al., 2022), antioxidant (Morales, 2020), and anti-inflammatory (Xu et al., 2022) effects. All these studies reveal that kombucha, as a functional fermented beverage, can provide multifaceted benefits for gut microbiota and systemic health.

Fermented Dairy Products

Fermented dairy products are foods that have become popular in the world, especially in the last few decades, and improve the quality of life thanks to their health benefits, such as nutritional value, high digestibility, antioxidant, antihypertensive, cholesterol-lowering, and immunoregulatory effects, as well as their appreciated organoleptic properties (García-Burgos et al., 2020). In addition to products such as yogurt, cheese, and kefir that are well-known worldwide, there are fermented milk products such as koumiss, skyr, shrikhand, gioddu, churpi, maasai, lassi, ayran, kurut, kashk, tarag, suusac, fene, and chhu that are traditionally consumed (Mathara et al., 2004; Lore et al., 2005; Ortu et al., 2007; Hemsworth et al., 2011; Güler & Park, 2011; Ispirli & Dertli, 2018; Liu et al., 2012; Bourrie et al., 2016; Chen et al., 2019; Pourjoula et al., 2020; Leksir et al., 2019; Wullschleger et al., 2013; Mangia et al., 2014; Shangpliang et al., 2018).

Fermented milk is a dairy product containing a specific culture or combination of cultures used for fermentation (Codex Alimentarius, 2011). Yogurt is a biologically active fermented dairy product that converts milk sugar into lactic acid through fermentation by lactic acid bacteria, thereby enhancing its nutritional properties and providing probiotic benefits. In addition to the traditional starter cultures (*Lactobacillus delbrueckii* subsp. *bulgaricus* and *Streptococcus thermophilus*) used in yogurt production, many other microorganisms can be added to the special culture to impart different flavors and characteristics (such as *Lactobacillus acidophilus*, *Bifidobacterium bifidum*, and *Lactobacillus casei*). Kefir is a beverage consumed for centuries for its health benefits. It differs from other fermented dairy products because it contains kefir grains and a large yeast population during fermentation. It has a sour, acidic, slightly alcoholic taste and a slightly viscous flow. Thanks to metabolites such as kefiran and exosaccharides, many beneficial effects on health, including gastrointestinal health, immune regulation, wound healing, anti-carcinogenic, antimicrobial, and cholesterol metabolism/angiotensin-converting enzyme inhibition, have been reported (Bourrie et al., 2016). Cheese is a probiotic-rich dairy product made by coagulating milk casein protein, and it is enjoyed in various forms worldwide. Cow, buffalo, goat, or sheep milk is usually used as raw material. According to Codex Alimentarius (2011), cheese should contain harmless bacteria that produce lactic acid and/or contribute to flavor and aroma, other harmless microorganisms as starter cultures, and appropriate and safe enzymes for coagulation (such as rennet) and sodium chloride. The positive effects of cheese consumption on muscle, bone, cholesterol, diabetes, obesity, and cancer have been proven (Tunick & Van Hekken, 2015).

The main components of fermented dairy products are proteins (casein, whey proteins, bioactive peptides, and bacteriocins), lipids (saturated, monounsaturated, and polyunsaturated fats), carbohydrates (lactose, oligosaccharides, polysaccharides, and exopolysaccharides), vitamins (A, B-complex (B1, B2, B3, B5, B6, B9, B12), and D), minerals (calcium, phosphorus, potassium, magnesium, and zinc), prebiotics (indigestible food components, lactose, exopolysaccharides), probiotics (*Bifidobacterium*, *Lactobacillus*, and *Streptococcus*), yeasts (*Saccharomyces boulardii*, *Kluyveromyces marxianus*) (García-Burgos et al., 2020). The probiotics and prebiotics they contain are reported to have many positive effects on the human digestive system, such as balancing intestinal microflora, promoting healthy bowel habits, and reducing diarrhea (Mansour et al., 2021).

To impart the desired properties during the production of fermented milk products, various genera of lactic acid bacteria, such as *Lactobacillus*, *Streptococcus*, *Leuconostoc*, *Pediococcus*, and *Lactococcus* from the Firmicutes phylum, and *Bifidobacterium* from the Actinobacteria phylum, are used (Goswami et al., 2017). However, regardless of the culture used, the primary purpose is for the culture to contribute to flavor development, remain alive in the product, and interact with the microbiota and cells in the intestinal wall (Granier et al., 2013). Lactic acid bacteria added to milk synthesize organic acids (mainly lactic acid), increasing the environment's acidity. In addition, acetic acid, ethanol, flavor compounds, flavor enhancers, and exosaccharides produced during fermentation enhance the product's microbial quality, functionality, and health benefits (Han et al., 2016). In the advanced stages of fermentation, macronutrients (protein, lipid, and carbohydrates) are digested by microorganisms and transformed into simpler structures such as peptides, free amino acids, free fatty acids, and lactic acid. Thanks to this transformation, the solubility of proteins in milk and the bioavailability of amino acids increase. According to Sharma et al. (2020), fermented dairy products contain casein proteins (α-, β-, and κ-casein) and whey proteins (e.g., α-lactoalbumin, β-lactoglobulin, lactoferrin, immunoglobulin, and enzymes) like milk, but the protein content, including free amino acids such as leucine, isoleucine, proline, arginine, and phenylalanine, is higher than milk at 85% (increasing up to 90–92%).

Derrien and van Hylckama Vlieg (2015) reported that fermented dairy products contain approximately 10^6–10^9 CFU of live bacteria and retain their viability while passing through the human digestive system. Regular kefir consumption has been reported to support digestive and immune health, regulate blood sugar levels, and alleviate symptoms of lactose intolerance (Şanlıer et al., 2019). Many researchers have found that consuming yogurt and other fermented dairy products positively changes the fecal microbial population (Uyeno et al., 2008; McNulty et al., 2011). Investigating the effects of consuming a fermented dairy product containing *Bifidobacterium animalis* on 28 subjects with irritable bowel syndrome, Veiga et al. (2014) reported that it increased colonic short-chain fatty acid production and decreased the pathogenic bacterium *Bilophila wadsworthia*. Researchers have stated that consumption of this fermented milk may support intestinal homeostasis. Alvarez et al. (2020) reported that in a 4-week study involving 96 healthy adults, subjects were given 1-3 bottles of yogurt containing the probiotic candidate strain *Lactobacillus paracasei* subsp. *paracasei* CNCM I-1518 and CNCM I-3689, and *Lactobacillus rhamnosus* CNCM I-3690, were daily and safely altered the intestinal microbiota in a dose-dependent manner.

The beneficial effects of fermented foods on intestinal microbiota and human health have been proven thanks to the probiotics, prebiotics, and various bioactive compounds they contain. Fermented foods are traditionally consumed in many countries and are available commercially. Many countries, such as Turkiye, India, Canada, the USA, England, Australia, Japan, Sweden, Portugal, and Italy, recognize the health benefits of fermented foods and include fermented foods among the recommended food groups in their national nutrition guides (some of which are included in the dairy products category). Thus, it is aimed to encourage healthy eating habits and draw attention to the positive effects of fermented foods on health. However, the probiotic bacteria in fermented foods may vary depending on individuals' dietary habits and other factors that affect intestinal regularity (Wu et al., 2019). Therefore, fermented foods should be consumed regularly to observe their long-term positive effects on human health.

Healing Foods in Traditional Medicine

In traditional medicine, some plants, spices, fruits, vegetables, and other food items have been recognized and used for centuries by different cultures and societies for their healing properties. Today, herbal medicines are used for a wide range of conditions, from the common cold to cardiovascular disorders and metabolic diseases. While the rate of herbal medicine use is approximately 33% in developed countries, it is estimated to be 80% in many developing countries (Zhou et al., 2007). More than 11,000 plant species are used for therapeutic purposes worldwide, and approximately 500 are widely known in Asia and other countries. However, very few of these plants have been studied scientifically and medically. Some of the plants mentioned for traditional treatments are discussed below.

Saffron (*Crocus sativus* L.) has been used traditionally in Chinese, Ayurvedic, Persian, and Unani medicine from ancient times to the present day. It was used mainly in Egypt and Rome to increase sexual desire, prevent mucosal inflammation, constipation, muscle pain, depression, respiratory difficulties, cough, menstrual irregularities, gas accumulation, support milk production, and reduce cramps (Hosseinzadeh et al., 2008; Pitsikas, 2016). It is known that Cleopatra used saffron with donkey milk to keep her skin young. Materia Medica, written by the Greek physician Pedanius Dioscorides in the 1st century AD, is a work containing the therapeutic properties of saffron (Christodoulou et al., 2015). For more detailed information about saffron, see the ''Spices: More Than Flavor'' section.

Ginseng is among the most recognized and valued plants in Chinese medicine and worldwide. It is widely used in health support products and daily nutrition. The species known as Asian Ginseng/Red Ginseng (*Ginseng radix rubra*) is used in Traditional Chinese Medicine for the treatment of fatigue, mental disorders, diabetes, digestive problems such as gastritis and stomach ulcers, and sexual dysfunctions (Zhou et al., 2007).

Safflower (*Carthamus tinctorius*) is a plant with many branches, thorns, herbaceous structures, and bright yellow, orange, or red flowers. It is cultivated for its flowers. Food coloring is obtained from its flowers (added to beverages). The colorless oil, obtained from safflower seeds, is used in meals, salads, and margarine production (Albaiz, 2022). More than 104 compounds are isolated from Safflower, including quinolones, flavonoids, alkaloids, polyacetylene, aromatic glycosides, and organic acids (Li & Yang, 2002). It is used in traditional Chinese medicine and other folk medicines to stimulate the menstrual cycle and relieve menstrual pain, breast pain, and traumatic injuries (Zhou et al., 2007). Today, studies have shown that the plant reveals a wide array of therapeutic effects, including correcting myocardial ischemia, regulating the immune system, preventing blood clotting, preventing thrombosis, exhibiting antioxidant properties, slowing down the aging process, protecting against oxygen deficiency, being effective against digestive system problems, reducing inflammation, preventing liver fibrosis, fighting cancer cells and pain-relieving effects thanks to the bioactive components it contains (Zhou et al., 2014).

Ginkgo biloba L. is a tree native to China, Japan, and Korea, where humans have used leaves and seeds for thousands of years. *G. biloba* is a herbal dietary supplement containing many biologically active components such as fatty acids, proanthocyanidins, and polysaccharides. The most dominant bioactive components are terpene trilactones (ginkgolides (A, B, C) and bilobalides) and flavonoids (Fang et al., 2020). Feng et al. (2019) stated that ginkgolides are used for neuroprotective treatment after cerebral infarction, a disorder with the potential to cause irreversible brain injury and neurological disorders. *G. biloba* L. leaf extracts have been shown to have positive results in the treatment of Alzheimer's disease, neurodegenerative diseases, neurosensory problems such as tinnitus and dizziness, dementia, and oxidative stress, and have antioxidant effects (Mohanta et al., 2014; Boateng, 2022). Traditional Chinese Medicine uses its seeds for chronic cough, asthma, leukemia, and urinary tract problems. At the same time, its leaves are reported to be used to treat cardiovascular disease, angina, hyperlipidemia, hypertension, and cerebral vascular spasm (Zhou et al., 2007). Studies are being conducted to develop products, such as fermented beverages (Biernacka et al., 2024) and food supplements (Czigle et al., 2018), produced from *G. biloba* leaves.

Turmeric (*Curcuma longa* L.) is an aromatic stimulant and carminative in ancient Hindu texts (Yadav et al., 2017; Akram, 2010). In China, saffron is used orally and externally for conditions such as urticaria, skin allergies, viral hepatitis, arthritis, sore throats, and wounds (Kapoor, 2018). Although the most common use of *C. longa* is oral, it can also be applied topically for skin conditions such as acne, wounds, eczema, and furuncles. It is also used in inhalation in the Ayurvedic tradition. For more information on turmeric, see the "Spices: More Than Flavor" section.

Helianthus annuus L. is one of approximately 25,000 species of the Asteraceae family. Species of this family are cultivated almost everywhere in the world except the polar regions. FAO estimates that global production of *H. annuus* will reach sixty million tons by 2050 (Fernández-Luqueño et al., 2014). *H. annuus*, which was cultivated in 2600 BC, is commonly known as the sunflower (Lenz et al., 2008). It is an annual plant with yellow flowers and a stem that is 1-3 m high (Bashir et al., 2021). Its high economic value is due to the phenolic compounds, including flavonoids and phenolic acids, alkaloids, terpenes, fixed oils, proteins, carbohydrates, and vitamins (B, D, E, and K) it contains (Subashini & Rakshitha, 2016). Sunflowers contain high amounts of polyunsaturated fatty acids, especially omega-3 and omega-6 (Van Nieuwenhove et al., 2019). The health benefits it exhibits, thanks to its bioactive components, are known, and its use in food and cosmetic products is widespread (Rolnik & Olas, 2021). Besides the above-mentioned uses, other plant parts are also quite valuable as animal feed. In addition, it is reported to be used in the production of biofuels, dyes, and surfactants (Adeleke & Babalola, 2020). Various studies have proven the antioxidant (Jiraungkoorskul, 2016), anti-inflammatory (Bashir et al., 2021), anticancer (Roy et al., 2015), anticholesterol (Rani et al., 2017) effects of sunflower, as well as its protective effect against cardiovascular diseases (Vijayakumar et al., 2016) and obesity (Islam, 2015). The use of the plant in cultural or ancestral remedies dates back about 3000 years. It is reported that its seeds, leaves, roots and flowers are used in Iraq, India, South Africa, Russia, America, Mexico, Morocco, Africa and Nigeria for their diuretic, expectorant, wound healing, hypoglycemic and gastrointestinal system regulating effects, as well as in the treatment of kidney diseases, fever, skin allergies, malaria and diabetes (Al-douri, 2000; Bnouham et al., 2002; Malangu, 2007; Onoja & Anaga, 2014; Bashir et al., 2021; Balkan et al., 2020; Oladeji et al., 2020).

Plantago major L. is a perennial herb, about 15 cm tall, with rosette-shaped leaves and small brown-green flowers. Seeds, leaves, flowers and roots of the plant contain flavonoids (luteolin, apigenin, baicalein, hispidulin, plantaginin, luteolin 7-glucoside, hispidulin 7-glucuronide, luteolin 7-diglucoside, apigenin 7-glucoside, nepetin 7-glucoside and luteolin 6-hydroxy 4′-methoxy 7-galactoside), alkaloids (indicain and plantagonin), terpenoids (loliolid, ursolic acid, oleanolic acid, sitosterol acid, and 18β-glycyrrhetinic), iridoid glycosides (such as aucubin, asperuloside, majoroside, catapol and geniposidic acid), various polysaccharides (e.g., xylose, arabinose, galacturonic acid, gardoside), fatty acids (lignoceric, palmitic, stearic, oleic, linoleic and linolenic acids) and minerals (Adom et al., 2017). It can be consumed by cooking the plant's leaves with lentils. *P. major* has been used in Iranian, Greek, European, Arab, Indian, and Chinese traditional medicine to treat various diseases for about 4000 years (Jonsson, 1983; Shahabi et al., 2008; Mozaffarian, 2013). It has been reported to be used in Iranian traditional medicine to reduce inflammation and bleeding (Najafian et al., 2018). Materia Medica, an encyclopaedia on herbal medicines and pharmacopoeias worldwide, states that nerve grass can treat dog bites (Roca-Garcia, 1972). Nielsen & Kehler (1969) reported that the Vikings used *P. major* for wound treatment in a Scandinavian text (Vølsunga saga). In the periods following the spread of Islam, the famous scientists named Razi, Ibn Sina, and Bıruni, who were trained in Islamic medical schools, presented information such as the growing environment of the plant, areas of use in treatment, dosage, and side effects in detail (Najafian et al., 2018). Various ethnopharmacological studies report that it is used against tumors, pain, and fever, in addition to skin, infection, digestive organs, respiratory organs, and reproductive and circulatory problems in many countries such as Norway, Turkiye, Iran, India, the USA, Brazil, Italy, Denmark, Chile, Thailand, and Peru (Samuelsen, 2000).

Cinnamon (*C. verum*) has long been used as a traditional folk remedy in Korea, China, and Russia (El-Desoky et al., 2012). It has been recorded in folk medicine with its anti-bronchitis properties (Ribeiro-Santos et al., 2017). It is also a source of healing frequently used by the public in the treatment of diseases such as sore throat, nausea, pain due to gas accumulation, sexual performance inadequacy, eye inflammation, and rheumatism (Abeysekera et al., 2013; Mariappan et al., 2013; Khasnavis & Pahan, 2014; Hamidpour et al., 2015; Gulcin et al., 2019). In traditional applications, cinnamon essential oil is used to alleviate bile irregularities, migraine-like headaches, hemorrhoidal complaints, dry mouth, lower respiratory tract infections, diarrhea, skin itching, cardiovascular problems, and urinary tract disorders (Farahpour & Habibi, 2012). For more detailed information on cinnamon, see the "Spices: More Than Flavor" section.

Petroselinum crispum (parsley) belongs to the Apiaceae family and is derived from the Latin words "petros" and "selinon". It is usually an annual plant, reaching 60-100 cm in height, glabrous, with bright green stems and leaves, and light yellow, umbrella-shaped flowers. It can be grown almost everywhere in the world. Although the chemical composition of parsley may vary according to the variety and environmental conditions in which it is grown, it can be said to be a good source of proteins, carbohydrates, and volatile oils (Stephens, 1994). Various studies have shown that the plant contains bioactive components such as flavonoids, coumarins, terpenes, vitamins C and A, and carotenoids (Larsen & Christensen, 2005; Cherng et al., 2008; Luthria, 2008; Heidari et al., 2018). In studies by Gadi et al. (2009) and Mekhfi et al. (2004), the effects of parsley on the cardiovascular system, particularly its antiplatelet (antiplatelet aggregation) properties, were stated. *P. crispum* is indispensable in salads and various dishes today. Ethnobotanical studies have determined that the seeds and leaves of parsley are traditionally used in countries such as Iran, Iraq, Turkiye, Morocco, Spain, Peru, and Serbia for the treatment of various skin diseases, urinary tract diseases, gastritis, heart and kidney diseases, diabetes, and blood pressure, as well as pain relief, diuretic, antitussive, and carminative purposes (Merzouki et al., 1997; Jouad et al., 2001; Eddouks et al., 2002; Bolkent et al., 2004; Moazedi et al., 2007; Montesano et al., 2012; Aljanaby, 2013; Šavikin et al., 2013; Soliman et al., 2015). In the encyclopaedia of medicinal substances called "Makhzan-al-Adviah" written by Aghili Khorasani, an Iranian writer, and in the book Muhit Azam written by Hakim Azam Khan, an Indian writer, it is reported that parsley is used in folk medicine as a carminative, diuretic, anti-rhinitis, skin abscess, anti-inflammatory, urinary tract, bladder cleanser, and digestive aid. It is also applied externally in the treatment of scorpion stings and scars (Bahramsoltani et al., 2024).

Taraxacum officinale, classified under the Asteraceae family, is commonly known as dandelion. It grows up to 60 cm and is a perennial plant with yellow-orange flowers. Dandelion flowers, roots, and leaves can be consumed raw, cooked, or brewed into tea (Escudero et al., 2003). Since the leaves taste bitter, they should be boiled before consumption. Dandelion and burdock, a traditional fizzy drink produced in England since the Middle Ages, also contains dandelion and burdock. Recent studies have proven the plant's antitumor, antioxidant, anti-inflammatory, and liver-protective effects (Ovadje et al., 2012; Colle et al., 2012; Koh et al., 2010; Park et al., 2010). It has a long history of use in traditional Chinese medicine to treat various infectious diseases. Schütz et al. (2006) reported that dandelion was traditionally used as a mild laxative, tonic, stimulant, and diuretic.

Ginger (*Zingiber officinale* Roscoe) is commonly used in traditional Ayurvedic, Chinese, Arabic, and African medicine. In traditional medicine, it is reported to be used for conditions such as anti-nausea, headache, nausea, digestive disorders, colds, arthritis, rheumatism, muscle pain, and inflammation (Baliga et al., 2011; Dehghani et al., 2011). For more detailed information on ginger, see the 'Spices: More Than Flavor' section.

Urtica dioica, a member of the Urticaceae family, is a herbaceous perennial plant with green leaves and stinging hairs (known as trichomes) covering the whole plant. The plant is mainly grown in temperate and tropical regions. However, it adapts to different climatic conditions. The nettles are used not only for cooking and brewing tea, but also in the pharmaceutical, cosmetic, biodynamic agriculture, feed, and textile industries (Hartl & Vogl, 2002; Vogl, 2003). Recent scientific studies have shown that the plant contains bioactive compounds such as flavonoids, phenolic and amino acids, fatty acids, and carotenoids, which are antioxidant, anti-kidney stone, antihypertensive, antidiabetic, anti-inflammatory, antiulcer, anticarcinogenic, antimutagenic, anti-hypoxic (Tahri et al., 2000; Guil-Guerrero et al., 2003; Wright et al., 2007; Khalili et al., 2015; Mzid et al., 2017; Mukundi et al., 2017; Afkari et al., 2019; Kasouni et al., 2021; Aziz et al., 2024). The long-standing traditional use of the plant in Turkiye and Iran has been reported by various sources (Hoseinynejad et al., 2022; Uzonur et al., 2013). It has been traditionally used for a very long time to reduce the effects of disorders such as iron deficiency, excessive menstruation, nosebleeds, hypertension, and asthma (Mikaeili et al., 2013; Di Virgilio et al., 2015). Current scientific research has reported positive effects of the plant on diabetes and rheumatoid arthritis (Namazi et al., 2012).

Black pepper (*P. nigrum*) is used in various traditional healing systems, including Ayurveda and Unani (Acharya et al., 2012; Ahmad et al., 2012). The flowers, seeds, fruits, and leaves of the plant have been traditionally employed to treat hypertension (Sethiya et al., 2018), cuts, wounds, fever, sinusitis, vomiting, stomach problems (Vijayakumar et al., 2015), epilepsy (Sharma et al., 2013), skin diseases (Policepatel & Manikrao, 2013), cough (Mallik et al., 2012), and diabetes (Lingaraju et al., 2013). For more detailed information on black pepper, see the 'Spices: More Than Flavor' section.

Although many plants we can use as food are used in traditional medicine, they may interact with drugs, posing a serious safety concern. To prevent these interactions, multiple studies should be conducted using appropriate in vitro and in vivo approaches. In addition, each person's metabolism is unique. Effects or side effects not observed in one person may be observed in another. Therefore, it is not recommended to use plants for therapeutic purposes without being sure of their effects and interactions with drugs.

Conclusion

In the last century, the direct use of herbal medicines, which have been used from the past to the present, has been dramatically reduced to crude extracts and pure components. Today, plants are generally utilized as raw materials or precursor molecules in drug design and manufacturing processes. Furthermore, advances in analytical methods and omics technologies have paved the way for the direct use of purified plant extracts as therapeutic products. Although most natural products do not meet the pharmacological and bioavailability criteria required in today's clinical trials, the therapeutic potential of herbal extracts may be effective, especially in complex chronic diseases that are difficult to treat with a single active substance.

References

Aarabi, F., Naake, T., Fernie, A. R., & Hoefgen, R. (2020). Coordinating sulfur pools under sulfate deprivation. *Trends in Plant Science*, *25*(12), 1227-1239.

Abadie, C., & Tcherkez, G. (2019). Plant sulphur metabolism is stimulated by photorespiration. *Communications Biology, 2*(1), 379.

Abdalla, M. A., & Mühling, K. H. (2019). Plant-derived sulfur containing natural products produced as a response to biotic and abiotic stresses: A review of their structural diversity and medicinal importance. *Journal of Applied Botany and Food Quality, 92*, 204-215.

Abeysekera, W. P. K. M., Premakumara, G. A. S., & Ratnasooriya, W. D. (2013). In vitro antioxidant properties of leaf and bark extracts of Ceylon cinnamon (Cinnamomum zeylanicum Blume). *Tropical Agricultural Research, 24*(2), 128–138.

Acharya, S. G., Momin, A. H., & Gajjar, A. V. (2012). Review of piperine as a bio-enhancer. *American Journal of PharmTech Research, 2*, 32-44.

Adeleke, B. S., & Babalola, O. O. (2020). Oilseed crop sunflower (Helianthus annuus) as a source of food: Nutritional and health benefits. *Food Science & Nutrition, 8*(9), 4666-4684.

Adom, M. B., Taher, M., Mutalabisin, M. F., Amri, M. S., Kudos, M. B. A., Sulaiman, M. W. A. W., ... & Susanti, D. (2017). Chemical constituents and medical benefits of Plantago major. *Biomedicine & pharmacotherapy, 96,* 348-360.

Afkari, R., Feizabadi, M. M., Ansari-Moghadam, A., Safari, T., & Bokaeian, M. (2019). Simultaneous use of oxalate-degrading bacteria and herbal extract to reduce the urinary oxalate in a rat model: A new strategy. *International Brazilian Journal of Urology, 45*(6), 1249-1259.

Afzal, M., Al-Hadidi, D., Menon, M., Pesek, J., & Dhami, M. S. I. (2001). Ginger: an ethnomedical, chemical and pharmacological review. *Drug metabolism and drug interactions*, 18(3-4), 159-190.

Ahmad, N., Fazal, H., Abbasi, B. H., Farooq, S., Ali, M., & Khan, M. A. (2012). Biological role of Piper nigrum L.(Black pepper): A review. *Asian Pacific Journal of Tropical Biomedicine, 2*(3), S1945-S1953.

Ahmed, E., Arshad, M., Khan, M. Z., Amjad, M. S., Sadaf, H. M., Riaz, I., Sabir, S., Ahmad, N. (2017). Secondary metabolites and their multidimensional prospective in plant life. *Journal of Pharmacognosy and Phytochemistry, 6*(2), 205-214.

Ahmed, R. S., Suke, S. G., Seth, V., Chakraborti, A., Tripathi, A. K., & Banerjee, B. D. (2008). Protective effects of dietary ginger (Zingiber officinales Rosc.) on lindane-induced oxidative stress in rats. *Phytotherapy Research, 22*(7), 902-906.

Ajayi, O. A., & Awe, O. T. (2022). Nutritional and Anti-Nutritional Composition of Fermented/Pickled Garden Egg (Solanum aethiopicum L.). *Food ScienTech Journal, 4*(1), 8-17.

Akram, M., Shahab-Uddin, A. A., Usmanghani, K. H. A. N., Hannan, A. B. D. U. L., Mohiuddin, E., & Asif, M. (2010). Curcuma longa and curcumin: a review article. *Romanian Journal of Biology – Plant Biology, 55*(2), 65-70.

Aksay, S., Arslan, R., Tokbaş, H., & Eroğlu, E. Ç. (2022). Changes in some biochemical properties of Brassica vegetables (Cabbage, Cauliflower and Broccoli) pickles and brines. *Journal of Raw Materials to Processed Foods, 3*(1), 26-34.

Albaiz, A. S. (2022). The Use of Safflower (Carthamus tinctorius) in treating depression and anxiety. *Cureus, 14*(2).

Al-douri, N. A. (2000). A survey of medicinal plants and their traditional uses in Iraq. *Pharmaceutical biology, 38*(1), 74-79.

Ali, B. H., Blunden, G., Tanira, M. O., & Nemmar, A. (2008). Some phytochemical, pharmacological and toxicological properties of ginger (Zingiber officinale Roscoe): a review of recent research. *Food and chemical Toxicology, 46*(2), 409-420.

Ali, S. M., Khan, A. A., Ahmed, I., Musaddiq, M., Ahmed, K. S., Polasa, H., ... & Ahmed, N. (2005). Antimicrobial activities of Eugenol and Cinnamaldehyde against the human gastric pathogen Helicobacter pylori. *Annals of clinical microbiology and antimicrobials, 4*, 1-7.

Aljanaby, A. A. J. J. (2013). Antibacterial activity of an aqueous extract of Petroselinum crispum leaves against pathogenic bacteria isolated from patients with burns infections in Al-najaf Governorate, Iraq. *Research on Chemical Intermediates, 39*, 3709-3714.

Alvarez, A. S., Tap, J., Chambaud, I., Cools-Portier, S., Quinquis, L., Bourlioux, P., ... & Derrien, M. (2020). Safety and functional enrichment of gut microbiome in healthy subjects consuming a multi-strain fermented milk product: a randomised controlled trial. *Scientific Reports, 10*(1), 15974.

Amin, A., & Hamza, A. A. (2006). Effects of Roselle and Ginger on cisplatin-induced reproductive toxicity in rats. *Asian journal of andrology, 8*(5), 607-612.

An, S. Y., Lee, M. S., Jeon, J. Y., Ha, E. S., Kim, T. H., Yoon, J. Y., ... & Lee, K. W. (2013). Beneficial effects of fresh and fermented kimchi in prediabetic individuals. *Annals of Nutrition and Metabolism, 63*(1-2), 111-119.

Aruoma, O. I., Spencer, J. P., Warren, D., Jenner, P., Butler, J., & Halliwell, B. (1997). Characterization of food antioxidants, illustrated using commercial garlic and ginger preparations. *Food chemistry, 60*(2), 149-156.

Azhari, H. N., Yap, S. S., & Nour, A. H. (2017). Extraction and chemical compositions of ginger (Zingiber officinale Roscoe) essential oils as cockroaches repellent. Australian *Journal of Basic and Applied Sciences, 11*(3), 1-8.

Aziz, N., Vinod, Y. R., Sarkar, D., Wal, A., Verma, V., & Wal, P. (2024). Urtica dioica (Stinging Nettle): A Comprehensive & Concise Review on its Nutritional Profile and Therapeutic Applications. *Current Traditional Medicine, 10*(7), 119-130.

Baba, S. A., A. H. Malik, Z. A. Wani, T. Mohiuddin, Z. Shah, N. Abbas, and N. Ashraf. 2015. Phytochemical analysis and antioxidant activity of different tissue types of Crocus sativus and oxidative stress alleviating potential of saffron extract in plants, bacteria, and yeast. *South African Journal of Botany 99*:80–7. doi: 10.1016/j.sajb.2015.03.194.

Baek, Y. H., Kwak, J. R., Kim, S. J., Han, S. S., & Song, Y. O. (2001). Effects of kimchi supplementation and/or exercise training on body composition and plasma lipids in obese middle school girls. *Journal of the Korean Society of Food Science and Nutrition,30*,906-912.

Bagheri, H., Manap, M. Y. B. A., & Solati, Z. (2014). Antioxidant activity of Piper nigrum L. essential oil extracted by supercritical CO2 extraction and hydro-distillation. *Talanta, 121*, 220-228.

Bahram-Parvar, M., & Lim, L. T. (2018). Fresh-cut onion: A review on processing, health benefits, and shelf-life. *Comprehensive Reviews in Food Science and Food Safety, 17*(2), 290-308.

Bahramsoltani, R., Ahmadian, R., Daglia, M., & Rahimi, R. (2024). Petroselinum crispum (Mill.) Fuss (Parsley): an updated review of the traditional uses, phytochemistry, and pharmacology. *Journal of Agricultural and Food Chemistry, 72*(2), 956-972.

Balasundram, N., Sundram, K., & Samman, S. (2006). Phenolic compounds in plants and agri-industrial by-products: Antioxidant activity, occurrence, and potential uses. *Food chemistry, 99*(1), 191-203.

Baliga, M. S., Haniadka, R., Pereira, M. M., D'Souza, J. J., Pallaty, P. L., Bhat, H. P., & Popuri, S. (2011). Update on the chemopreventive effects of ginger and its phytochemicals. *Critical reviews in food science and nutrition, 51*(6), 499-523.

Balkan, A. A., Yıldız, A., Berber, D., Gökalsın, B., & Sesal, C. (2020). Infusion or decoction extracts of Helianthus annuus leaves: potential inhibitors for QS system and biofilm formation in Pseudomonas aeruginosa. *International Journal of Advances in Engineering and Pure Sciences, 32*(4), 499-506.

Baschali, A., Tsakalidou, E., Kyriacou, A., Karavasiloglou, N., & Matalas, A. L., (2017). Traditional low-alcoholic and non-alcoholic fermented beverages consumed in european countries: a neglected food group. 10.1017/S0954422416000202.

Bashir, T., Zia-Ur-Rehman Mashwani, K. Z., Haider, S., & Shaista Tabassum, M. (2021). Chemistry, pharmacology and ethnomedicinal uses of Helianthus annuus (Sunflower): A Review. *Pure and Applied Biology (PAB), 4*(2), 226-235.

Batt, C. A. Microbiology of fermentations. In *Reference Module in Food Science*; Elsevier: Amsterdam, The Netherlands, 2016.

Behera, S. S., Ray, R. C., & Zdolec, N. (2018). Lactobacillus plantarum with functional properties: an approach to increase safety and shelf-life of fermented foods. *BioMed Research International, 2018*(1), 9361614.

Bhandari, U., & Pillai, K. K. (2005). Effect of ethanolic extract of Zingiber officinale on dyslipidaemia in diabetic rats. *Journal of ethnopharmacology, 97*(2), 227-230.

Biernacka, P., Felisiak, K., & Adamska, I. (2024). The potential of dried Ginkgo Biloba leaves as a novel ingredient in fermented beverages of enhanced flavour and antioxidant properties. *Food Chemistry*, 461, 141018.

Bnouham, M., Legssyer, A., Mekhfi, H., & Ziyyat, A. (2002). Medicinal plants used in the treatment of diabetes in Morocco. *International Journal of Diabetes and Metabolism, 10*(1), 33-50.

Boateng, I. D. (2022). A critical review of current technologies used to reduce ginkgotoxin, ginkgotoxin-5′-glucoside, ginkgolic acid, allergic glycoprotein, and cyanide in Ginkgo biloba L. seed. *Food Chemistry, 382*, 132408.

Bolkent, S., Yanardag, R., Ozsoy-Sacan, O., & Karabulut-Bulan, O. (2004). Effects of parsley (Petroselinum crispum) on the liver of diabetic rats: a morphological and biochemical study. *Phytotherapy Research, 18*(12), 996-999.

Borquaye, L. S., Gasu, E. N., Ampomah, G. B., Kyei, L. K., Amarh, M. A., Mensah, C. N., ... & Aboagye, C. I. (2020). Alkaloids from Cryptolepis sanguinolenta as potential inhibitors of SARS-CoV-2 viral proteins: An in silico study. *BioMed Research International, 2020*(1), 5324560.

Bourrie, B. C. T., Willing, B. P., & Cotter, P. D. (2016). The microbiota and health promoting characteristics of the fermented beverage kefir. *Frontiers in Microbiology, 7*, 1–17.

Brglez Mojzer, E., Knez Hrnčič, M., Škerget, M., Knez, Ž., & Bren, U. (2016). Polyphenols: Extraction methods, antioxidative action, bioavailability and anticarcinogenic effects. *Molecules, 21*(7), 901.

Brugnerotto, P., Seraglio, S. K. T., Schulz, M., Gonzaga, L. V., Fett, R., & Costa, A. C. O. (2021). Pyrrolizidine alkaloids and beehive products: A review. *Food chemistry, 342*, 128384.

Bui, T. T., Piao, C. H., Song, C. H., Shin, H. S., Shon, D. H., & Chai, O. H. (2017). Piper nigrum extract ameliorated allergic inflammation through inhibiting Th2/Th17 responses and mast cells activation. *Cellular Immunology, 322*, 64-73.

Bukhari, I. A., Alhumayyd, M. S., Mahesar, A. L., & Gilani, A. H. (2013). The analgesic and anticonvulsant effects of piperine in mice. *Journal of physiology and pharmacology, 64*(6), 789-794.

Cadenas, E., Packer, L., 2002. *Handbook of antioxidants*, Marcel Dekker, Inc., 2nd Edition, New York, 279-337.

Cardoso-Ugarte, G. A., Lopez-Malo, A., & Sosa-Morales, M. E. (2016). Cinnamon (Cinnamomum zeylanicum) essential oils. In *Essential Oils in Food Preservation*, Flavor and Safety, 339-347.

Castañeda-Ovando, A., de Lourdes Pacheco-Hernández, M., Páez-Hernández, M. E., Rodríguez, J. A., & Galán-Vidal, C. A. (2009). Chemical studies of anthocyanins: A review. *Food chemistry, 113*(4), 859-871.

Castellone, V., Bancalari, E., Rubert, J., Gatti, M., Neviani, E., & Bottari, B. (2021). Eating fermented: Health benefits of LAB-fermented foods. *Foods, 10*(11), 2639.

Cemeroğlu, B. S. (2011). *Meyve ve sebze işleme teknolojisi*. Nobel Akademik Yayıncılık.

Chan, E. W. C., Lim, Y. Y., Wong, S. K., Lim, K. K., Tan, S. P., Lianto, F. S., & Yong, M. Y. (2009). Effects of different drying methods on the antioxidant properties of leaves and tea of ginger species. *Food Chemistry, 113*(1), 166-172.

Chan, M., Liu, D., Wu, Y., Yang, F., & Howell, K. (2021). Microorganisms in whole botanical fermented foods survive processing and simulated digestion to affect gut microbiota composition. *Frontiers in Microbiology, 12*, 759708.

Chast, F. A. (2008). Part I: General aspects of medicinal chemistry. In C. G. Wermuth (Ed.), The practice of medicinal chemistry (3rd ed., pp. 4–52). Amsterdam: Elsevier.

Chen, J. C., Huang, L. J., Wu, S. L., Kuo, S. C., Ho, T. Y., & Hsiang, C. Y. (2007). Ginger and its bioactive component inhibit enterotoxigenic Escherichia coli heat-labile enterotoxin-induced diarrhea in mice. *Journal of Agricultural and Food Chemistry, 55*(21), 8390-8397.

Chen, M. S. (2008). Inducible direct plant defense against insect herbivores: a review. *Insect science, 15*(2), 101-114.

Chen, M., Ye, X., Shen, D., & Ma, C. (2019). Modulatory effects of gut microbiota on constipation: The commercial beverage yakult shapes stool consistency. *Journal of Neurogastroenterology and Motility, 25*(3), 475–477.

Chen, Y., Chen, L., Liu, L., Bi, X., & Liu, X. (2023). Characteristics of microbial communities in fermentation of pickled ginger and their correlation with its volatile flavors. *Food Bioscience, 53*, 102736.

Cherng, J. M., Chiang, W., & Chiang, L. C. (2008). Immunomodulatory activities of common vegetables and spices of Umbelliferae and its related coumarins and flavonoids. *Food chemistry, 106*(3), 944-950.

Chiocchio, I., Mandrone, M., Tomasi, P., Marincich, L., & Poli, F. (2021). Plant secondary metabolites: An opportunity for circular economy. *Molecules, 26*(2), 495.

Choi, I. H., Noh, J. S., Han, J. S., Kim, H. J., Han, E. S., & Song, Y. O. (2013). Kimchi, a fermented vegetable, improves serum lipid profiles in healthy young adults: randomized clinical trial. *Journal of Medicinal Food, 16*(3), 223-229.

Choi, S. H., Kim, H. J., Kwon, M. J., Baek, Y. H., & Song, Y. O. (2001). The effect of kimchi pill supplementation on plasma lipid concentration in healthy people. *Journal of the Korean Society of Food Science and Nutrition, 30*, 913–920.

Christodoulou, E., Kadoglou, N. P., Kostomitsopoulos, N., & Valsami, G. (2015). Saffron: a natural product with potential pharmaceutical applications. *Journal of Pharmacy and Pharmacology, 67*(12), 1634-1649.

Chrubasik, S., Pittler, M. H., & Roufogalis, B. D. (2005). Zingiberis rhizoma: a comprehensive review on the ginger effect and efficacy profiles. *Phytomedicine, 12*(9), 684-701.

Cinelli, M. A., & Jones, A. D. (2021). Alkaloids of the genus datura: Review of a rich resource for natural product discovery. *Molecules, 26*(9), 2629.

Codex Alimentarius, (2011). *Codex standard. Rome, Food Agriculture Organization*, 33-198.

Colle, D., Arantes, L. P., Gubert, P., da Luz, S. C. A., Athayde, M. L., Teixeira Rocha, J. B., & Soares, F. A. A. (2012). Antioxidant properties of Taraxacum officinale leaf extract are involved in the protective effect against hepatoxicity induced by acetaminophen in mice. *Journal of Medicinal Food, 15*(6), 549-556.

Cortright, D. N., Krause, J. E., & Broom, D. C. (2007). TRP channels and pain. *Biochimica et Biophysica Acta (BBA)-Molecular Basis of Disease, 1772*(8), 978-988.

Cragg, G. M., & Newman, D. J. (2013). Natural products: a continuing source of novel drug leads. *Biochimica et Biophysica Acta (BBA)-General Subjects, 1830*(6), 3670-3695.

Czigle, S., Tóth, J., Jedlinszki, N., Háznagy-Radnai, E., Csupor, D., & Tekeľová, D. (2018). Ginkgo biloba food supplements on the European market–adulteration patterns revealed by quality control of selected samples. *Planta medica, 84*(06/07), 475-482.

D'Archivio, A. A., and M. A. Maggi. 2017. Geographical identification of saffron (Crocus sativus L.) by linear discriminant analysis applied to the UV-visible spectra of aqueous extracts. *Food Chemistry 219*, 408–413.

Dai, J., & Mumper, R. J. (2010). Plant phenolics: extraction, analysis and their antioxidant and anticancer properties. *Molecules, 15*(10), 7313-7352.

Daniel, A. N., Sartoretto, S. M., Schmidt, G., Caparroz-Assef, S. M., Bersani-Amado, C. A., & Cuman, R. K. N. (2009). Anti-inflammatory and antinociceptive activities A of eugenol essential oil in experimental animal models. *Revista Brasileira de Farmacognosia, 19*, 212-217.

Das, A., Sarkar, S., Bhattacharyya, S., & Gantait, S. (2020). Biotechnological advancements in Catharanthus roseus (L.) G. Don. *Applied Microbiology and Biotechnology, 104*, 4811-4835.

De Campos Costa, M. A., da Silva Duarte, V., Fraiz, G. M., Cardoso, R. R., da Silva, A., Martino, H. S. D., ... & de Barros, F. A. R. (2024). Regular Consumption of Black Tea Kombucha Modulates the Gut Microbiota in Individuals with and without Obesity. *The Journal of Nutrition, 155*(5), 1331-1349.

Dehghani I, Mostajeran A, Asghari G. 2011. In vitro and in vivo production of gingerols and zingiberene in ginger plant (Zingiber officinale Roscoe). *Iranian Journal of Pharmaceutical Sciences, 7*(2), 117-121.

Derrien, M., & van Hylckama Vlieg, J. E. (2015). Fate, activity, and impact of ingested bacteria within the human gut microbiota. *Trends in Microbiology, 23*(6), 354-366.

Di Virgilio, N., Papazoglou, E. G., Jankauskienc, Z., Di Lonardo, S., Praczyk, M., & Wielgusz, K. (2015). The potential of stinging nettle (Urtica dioica L.) as a crop with multiple uses. *Industrial Crops and Products, 68*, 42-49.

Diez-Ozaeta, I., & Astiazaran, O. J. (2022). Fermented foods: An update on evidence-based health benefits and future perspectives. *Food Research International, 156*, 111133.

Dimidi, E., Cox, S.R., Rossi, M. & Whelan, K. (2019). Fermented foods: definitions and characteristics, impact on the gut microbiota and effects on gastrointestinal health and disease. *Nutrients, 11*, 1806.

Divekar, P. A., Narayana, S., Divekar, B. A., Kumar, R., Gadratagi, B. G., Ray, A., ... & Behera, T. K. (2022). Plant secondary metabolites as defense tools against herbivores for sustainable crop protection. *International Journal of Molecular Sciences, 23*(5), 2690.

Dorman, H. D., & Deans, S. G. (2000). Antimicrobial agents from plants: antibacterial activity of plant volatile oils. *Journal of Applied Microbiology, 88*(2), 308-316.

Ecklu-Mensah, G., Miller, R., Maseng, M. G., Hawes, V., Hinz, D., Kim, C., & Gilbert, J. A. (2024). Modulating the human gut microbiome and health markers through kombucha consumption: a controlled clinical study. *Scientific Reports, 14*(1), 31647.

Eddouks, M., Maghrani, M., Lemhadri, A., Ouahidi, M. L., & Jouad, H. (2002). Ethnopharmacological survey of medicinal plants used for the treatment of diabetes mellitus, hypertension and cardiac diseases in the south-east region of Morocco (Tafilalet). *Journal of Ethnopharmacology, 82*(2-3), 97-103.

Ee, G. C., C. M. Lim, C. K. Lim, M. Rahmani, K. Shaari, and C. F. Bong. 2009. Alkaloids from Piper sarmentosum and Piper nigrum. *Natural Product Research, 23*(15),1416–1423.

El Sayed, S. M., & Moustafa, R. A. (2016). Effect of combined administration of ginger and cinnamon on high fat diet induced hyperlipidemia in rats. *Journal of Pharmaceutical, Chemical and Biological Sciences, 3*(4), 561-572.

El Sheikha, A. F., & Hu, D. M. (2020). Molecular techniques reveal more secrets of fermented foods. *Critical Reviews in Food Science and Nutrition, 60*(1), 11-32.

El-Desoky, G. E., Aboul-Soud, M. A., & Al-Numair, K. S. (2012). Antidiabetic and hypolipidemic effects of Ceylon cinnamon (Cinnamomum verum) in alloxan-diabetic rats. *Journal of Medicinal Plants Research, 6*(9), 1685-91.

Escudero, N. L., De Arellano, M. L., Fernández, S., Albarracín, G., & Mucciarelli, S. (2003). Taraxacum officinale as a food source. *Plant Foods for Human Nutrition, 58*, 1-10.

Estevinho, M. M., Yuan, Y., Rodríguez-Lago, I., Sousa-Pimenta, M., Dias, C. C., Barreiro-de Acosta, M., ... & Magro, F. (2024). Efficacy and safety of probiotics in IBD: An overview of systematic reviews and updated meta-analysis of randomized controlled trials. *United European Gastroenterology Journal, 12*(7), 960-981.

Faghani, M., Rahimian, Y., Rafiee, A., & Namjoo, A. R. (2014). Effect of garlic and cinnamon in comparison to virginiamycin on performance and some haematological parameters in broiler chicks. *Research Opinions in Animal & Veterinary Sciences,* 4(9), 504-507.

Fang, J., Wang, Z., Wang, P., & Wang, M. (2020). Extraction, structure and bioactivities of the polysaccharides from Ginkgo biloba: A review. *International Journal of Biological Macromolecules, 162*, 1897-1905.

Farahpour, M. R., & Habibi, M. (2012). Evaluation of the wound healing activity of an ethanolic extract of Ceylon cinnamon in mice. *Veterinarni Medicina, 57*(1), 53–57.

Fauvelle, C., Lambotin, M., Heydmann, L., Prakash, E., Bhaskaran, S., Vishwaraman, M., ... & Moog, C. (2017). A cinnamon-derived procyanidin type A compound inhibits hepatitis C virus cell entry. *Hepatology international, 11*, 440-445.

Fayek, N. M., Farag, M. A., & Saber, F. R. (2021). Metabolome classification via GC/MS and UHPLC/MS of olive fruit varieties grown in Egypt reveal pickling process impact on their composition. *Food Chemistry, 339*, 127861.

Feng, Z., Sun, Q., Chen, W., Bai, Y., Hu, D., & Xie, X. (2019). The neuroprotective mechanisms of ginkgolides and bilobalide in cerebral ischemic injury: a literature review. *Molecular Medicine, 25*, 1-8.

Fernández-Luqueño, F., López-Valdez, F., Miranda-Arámbula, M., Rosas-Morales, M., Pariona, N., & Espinoza-Zapata, R. (2014). An introduction to the sunflower crop. Sunflowers: Growth and Development, Environmental Influences and Pests/Diseases. Valladolid, Spain: Nova Science Publishers, 1-18.

Fich, E. A., Segerson, N. A., & Rose, J. K. (2016). The plant polyester cutin: biosynthesis, structure, and biological roles. *Annual Review of Plant Biology, 67*(1), 207-233.

Fijan, S., Fijan, P., Wei, L., & Marco, M. L. (2024). Health benefits of kimchi, sauerkraut, and other fermented foods of the genus brassica. *Applied Microbiology, 4*(3), 1165-1176.

Fike, D. A., Bradley, A. S., & Rose, C. V. (2015). Rethinking the ancient sulfur cycle. *Annual Review of Earth and Planetary Sciences, 43*(1), 593-622.

Fraiz, G. M., Bonifácio, D. B., Lacerda, U. V., Cardoso, R. R., Corich, V., Giacomini, A., Martino, H. S. D., Esteban-Echeverría, S., Romo-Hualde, A., Muñoz-Prieto, D., Barros, F. A. R. d., Milagro, F. I., & Bressan, J. (2024). The Impact of Green Tea Kombucha on the Intestinal Health, Gut Microbiota, and Serum Metabolome of Individuals with Excess Body Weight in a Weight Loss Intervention: A Randomized Controlled Trial. *Foods, 13*(22), 3635.

Gadi, D., Bnouham, M., Aziz, M., Ziyyat, A., Legssyer, A., Legrand, C., ... & Mekhfi, H. (2009). Parsley extract inhibits in vitro and ex vivo platelet aggregation and prolongs bleeding time in rats. *Journal of Ethnopharmacology, 125*(1), 170-174.

Gaggìa, F., Baffoni, L., Galiano, M., Nielsen, D. S., Jakobsen, R. R., Castro-Mejía, J. L., ... & Di Gioia, D. (2018). Kombucha beverage from green, black and rooibos teas: A comparative study looking at microbiology, chemistry and antioxidant activity. *Nutrients, 11*(1), 1.

Garas, N. A., & Waiss Jr, A. C. (1986). Differential accumulation and distribution of antifungal sesquiterpenoids in cotton stems inoculated with Verticillium dahliae. *Phytopathology, 76*(10), 1011-1017.

García-Burgos, M., Moreno-Fernández, J., Alférez, M. J., Díaz-Castro, J., & López-Aliaga, I. (2020). New perspectives in fermented dairy products and their health relevance. *Journal of Functional Foods, 72*, 104059.

Garg, R., Gupta, S., & Maru, G. B. (2008). Dietary curcumin modulates transcriptional regulators of phase I and phase II enzymes in benzo [a] pyrene-treated mice: mechanism of its anti-initiating action. *Carcinogenesis, 29*(5), 1022-1032.

Garg, S., & Roy, A. (2020). In silico analysis of selected alkaloids against main protease (Mpro) of SARS-CoV-2. *Chemico-Biological Interactions, 332*, 109309.

Ghasemzadeh, A., Jaafar, H. Z., & Rahmat, A. (2010). Antioxidant activities, total phenolics and flavonoids content in two varieties of Malaysia young ginger (Zingiber officinale Roscoe). *Molecules, 15*(6), 4324-4333.

Ghayur, M. N., Gilani, A. H., Afridi, M. B., & Houghton, P. J. (2005). Cardiovascular effects of ginger aqueous extract and its phenolic constituents are mediated through multiple pathways. *Vascular Pharmacology, 43*(4), 234-241.

Goel, A., Kunnumakkara, A. B., & Aggarwal, B. B. (2008). Curcumin as "Curecumin": from kitchen to clinic. *Biochemical Pharmacology, 75*(4), 787-809.

González-Vallinas, M., González-Castejón, M., Rodríguez-Casado, A., & Ramírez de Molina, A. (2013). Dietary phytochemicals in cancer prevention and therapy: a complementary approach with promising perspectives. *Nutrition Reviews, 71*(9), 585-599.

Gopinath, S. M., Vidya, G. M., & Narasimha Murthy, T. P. (2014). Isolation, characterization and purification of bio-active compounds from Andrographis paniculata and Cinnamon verum for anti-HIV activity. *International Journal of Innovative Research in Science, Engineering and Technology, 3*(6), 13777–13783.

Goswami, G., Bora, S. S., Parveen, A., Boro, R. C., & Barooah, M. (2017). Identification and functional properties of dominant lactic acid bacteria isolated from Kahudi, a traditional rapeseed fermented food product of Assam, India. *Journal of Ethnic Foods*, 4(3), 187-197.

Granier, A., Goulet, O., & Hoarau, C. (2013). Fermentation products: Immunological effects on human and animal models. *Pediatric Research, 74*, 238–244.

Griffiths, M. R., Strobel, B. W., Hama, J. R., & Cedergreen, N. (2021). Toxicity and risk of plant-produced alkaloids to Daphnia magna. *Environmental Sciences Europe, 33*(1), 1-12.

Gross, G. G. (2008). From lignins to tannins: Forty years of enzyme studies on the biosynthesis of phenolic compounds. *Phytochemistry, 69*(18), 3018-3031.

Gruenwald, J., Freder, J., & Armbruester, N. (2010). Cinnamon and health. *Critical Reviews in Food Science and Nutrition, 50*(9), 822-834.

Grzanna, R., Lindmark, L., & Frondoza, C. G. (2005). Ginger—an herbal medicinal product with broad anti-inflammatory actions. *Journal of Medicinal Food, 8*(2), 125-132.

Guil-Guerrero, J. L., Rebolloso-Fuentes, M. M., & Isasa, M. T. (2003). Fatty acids and carotenoids from Stinging Nettle (Urtica dioica L.). *Journal of Food Composition and Analysis, 16*(2), 111-119.

Gukovsky, I., Reyes, C. N., Vaquero, E. C., Gukovskaya, A. S., & Pandol, S. J. (2003). Curcumin ameliorates ethanol and nonethanol experimental pancreatitis. *American Journal of Physiology-Gastrointestinal and Liver Physiology, 284*(1), 85-95.

Gulcin, I., Kaya, R., Goren, A. C., Akincioglu, H., Topal, M., Bingol, Z., ... & Alwasel, S. (2019). Anticholinergic, antidiabetic and antioxidant activities of cinnamon (Cinnamomum verum) bark extracts: Polyphenol contents analysis by LC-MS/MS. *International Journal of Food Properties, 22*(1), 1511-1526.

Gülçin, İ. (2011). Antioxidant activity of eugenol: A structure–activity relationship study. *Journal of Medicinal Food, 14*(9), 975-985.

Gülçin, İ., Şat, İ. G., Beydemir, Ş., Elmastaş, M., & Küfrevioğlu, Ö. İ. (2004). Comparison of antioxidant activity of clove (Eugenia caryophylata Thunb) buds and lavender (Lavandula stoechas L.). *Food chemistry, 87*(3), 393-400.

Güler, Z., & Park, Y. W. (2011). Characteristics of physico-chemical properties, volatile compounds and free fatty acid profiles of commercial set-type Turkish yoghurts. *Open Journal of Animal Sciences, 1*(1), 1-8.

Gwee, K. A., Lee, W. R. W., Chua, Q., Chiou, F. K., Aw, M. M., & Koh, Y. H. (2025). The evidence for probiotics in the treatment of digestive disorders in the pediatric population. *Journal of Gastroenterology and Hepatology, 40*(1), 41-47.

Hamidpour, R., Hamidpour, M., Hamidpour, S., & Shahlari, M. (2015). Cinnamon from the selection of traditional applications to its novel effects on the inhibition of angiogenesis in cancer cells and prevention of Alzheimer's disease, and a series of functions such as antioxidant, anticholesterol, antidiabetes, antibacterial, antifungal, nematicidal, acaracidal, and repellent activities. *Journal of Traditional and Complementary Medicine, 5*(2), 66-70.

Han, X., Yang, Z., Jing, X., Yu, P., Zhang, Y., Yi, H., & Zhang, L. (2016). Improvement of the texture of yogurt by use of exopolysaccharide producing lactic acid bacteria. *BioMed Research International, 2016*(1), 7945675.

Hansel, R., Keller, K., Rimpler, H., & Schneider, G. (1992). Uvae ursi folium (Bärentraubenblätter)" In: *Hagers Handbuch der pharmazeutischen Praxis*: 4, Drogen AD.

Hao, C. Y., Rui, F. A. N., Ribeiro, M. C., Tan, L. H., Wu, H. S., Yang, J. F., ... & Huan, Y. (2012). Modeling the potential geographic distribution of black pepper (Piper nigrum) in Asia using GIS tools. *Journal of Integrative Agriculture, 11*(4), 593-599.

Hartl, A., & Vogl, C. R. (2002). Dry matter and fiber yields, and the fiber characteristics of five nettle clones (Urtica dioica L.) organically grown in Austria for potential textile use. *American Journal of Alternative Agriculture, 17*(4), 195-200.

Hashemi, P., & Erim, F. B. (2016). Analysis of vitamin B2 in saffron stigmas (Crocus sativus L) by capillary electrophoresis coupled with laser-induced fluorescence detector. *Food Analytical Methods, 9*, 2395-2399.

Hashizume, K., Tsukahara, T., Yamada, K., Koyama, H., & Ushida, K. (2003). Megasphaera elsdenii JCM1772T normalizes hyperlactate production in the large intestine of fructooligosaccharide-fed rats by stimulating butyrate production. *The Journal of Nutrition, 133*(10), 3187-3190.

Hawksworth, D. L. (2004). Fungal diversity and its implications for genetic resource collections. *Studies in Mmycology, 50*(1), 9-17.
Heath, D. D., Khwaja, F., & Rock, C. L. (2004, March). Curcumin content of turmeric and curry powders. *Faseb Journal, 18*,(4)125-125
Heidari, T., Moazedi, A. A., Seyyednejad, S. M., & Borojeni, M. P. (2017). The role of histaminergic H2 receptors on spasmolytic activity of hydroalcoholic extract of parsley (Petroselinum crispum) Seeds in Isolated Rat's Ileum. *Journal of Natural Remedies*, 114-124.
Hemsworth, J., Hekmat, S., & Reid, G. (2011). The development of micronutrient supplemented probiotic yogurt for people living with HIV: Laboratory testing and sensory evaluation. *Innovative Food Science & Emerging Technologies, 12*(1), 79-84.
Hendrich, A. B. (2006). Flavonoid-membrane interactions: possible consequences for biological effects of some polyphenolic compounds 1. *Acta Pharmacologica Sinica, 27*(1), 27-40.
Holzapfel, W., Schillinger, U., & Buckenhüskes, H. J. (2003). 14 Sauerkraut. *Handbook of fermented functional foods*, 343.
Hong, J., Bose, M., Ju, J., Ryu, J. H., Chen, X., Sang, S., ... & Yang, C. S. (2004). Modulation of arachidonic acid metabolism by curcumin and related β-diketone derivatives: effects on cytosolic phospholipase A 2, cyclooxygenases and 5-lipoxygenase. *Carcinogenesis, 25*(9), 1671-1679.
Hong, S. H., Ismail, I. A., Kang, S. M., Han, D. C., & Kwon, B. M. (2016). Cinnamaldehydes in cancer chemotherapy. *Phytotherapy Research, 30*(5), 754-767.
Hoseinynejad, K., Amini, F., Hasanloo, E., Shayanpour, S., Pouladzadeh, M., & Nazer, M. (2022). Phytotherapy in Renal Failure Due to Blood Pressure and Diabetes: A Systematic Review Study in Iram Ethnobotanical Documents. *Indian Journal of Forensic Medicine & Toxicology, 16*(1).
Hosseinzadeh, H., Ziaee, T., & Sadeghi, A. (2008). The effect of saffron, Crocus sativus stigma, extract and its constituents, safranal and crocin on sexual behaviors in normal male rats. *Phytomedicine, 15*(6-7), 491-495.
Hritcu, L., Noumedem, J. A., Cioanca, O., Hancianu, M., Postu, P., & Mihasan, M. (2015). Anxiolytic and antidepressant profile of the methanolic extract of Piper nigrum fruits in beta-amyloid (1–42) rat model of Alzheimer's disease. *Behavioral and Brain Functions, 11*, 1-13.
Islam, A. T., Islam, A. T., & Mazumder, K. (2015). In vivo anti-obesity activity of methanolic extract of Helianthus annuus seeds. *International Journal of Applied Research, 1*(13), 518-22.
Islam, M. S., & Choi, H. (2008). Comparative effects of dietary ginger (Zingiber officinale) and garlic (Allium sativum) investigated in a type 2 diabetes model of rats. *Journal of Medicinal Food, 11*(1), 152-159.
Ispirli, H., & Dertli, E. (2017). Isolation and characterisation of lactic acid bacteria from traditional koumiss and kurut. *International journal of food properties*, 20(3), 2441-2449.
Jadouali, S. M., H. Atifi, R. Mamouni, K. Majourhat, Z. Bouzoubaâ, A. Laknifli, and A. Faouzi. 2019. Chemical characterization and antioxidant compounds of flower parts of Moroccan Crocus sativus L. *Journal of the Saudi Society of Agricultural Sciences 18*(4),476–80.
Jeena, K., Liju, V. B., Umadevi, N. P., & Kuttan, R. (2014). Antioxidant, anti-inflammatory and antinociceptive properties of black pepper essential oil (Piper nigrum Linn). *Journal of Essential Oil Bearing Plants, 17*(1), 1-12.
Jiraungkoorskul, W. A. N. N. E. E. (2016). Review of nutraceutical uses of an antioxidant sunflower sprout, Helianthus annuus. *Asian Journal of Pharmaceutical and Clinical Research, 9*(6), 21-23.
Jirovetz, L., Buchbauer, G., Stoilova, I., Stoyanova, A., Krastanov, A., & Schmidt, E. (2006). Chemical composition and antioxidant properties of clove leaf essential oil. *Journal of Agricultural and Food Chemistry, 54*(17), 6303-6307.
Jonsson, S. (1983). Blomsterboken: markens urter, lyng og trær. Teknologisk forl..
Jouad, H., Haloui, M., Rhiouani, H., El Hilaly, J., & Eddouks, M. (2001). Ethnobotanical survey of medicinal plants used for the treatment of diabetes, cardiac and renal diseases in the North centre region of Morocco (Fez–Boulemane). *Journal of Ethnopharmacology, 77*(2-3), 175-182.
Jung, H. W., Yoon, C. H., Park, K. M., Han, H. S., & Park, Y. K. (2009). Hexane fraction of Zingiberis Rhizoma Crudus extract inhibits the production of nitric oxide and proinflammatory cytokines in LPS-stimulated BV2 microglial cells via the NF-kappaB pathway. *Food and Chemical Toxicology, 47*(6), 1190-1197.

Jung, J. Y., Lee, S. H., Kim, J. M., Park, M. S., Bae, J. W., Hahn, Y., ... & Jeon, C. O. (2011). Metagenomic analysis of kimchi, a traditional Korean fermented food. *Applied and Environmental Microbiology, 77*(7), 2264-2274.

Jung, Y., Kim, I., Mannaa, M., Kim, J., Wang, S., Park, I., ... & Seo, Y. S. (2019). Effect of Kombucha on gut-microbiota in mouse having non-alcoholic fatty liver disease. *Food Science and Biotechnology, 28*, 261-267.

Kabak, B., & Dobson, A. D. (2011). An introduction to the traditional fermented foods and beverages of Turkey. *Critical Reviews in Food Science and Nutrition, 51*(3), 248-260.

Kampa, M., Nifli, A. P., Notas, G., & Castanas, E. (2007). Polyphenols and cancer cell growth. *Reviews of Physiology, Biochemistry and Pharmacology*, 79-113.

Kapoor, L. D. (2018). *CRC handbook of Ayurvedic medicinal plants*. CRC press.

Karimi, E., E. Oskoueian, R. Hendra, and H. Z. Jaafar. 2010. Evaluation of Crocus sativus L. stigma phenolic and flavonoid compounds and its antioxidant activity. *Molecules (Basel, Switzerland), 15* (9),6244–6256.

Kasouni, A. I., Chatzimitakos, T. G., Stalikas, C. D., Trangas, T., Papoudou-Bai, A., & Troganis, A. N. (2021). The unexplored wound healing activity of urtica dioica L. extract: an *in vitro* and *in vivo* study. *Molecules, 26*(20), 6248.

Kaundal, R., Kumar, M., Kumar, S., Singh, D., & Kumar, D. (2022). Polyphenolic profiling, antioxidant, and antimicrobial activities revealed the quality and adaptive behavior of viola species, a dietary spice in the himalayas. *Molecules, 27*(12), 3867.

Khalili, M., Dehdar, T., Hamedi, F., Ebrahimzadeh, M., & Karami, M. (2015). Antihypoxic activities of Eryngium caucasicum. *European Review for Medical and Pharmacological Sciences, 19*(17), 3282-5.

Khasnavis, S., & Pahan, K. (2014). Cinnamon treatment upregulates neuroprotective proteins Parkin and DJ-1 and protects dopaminergic neurons in a mouse model of Parkinson's disease. *Journal of Neuroimmune Pharmacology, 9*(4), 569–581.

Khattak, S., Shah, H. U., Ahmad, W., & Ahmad, M. (2005). Biological effects of indigenous medicinal plants Curcuma longa and Alpinia galanga. *Fitoterapia, 76*(2), 254-257.

Kim, D. C., Kim, S. H., Choi, B. H., Baek, N. I., Kim, D., Kim, M. J., & Kim, K. T. (2005). Curcuma longa extract protects against gastric ulcers by blocking H2 histamine receptors. *Biological and Pharmaceutical Bulletin, 28*(12), 2220-2224.

Kim, E. K., An, S. Y., Lee, M. S., Kim, T. H., Lee, H. K., Hwang, W. S., ... & Lee, K. W. (2011). Fermented kimchi reduces body weight and improves metabolic parameters in overweight and obese patients. *Nutrition Research, 31*(6), 436-443.

Kim, H. Y., & Park, K. Y. (2018). Clinical trials of kimchi intakes on the regulation of metabolic parameters and colon health in healthy Korean young adults. *Journal of Functional Foods, 47*, 325-333.

Kim, H. Y., Park, E. S., Choi, Y. S., Park, S. J., Kim, J. H., Chang, H. K., & Park, K. Y. (2022). Kimchi improves irritable bowel syndrome: results of a randomized, double-blind placebo-controlled study. *Food & Nutrition Research, 66*(10), 29219.

Kim, M., & Chun, J. (2005). Bacterial community structure in kimchi, a Korean fermented vegetable food, as revealed by 16S rRNA gene analysis. *International Journal of Food Microbiology, 103*(1), 91-96.

Kim, M., Miyamoto, S., Yasui, Y., Oyama, T., Murakami, A., & Tanaka, T. (2009). Zerumbone, a tropical ginger sesquiterpene, inhibits colon and lung carcinogenesis in mice. *International Journal of Cancer, 124*(2), 264-271.

Kinghorn, A. D., Pan, L., Fletcher, J. N., & Chai, H. (2011). The relevance of higher plants in lead compound discovery programs. *Journal of Natural Products, 74*(6), 1539-1555.

Kiokias, S., & Oreopoulou, V. (2021). A review of the health protective effects of phenolic acids against a range of severe pathologic conditions (including coronavirus-based infections). *Molecules, 26*(17), 5405.

Klotz, U. (1991). Occurrence of "natural" benzodiazepines. *Life Sciences, 48*(3), 209-215.

Knez, E., Hałasa, R., Turecka, K., Ośko, J., Kadac-Czapska, K., Waleron, K., & Grembecka, M. (2024). Microbiological Safety and Health Properties of Marketed Fermented Root Vegetables. *Applied Sciences, 15*(1), 121.

Koh, A., De Vadder, F., Kovatcheva-Datchary, P., & Bäckhed, F. (2016). From dietary fiber to host physiology: short-chain fatty acids as key bacterial metabolites. *Cell, 165*(6), 1332–1345.

Koh, Y. J., Cha, D. S., Ko, J. S., Park, H. J., & Choi, H. D. (2010). Anti-inflammatory effect of Taraxacum officinale leaves on lipopolysaccharide-induced inflammatory responses in RAW 264.7 cells. *Journal of Medicinal Food, 13*(4), 870-878.

Kopriva, S., Malagoli, M., & Takahashi, H. (2019). Sulfur nutrition: impacts on plant development, metabolism, and stress responses. *Journal of Experimental Botany, 70*(16), 4069-4073.

Kortbeek, R. W., van der Gragt, M., & Bleeker, P. M. (2019). Endogenous plant metabolites against insects. *European Journal of Plant Pathology, 154*, 67-90.

Kulkarni, S. K., & Dhir, A. (2010). An overview of curcumin in neurological disorders. *Indian Journal of Pharmaceutical Sciences, 72*(2), 149.

Kumar, N., & Goel, N. (2019). Phenolic acids: Natural versatile molecules with promising therapeutic applications. *Biotechnology Reports, 24*, e00370.

Kundu, S., Biswas, T. K., Das, P., Kumar, S., & De, D. K. (2005). Turmeric (Curcuma longa) rhizome paste and honey show similar wound healing potential: a preclinical study in rabbits. *The International Journal of Lower Extremity Wounds, 4*(4), 205-213.

Kurokawa, M., Hozumi, T., Basnet, P., Nakano, M., Kadota, S., Namba, T., ... & Shiraki, K. (1998). Purification and characterization of eugeniin as an anti-herpesvirus compound from Geum japonicum and Syzygium aromaticum. *The Journal of Pharmacology and Experimental Therapeutics, 284*(2), 728-735.

Künstler, A., Gullner, G., Ádám, A. L., Kolozsváriné Nagy, J., & Király, L. (2020). The versatile roles of sulfur-containing biomolecules in plant defense—A road to disease resistance. *Plants, 9*(12), 1705.

Lall, R. K., Syed, D. N., Adhami, V. M., Khan, M. I., & Mukhtar, H. (2015). Dietary polyphenols in prevention and treatment of prostate cancer. *International Journal of Molecular Sciences, 16*(2), 3350-3376.

Langner, E., Greifenberg, S., & Gruenwald, J. (1998). Ginger: history and use. *Advances in Therapy, 15*(1), 25-44.

Larsen, E., & Christensen, L. P. (2005). Simple saponification method for the quantitative determination of carotenoids in green vegetables. *Journal of Agricultural and Food Chemistry, 53*(17), 6598-6602.

Lattanzio, V., Cardinali, A., & Linsalata, V. (2012). Plant phenolics: a biochemical and physiological perspective. *Recent Advances in Polyphenol Research, 3*, 1-39.

Lattanzio, V., Kroon, P. A., Quideau, S., & Treutter, D. (2009). Plant phenolics—secondary metabolites with diverse functions. *Recent Advances in Polyphenol Research, 1*, 1-35.

Lee, H. S. (2006). Antiplatelet property of Curcuma longa L. rhizome-derived ar-turmerone. *Bioresource Technology, 97*(12), 1372-1376.

Lee, H., Yoon, H., Ji, Y., Kim, H., Park, H., Lee, J., ... & Holzapfel, W. (2011). Functional properties of Lactobacillus strains isolated from kimchi. *International Journal of Food Microbiology, 145*(1), 155-161.

Lee, S. H., Whon, T. W., Roh, S. W., & Jeon, C. O. (2020). Unraveling microbial fermentation features in kimchi: from classical to meta-omics approaches. *Applied Microbiology and Biotechnology, 104*, 7731-7744.

Lee, Y. K., Yuk, D. Y., Lee, J. W., Lee, S. Y., Ha, T. Y., Oh, K. W., ... & Hong, J. T. (2009). (−)-Epigallocatechin-3-gallate prevents lipopolysaccharide-induced elevation of beta-amyloid generation and memory deficiency. *Brain Research, 1250*, 164-174.

Leksir, C., Boudalia, S., Moujahed, N., & Chemmam, M. (2019). Traditional dairy products in Algeria: case of Klila cheese. *Journal of Ethnic Foods, 6*(1), 1-14.

Li, A. N., Li, S., Zhang, Y. J., Xu, X. R., Chen, Y. M., & Li, H. B. (2014). Resources and biological activities of natural polyphenols. *Nutrients, 6*(12), 6020-6047.

Li, D. P., & Yang, S. L. (2002). Modern Chinese Material Medica.

Li, H. Y., Lee, B. K., Kim, J. S., Jung, S. J., & Oh, S. B. (2008). Eugenol inhibits ATP-induced P2X currents in trigeminal ganglion neurons. *The Korean Journal of Physiology & Pharmacology, 12*(6), 315.

Li, R., Liang, T., Xu, L., Li, Y., Zhang, S., & Duan, X. (2013). Protective effect of cinnamon polyphenols against STZ-diabetic mice fed high-sugar, high-fat diet and its underlying mechanism. *Food and Chemical Toxicology, 51*, 419-425.

Lim, T. K. 2012. *Edible medicinal and non-medicinal plants*. Vol. 1. Dordrecht: Springer.

Lingaraju, D. P., Sudarshana, M. S., & Rajashekar, N. (2013). Ethnopharmacological survey of traditional medicinal plants in tribal areas of Kodagu district, Karnataka, India. *Journal of pharmacy research, 6*(2), 284-297.

Liu, W., Bao, Q., Qing, M., Chen, X., Sun, T., Li, M., ... & Zhang, H. (2012). Isolation and identification of lactic acid bacteria from Tarag in Eastern Inner Mongolia of China by 16S rRNA sequences and DGGE analysis. *Microbiological Research, 167*(2), 110-115.

Liu, W., Feng, Y., Yu, S., Fan, Z., Li, X., Li, J., & Yin, H. (2021). The flavonoid biosynthesis network in plants. *International Journal of Molecular Sciences, 22*(23), 12824.

Llorca, E., Puig, A., Hernando, I., Salvador, A., Fiszman, S. M., & Lluch, M. A. (2001). Effect of fermentation time on texture and microstructure of pickled carrots. *Journal of the Science of Food and Agriculture, 81*(15), 1553-1560.

Lore, T. A., Mbugua, S. K., & Wangoh, J. (2005). Enumeration and identification of microflora in suusac, a Kenyan traditional fermented camel milk product. *LWT Food Science and Technology, 38*(2), 125-130.

Luthria, D. L. (2008). Influence of experimental conditions on the extraction of phenolic compounds from parsley (Petroselinum crispum) flakes using a pressurized liquid extractor. *Food Chemistry, 107*(2), 745-752.

Major, N., Bažon, I., Išić, N., Kovačević, T. K., Ban, D., Radeka, S., & Goreta Ban, S. (2022). Bioactive properties, volatile compounds, and sensory profile of sauerkraut are dependent on cultivar choice and storage conditions. *Foods, 11*(9), 1218.

Malangu, N. (2007). Self-reported use of traditional, complementary and over-the-counter medicines by HIV-infected patients on antiretroviral therapy in Pretoria, South Africa. *African Journal of Traditional, Complementary and Alternative Medicines, 4*(3), 273-278.

Mallik, B. K., Panda, T., & Padhy, R. N. (2012). Traditional herbal practices by the ethnic people of Kalahandi district of Odisha, India. *Asian Pacific Journal of Tropical Biomedicine, 2*(2), 988-994.

Mallikarjuna, K., Chetan, P. S., Reddy, K. S., & Rajendra, W. (2008). Ethanol toxicity: Rehabilitation of hepatic antioxidant defense system with dietary ginger. *Fitoterapia, 79*(3), 174-178.

Manach, C., Williamson, G., Morand, C., Scalbert, A., & Rémésy, C. (2005). Bioavailability and bioefficacy of polyphenols in humans. I. Review of 97 bioavailability studies. *The American Journal of Clinical Nutrition, 81*(1), 230-242.

Mangia, N. P., Garau, G., Murgia, M. A., Bennani, A., & Deiana, P. (2014). Influence of autochthonous lactic acid bacteria and enzymatic yeast extracts on the microbiological, biochemical and sensorial properties of Lben generic products. *Journal of Dairy Research, 81*(2), 193-201.

Mansour, S. R., Moustafa, M. A. A., Saad, B. M., Hamed, R., & Moustafa, A. R. (2021). Impact of diet on human gut microbiome and disease risk. *New Microbes and New Infections, 41*, 100845.

Marco, M. L., Heeney, D., Binda, S., Cifelli, C. J., Cotter, P. D., Foligné, B., ... & Hutkins, R. (2017). Health benefits of fermented foods: microbiota and beyond. *Current Opinion in Biotechnology, 44*, 94-102.

Marco, M. L., Sanders, M. E., Gänzle, M., Arrieta, M. C., Cotter, P. D., De Vuyst, L., ... & Hutkins, R. (2021). The International Scientific Association for Probiotics and Prebiotics (ISAPP) consensus statement on fermented foods. *Nature Reviews Gastroenterology & Hepatology, 18*(3), 196-208.

Mariappan, P. M., Sabesan, G., Koilpillai, B., Janakiraman, S., & Sharma, N. K. (2013). Chemical characterisation and antifungal activity of methanolic extract of Cinnamomum verum J. Presl bark against Malassezia spp. *Pharmacognosy Journal, 5*(5), 197–204.

Marinova, E. M., & Yanishlieva, N. V. (2003). Antioxidant activity and mechanism of action of some phenolic acids at ambient and high temperatures. *Food Chemistry, 81*(2), 189-197.

Martínez Leal, J., Valenzuela Suárez, L., Jayabalan, R., Huerta Oros, J., & Escalante-Aburto, A. (2018). A review on health benefits of kombucha nutritional compounds and metabolites. *CyTA-Journal of Food, 16*(1), 390-399.

Mathara, J. M., Schillinger, U., Kutima, P. M., Mbugua, S. K., & Holzapfel, W. H. (2004). Isolation, identification and characterisation of the dominant microorganisms of kule naoto: the Maasai traditional fermented milk in Kenya. *International Journal of Food Microbiology, 94*(3), 269-278.

McNulty, N. P., Yatsunenko, T., Hsiao, A., Faith, J. J., Muegge, B. D., Goodman, A. L., ... & Gordon, J. I. (2011). The impact of a consortium of fermented milk strains on the gut microbiome of gnotobiotic mice and monozygotic twins. *Science Translational Medicine, 3*(106), 106ra106-106ra106.

Mekhfi, H., El Haouari, M., Legssyer, A., Bnouham, M., Aziz, M., Atmani, F., ... & Ziyyat, A. (2004). Platelet anti-aggregant property of some Moroccan medicinal plants. *Journal of Ethnopharmacology, 94*(2-3), 317-322.

Melnyk, J. P., Wang, S., & Marcone, M. F. (2010). Chemical and biological properties of the world's most expensive spice: Saffron. *Food Research International, 43*(8), 1981-1989.

Menia, M., Iqbal, S., Zahida, R., Tahir, S., Kanth, R. H., Saad, A. A., & Hussian, A. (2018). Production technology of saffron for enhancing productivity. J. Pharmacognos. *Phytochem, 7*(1), 1033-1039.

Merzouki, A., Ed-Derfoufi, F., El Aallali, A., & Molero-Mesa, J. (1997). Wild medicinal plants used by local Bouhmed population (Morocco). *Fitoterapia, 68*(5): 444-460.

Metodiewa, D., Kochman, A., & Karolczak, S. (1997). Evidence for antiradical and antioxidant properties of four biologically active N, N-Diethylaminoethyl ethers of flavaone oximes: A comparison with natural polyphenolic flavonoid rutin action. *International Union of Biochemistry and Molecular Biology Life, 41*(5), 1067-1075.

Miękus, N., Marszałek, K., Podlacha, M., Iqbal, A., Puchalski, C., & Świergiel, A. H. (2020). Health benefits of plant-derived sulfur compounds, glucosinolates, and organosulfur compounds. *Molecules, 25*(17), 3804.

Mikaeili, A., Karimi, I., Modaresi, M., & Bagherinasab, Z. (2013). Assessment of antidermatophytic activities of Urtica dioica L against Microsporum canis in a guinea pig model. *Tropical Journal of Pharmaceutical Research, 12*(6), 997-1002.

Moazedi, A. A., Mirzaie, D. N., Seyyednejad, S. M., Zadkarami, M. R., & Amirzargar, A. (2007). Spasmolytic effect of Petroselinum crispum (Parsley) on rat's ileum at different calcium chloride concentrations. *Pakistan Journal of Biological Sciences, 10*(22), 4036-4042.

Mohanta, T. K., Tamboli, Y., & Zubaidha, P. K. (2014). Phytochemical and medicinal importance of Ginkgo biloba L. *Natural Product Research, 28*(10), 746–752.

Molina, R. V., Valero, M., Navarro, Y., Guardiola, J. L., & Garcia-Luis, A. J. S. H. (2005). Temperature effects on flower formation in saffron (Crocus sativus L.). *Scientia Horticulturae, 103*(3), 361-379.

Montesano, V., Negro, D., Sarli, G., De Lisi, A., Laghetti, G., & Hammer, K. (2012). Notes about the uses of plants by one of the last healers in the Basilicata Region (South Italy). *Journal of Ethnobiology and Ethnomedicine, 8*, 1-10.

Morales, D. (2020). Biological activities of kombucha beverages: The need of clinical evidence. *Trends in Food Science & Technology, 105*, 323-333.

Moreira, G. V., Araujo, L. C., Murata, G. M., Matos, S. L., & Carvalho, C. R. (2022). Kombucha tea improves glucose tolerance and reduces hepatic steatosis in obese mice. *Biomedicine & Pharmacotherapy, 155*, 113660.

Mozaffarian, V. (2013). *Identification of medicinal and aromatic plants of Iran.*

Mukundi, M. J., Mwaniki, N. E., Piero, N. M., Murugi, N. J., Juma, K. K., Yusuf, A. A., ... & Muchugi, A. N. (2017). Potential anti-diabetic effects and safety of aqueous extracts of Urtica dioica collected from Narok County, *Kenya, 8*(5), 1000548.

Muthaura, C. N., Rukunga, G. M., Chhabra, S. C., Mungai, G. M., & Njagi, E. N. M. (2007). Traditional phytotherapy of some remedies used in treatment of malaria in Meru district of Kenya. *South African Journal of Botany, 73*(3), 402-411.

Mzid, M., Ben Khedir, S., Ben Salem, M., Regaieg, W., & Rebai, T. (2017). Antioxidant and antimicrobial activities of ethanol and aqueous extracts from Urtica urens. *Pharmaceutical Biology, 55*(1), 775-781.

Na, G., & Salt, D. E. (2011). The role of sulfur assimilation and sulfur-containing compounds in trace element homeostasis in plants. *Environmental and Experimental Botany, 72*(1), 18-25.

Nahak, G., & Sahu, R. K. (2011). Phytochemical evaluation and antioxidant activity of Piper cubeba and Piper nigrum. *Journal of Applied Pharmaceutical Science, 1*(8), 153-157.

Najafian, Y., Hamedi, S. S., Farshchi, M. K., & Feyzabadi, Z. (2018). Plantago major in Traditional Persian Medicine and modern phytotherapy: a narrative review. *Electronic Physician, 10*(2), 6390.

Namazi, N., Tarighat, A., & Bahrami, A. (2012). The effect of hydro alcoholic nettle (Urtica dioica) extract on oxidative stress in patients with type 2 diabetes: a randomized double-blind clinical trial. *Pakistan Journal of Biological Sciences, 15*(2), 98-102.

Naveed, R., Hussain, I., Tawab, A., Tariq, M., Rahman, M., Hameed, S., & Iqbal, M. (2013). Antimicrobial activity of the bioactive components of essential oils from Pakistani spices against Salmonella and other multi-drug resistant bacteria. *BMC Complementary and Alternative Medicine, 13*(1), 265.

Nematollahi, Z., Ebrahimi, M., Raeisi, M., Dadban Shahamat, Y., Ghodsi Moghadam, M., Hashemi, M., ... & Shirzad, H. (2020). The antibacterial activity of cinnamon essential oil against foodborne bacteria: a mini-review. *Journal of Human Environment and Health Promotion, 6*(3), 101-105.

Neveu, V., Perez-Jiménez, J., Vos, F., Crespy, V., du Chaffaut, L., Mennen, L., ... & Scalbert, A. (2010). Phenol-Explorer: an online comprehensive database on polyphenol contents in foods. *Database,* 2010.

Nielsen, E. S., Garnås, E., Jensen, K. J., Hansen, L. H., Olsen, P. S., Ritz, C., ... & Nielsen, D. S. (2018). Lacto-fermented sauerkraut improves symptoms in IBS patients independent of product pasteurisation–a pilot study. *Food & Function, 9*(10), 5323-5335.

Nielsen, H., & Kehler, S. (1969). Lægeplanter og trolddomsurter. In: Kehler, S. (Ed.). Politikens Forlag, København, pp. 321-324.

Oladeji, O. S., Oluyori, A. P., Bankole, D. T., & Afolabi, T. Y. (2020). Natural products as sources of antimalarial drugs: ethnobotanical and ethnopharmacological studies. *Scientifica, 2020*(1), 7076139.

Oliviero, F., Scanu, A., Zamudio-Cuevas, Y., Punzi, L., & Spinella, P. (2018). Anti-inflammatory effects of polyphenols in arthritis. *Journal of the Science of Food and Agriculture, 98*(5), 1653-1659.

Onoja, S. O., & Anaga, A. O. (2014). Evaluation of the antidiabetic and antioxidant potentials of methanolic leaf extract of Helianthus annuus L. on alloxan-induced hyperglycemic rats. *Comparative Clinical Pathology, 23*, 1565-1573.

Onyesife, C. O., Ogugua, V. N., & Anaduaka, E. G. (2014). Hypoglycemic potentials of ethanol leaves extract of black pepper (Piper nigrum) on alloxan-induced diabetic rats. *Annals of Biological Research, 5*(6), 26-31.

Ortu, S., Felis, G. E., Marzotto, M., Deriu, A., Molicotti, P., Sechi, L. A., ... & Zanetti, S. (2007). Identification and functional characterization of Lactobacillus strains isolated from milk and Gioddu, a traditional Sardinian fermented milk. *International Dairy Journal, 17*(11), 1312-1320.

Ovadje, P., Hamm, C., & Pandey, S. (2012). Efficient induction of extrinsic cell death by dandelion root extract in human chronic myelomonocytic leukemia (CMML) cells. *PloS one, 7*(2), e30604.

Ozgoli, G., Goli, M., & Simbar, M. (2009). Effects of ginger capsules on pregnancy, nausea, and vomiting. The *Journal of Alternative and Complementary Medicine, 15*(3), 243-246.

Panche, A. N., Diwan, A. D., & Chandra, S. R. (2016). Flavonoids: an overview. *Journal of Nutritional Science, 5*, e47.

Park, C. M., Youn, H. J., Chang, H. K., & Song, Y. S. (2010). TOP1 and 2, polysaccharides from Taraxacum officinale, attenuate CCl4-induced hepatic damage through the modulation of NF-κB and its regulatory mediators. *Food and Chemical Toxicology, 48*(5), 1255-1261.

Park, E. J., Jeon, C. H., Ko, G., Kim, J., & Sohn, D. H. (2000). Protective effect of curcumin in rat liver injury induced by carbon tetrachloride. *Journal of Pharmacy and Pharmacology, 52*(4), 437-440.

Parmar, V. S., Jain, S. C., Bisht, K. S., Jain, R., Taneja, P., Jha, A., ... & Boll, P. M. (1997). Phytochemistry of the genus Piper. *Phytochemistry, 46*(4), 597-673.

Pereira, A. G., Cassani, L., Garcia-Oliveira, P., Otero, P., Mansoor, S., Echave, J., ... & Prieto, M. A. (2023). Plant alkaloids: production, extraction, and potential therapeutic properties. In *Natural secondary metabolites: from nature, through science, to industry* (pp. 157-200). Cham: Springer International Publishing.

Perez-Conesa, D., McLandsborough, L., & Weiss, J. (2006). Inhibition and inactivation of Listeria monocytogenes and Escherichia coli O157: H7 colony biofilms by micellar-encapsulated eugenol and carvacrol. *Journal of Food Protection, 69*(12), 2947-2954.

Pérez-Jiménez, J., Neveu, V., Vos, F., & Scalbert, A. (2010). Identification of the 100 richest dietary sources of polyphenols: an application of the Phenol-Explorer database. *European Journal of Clinical Nutrition, 64*(3), 112-120.

Pitsikas, N. (2016). Constituents of saffron (Crocus sativus L.) as potential candidates for the treatment of anxiety disorders and schizophrenia. *Molecules (Basel, Switzerland), 21*(3):303.

Pittman, S. (2011). Cinnamon: It's not just for making cinnamon rolls. *Ethnobotanical Leaflets, 2000*(2), 11.

Policepatel, S. S., & Manikrao, V. G. (2013). Ethnomedicinal plants used in the treatment of skin diseases in Hyderabad Karnataka region, Karnataka, India. *Asian Pacific Journal of Tropical Biomedicine, 3*(11), 882-886.

Pourjoula, M., Picariello, G., Garro, G., D'Auria, G., Nitride, C., Ghaisari, A. R., & Ferranti, P. (2020). The protein and peptide fractions of kashk, a traditional Middle East fermented dairy product. *Food Research International, 132*, 109107.

Prasad, R., & Shivay, Y. S. (2018). Sulphur in soil, plant and human nutrition. Proceedings of the National Academy of Sciences, India Section B: *Biological Sciences, 88*, 429-434.

Prashant, A., Rangaswamy, C., Yadav, A. K., Reddy, V., Sowmya, M. N., & Madhunapantula, S. (2017). In vitro anticancer activity of ethanolic extracts of Piper nigrum against colorectal carcinoma cell lines. *International Journal of Applied and Basic Medical Research, 7*(1), 67-72.

Price, A. H., Atherton, N. M., & Hendry, G. A. (1989). Plants under drought-stress generate activated oxygen. *Free Radical Research Communications, 8*(1), 61-66.

Rahemi, D., Babaee, N., Kazemi, S., Sefidgar, S. A. A., & Moghadamnia, A. A. (2015). An in vitro study of the effect of cinnamaldehyde on the growth of Candida albicans compared to nystatin and fluconazole. *Crescent Journal of Medical and Biological Science,2*,70-80.

Ramos, S. (2008). Cancer chemoprevention and chemotherapy: dietary polyphenols and signalling pathways. *Molecular Nutrition & Food Research, 52*(5), 507-526.

Rana, I. S., Rana, A. S., & Rajak, R. C. (2011). Evaluation of antifungal activity in essential oil of the Syzygium aromaticum (L.) by extraction, purification and analysis of its main component eugenol. *Brazilian Journal of Microbiology, 42*, 1269-1277.

Rani, R., Sheoran, R., & Sharma, B. (2017). Perspectives of breeding for altering sunflower oil quality to obtain novel oils. *International Journal of Current Microbiology and Applied Sciences, 6*(8), 949-962.

Ravindran, P. N., & Kallupurackal, J. A. (2012). Black pepper. In Han, X., Z. Yang, X. Jing, P. Yu, Y. Zhang, H. Yi, and L. Zhang. Book of herbs and spices (pp. 86-115). Woodhead Publishing.

Razak, S. I. A., M. S. Anwar Hamzah, F. C. Yee, M. R. A. Kadir, and N. H. M. Nayan. 2017. A review on medicinal properties of saffron toward major diseases. *Journal of Herbs, Spices & Medicinal Plants, 23* (2):98–116.

Ribeiro-Santos, R., Andrade, M., Madella, D., Martinazzo, A. P., Moura, L. D. A. G., de Melo, N. R., & Sanches-Silva, A. (2017). Revisiting an ancient spice with medicinal purposes: Cinnamon. *Trends in Food Science & Technology, 62*, 154-169.

Rios, J. L., Simeon, S., & Villar, A. (1989). Pharmacological activity of aporphinoid alkaloids. A review. Pharmacological activity of aporphinoid alkaloids. *Fitoterapia, 60*(5), 387-412.

Roca-Garcia, H. (1970). Weeds: a link with the past. *Arnoldia, 30*(3), 114-115.

Rolnik, A., & Olas, B. (2021). The plants of the Asteraceae family as agents in the protection of human health. *International Journal of Molecular Sciences, 22*(6), 3009.

Roy, R. D., Hossan, M. S., & Rahmatullah, M. (2015). A review of anticancer potential of elephantopus scaber and its phytoconstituents. *World Journal of Pharmacology and Pharmaceutical L Science, 4*(10), 86-94.

Ruwizhi, N., & Aderibigbe, B. A. (2020). Cinnamic acid derivatives and their biological efficacy. *International Journal of Molecular Sciences, 21*(16), 5712.

Sadeghi, A. (2016). In vitro assessment of some probiotic properties of Lactobacillus fermentum isolated from pickled garlic. *Journal of Food Quality and Hazards Control, 3*(2), 67-72.

Samuelsen, A. B. (2000). The traditional uses, chemical constituents and biological activities of Plantago major L. A review. *Journal of Ethnopharmacology, 71*(1-2), 1-21.

Šavikin, K., Zdunić, G., Menković, N., Živković, J., Ćujić, N., Tereščenko, M., & Bigović, D. (2013). Ethnobotanical study on traditional use of medicinal plants in South-Western Serbia, Zlatibor district. *Journal of Ethnopharmacology, 146*(3), 803-810.

Schütz, K., Carle, R., & Schieber, A. (2006). Taraxacum—a review on its phytochemical and pharmacological profile. *Journal of Ethnopharmacology, 107*(3), 313-323.

Selhub, E. M., Logan, A. C., & Bested, A. C. (2014). Fermented foods, microbiota, and mental health: ancient practice meets nutritional psychiatry. *Journal of Physiological Anthropology, 33*, (1) 1-12.

Semwal, R. B., Semwal, D. K., Combrinck, S., & Viljoen, A. M. (2015). Gingerols and shogaols: Important nutraceutical principles from ginger. *Phytochemistry, 117*, 554-568.

Sethiya, N. K., Ahmed, N. M., Shekh, R. M., Kumar, V., Singh, P. K., & Kumar, V. (2018). Ethnomedicinal, phytochemical and pharmacological updates on Hygrophila auriculata (Schum.) Hiene: an overview. *Journal of Integrative Medicine, 16*(5), 299-311.

Shahabi, S., Hassan, Z. M., Mahdavi, M., Dezfouli, M., Rahvar, M. T., Naseri, M., ... & Khalkhali, H. R. (2008). Hot and Cold natures and some parameters of neuroendocrine and immune systems in traditional Iranian medicine: a preliminary study. *The Journal of Alternative and Complementary Medicine, 14*(2), 147-156.

Shahbazi, R., Sharifzad, F., Bagheri, R., Alsadi, N., Yasavoli-Sharahi, H., & Matar, C. (2021). Anti-inflammatory and immunomodulatory properties of fermented plant foods. *Nutrients, 13*(5), 1516.

Shahrajabian, M. H., Sun, W., & Cheng, Q. (2019). Clinical aspects and health benefits of ginger (Zingiber officinale) in both traditional Chinese medicine and modern industry. *Acta Agriculturae Scandinavica, Section b—Soil & Plant Science, 69*(6), 546-556.

Shan, B., Cai, Y. Z., Sun, M., & Corke, H. (2005). Antioxidant capacity of 26 spice extracts and characterization of their phenolic constituents. *Journal of Agricultural and Food Chemistry, 53*(20), 7749-7759.

Shangpliang, H. N. J., Rai, R., Keisam, S., Jeyaram, K., & Tamang, J. P. (2018). Bacterial community in naturally fermented milk products of Arunachal Pradesh and Sikkim of India analysed by high-throughput amplicon sequencing. *Scientific Reports, 8*(1), 1532.

Shanmugapriya, K. S. P. S., Saravana, P. S., Payal, H., Mohammed, S., & Williams, B. (2012). Antioxidant potential of pepper (Piper nigrum Linn.) leaves and its antimicrobial potential against some pathogenic microbes. *Indian Journal of Natural Products and Resources,3*(4),570-577.

Sharma, J., Gairola, S., Gaur, R. D., Painuli, R. M., & Siddiqi, T. O. (2013). Ethnomedicinal plants used for treating epilepsy by indigenous communities of sub-Himalayan region of Uttarakhand, India. *Journal of Ethnopharmacology, 150*(1), 353-370.

Sharma, R., Garg, P., Kumar, P., Bhatia, S. K., & Kulshrestha, S. (2020). Microbial fermentation and its role in quality improvement of fermented foods. *Fermentation, 6*(4), 106.

Sharma, S. S., Kochupillai, V., Gupta, S. K., Seth, S. D., & Gupta, Y. K. (1997). Antiemetic efficacy of ginger (Zingiber officinale) against cisplatin-induced emesis in dogs. *Journal of ethnopharmacology*, 57(2), 93-96.

Sharma, U. K., Sharma, A. K., & Pandey, A. K. (2016). Medicinal attributes of major phenylpropanoids present in cinnamon. *BMC Complementary and Alternative Medicine, 16*, 1-11.

Shetty, S. A., Marathe, N. P., Lanjekar, V., Ranade, D., & Shouche, Y. S. (2013). Comparative genome analysis of Megasphaera sp. reveals niche specialization and its potential role in the human gut. *PloS One, 8*(11), e79353.

Shimojima, M. (2011). Biosynthesis and functions of the plant sulfolipid. *Progress in Lipid Research, 50*(3), 234-239.

Shoskes, D., Lapierre, C., & Cruz-Corerra, M. (2006). Beneficial effects of the bioflavonoids curcumin and quercetin on early function in cadaveric renal transplantation: a randomized placebo controlled trial. *Alternative Medicine Review, 11*(1), 71-72.

Siddiqui, S. A., Erol, Z., Rugji, J., Taşçı, F., Kahraman, H. A., Toppi, V., ... & Castro-Muñoz, R. (2023). An overview of fermentation in the food industry-looking back from a new perspective. *Bioresources and Bioprocessing, 10*(1), 85.

Šikić-Pogačar, M., Turk, D. M., & Fijan, S. (2022). Knowledge of fermentation and health benefits among general population in North-eastern Slovenia. *BMC Public Health, 22*(1), 1695.

Singhal, P., Satya, S., & Naik, S. N. (2021). Fermented bamboo shoots: A complete nutritional, anti-nutritional and antioxidant profile of the sustainable and functional food to food security. Food chemistry: *Molecular Sciences, 3*, 100041.

Slavin, J. (2013). Fiber and prebiotics: mechanisms and health benefits. *Nutrients, 5*(4), 1417-1435.

Sneader, W. (1997). Drug prototypes and their exploitation. *European Journal of Medicinal Chemistry, 1*(32), 91.

Sofia, P. K., Prasad, R., Vijay, V. K., & Srivastava, A. K. (2007). Evaluation of antibacterial activity of Indian spices against common foodborne pathogens. *International Journal of Food Science and Technology*, 42(8), 910-915.

Soliman, H. A., Eltablawy, N. A., & Hamed, M. S. (2015). The ameliorative effect of Petroselinum crispum (parsley) on some diabetes complications. *Journal of Medicinal Plants Studies*, 3(4), 92-100.

Sonkusare, S. (2008). Hyperemesis gravidarum: a review. *Med J Malaysia*, 63(3), 272-6.

Sova, M., & Saso, L. (2020). Natural sources, pharmacokinetics, biological activities and health benefits of hydroxycinnamic acids and their metabolites. *Nutrients*, 12(8), 2190.

Sreejayan, N., Rao, M. N. A., Priyadarsini, K. I., & Devasagayam, T. P. A. (1997). Inhibition of radiation-induced lipid peroxidation by curcumin. *International Journal of Pharmaceutics*, 151(1), 127-130.

Srivastava, S., & Singh, K. (2016). Changes occur on nutritional value of beetroot ("Beta vulgaris") after pickling. *Current Research in Nutrition and Food Science*, 4(3), 217.

Sriwiriyajan, S., Ninpesh, T., Sukpondma, Y., Nasomyon, T., & Graidist, P. (2014). Cytotoxicity screening of plants of genus Piper in breast cancer cell lines. *Tropical Journal of Pharmaceutical Research*, 13(6), 921-928.

Sroka, Z., & Cisowski, W. H. P. S. (2003). Hydrogen peroxide scavenging, antioxidant and anti-radical activity of some phenolic acids. *Food and Chemical Toxicology*, 41(6), 753-758.

Stephens, J. M. (1994). *Parsley--Petroselinum Crispum* (Mill.) Nym. University of Florida Cooperative Extension Service, Institute of Food and Agriculture Sciences, EDIS.

Stoner, G. D., & Mukhtar, H. (1995). Polyphenols as cancer chemopreventive agents. *Journal of Cellular Biochemistry, 59*(S22), 169-180.

Subashini, R., & Rakshitha, S. U. (2012). Phytochemical screening, antimicrobial activity and in vitro antioxidant investigation of methanolic extract of seeds from Helianthus annuus L. *Chemical Science Review and Letters, 1*(1), 30-34.

Sung, B., Murakami, A., Oyajobi, B. O., & Aggarwal, B. B. (2009). Zerumbone abolishes RANKL-induced NF-κB activation, inhibits osteoclastogenesis, and suppresses human breast cancer–induced bone loss in athymic nude mice. *Cancer Research, 69*(4), 1477-1484.

Şanlier, N., Gökcen, B. B., & Sezgin, A. C. (2019). Health benefits of fermented foods. *Critical Reviews in Food Science and Nutrition, 59*(3), 506-527.

Tahri, A., Yamani, S., Legssyer, A., Aziz, M., Mekhfi, H., Bnouham, M., & Ziyyat, A. (2000). Acute diuretic, natriuretic and hypotensive effects of a continuous perfusion of aqueous extract of Urtica dioica in the rat. *Journal of Ethnopharmacology, 73*(1-2), 95-100.

Tamang, J. P., & Kailasapathy, K. (Eds.). (2010). Fermented foods and beverages of the world. CRC press, Taylor & Francis, Boca Raton. https://doi.org/10.1111/j.1471- 0307.2011.00717.x.

Tasdemir, S. S. & Sanlier, N. (2020). An insight into the anticancer effects of fermented foods: a review. *Journal of Functional Foods, 75*, 104281.

Tasleem, F., Azhar, I., Ali, S. N., Perveen, S., & Mahmood, Z. A. (2014). Analgesic and anti-inflammatory activities of Piper nigrum L. *Asian Pacific Jjournal of Tropical Medicine, 7*, 461-468.

Teoh, E. S., & Teoh, E. S. (2016). Secondary metabolites of plants. *Medicinal Orchids of Asia,* 59-73.

Tiku, A. R. (2018). Antimicrobial compounds and their role in plant defense. *Molecular aspects of plant-pathogen interaction,* Singh, A., Singh, I. K., Eds.; Springer: Singapore, 2018; pp. 283–307.

Touret, T., Oliveira, M., & Semedo-Lemsaddek, T. (2018). Putative probiotic lactic acid bacteria isolated from sauerkraut fermentations. *PloS One, 13*(9), e0203501.

Tunick, M. H., & Van Hekken, D. L. (2015). Dairy products and health: recent insights. *Journal of Agricultural and Food Chemistry, 63*(43), 9381-9388.

Umar, O. B., Ranti, L. A., Abdulbaki, A. S., Bola, A. U. L., Abdulhamid, A. K., Biola, M. R., & Victor, K. O. (2021). Stresses in plants: Biotic and abiotic. In *Current trends in wheat research.* IntechOpen.

Upadhyay, S. U., Patel, V. B., Patel, A. A., Upadhyay, U. M., & Patel, N. M. (2012). Ajuga Bracteosa: A Promısıng Herb. *Pharma Science Monitor, 3*(4).

USDA, U. (2013). National nutrient database for standard reference, release 28. US Department of Agriculture, Agricultural Research Service, Nutrient Data Laboratory. The United States Department of Agriculture, Beltsville, MD, USA.

Uyeno, Y., Sekiguchi, Y., & Kamagata, Y. (2008). Impact of consumption of probiotic lactobacilli-containing yogurt on microbial composition in human feces. *International Journal of Food Microbiology, 122*(1-2), 16-22.

Uzonur, I., Akdeniz, G., Katmer, Z., & Ersoy, S. K. (2013). RAPD-PCR and real-time PCR HRM based genetic variation evaluations of Urtica dioica parts, ecotypes and evaluations of morphotypes in Turkey. *African Journal of Traditional, Complementary and Alternative Medicines, 10*(2), 232-245.

Van Nieuwenhove, E., Lagou, V., Van Eyck, L., Dooley, J., Bodenhofer, U., Roca, C., ... & Liston, A. (2019). Machine learning identifies an immunological pattern associated with multiple juvenile idiopathic arthritis subtypes. *Annals of the Rheumatic Diseases, 78*(5), 617-628.

Veiga, P., Pons, N., Agrawal, A., Oozeer, R., Guyonnet, D., Brazeilles, R., ... & Kennedy, S. P. (2014). Changes of the human gut microbiome induced by a fermented milk product. *Scientific Reports, 4*(1), 6328.

Verma, S., Nizam, S., & Verma, P. K. (2013). Biotic and abiotic stress signaling in plants. Stress signaling in plants: genomics and proteomics perspective, Springer Science, New York, USA

Vijayakumar, M., Vasudevan, D. M., Sundaram, K. R., Krishnan, S., Vaidyanathan, K., Nandakumar, S., ... & Mathew, N. (2016). A randomized study of coconut oil versus sunflower oil on cardiovascular risk factors in patients with stable coronary heart disease. *Indian Heart Journal, 68*(4), 498-506.

Vijayakumar, R. S., & Nalini, N. (2006). Efficacy of piperine, an alkaloidal constituent from Piper nigrum on erythrocyte antioxidant status in high fat diet and antithyroid drug induced hyperlipidemic rats. *Cell Biochemistry and Function, 24*(6), 491-498.

Vijayakumar, S., Yabesh, J. M., Prabhu, S., Manikandan, R., & Muralidharan, B. (2015). Quantitative ethnomedicinal study of plants used in the Nelliyampathy hills of Kerala, India. *Journal of Ethnopharmacology, 161*, 238-254.

Vīna, I., Semjonovs, P., Linde, R., & Deniņa, I. (2014). Current evidence on physiological activity and expected health effects of kombucha fermented beverage. *Journal of Medicinal Food, 17*(2), 179-188.

Vogl, C. R., & Hartl, A. (2003). Production and processing of organically grown fiber nettle (Urtica dioica L.) and its potential use in the natural textile industry: A review. *American Journal of Alternative Agriculture, 18*(3), 119-128.

Voidarou, C., Antoniadou, M., Rozos, G., Tzora, A., Skoufos, I., Varzakas, T., ... & Bezirtzoglou, E. (2020). Fermentative foods: Microbiology, biochemistry, potential human health benefits and public health issues. *Foods, 10*(1), 69.

Walker, E. H., Pacold, M. E., Perisic, O., Stephens, L., Hawkins, P. T., Wymann, M. P., & Williams, R. L. (2000). Structural determinants of phosphoinositide 3-kinase inhibition by wortmannin, LY294002, quercetin, myricetin, and staurosporine. *Molecular Cell, 6*(4), 909-919.

Wickenberg, J., Ingemansson, S. L., & Hlebowicz, J. (2010). Effects of Curcuma longa (turmeric) on postprandial plasma glucose and insulin in healthy subjects. *Nutrition Journal, 9*, 1-5.

Wilburn, J. & Ryan, E. (2017). Fermented foods in health promotion and disease prevention: an overview. *Fermented Foods in Health and Disease Prevention*, 3–19. Cambridge, MA: Academic Press.

Wilson, E. M., Johanningsmeier, S. D., & Osborne, J. A. (2015). Consumer acceptability of cucumber pickles produced by fermentation in calcium chloride brine for reduced environmental impact. *Journal of Food Science, 80*(6), 1360-1367.

Wright, C. I., Van-Buren, L., Kroner, C. I., & Koning, M. M. G. (2007). Herbal medicines as diuretics: a review of the scientific evidence. *Journal of Ethnopharmacology, 114*(1), 1-31.

Wu, J. C., Lai, C. S., Lee, P. S., Ho, C. T., Liou, W. S., Wang, Y. J., & Pan, M. H. (2016). Anti-cancer efficacy of dietary polyphenols is mediated through epigenetic modifications. *Current Opinion in Food Science, 8*, 1-7.

Wu, Y., Wan, J., Choe, U., Pham, Q., Schoene, N. W., He, Q., ... & Wang, T. T. (2019). Interactions between food and gut microbiota: impact on human health. *Annual Review of Food Science and Technology, 10*(1), 389-408.

Wullschleger, S., Lacroix, C., Bonfoh, B., Sissoko-Thiam, A., Hugenschmidt, S., Romanens, E., ... & Meile, L. (2013). Analysis of lactic acid bacteria communities and their seasonal variations in a spontaneously fermented dairy product (Malian fènè) by applying a cultivation/genotype-based binary model. *International Dairy Journal, 29*(1), 28-35.

Xiao, Z. B., Zhu, J. C., Feng, T., Tian, H. X., Yu, H. Y., Niu, Y. W., & Zhang, X. M. (2010). Comparison of volatile components in Chinese traditional pickled peppers using HS–SPME–GC–MS, GC–O and multivariate analysis. *Natural Product Research, 24*(20), 1939-1953.

Xu, G. L., Li, G., Ma, H. P., Zhong, H., Liu, F., & Ao, G. Z. (2009). Preventive effect of crocin in inflamed animals and in LPS-challenged RAW 264.7 cells. *Journal of Agricultural and Food Chemistry, 57*(18), 8325-8330.

Xu, S., Wang, Y., Wang, J., & Geng, W. (2022). Kombucha reduces hyperglycemia in type 2 diabetes of mice by regulating gut microbiota and its metabolites. *Foods, 11*(5), 754.

Xu, Y., Ku, B., Tie, L., Yao, H., Jiang, W., Ma, X., & Li, X. (2006). Curcumin reverses the effects of chronic stress on behavior, the HPA axis, BDNF expression and phosphorylation of CREB. *Brain Research, 1122*(1), 56-64.

Yadav, R. P., Tarun, G., Roshan, C., & Yadav, P. (2017). Versatility of turmeric: A review the golden spice of life. *Journal of Pharmacognosy and Phytochemistry, 6*(1), 41-46.

Yang, C. S., Landau, J. M., Huang, M. T., & Newmark, H. L. (2001). Inhibition of carcinogenesis by dietary polyphenolic compounds. *Annual Review of Nutrition, 21*(1), 381-406.

Yang, W., Chen, Y. H., Liu, H., & Qu, H. D. (2015). Neuroprotective effects of piperine on the 1-methyl-4-phenyl-1, 2, 3, 6-tetrahydropyridine-induced Parkinson's disease mouse model. *International Journal of Molecular Medicine, 36*(5), 1369-1376.

Yang, Y., Fan, Y., Li, T., Yang, Y., Zeng, F., Wang, H., ... & Zhang, Y. (2022). Microbial composition and correlation between microbiota and quality-related physiochemical characteristics in chongqing radish paocai. *Food Chemistry, 369*, 130897.

Yashin, A., Yashin, Y., Xia, X., & Nemzer, B. (2017). Antioxidant activity of spices and their impact on human health: A review. *Antioxidants, 6*(3), 70.

Young, H. Y., Liao, J. C., Chang, Y. S., Luo, Y. L., Lu, M. C., & Peng, W. H. (2006). Synergistic effect of ginger and nifedipine on human platelet aggregation: a study in hypertensive patients and normal volunteers. *The American Journal of Chinese Medicine, 34*(04), 545-551.

Zhang, C., Zhang, J., & Liu, D. (2021). Biochemical changes and microbial community dynamics during spontaneous fermentation of Zhacai, a traditional pickled mustard tuber from China. *International Journal of Food Microbiology, 347*, 109199.

Zhou, S. F., Zhou, Z. W., Li, C. G., Chen, X., Yu, X., Xue, C. C., & Herington, A. (2007). Identification of drugs that interact with herbs in drug development. *Drug Discovery Today, 12*(15-16), 664-673.

Zhou, X., Tang, L., Xu, Y., Zhou, G., & Wang, Z. (2014). Towards a better understanding of medicinal uses of Carthamus tinctorius L. in traditional Chinese medicine: a phytochemical and pharmacological review. *Journal of Ethnopharmacology, 151*(1), 27-43.

Chapter 6

The Role and Impact of Food Supplements in Modern Health

Seda Ufuk[1]
Emel Uçarer[2,*]
and Memnune Şengül[1]
[1]Department of Food Engineering, Ataturk University, Erzurum, Turkiye
[2]Department of Nutrition and Dietetics, Igdir University, Igdir, Turkiye

Abstract

Food is essential for human survival, providing energy, structural support, and regulation of bodily functions. Nutrition—the biochemical process by which the body consumes and utilizes food—-is vital for growth, health, and disease prevention. A balanced diet that supplies all necessary nutrients is crucial for proper functioning. However, many individuals today experience nutritional deficiencies due to changing lifestyles, leading to an increased reliance on food supplements to address these gaps. Food supplements, which often contain vitamins, minerals, proteins, and herbal ingredients, are used to complement the diet and support health. While they can provide benefits, these products are not substitutes for a balanced diet and should be used under professional guidance. In Turkey, food supplements are regulated under the Turkish Food Codex, which enforces strict standards to ensure safety, accurate labeling, and the prevention of misleading claims. Unlike pharmaceutical drugs, food supplements are not intended to treat, prevent, or cure diseases. Despite their potential to support health, supplements can pose risks, especially when misused or when they interact with medications. The growing use of food supplements has raised concerns about safety, appropriate dosing, and potential side effects. This chapter emphasizes the importance of using supplements responsibly, understanding their potential risks, and maintaining strong regulatory oversight to protect consumer health.

Keywords: food supplements, supplement regulation, nutrition, consumer safety, nutritional deficiencies

In: Let Food Be Your Medicine, and Medicine Be Your Food
Editors: Gulcin Yildiz and Arzu Imece
ISBN: 979-8-90134-204-6

Introduction

Food is any substance that meets the body's physiological needs for energy, structural support, regulation, and protection (Tripathi and Patni, 2024). It exists in various forms and is essential for sustaining life. Food includes edible tissues derived from animals and plants and is composed primarily of water, along with organic and inorganic compounds collectively known as nutrients. These nutrients are vital for proper development, growth, and the maintenance of good health. Because the human body cannot synthesize essential nutrients on its own, they must be obtained through dietary intake (Kanojia and Bano, 2024).

Research in nutritional science has shown that humans require more than 50 different nutrients for growth, development, and overall health. These nutrients are classified into six main groups: proteins, carbohydrates, fats, water, minerals, and vitamins (Baysal, 2004). Nutrition is the biochemical and physiological process through which organisms consume and utilize food to support growth, sustain life, and maintain health (Bano and Kanojia, 2024). Nutrition represents a fundamental human need that underpins physiological functions and overall well-being. It is a scientific discipline that examines the complex relationship between humans and food, encompassing all stages from nutrient production to their cellular utilization. Consequently, nutrition intersects with medical and social sciences, which address human health and behavior, as well as with the physical sciences, which investigate the properties and processes of food and nutrients (Baysal, 2004).

A balanced diet provides adequate amounts of each nutrient in the proper proportions, as outlined by the Recommended Dietary Allowances (RDA). Adequate and balanced nutrition involves consuming the appropriate quantities of nutrients required for growth, repair, and optimal body function, as well as ensuring their effective utilization within the body (Unsal, 2020). Inadequate nutrition directly affects all aspects of human functioning, particularly during illness and disease (Baker, 2025). Research indicates that insufficient and unbalanced nutrition can negatively affect both mental and physical development (Özbekler, 2019).

Malnutrition refers to an insufficient, excessive, or imbalanced intake of nutrients that can adversely affect body composition, physical function, and overall health. It influences the functioning and healing processes of all organ systems (Saunders and Smith, 2010) and encompasses both undernutrition and overnutrition (Siddiqui, 2020). Although malnutrition encompasses both conditions, the term is often used to describe undernutrition, as in this chapter. The main factors contributing to malnutrition within society are illustrated in Figure 1 below. Malnutrition may result from inadequate food intake, difficulty in eating, or problems with nutrient absorption. Health-related factors such as vomiting, loss of appetite, mental health disorders, and the use of certain medications can also contribute to its development (Ersado, 2022).

Changes in eating habits, driven by shifts in daily living and working conditions, have contributed to an increase in malnutrition-related problems. Consequently, many individuals turn to food supplements to obtain essential nutrients that may be lacking in their diets (Savlak et al., 2022). While food supplements cannot replace a balanced diet, they can complement nutritional intake when regular diets are insufficient or when nutrient needs increase at various stages of life (Isık, 2024). People use food supplements for a range of purposes, including maintaining proper nutrition, supporting tissue health, reducing the risk of certain diseases and age-related decline, and enhancing physical performance. However, the scientific evidence supporting these benefits varies depending on the specific ingredients of each supplement (Hathcock, 2001).

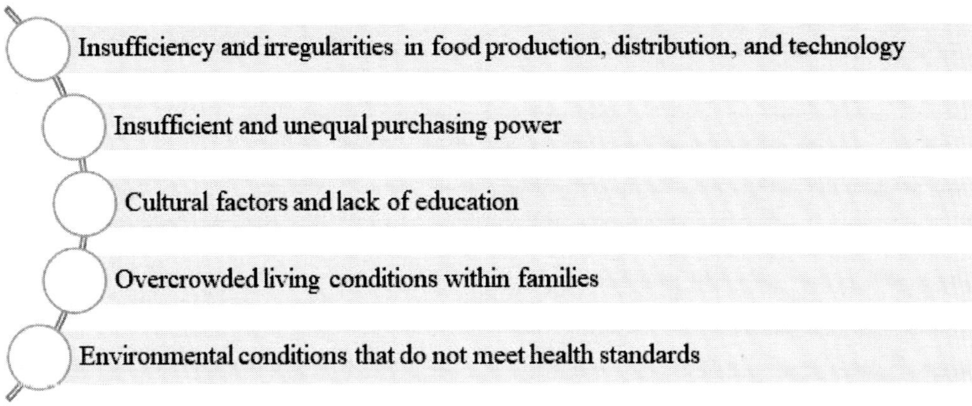

Figure 1. Causes of Malnutrition in Society (Baysal, 2004).

According to the Turkish Food Codex Supplementary Food Regulation (2013/49), food supplements are products in forms such as capsules, tablets, powders, or liquids, designed to complement a regular diet. They may contain vitamins, minerals, proteins, carbohydrates, fibers, fatty acids, amino acids, or bioactive substances derived from plant, animal, or herbal sources. These substances can be either isolated or combined and are subject to defined daily intake limits (Turkish Ministry of Agriculture and Forestry, 2013). Herbs and botanicals are commonly used as functional ingredients in both functional foods and food supplements. Additionally, many pharmaceutical formulations incorporate ingredients derived from plant sources (Eussen et al., 2011). The production of supplements generally involves multiple stages, including recipe development, mixing, milling, granulation, drying, final mixing, tablet formation, tablet pressing, coating, and encapsulation (Tousel, 2015). These products are distributed through various retail channels, including pharmacies, supermarkets, and online platforms (Rusu et al., 2023; Poniedziałek et al., 2023).

The form of a supplement is influenced by factors such as convenience and nutrient composition. Powders can be easily mixed with food or beverages and may include added sugars to enhance taste. Liquid and chewable supplements can provide a broad range of nutrients; however, their flavors can be unpalatable, and they may not offer a complete nutrient profile. It is essential to evaluate whether a supplement adequately meets an individual's nutritional needs (Ertaş, 2022).

Globally, food supplements are regulated as food products. Although they may resemble drugs in appearance or formulation, they are not subject to the exact stringent testing requirements. Manufacturers are required to adhere to Current Good Manufacturing Practices (cGMP) for production, packaging, and labeling (Vovk and Glavnik, 2015; Eussen et al., 2011). Unlike pharmaceutical drugs, supplements are not intended to diagnose, treat, or cure diseases and do not undergo rigorous pre-market approval processes (Halsted, 2003).

In Turkey, the Ministry of Health is responsible for approving drugs, while the Ministry of Agriculture and Forestry regulates food supplements. Food supplements must comply with legally established labeling regulations, and the information provided on labels is subject to strict requirements. Unlike drugs, however, food supplements may be advertised. The classification of a product as either a food or a drug depends on its presentation and the expected pharmacological, immunological, or metabolic effects. As a result, herbs and botanicals can be used both as food supplements and as medicines within a Member State, depending on their dosage and formulation (Eussen et al., 2011).

The most widely used food supplements are primarily vitamins and minerals, including multivitamins, multimineral, B vitamins, antioxidant vitamins, calcium (Ca), iron (Fe), omega-3 fatty acids (fish oil), vitamin D, zinc (Zn), magnesium (Mg), chromium (Cr), iodine (I), copper (Cu), manganese (Mn), potassium (K), and silicon (Si). Less common supplements include various herbal extracts such as ginseng, ginkgo biloba, garlic, royal jelly, wheat germ, gamma-linolenic acid, ginger, chlorophyll, algae, gut-regulating supplements (primarily insoluble fibers), brewer's yeast, valerian, guar gum, lecithin, propolis, and other functional food supplements (Kofoed et al., 2015). According to Deng (2012), five commonly used herbal food supplements claimed to have hypoglycemic effects are *Emblica officinalis* (gooseberry), fenugreek, green tea, *Momordica charantia* (bitter melon), and cinnamon. A study by Dickinson et al. (2014) reported that multivitamins were the most frequently used supplements among users (71%), followed by omega-3 (33%), calcium (32%), vitamin D (32%), and vitamin C (32%). Additionally, there is a growing interest in fish oils, probiotics, other non-nutritive components, and herbal supplements (Atalay and Erge, 2018). The primary reasons for supplement use were overall health and wellness (58%) and filling dietary nutrient gaps (42%).

The use of food supplements has increased significantly in recent years, often to compensate for poor dietary habits, manage health conditions, or enhance overall well-being. While supplements can help prevent nutrient deficiencies and support bodily functions, their growing popularity has raised concerns. Regulations vary widely across countries, and issues such as safety, quality control, and misleading claims are becoming increasingly common. Many individuals also rely on supplements rather than improving their diets, raising questions about whether supplementation is necessary for most people. This chapter explores the reasons for supplement use, their benefits and risks, and the challenges associated with their regulation and responsible consumption.

Regulation of Food Supplements

Food supplements must be safe for consumers worldwide; however, regulations differ significantly between countries. Definitions and classifications of terms such as nutraceuticals vary widely, and no universal regulatory standard exists. A product classified as a food supplement in one country may be considered as a medicinal product in another, depending on the respective national legal framework (Zovi et al., 2025).

In the United States, the Food and Drug Administration (FDA) regulates food supplements under the Dietary Supplement Health and Education Act (DSHEA) of 1994 (Starr, 2015; Wallace, 2015). According to the Dietary Supplement Health and Education Act, food supplements encompass a broad and diverse category of products intended to supplement the diet, but not more appropriately classified as drugs, conventional foods, or food additives. These supplements may consist, either wholly or in concentrated, metabolized, constituent, or extract form, of one or more vitamins, minerals, amino acids, herbs, or other botanicals, as well as other substances aimed at enhancing overall dietary intake, such as enzymes, organ tissues, and oils. To qualify as a supplement, a product must be intended for ingestion, sold in forms such as capsules, tablets, soft gels, gel caps, powders, or liquids, and not marketed as a conventional food (DSHEA, 1994). Under DSHEA, manufacturers are responsible for ensuring their products are safe prior to marketing, but they are not required to demonstrate efficacy. The FDA monitors post-market, enforcing Good Manufacturing Practices (GMP) and tracking adverse events (Bose and Sharma, 2025).

In the European Union (EU), food supplements are classified as a specific category of food under existing legislation. They are defined as products intended to complement the regular diet and contain concentrated sources of nutrients or other substances with nutritional or physiological effects. These supplements are sold in measured doses, such as capsules, tablets, pills, powders, liquid ampoules, dropper bottles, and other small, portioned forms, as outlined in Directive 2002/46/EC (Coppens, 2020). Within the EU, food supplements are subject to a notification procedure that requires manufacturers to inform the relevant national authority before marketing their products. However, suppose a food supplement contains novel or innovative ingredients (as defined under the Novel Foods Regulation) or makes health claims (as defined under Regulation 1924/2006). In that case, the manufacturer must obtain pre-market authorization before the product can be sold (Zovi et al., 2025). Food supplements in the European Union (EU) are regulated by a comprehensive framework of EU and national laws governing their production, composition, and labeling. Directive 2002/46/EC defines food supplements and establishes specific rules for vitamins and minerals, as well as labeling requirements. Regulation (EC) No 178/2002, known as the General Food Law, sets overarching food safety standards, outlines producer responsibilities, and defines procedures for notification and product recall in the event of safety concerns. The European Food Safety Authority (EFSA), which also evaluates the safety of additives, novel ingredients, and health claims, states that food supplements (FS) can be used to address nutrient deficiencies and help ensure adequate intake (Demir et al., 2021). Labeling requirements for food supplements are detailed in Regulation (EU) No 1169/2011, while Regulation (EC) No 1924/2006 governs health claims on food products, including supplements. Novel ingredients not commonly used before May 15, 1997, require assessment by the European Food Safety Authority (EFSA) and pre-market authorization under Regulation (EU) 2015/2283. Regulation (EC) 1333/2008 specifies permitted food additives and their conditions of use. Regulation (EC) No 1881/2006 establishes maximum levels for contaminants such as heavy metals, mycotoxins, and dioxins, while limits for pyrrolizidine alkaloids are still under development. Regulation (EC) No 396/2005 sets thresholds for pesticide residues. Finally, Regulations (EC) No 852/2004 and 853/2004, together with Regulation (EU) 2017/625, ensure hygiene standards and official controls are applied to food supplement manufacturers (Savlak et al., 2022; Coppens, 2020).

In the European Union, food supplements are regulated as foodstuffs and do not require clinical studies for registration. However, they must meet specific labeling requirements, including information on nutrient content, recommended daily intake, dosage warnings, dietary disclaimers, and child safety notices (Poniedziałek et al., 2023).

Terminology also differs: in the United States, these products are referred to as dietary supplements, whereas in the European Union, they are called food supplements. While both regions categorize products by function — such as vitamins, minerals, and probiotics — the US places greater emphasis on herbs, enzymes, and amino acids. In contrast, the EU focuses more on herbal and fatty acid supplements. These distinctions reflect differences in consumer trends and legal interpretations (Bose and Sharma, 2025).

Table 1. Food supplements regulations in Turkiye

Regulation on Import, Production, Processing, and Placement on the Market of Supplementary Foods
Approval Procedures for Supplementary Foods Application Instructions
Turkish Food Codex Communiqué on Supplementary Foods
Turkish Food Codex List of Restricted Substances in Food Supplements
Appropriate Plant List
Turkish Food Codex Regulation on the addition of vitamins and minerals and of certain other substances to foods
Turkish Food Codex Regulation on nutrition and health claims made on food
Turkish Food Codex Regulation on labeling and provision of food information to consumers
Guide for Turkish Food Codex Food Labeling and Consumer Information Regulation
Turkish Food Codex Regulation on contaminants

In Turkey, food supplements are primarily regulated by the Turkish Food Codex, administered by the Ministry of Agriculture and Forestry, which oversees their import, production, processing, and marketing. These regulations aim to ensure consumer safety and public health. Effective since May 2, 2013, under Law No. 5996 on Veterinary Services, Plant Health, Food and Feed, all food supplements must be approved prior to production, import, or sale. Products must undergo testing, meet safety standards, and ensure marketing aligns with scientifically supported claims. Misleading advertisements, such as those claiming weight loss, performance enhancement, or disease treatment, are strictly prohibited.

Furthermore, businesses must be registered and comply with stringent labeling requirements to provide accurate information (Savlak et al., 2022). Approval of food supplements is not based solely on their content declarations; instead, a detailed evaluation process is followed, ensuring that products are authorized only after necessary analyses are completed. After approval, supplements are subject to strict post-market monitoring (Şimşek, 2021). In Turkey, regulations regarding supplements are structured through laws, regulations, communiqués, and instructions. Table 1 presents a list of key regulations and guidelines for food supplements.

Companies intending to sell imported or domestically produced food supplements in Turkey must first register with the Provincial Directorates of the Ministry of Agriculture and Forestry and obtain product approval before placing their products on the market. The approval procedures are conducted in accordance with the relevant food supplement regulations published by the Ministry. Laws, regulations, communiqués, and official instructions have been established to standardize the production, preparation, processing, storage, transportation, and marketing of food supplements, ensuring compliance with hygiene and sanitation standards. The daily recommended intake of vitamins and minerals in food supplements must not exceed the limits specified in the Turkish Food Codex Communiqué on Supplementary Foods. Additionally, the botanical species used in the formulation of food supplements must comply with the Appropriate Plant List issued by the Ministry of Agriculture and Forestry. Other ingredients in the final product must also comply with the Turkish Food Codex List of Restricted Substances in Food Supplements, which constitutes a critical legislative requirement. To ensure that food supplements are safe and suitable for consumption, all relevant regulations must be strictly implemented. The Ministry of Agriculture and Forestry, which is responsible for oversight under the Regulation on Import, Production, Processing, and Placement on the Market of Supplementary Foods, does not permit the production, processing, sale, or import of supplementary foods that have not been approved in accordance with the Supplementary Food Approval Procedures. Additionally, unregistered businesses and false advertisements, such as claims of performance enhancement, weight loss or gain, smoking cessation, or the treatment of various diseases, are strictly prohibited (Savlak et al., 2022).

The regulation of food supplements varies considerably across regions, with the United States, European Union, and Turkey each implementing distinct legal frameworks to ensure the safety and quality of these products. In the United States, the Dietary Supplement Health and Education Act (DSHEA) grants manufacturers greater flexibility regarding product efficacy requirements. In contrast, the European Union and Turkey apply more stringent regulatory processes, particularly regarding product approval, labeling, and health claims. These differences illustrate the diverse regulatory approaches taken to govern food supplements and reflect each region's commitment to protecting consumer health and safety.

Why Do People Need Food Supplements?

Throughout human history, diets have been based mainly on plant foods and seafood. Hunter-gatherer societies gradually incorporated meat from the large animals they hunted. By around 10,000 BC, the development of agriculture and animal husbandry enabled the regular consumption of meat and cereals. At that time, there was no systematic knowledge of the roles of vitamins, minerals, proteins, carbohydrates, and fats in human health. However, local communities in different geographical areas had, through experience, developed a variety of foods and cooking methods that promoted good health. Traditional knowledge has led to the selection of specialized plants and foods for nutrition, particularly for children, pregnant women, and the elderly. For example, Native Americans knew that tea made from vitamin C-rich pine bark and needles could treat scurvy. This condition was later scientifically recognized as a vitamin C deficiency disease. In 1749, Dr. James Lind's discovery that citrus fruits prevented scurvy marked an important milestone in nutritional science (Shils and Shike, 2006; Fraga, 2009). Significant progress was made in vitamin research during the 1920s and 1930s, and in 1935, the first synthetic vitamin C was introduced to the market (Kourkouta et al., 2016). In the following period, research on vitamins and other food supplements accelerated, and particularly in the 1980s, intensive studies were conducted on the protective effects of antioxidants such as vitamin C, A, E, β-carotene, and selenium against free radical damage. In addition, epidemiological findings indicate that diets rich in vegetables and fruits, which contain antioxidants, may reduce the risk of cardiovascular disease and cancer (Alissa and Ferns, 2017; Steck and Murphy, 2020).

Food supplements are defined as concentrated sources of vitamins, minerals, or other substances with nutritional or physiological benefits, intended to complement the regular diet (DSHEA, 1994). Despite their pharmaceutical-like presentation, they are legally classified as food products rather than medical drugs, which positions users as consumers rather than patients. However, debates continue over whether their use should be restricted to healthy individuals or extended to those with medical conditions. The food supplement market continues to grow rapidly, with products targeting a wide range of health concerns. Public concern about the perceived nutrient depletion of modern foods, along with a growing emphasis on preventive health, are significant factors driving the normalization of supplement use (Wierzejska, 2021).

Food supplements are often used to support body functions and prevent illness (Hys, 2020). The most common purposes for individuals using food supplements are stress relief, digestive care, physical and sport performance, and immunity boosting (Chaloupkova et al., 2020). It is known that the use of food supplements is widespread, particularly in both developed and developing countries. The consumption of individual nutrient supplements has increased significantly in recent years; for instance, vitamin D supplement use among US adults quadrupled between 1999 and 2012, while omega-3 fatty acid supplement use grew sevenfold (Kantor et al., 2016). Although supplement consumption is generally lower outside North America, prevalence varies significantly across countries, for example, 51% in Denmark, 43% in Australia, 36% in the UK, and considerably lower in Spain (6%) and Greece (2%) (Skeie et al., 2009; O'Brien et al., 2017). In the US, over 70% of adults aged 65 and older use supplements, compared to approximately one-third of children and adolescents (Blumberg et al., 2017a; Qato et al., 2018). Supplement use is more common among women and is positively associated with higher educational attainment and socioeconomic status (Blumberg et al., 2017a; Chen et al., 2019). Supplement users also tend to exhibit healthier lifestyle behaviors, such as non-smoking, moderate alcohol consumption, maintaining a healthy weight, and engaging in regular physical activity, and they typically achieve better diet quality compared to non-users (Bailey et al., 2011; Chen et al., 2019; Harrison et al., 2004). Additionally, there was a significant increase in the use of vitamin C, vitamin D, multivitamins, zinc, fish oil, probiotics, and propolis supplements to support the immune system during the COVID-19 period (Aysin and Urhan, 2021).

Supplement use significantly reduces the prevalence of inadequate nutrient intake in the general population, particularly for calcium and vitamin D (Blumberg et al., 2017b). However, despite widespread supplement use, micronutrient deficiencies remain common in high-income countries due to energy-dense but nutrient-poor diets. In low- and middle-income countries, where deficiencies in iodine, iron, zinc, and vitamin A are prevalent, supplementation is recommended when dietary modification, food fortification, or food provision strategies are insufficient (Bailey et al., 2015). In Turkey, chronic nutritional deficiencies —including underweight and stunting, vitamin and mineral deficiencies (vitamin B_1, B_{12}, and D; iron, zinc, calcium, iodine, phosphate), and chronic diseases such as diabetes and obesity — are reported as the leading nutrition-related problems (TNHS, 2017). When necessary, it is vital for individuals to obtain nutrients, such as vitamins and minerals/, through dietary supplements to maintain a healthy, high-quality life (Özbekler, 2019). In general, adults who consume a balanced, adequate diet do not require vitamin or mineral supplements. However, even if they are healthy, individuals with certain special conditions may need to use food supplements in the situations specified in Table 2.

Food supplements are designed to complement the diet and address basic nutritional needs when necessary. They provide concentrated amounts of nutrients with minimal or no caloric contribution relative to total energy intake, allowing for improvements in dietary adequacy without significantly increasing energy consumption. This makes food supplements particularly valuable for populations with inadequate energy intake and for individuals at nutritional risk. Supplements are beneficial across specific life stages and medical conditions. For instance, they support pregnant women, for whom meeting nutrient requirements through diet alone can be challenging. Infants may require additional iron, vitamin K, vitamin D, and potentially choline, while elderly individuals at risk of achlorhydria are recommended to take vitamin B_{12} supplements. Individuals who have undergone intestinal bypass surgery often require daily supplementation with micronutrients, such as iron, calcium, and vitamins A, B, C, and D, to prevent deficiencies caused by malabsorption. Furthermore, food supplements have been widely used to prevent and treat malnutrition, with micronutrient powders used in developed countries to address deficiencies in children and pregnant women (Binns et al., 2018; Dwyer et al., 2022).

Table 2. Situations where the use of food supplements may be necessary (Tek and Pekcan, 2008)

Individuals following low-energy diets
Those unable to provide adequate and balanced nutrition due to psychological or economic factors
Vegetarians, especially vegans who avoid all animal products
Infants (to prevent vitamin D deficiency) and children
Pregnant and breastfeeding women (to support iron, folate, vitamin B12, etc.)
Women experiencing significant bone loss after menopause
The elderly
Individuals on long-term medication (such as antacids, antibiotics, laxatives, diuretics, etc.)
Those with allergic conditions that interfere with nutrient absorption
Patients receiving medical nutrition therapy due to illness
People undergoing dialysis treatment, and others in similar situations

Advantages and Benefits of Food Supplements

Maintaining overall health and long-term well-being often requires a balanced diet, sometimes supported by appropriate supplementation. Food supplements include a wide range of products containing dietary ingredients such as vitamins, minerals, amino acids, fatty acids, herbs, and botanicals. These ingredients can be taken individually or in combination to meet essential nutritional needs, particularly for individuals with restrictive diets, such as vegans or those with food allergies, who may be at risk for deficiencies (Brown and Green, 2021).

While people use supplements for various reasons, the most common motivations include ensuring adequate nutrient intake, reducing age-related health risks, and maintaining healthy body tissues (Djaoudene et al., 2023). Supplements are also widely used for disease prevention, diet balancing, enhancing physical appearance, overall wellness, and sports performance (Boggia et al., 2020). In recent years, food supplements have gained global attention for their role in supporting health and managing specific conditions. They are generally easy to use and often provide a cost-effective alternative to pharmaceutical compounds. Among supplement types, botanicals have become particularly popular. These plant-based supplements are made from entire herbs or specific parts, such as flowers, leaves, roots, berries, or exudates, and may be used alone or in blended formulations (Durazzo et al., 2022).

Food supplements provide concentrated nutrients with minimal or no caloric contribution, making them a practical option for improving nutritional status without increasing energy intake. They are beneficial for individuals who do not obtain enough nutrients from food alone, such as those with low energy intake, pregnant women, infants, the elderly, or people with certain health conditions. In both developing and developed countries, food supplements are widely used to help prevent and treat malnutrition, especially in children and pregnant women (Djaoudene et al., 2023). In addition to general nutrition support, some food supplements are marketed to assist with weight management. They may help compensate for nutritional gaps in calorie-restricted diets and, in some cases, are believed to support weight loss by suppressing appetite or increasing metabolic rate (Dwyer et al., 2005).

Food supplements may benefit specific groups, such as competitive athletes, particularly when their diets lack essential nutrients. In such cases, increasing nutrient intake through food or supplements is recommended once a deficiency is identified (Verma et al., 2022). The benefits of food supplements are illustrated in Figure 2. Building on these benefits, nutritional supplements combine advanced food science with essential nutrients to deliver holistic nutrition that supports a healthy lifestyle (Rajakumari et al., 2018).

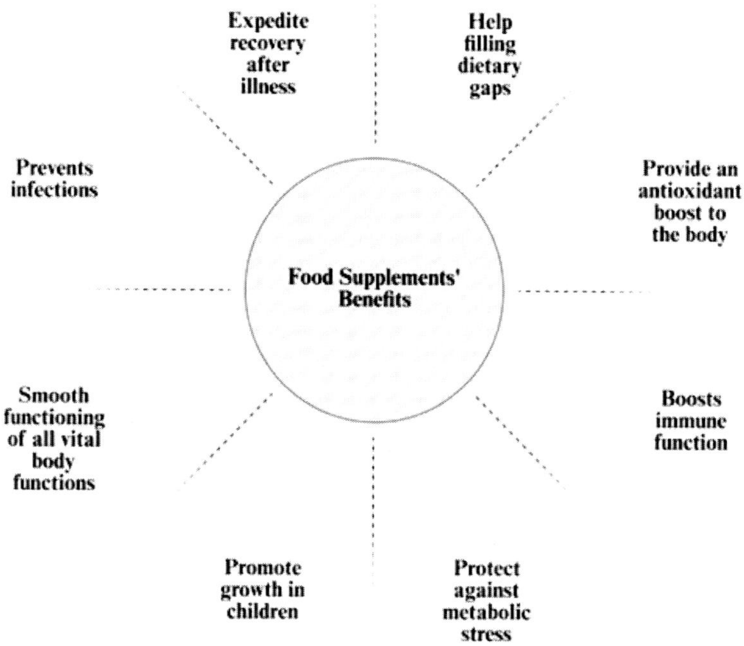

Figure 2. Summary of the Benefits of Food Supplements (Rajakumari et al., 2018).

Many food supplements are marketed for benefits such as weight loss, fertility, brain function, or appearance. However, in the European Union, health claims must be approved by the European Food Safety Authority (EFSA), and disease-related claims are prohibited under Regulation (EU) No 1169/2011. Nevertheless, many users continue to perceive supplements as medicinal or therapeutic alternatives (Poniedziałek et al., 2023).

Growing Issues with the Increased Consumption of Food Supplements

Food supplements play a physiological and nutritional role in disease prevention, health maintenance, the alleviation of minor ailments, and the restoration of physiological balance. Their use has expanded into areas such as wellness, beauty, and general health, contributing to a global increase in consumption (Dwyer et al., 2021). Use of food supplements is increasing worldwide, driven by several factors, as shown in Table 3. The largest markets for food supplements are the United States, Europe, and Japan, followed by Asia, Australia, and

Oceania, with growing demand in the Middle East and Africa. Socio-demographic trends, cultural factors, consumer behavior, and aging populations shape regional differences in consumption. Globally, the food supplement market expanded from USD 82 billion to USD 149.5 billion by 2021 and is projected to reach USD 308 billion by 2028, with a compound annual growth rate of 8.9% (Hys, 2020).

Table 3. Factors Affecting the Increased Use of Food Supplements

An aging baby boomer population prioritizing wellness
A preference for natural substances over synthetic chemicals
Skepticism towards pharmaceutical companies
Increased supplement use among individuals with chronic diseases

Despite this rapid growth, many consumers remain unaware of the potential health risks associated with misuse. Scientific evidence regarding the safety and efficacy of numerous supplements is still limited, and misleading advertising, through exaggeration, omission, or vague wording, can lead to widespread misconceptions (Petty and Kopp, 1995; Kreth et al., 2000). While middle-aged adults are generally better at detecting deceptive claims, older adults are more susceptible to misleading visual cues, making them more vulnerable to misinformation and related health risks (Sun et al., 2024).

Quality and safety issues are also a concern, particularly with botanical supplements. These products are often susceptible to misidentification, contamination, or adulteration. Supplements marketed for weight loss or athletic performance are especially prone to containing undeclared synthetic ingredients (Dwyer et al., 2018). Variability between actual and labeled nutrient content is common; studies have shown that vitamin D levels in supplements can range from 8% to 177% of declared values (Verkaik-Kloosterman et al., 2017), with similar discrepancies reported for iron and iodine supplements (Teas et al., 2004). Fish oil capsules, widely consumed for omega-3 intake, were found in a New Zealand study to be heavily oxidized and metabolically inactive by the time they were consumed. In addition to quality concerns, supplement safety is dose-dependent (Albert et al., 2015). High doses, particularly of nutrients such as calcium and vitamin D, can lead to adverse health outcomes, including soft-tissue calcification, even in individuals with normal kidney function (Drüeke and Massy, 2012; Rooney et al., 2017).

Consuming food supplements can negatively affect consumers financially, as many products fail to deliver on their health claims. More critically, reliance on supplements for treating serious illnesses can delay or replace necessary medical care, posing significant health risks. This underscores the importance of regulatory measures that prohibit unverified health claims. Consequently, stronger regulatory mechanisms are needed to protect consumers (Muela-Molina et al., 2021).

Challenges in the Use of Food Supplements

Health-conscious consumers often turn to food supplements to enhance their quality of life, yet many are unaware that these products are not always natural or safe (Chiba and Tanemura, 2022). The growing popularity and widespread use of food supplements have raised significant safety concerns, particularly because patients often do not disclose their use to healthcare providers (Egan et al., 2011). Risks associated with food supplements include organ toxicity, drug interactions, contamination, and the false assumption that supplements can replace a healthy diet and lifestyle. The US Food and Drug Administration (FDA) estimates that around 50,000 adverse events related to supplements occur annually, though the actual number is likely higher due to underreporting (Woo, 2007; Vatistas and Samules, 2012).

Microbial contamination is a significant issue, particularly in herbal supplements. One study examined 183 food supplement products and identified toxigenic fungi, molds, and yeasts in products such as alfalfa, garlic, and St. John's Wort. In contrast, vitamin and mineral supplements showed no contamination (Tournas, 2009). Additionally, heavy metal contamination poses long-term health risks, including cognitive decline and cancer. In a study of 121 supplements in Canada, 5% exceeded arsenic limits, and smaller percentages exceeded limits for lead, cadmium, aluminum, and mercury, exposing consumers to significant toxicological hazards (Genuis et al., 2012).

Many supplements contain ingredients commonly found in prescription drugs and may not be appropriately regulated. It is therefore essential to understand what food supplements are, whether they are truly necessary, their potential benefits and risks, and consumers' level of awareness about them. Although it is typically more important to use single-ingredient food supplements consciously, high doses of certain food supplements can have significant biological effects on the body (Özbekler, 2019). Additionally, misidentification of medicinal plants, improper labeling, and intentional adulteration with pharmaceutical compounds for economic gain further increase the risk of adverse reactions (Egan et al., 2011; Şimşek, 2021; Walker, 2004). Furthermore, herbs can become contaminated with pesticides, heavy metals, and radioactivity from the soil, as well as with organic solvents and microbial contaminants resulting from improper processing. Deliberate substitution or adulteration of herbs with other plants or chemical drugs may also occur for economic reasons, either to replace a more expensive herb with a counterfeit or to enhance the effectiveness of an herbal mixture by adding a drug with well-known activity for the same indication (Eichhorn et al., 2011).

Many dietary supplements implicitly claim efficacy in treating diseases, thereby misleading consumers into using them for therapeutic purposes. Furthermore, numerous ingredients in food supplements, particularly botanicals, may interact with pharmaceutical drugs and cause serious adverse effects. Therefore, particular attention should be given to the use of food supplements by patients (Chiba and Tanemura, 2022). Many food supplements contain bioactive ingredients that can exert potent biological effects on the body. However, these effects can be detrimental or even life-threatening under certain circumstances. The combination of different supplements or the concurrent use of supplements with prescription or over-the-counter medications can lead to harmful interactions (Eichhorn et al., 2011). Adverse interactions are particularly concerning with commonly used medications, and substances like St. John's Wort have been shown to interfere with drug metabolism, notably in HIV/AIDS, anticoagulation, and cardiovascular treatments (Gardiner et al., 2008; Tsai et al., 2012). Furthermore, substituting supplements for prescribed medications or exceeding the recommended dosage of specific nutrients, such as vitamin A, vitamin D, or iron, can result in significant adverse health outcomes (Eichhorn et al., 2011). Therefore, enhanced quality control procedures and monitoring programs should be implemented to prevent the production of unsafe food supplements and ensure their safety (Costa et al., 2019).

Although one of the primary purposes of food supplements is to reduce the risk of chronic diseases, their effectiveness in doing so is controversial. Food supplements are generally ineffective in reducing the risk of cardiovascular diseases, cancer and type 2 diabetes in healthy individuals without nutritional deficiencies. Evidence suggests that omega-3 supplements do not confer the same benefits as consuming fish directly, indicating that capsules cannot effectively substitute for fish in the diet (Wierzejska, 2021). Similarly, meta-analyses and extensive cohort studies have found no significant association between vitamin–mineral supplementation and reduced risk of stroke or cardiovascular mortality (Kim et al., 2018). Some experts argue that antioxidant supplementation may only benefit individuals with existing deficiencies, and excessive doses could be harmful, particularly in those with an undiagnosed phase of cancer (Vernieri et al., 2018). While vitamin D has attracted interest for its potential role in cancer prevention due to its anti-inflammatory and immunomodulatory properties, meta-analyses of randomized trials have not confirmed a reduction in cancer incidence with supplementation (Goulão et al., 2018; Keum et al., 2019). Although nutrients such as chromium, omega-3 fatty acids, vitamin D, zinc, and magnesium may show some promise in the prevention and management of type 2 diabetes, overall clinical evidence supporting the use of supplements for diabetes control remains limited (Behrouz et al., 2020; Hannon et al., 2020). Consumers of food supplements often do not consult healthcare professionals, which can lead to adverse health effects from excessive intake of vitamins or minerals, inappropriate supplementation, or interactions with medications. Furthermore, many physicians are unaware of food supplement properties, and consumers may mistakenly view them as drugs for disease treatment (Poniedziałek et al., 2023).

In many countries, regulatory frameworks for food supplements remain relatively lax, prioritizing market access and industry profits over public health. Given the size of the food supplements market, valued at approximately US$180 billion, and its potential risks, stricter regulation is warranted. This should include tighter controls on product labeling, clear warnings about side effects, and verified efficacy for any health claims. Supplements recommended for specific life stages, such as folate, iodine, iron, vitamin D, and vitamin K during pregnancy and infancy, should be regulated separately from general products (Binns et al., 2018). Effective regulation should include (Nowak, 2010):

- A permitted substances list, rapid removal of toxic products, and enhanced monitoring of adverse events, with mandatory recording of supplement use in hospital admissions.
- A comprehensive registration system with strict compositional standards.
- Regular manufacturing inspections and licensing requirements.
- Enforcement of upper limits for nutrient contents based on established dietary reference intakes (RDIs).
- Clear, accurate labeling produced through standardized laboratory analyses, including allergen warnings, to guide both experts and lay consumers.
- Prohibition of off-label use and promotion of health claims only when substantiated by robust clinical evidence.
- Restriction to cultivated botanical sources and enforcement of bans on endangered species in supplements.
- Standardized international definitions and classification systems to support effective surveillance and trade regulation through Codex Alimentarius initiatives.

- Enhanced oversight of online food supplement sales through the creation of an international regulatory task force.

Comprehensive regulatory reforms are essential to safeguard public health amid the growing, largely unregulated global supplement market.

Are Food Supplements Really Necessary?

Although food supplements are widely used, their health benefits remain a topic of debate. While deficiencies in vitamins can cause diseases such as scurvy, beriberi, pellagra, and rickets, most well-balanced diets provide sufficient nutrients to prevent these conditions. The use of supplements is not consistently supported for preventing disease in well-nourished individuals (Moyer, 2014). Associations between supplement use and reduced disease risk have been reported; however, these findings are often confounded by factors such as socioeconomic status and overall healthier lifestyles. Furthermore, a focus on high-risk groups limits the applicability of the results to the general population (Zhang et al., 2020).

Nonetheless, food supplements can be beneficial for certain groups, such as individuals with insufficient dietary intake. For instance, during pregnancy, nutrients like folic acid, iron, and vitamin B12 are essential for fetal development. They can reduce the risk of complications, including pre-eclampsia, gestational diabetes, and low birth weight (Brown and Wright, 2020). Omega-3 supplementation during pregnancy and lactation has also been associated with improved cognitive outcomes in children (Nevins et al., 2021). Similarly, breastfed infants require vitamin D supplementation until the introduction of solid foods. Parental use of supplements for managing conditions such as attention-deficit/hyperactivity disease (ADHD), asthma, and colds in children is common. While some positive outcomes have been reported, there is also a risk of overdose and adverse effects (Smolinske, 2017). Older adults, due to decreased nutrient absorption, may benefit from supplements, although more research is necessary to assess long-term safety and efficacy (Walrand, 2018). Among athletes, supplements such as creatine, β-alanine, bicarbonate, caffeine, and omega-3 fatty acids have demonstrated performance benefits, highlighting the need for standardized international regulations to ensure safety in the growing sports supplement market (Alonso and Fernandez-Garcia, 2020; Cannataro et al., 2022).

The increasing prevalence of restrictive diets (e.g., ketogenic and vegan) underscores the importance of evaluating the role of supplements in addressing potential nutrient gaps. In low- and middle-income countries, fortification and supplementation are critical to preventing severe nutrient deficiencies, particularly among children under five years, for whom malnutrition remains a leading cause of mortality. Current evidence does not support routine supplementation with vitamins or fish oil to prevent non-communicable diseases in the general, well-nourished population. Further research is needed to better distinguish the roles of nutrients from food versus those from supplements, particularly to address the needs of vulnerable populations and inform clinical and public health guidelines (Zhang et al., 2020).

It should be emphasized that food supplements are not universally recommended for general health maintenance or disease prevention. Instead, their use should be limited to individuals with prolonged dietary inadequacies or clinically diagnosed nutrient deficiencies.

Greater emphasis should be placed on promoting dietary improvements, particularly increasing the intake of fruits and vegetables, which naturally provide vitamins, minerals, and other bioactive compounds that supplements cannot fully replicate. Individuals considering supplement use on their own should avoid products with excessively high nutrient concentrations, as, unlike nutrients from whole foods, excessive intake from supplements can lead to serious adverse health outcomes (Wierzejska, 2021).

Conclusion

Food supplements serve as tools to support physiological functions and enhance overall well-being by addressing dietary gaps and correcting nutritional imbalances. The increasing accessibility of food supplements, combined with relatively limited regulatory and labeling requirements compared with pharmaceuticals, has contributed to their widespread use. The supplement industry, driven by major corporations and substantial profits, operates under relatively lenient production standards, often with minimal oversight regarding safety, quality, and efficacy.

Although numerous food supplements have been developed in recent years, regulatory progress has been slow. Growing awareness of the need for stricter oversight has prompted food and drug authorities to begin standardizing regulations to enhance consumer safety. Given the global surge in food supplement consumption, there is an urgent need for large-scale studies to assess their efficacy and safety accurately. Moreover, the development of comprehensive compositional databases is essential to tracking nutrient and bioactive compound intake from supplements and for understanding their impact on human health. Stronger collaboration among governmental agencies is also necessary to ensure adequate oversight.

As societal interest in food supplements continues to rise, accompanying research must address their broader implications. Ultimately, supplements should be reserved for clearly defined public health or medical needs, such as deficiencies in iron, folic acid, or iodine. To promote long-term health and longevity, the emphasis should remain on a balanced diet rich in fruits and vegetables rather than on routine supplementation.

References

Albert, B. B., Derraik, J. G., Cameron-Smith, D., Hofman, P. L., Tumanov, S., Villas-Boas, S. G., & Cutfield, W. S. 2015. Fish oil supplements in New Zealand are highly oxidized and do not meet the label claims for n-3 PUFA content. *Scientific reports*, 5(1), 7928.

Alissa, E. M., and Ferns, G. A. 2017. Dietary fruits and vegetables and cardiovascular diseases risk. *Critical reviews in food science and nutrition*, 57(9), 1950-1962.

Alonso, M. R., and Fernández-García, B. 2020. Evolution of the use of sports supplements. *PharmaNutrition,* 14, 100239.

Atalay, D. and Erge, H. S. 2018. Dietary Supplements and Their Effects on Health. *Food and Health*, 4(2), 98-111. doi: 10.3153/FH18010.

Aysin, E., and Urhan, M. 2021. Dramatic increase in dietary supplement use during COVID-19. *Current Developments in Nutrition*, 5, 207.

Baker, S. S. 2025. Nutrition. Pediatric Board Study Guide. doi: 10.1007/978-3-031-65148-9_21.

Bailey, R. L., Fulgoni III, V. L., Keast, D. R., & Dwyer, J. T. 2011. Dietary supplement use is associated with higher intakes of minerals from food sources. *The American journal of clinical nutrition*, 94(5), 1376-1381.

Bailey, R. L., West Jr, K. P., & Black, R. E. 2015. The epidemiology of global micronutrient deficiencies. *Annals of nutrition and metabolism*, 66(Suppl. 2), 22-33.

Bano, F., and Kanojia, S. 2024. Definition and Terminology related to Food Science and Nutrition. *Basics of Community Science*, 10.

Baysal, A. 2004. *Beslenme*. Hatiboğlu Yayınevi. [Nutrition. Hatiboğlu Publishing House.]

Behrouz, V., Dastkhosh, A., & Sohrab, G. 2020. Overview of dietary supplements on patients with type 2 diabetes. Diabetes & Metabolic Syndrome: *Clinical Research & Reviews*, 14(4), 325-334.

Binns, C. W., Lee, M. K., & Lee, A. H. 2018. Problems and prospects: public health regulation of dietary supplements. *Annual review of public health*, 39(1), 403-420.

Blumberg, J. B., Frei, B., Fulgoni Iii, V. L., Weaver, C. M., & Zeisel, S. H. 2017a. Contribution of dietary supplements to nutritional adequacy in various adult age groups. *Nutrients*, 9(12), 1325.

Blumberg, J. B., Frei, B. B., Fulgoni Iii, V. L., Weaver, C. M., & Zeisel, S. H. 2017b. Impact of frequency of multi-vitamin/multi-mineral supplement intake on nutritional adequacy and nutrient deficiencies in US adults. *Nutrients*, 9(8), 849.

Boggia, R., Zunin, P., & Turrini, F. 2020. Functional foods and food supplements. *Applied Sciences*, 10(23), 8538.

Bose, A., and Sharma, S. 2025. Global Regulatory Trends and Comparative Insights: Nutraceuticals in the USA, India, and Europe. *PharmaNutrition*, 100430.

Brown, B., and Wright, C. 2020. Safety and efficacy of supplements in pregnancy. *Nutrition Reviews*, 78(10), 813-826.

Cannataro, R., Straface, N., & Cione, E. 2022. Nutritional supplements in combat sports: What we know and what we do. *Human Nutrition & Metabolism*, 29, 200155.

Chaloupkova, P., Petrtyl, M., Verner, V., & Kokoska, L. 2020. Dietary supplements versus functional foods: consumers' attitudes to their consumption. *British Food Journal*, 122(12), 3853-3868.

Chen, F., Du, M., Blumberg, J. B., Chui, K. K. H., Ruan, M., Rogers, G., & Zhang, F. F. 2019. Association among dietary supplement use, nutrient intake, and mortality among US adults: a cohort study. *Annals of internal medicine,* 170(9), 604-613.

Chiba, T., and Tanemura, N. 2022. Differences in the perception of dietary supplements between dietary supplement/medicine users and non-users. *Nutrients,* 14(19), 4114.

Coppens, P. 2020. The importance of food supplements for public health and well-being. *World Rev. Nutr.* Diet, 121, 66-72.

Costa, J. G., Vidovic, B., Saraiva, N., do Céu Costa, M., Del Favero, G., Marko, D., & Fernandes, A. S. 2019. Contaminants: a dark side of food supplements? *Free radical research*, 53(sup1), 1113-1135.

Demir, G., Kılıçkalkan, B., & Takak, M. K. 2021. Investigation of Adults' Use of Dietary Supplements during the COVID-19 Pandemic. *The Journal of General Medicine*, 31(4), 430-439.

Deng, R. 2012. A review of the hypoglycemic effects of five commonly used herbal food supplements. *Recent patents on food, nutrition & agriculture*, 4(1), 50-60.

Dickinson, A., Blatman, J., El-Dash, N., & Franco, J. C. 2014. Consumer usage and reasons for using dietary supplements: report of a series of surveys. *Journal of the American College of Nutrition*, 33(2), 176-182.

Dietary Supplement Health and Education Act of 1994, Pub L 103–417, 108 Stat. 4325.

Djaoudene, O., Romano, A., Bradai, Y. D., Zebiri, F., Ouchene, A., Yousfi, Y., & Madani, K. 2023. A global overview of dietary supplements: regulation, market trends, usage during the COVID-19 pandemic, and health effects. *Nutrients,* 15(15), 3320.

Doğan, S., Okumuş, E., Bakkalbaşı, E., Cavidoğlu, İ. 2020. Tüketicilerin takviye edici gıdaları kullanım amacı, satın alma tercihleri, ürünlere olan güveni ve yasal düzenlemeler hakkındaki düşünceleri: Van ili örneği. Yüzüncü Yıl Üniversitesi Tarım Bilimleri Dergisi, [Consumption Purposes, Purchasing Preference, Trust towards Food Supplements and Consumer's Opinions about Legal Regulations: Case of Van Province. Yuzuncu Yıl University *Journal of Agriculture Sciences*], 30(4): 821831.

Drüeke, T. B., and Massy, Z. A. 2012. Role of vitamin D in vascular calcification: bad guy or good guy?. Nephrology Dialysis Transplantation, 27(5), 1704-1707.

Durazzo, A., Arcanjo, D. D. R., & Lucarini, M. 2022. The Health Effects of Dietary Supplements. *Evidence-based Complementary and Alternative Medicine: eCAM*, 2022, 9851048.

Dwyer, J. T., Allison, D. B., & Coates, P. M. 2005. Dietary supplements in weight reduction. *Journal of the American Dietetic Association*, 105(5), 80-86.

Dwyer, J. T., Coates, P. M., & Smith, M. J. 2018. Dietary supplements: regulatory challenges and research resources. *Nutrients*, 10(1), 41.

Dwyer, J., Saldanha, L., Bailen, R., Durazzo, A., Le Donne, C., Piccinelli, R., & Turrini, A. 2021. Commentary: an impossible dream? Integrating dietary supplement label databases: needs, challenges, next steps. *Journal of Food Composition and Analysis*, 102, 103882.

Dwyer, J. T., Saldanha, L. G., & Bailen, R. 2022. Dietary supplement databases: Public health tools. *Journal of Food Composition and Analysis*, 105, 104244.

Egan, B., Hodgkins, C., Shepherd, R., Timotijevic, L., & Raats, M. 2011. An overview of consumer attitudes and beliefs about plant food supplements. *Food & function*, 2(12), 747-752.

Eichhorn, T., Greten, H. J., & Efferth, T. 2011. Self-medication with nutritional supplements and herbal over-the counter products. *Natural Products and Bioprospecting*, 1, 62-70.

Ersado, T. L. 2022. Causes of malnutrition. In Combating malnutrition through sustainable approaches. *IntechOpen*.

Ertaş, M. 2022. Importance of the Conscious Consumption of Nutritional Supplements. Istanbul Sabahattin Zaim University *Journal of the Institute of Science and Technology*, 4(1), 48-53.

Eussen, S. R., Verhagen, H., Klungel, O. H., Garssen, J., van Loveren, H., van Kranen, H. J., & Rompelberg, C. J. 2011. Functional foods and dietary supplements: products at the interface between pharma and nutrition. *European journal of pharmacology*, 668, S2-S9.

Fraga, C. G. 2009. *Plant phenolics and human health: biochemistry, nutrition and pharmacology*, vol 1. Wiley, Hoboken.

Gardiner, P., Phillips, R., & Shaughnessy, A. F. 2008. Herbal and dietary supplement–drug interactions in patients with chronic illnesses. *American family physician*, 77(1), 73-78.

Genuis, S. J., Schwalfenberg, G., Siy, A. K. J., & Rodushkin, I. 2012. Toxic element contamination of natural health products and pharmaceutical preparations. *PLoS One*, 7(11), e49676.

Goulão, B., Stewart, F., Ford, J. A., MacLennan, G., & Avenell, A. 2018. Cancer and vitamin D supplementation: a systematic review and meta-analysis. *The American journal of clinical nutrition*, 107(4), 652-663.

Halsted, C. H. 2003. Dietary supplements and functional foods: 2 sides of a coin?. *The American journal of clinical nutrition*, 77(4), 1001S-1007S.

Hannon, B. A., Fairfield, W. D., Adams, B., Kyle, T., Crow, M., & Thomas, D. M. 2020. Use and abuse of dietary supplements in persons with diabetes. *Nutrition & diabetes*, 10(1), 14.

Harrison, R. A., Holt, D., Pattison, D. J., & Elton, P. J. 2004. Are those in need taking dietary supplements? A survey of 21 923 adults. *British Journal of Nutrition*, 91(4), 617-623.

Hassan, S., Egbuna, C., Tijjani, H., Ifemeje, J. C., Olisah, M. C., Patrick-Iwuanyanwu, K. C., & Ephraim-Emmanuel, B. C. 2020. Dietary supplements: Types, health benefits, industry and regulation. *Functional Foods and Nutraceuticals: Bioactive Components, Formulations and Innovations*, 23-38.

Hathcock, J. 2001. Dietary supplements: how they are used and regulated. *The Journal of nutrition*, 131(3), 1114S-1117S.

Hys, K. 2020. Identification of the reasons why individual consumers purchase dietary supplements. Perspectives on Consumer Behaviour: *Theoretical Aspects and Practical Applications*, 193-209.

Isık F. 2024. Food Supplements: Legal Regulations and Health Risks. *J One Health Res,* 2(2):1-10.

Kanojia, S., & Bano, F. 2024. Meal Planning and Recommended Dietary Allowance. *Basics of Community Science*, 37.

Kantor, E. D., Rehm, C. D., Du, M., White, E., & Giovannucci, E. L. 2016. Trends in dietary supplement use among US adults from 1999-2012. *Jama,* 316(14), 1464-1474.

Keum, N., Lee, D. H., Greenwood, D. C., Manson, J. E., & Giovannucci, E. 2019. Vitamin D supplementation and total cancer incidence and mortality: a meta-analysis of randomized controlled trials. *Annals of Oncology,* 30(5), 733-743.

Kim, J., Choi, J., Kwon, S. Y., McEvoy, J. W., Blaha, M. J., Blumenthal, R. S., & Michos, E. D. 2018. Association of multivitamin and mineral supplementation and risk of cardiovascular disease: a systematic review and meta-analysis. *Circulation: Cardiovascular Quality and Outcomes*, 11(7), e004224.

Kofoed, C. L., Christensen, J., Dragsted, L. O., Tjønneland, A., & Roswall, N. 2015. Determinants of dietary supplement use–healthy individuals use dietary supplements. *British Journal of Nutrition*, 113(12), 1993-2000.

Kourkouta L, Iliadis C, Frantzana E, Monios A, Dimitriadou A, Papathanassiou IV. 2016. Health and dietary supplements. *Int J Eng Appl Sci* 3(9):2394–3661

Kreth, M. L. 2000. Exploring" Responsibility" in Advertising: Health Claims about Dietary Supplements. *Business Communication Quarterly*, 63(3), 66-70.

Moyer, M. W. 2014. Vitamins on trial. Nature, 510(7506), 462.

Muela-Molina, C., Perelló-Oliver, S., & García-Arranz, A. 2021. False and misleading health-related claims in food supplements on Spanish radio: an analysis from a European Regulatory Framework. *Public health nutrition*, 24(15), 5156-5165.

Nevins, J. E., Donovan, S. M., Snetselaar, L., Dewey, K. G., Novotny, R., Stang, J., & De Jesus, J. 2021. Omega-3 fatty acid dietary supplements consumed during pregnancy and lactation and child neurodevelopment: a systematic review. *The Journal of nutrition*, 151(11), 3483-3494.

Nowak, R. E. 2010. DSHEA's failure: why a proactive approach to dietary supplement regulation is needed to effectively protect consumers. *U. Ill. L. Rev.*, 1045.

O'Brien, S. K., Malacova, E., Sherriff, J. L., & Black, L. J. 2017. The prevalence and predictors of dietary supplement use in the Australian population. *Nutrients,* 9(10), 1154.

Özbekler, T. M. 2019. "In The Perspective of Changing Nutritional Habits Through Supplementary Foods: How Conscious Are Consumers?", *International Social Sciences Studies Journal*, 5(51): 6866-6882.

Petty, R. D., and Kopp, R. J. 1995. Advertising challenges: A strategic framework and current review. *Journal of Advertising Research,* 35(2), 41-56.

Poniedziałek, B., Niedzielski, P., Kozak, L., Rzymski, P., Wachelka, M., Rzymska, I., & Rzymski, P. 2018. Monitoring of essential and toxic elements in multi-ingredient food supplements produced in European Union. *Journal of Consumer Protection and Food Safety,* 13, 41-48.

Rooney, M. R., Harnack, L., Michos, E. D., Ogilvie, R. P., Sempos, C. T., & Lutsey, P. L. 2017. Trends in use of high-dose vitamin D supplements exceeding 1000 or 4000 international units daily, 1999-2014. *Jama,* 317(23), 2448-2450.

Qato, D. M., Alexander, G. C., Guadamuz, J. S., & Lindau, S. T. 2018. Prevalence of dietary supplement use in US children and adolescents, 2003-2014. *Jama Pediatrics*, 172(8), 780-782.

Rajakumari, R., Oluwafemi, O. S., Thomas, S., & Kalarikkal, N. 2018. Dietary supplements containing vitamins and minerals: Formulation, optimization and evaluation. *Powder Technology*, 336, 481-492.

Republic of Turkey Ministry of Health General Director of Public Health. *Turkey Nutrition and Health Survey 2017 (TNHS 2017)*; Tiraj Basım ve Yayın Sanayi Ticaret Ltd. Şti.: Ankara, Turkey, 2019; ISBN 978-975-590-722-2.

Rusu, A. V., Trif, M., & Rocha, J. M. 2023. Microbial secondary metabolites via fermentation approaches for dietary supplementation formulations. *Molecules,* 28(16), 6020.

Saunders, J., and Smith, T. 2010. Malnutrition: causes and consequences. *Clinical medicine*, 10(6), 624.

Savlak, N., Çağındı, Ö., Dedeoğlu, M., İnce, C., & Köse, E. 2022. An overview of the dietary supplements in Turkey and the world. *Gıda* (2022) 47 (4) 576-590 doi: 10.15237/ gida.GD22021.

Shils M. E, Shike M (eds). 2006. Modern nutrition in health and disease. Lippincott Williams and Wilkins, Philadelphia

Siddiqui, F., Salam, R. A., Lassi, Z. S., & Das, J. K. 2020. The intertwined relationship between malnutrition and poverty. *Frontiers in public health*, 8, 453.

Skeie, G., Braaten, T., Hjartåker, A., Lentjes, M., Amiano, P., Jakszyn, P., & Slimani, N. 2009. Use of dietary supplements in the European Prospective Investigation into Cancer and Nutrition calibration study. *European journal of clinical nutrition*, 63(4), S226-S238.

Smolinske, S. C. 2017. Dietary supplements in children. *Pediatr. Clin. N. Am*, 64, 1243-1255.

Starr, R. R. 2015. Too little, too late: ineffective regulation of dietary supplements in the United States. *American journal of public health*, 105(3), 478-485.

Steck, S. E., and Murphy, E. A. 2020. Dietary patterns and cancer risk. *Nature Reviews Cancer*, 20(2), 125-138.

Sun, Y., Zhang, J., & Li, Y. 2024. Realisation that online advertisements are misleading: Involvement of middle-aged and older adults with botanical dietary supplements. *Journal of Retailing and Consumer Services,* 79, 103832.

Şimşek, S. 2021. Approval Procedures and Administrative Inspection of Food Supplements. *Social Sciences University of Ankara*, 3(1), 125-191.

Teas, J., Pino, S., Critchley, A., & Braverman, L. E. 2004. Variability of iodine content in common commercially available edible seaweeds. *Thyroid*, 14(10), 836-841.

Tek, N. A., and Pekcan, G. 2008. Besin Destekleri Kullanılmalı mı?, Türkiye Sağlık Bakanlığı, Klasmat Matbaacılık, [Should Nutritional Supplements Be Used?, Republic of Turkey Ministry of Health, Klasmat Printing] Ankara.

Tournas, V. H. 2009. Microbial contamination of select dietary supplements. *Journal of food safety*, 29(3), 430-442.

Tousel, M. D. 2015. The manufacturing process, tablet and capsule manufacturing. *Techceuticals,* 15, 1-12.

Tripathi, P. and Patni, K. 2024. Nutrition and Food in Contemporary Science. *International Journal of Advance and Applied Research.* Vol. 11 No. 5, 320-324.

Turkish Ministry of Agriculture and Forestry. 2013. *Turkish Food Codex Supplementary Food Regulation* (Regulation No: 2013/49).

Tsai, H. H., Lin, H. W., Simon Pickard, A., Tsai, H. Y., & Mahady, G. B. 2012. Evaluation of documented drug interactions and contraindications associated with herbs and dietary supplements: a systematic literature review. *International journal of clinical practice*, 66(11), 1056-1078.

Unsal, A. 2020. Nutrition and Basic Food Items. Kırşehir Ahi Evran University *Journal of Health Sciences*, 2(3), 1-10.

Vatistas, T. J., and Samuels, J. G. 2012. The regulation of dietary supplements in the United States: advocating for a reasonable approach, protecting patient safety, and the role of nursing. *Policy, Politics, & Nursing Practice,* 13(2), 113-116.

Verkaik-Kloosterman, J., Seves, S. M., & Ocké, M. C. 2017. Vitamin D concentrations in fortified foods and dietary supplements intended for infants: Implications for vitamin D intake. *Food chemistry*, 221, 629-635.

Verma, N. K., Singh, A. K., Rao, P. K., Singh, S., Mishra, S., & Singh, J. 2024. Dietary Supplements and Its Health Benefits: An Overview. *Sch Acad J Pharm*, 5, 181-188.

Vernieri, C., Nichetti, F., Raimondi, A., Pusceddu, S., Platania, M., Berrino, F., & de Braud, F. 2018. Diet and supplements in cancer prevention and treatment: Clinical evidences and future perspectives. *Critical reviews in oncology/hematology*, 123, 57-73.

Vovk, I., and Glavnik, V. 2015. Analysis of dietary supplements. In Instrumental thin-layer chromatography (pp. 589-635). Elsevier.

Walker, R. 2004. Criteria for risk assessment of botanical food supplements. *Toxicology letters*, 149(1-3), 187-195.

Wallace, T. C. 2015. Twenty years of the dietary supplement health and education act—how should dietary supplements be regulated?. *The Journal of Nutrition*, 145(8), 1683-1686.

Walrand, S. 2018. Dietary supplement intake among the elderly: hazards and benefits. *Current Opinion in Clinical Nutrition & Metabolic Care*, 21(6), 465-470.

Wierzejska, R. E. 2021. Dietary supplements—for whom? The current state of knowledge about the health effects of selected supplement use. *International journal of environmental research and public health*, 18(17), 8897.

Woo, J. J. 2007. Adverse event monitoring and multivitamin-multimineral dietary supplements. *The American journal of clinical nutrition*, 85(1), 323S-324S.

Zhang, F. F., Barr, S. I., McNulty, H., Li, D., & Blumberg, J. B. 2020. Health effects of vitamin and mineral supplements. BMJ, 369.

Zovi, A., Vitiello, A., Sabbatucci, M., Musazzi, U. M., Sagratini, G., Cifani, C., & Vittori, S. 2025. Food supplements marketed worldwide: A comparative analysis between the European and the US regulatory frameworks. *Journal of Dietary Supplements*, 22(1), 25-40.

Chapter 7

The Future of Food and Nutrition

Ayse Betul Demirbas Bilen[1]
Umut Uyar[2,*]
and Gulen Ecem Kalkan[1]
[1]Department of Nutrition and Dietetics, Istanbul Atlas University, Istanbul, Turkiye
[1]Department of Nutrition and Dietetics, Igdir University, Iğdır, Turkiye

Abstract

This chapter provides a comprehensive examination of the biotechnological and engineering innovations shaping the future of food systems. Globally, the increasing population, climate change, depletion of natural resources, and growing food security challenges necessitate radical transformations across all stages of production and consumption. First, the chapter highlights genetic engineering and genome editing (particularly CRISPR-Cas9) as tools for developing high-yield, nutrient-enhanced, and stress-resistant crops. These technologies hold promise in reducing pesticide use, preventing yield losses, and strengthening food security, although they remain constrained by ethical debates, regulatory frameworks, and public acceptance. Second, the contribution of biotechnology to nutrition and public health is emphasized. Biofortification strategies, such as Golden Rice, are discussed as sustainable solutions to "hidden hunger" caused by widespread micronutrient deficiencies. In addition, targeted modifications in macronutrients and reductions of harmful compounds underscore biotechnology's capacity to improve long-term metabolic health and food safety. The chapter also explores fermentation technologies, especially precision fermentation, which enable the production of animal-free proteins (dairy, eggs, and meat analogues) with lower environmental impacts while supporting dietary diversity and sustainability. Synthetic biology is presented as a frontier for producing high-value compounds such as natural flavorings, vitamins, and functional probiotics, strengthening the concept of "food as medicine". Furthermore, the discussion expands to enzyme engineering, nanotechnology and smart packaging, 3D/4D food printing, and smart food systems integrating biosensors and blockchain, each offering new opportunities to enhance efficiency, transparency, and consumer trust in modern food chains. Critical attention is also given to ethical, regulatory, and societal aspects, stressing that the success of biotechnological innovations depends not only on technical advances but also on transparency, equitable access, and consumer

* Corresponding Author's Email: umut.uyar@igdir.edu.tr

In: Let Food Be Your Medicine, and Medicine Be Your Food
Editors: Gulcin Yildiz and Arzu Imece
ISBN: 979-8-90134-204-6

confidence. From a sustainability perspective, biotechnological foods are shown to play a central role in reducing greenhouse gas emissions, conserving resources, preventing food waste, and supporting climate resilience. Finally, the chapter emphasizes that plant-based dietary models, alternative protein sources, and the Planetary Health Diet (EAT–Lancet) will be indispensable in advancing both human health and environmental sustainability.

Keywords: biotechnology, genetic engineering, biofortification, precision fermentation, sustainable food systems

Introduction

The global food system is currently under unprecedented pressure. Rising population levels, climate instability, and the depletion of fertile soils and freshwater resources create a complex set of challenges for agriculture (Godfray et al., 2010). For centuries, farming relied on traditional practices such as selective breeding, irrigation, and soil management. While these approaches shaped civilizations, they now prove insufficient to meet the urgent demands of food security in a rapidly changing world. Farmers today face a dual challenge: producing enough to feed billions, while also ensuring nutritional quality and food safety.

In this context, biotechnology and food engineering emerge as crucial tools. Conventional breeding methods may require decades to produce significant improvements, but biotechnological approaches operate with remarkable precision and speed. Tools such as CRISPR-Cas9, synthetic biology, and precision fermentation enable scientists to reconfigure crops or even design food components in controlled environments (Sufficiency, et al., 2022). These innovations go far beyond yield optimization. Their goals include creating foods that are more nutrient-rich, environmentally sustainable, and compatible with human health. A central dimension of this transformation is sustainability. At present, food systems contribute nearly one-third of global greenhouse gas emissions, with animal agriculture responsible for a large share of methane release and land use (Xu et al., 2021). Biotechnology provides alternatives that may reduce this burden: engineered crops requiring fewer inputs, new protein sources that bypass intensive livestock production, and packaging solutions that extend shelf life. Collectively, these innovations serve two pressing aims simultaneously—feeding a growing population and reducing ecological damage.

Nevertheless, the trajectory of biotechnology will not be determined by scientific progress alone. Public trust, ethical debates, and regulatory frameworks are equally decisive. As emphasized by scholars, no matter how advanced, technological solutions can only succeed if societies accept them as safe, fair, and beneficial (Muha-Ud-Din, 2024; Walait et al., 2024). Thus, the "future of food" is not only shaped in laboratories and on farms, but also in the dialogue between scientists, policymakers, and consumers.

Genetic Engineering and Genome Editing

Genetic engineering has marked one of the most influential milestones in the field of food biotechnology. During the 1990s, the first generation of genetically modified (GM) crops entered global markets, including varieties such as herbicide-tolerant soybeans and pest-resistant cotton. These innovations provided farmers with concrete benefits, particularly in terms of reducing crop damage and lowering production costs (Abdul Aziz et al., 2022). Yet, despite these advantages, they also triggered intense debates worldwide, centered on health risks, ethics, and long-term ecological consequences. The emergence of genome editing dramatically shifted this landscape. Among these tools, CRISPR-Cas9, developed in the last decade, stands out as both more precise and more accessible than earlier genetic engineering methods (Camerlengo et al., 2022). Unlike traditional transgenic approaches that introduce foreign DNA, CRISPR works by making targeted, small-scale modifications within an organism's own genome. This capacity has significant implications: in major crops such as wheat, rice, and maize, researchers have used CRISPR to enhance tolerance to drought and salinity, and to build resistance against bacterial and fungal diseases (Ndudzo et al., 2024). In practice, these adjustments not only improve yields but also reduce reliance on pesticides and fertilizers, aligning food production with sustainability goals. Rice provides a striking example. In Asia, bacterial blight has historically caused devastating yield losses. With CRISPR-based edits, scientists modified specific genes to block the pathogen, creating plants that are both more resilient and more productive (Raffan, et al., 2020). Such targeted interventions show how modern tools can offer solutions to problems that traditional breeding could not address quickly enough.

Nevertheless, genome editing is not without technical and social barriers. Issues such as unintended off-target mutations and the heavy reliance on tissue culture techniques remain scientific challenges (Biswas et al., 2021). Beyond the laboratory, consumer skepticism is significant. Regulatory responses differ by country: while some governments apply lighter rules to genome-edited crops compared to GMOs, others enforce identical regulations, leaving producers uncertain about how to proceed. At the same time, novel techniques are emerging. Base editing and prime editing allow precise single-nucleotide changes or insertions without introducing DNA double-strand breaks, thus reducing risks of unintended errors (Tian et al., 2025). In parallel, multiplexed editing—altering multiple genes simultaneously—is gaining ground. By combining CRISPR with systems biology and AI-based modeling, researchers envision crops that combine climate resilience, nutritional enhancement, and high productivity (Sagawa et al., 2021).

In summary, genetic engineering has evolved from its controversial beginnings into a rapidly expanding toolkit for food security. Its success, however, will depend not only on technical progress but also on transparent governance and the readiness of societies to accept such technologies.

Nutritional Enhancement and Public Health Applications

The role of biotechnology in improving human nutrition is often overshadowed by discussions about yield and sustainability, yet its contribution to health is equally critical. Around the world, so-called "hidden hunger" — widespread micronutrient deficiencies that may not be visible but have serious long-term effects — remains a major public health concern. Deficiencies in iron, zinc, and vitamin A continue to affect billions of people, leading to weakened immune function, impaired growth, and in severe cases, preventable blindness (WHO, 2021). Biotechnology provides promising ways to address these deficiencies. One of the most recognized strategies is biofortification. Instead of relying on supplements that may not reach all populations consistently, crops can be engineered to naturally provide essential nutrients. Golden Rice, developed to produce beta-carotene, was designed to combat vitamin A deficiency in areas where rice is the staple food but diets lack sufficient fruits and vegetables (Palmer et al., 2025). Similar approaches have been applied to maize and cassava, aiming to boost zinc and iron levels and improve the nutritional status of communities in Sub-Saharan Africa and Latin America.

Beyond micronutrients, biotechnological tools can also improve the quality of macronutrients. Gene editing allows for modifications in the fatty acid composition of vegetable oils, enhancing levels of heart-protective unsaturated fats while decreasing harmful trans fats. In cereals, targeted changes can increase resistant starch content, supporting better glycemic control and helping reduce the risk of type 2 diabetes (Augustin et al., 2024). Such improvements reflect the growing recognition of food as a driver of long-term metabolic health rather than just a source of calories. Biotechnology is also being applied to minimize harmful compounds in foods. For example, researchers have used CRISPR to modify wheat in a way that lowers acrylamide formation during baking — a compound classified as potentially carcinogenic (Raffan, et al., 2020). This demonstrates how biotechnology not only enhances nutritional value but also contributes to food safety. Moreover, these advances increasingly intersect with the principles of precision nutrition. As genomic and microbiome research expands, it becomes feasible to tailor diets to individual biological needs. In the future, biotechnology may enable crops designed with nutrient profiles targeted at populations with specific genetic risks, or even foods customized for personalized medical diets (Liu et al., 2025). For instance, wheat could be engineered with a lower glycemic index for populations prone to diabetes, or maize could be enriched with antioxidants for communities with high cardiovascular disease prevalence.

Despite these advances, scaling nutritional innovations presents challenges. Accessibility remains the greatest concern. For biofortified crops to make an impact, they must be safe, affordable, culturally acceptable, and trusted by consumers. If they are too costly for smallholder farmers or face public resistance due to lack of confidence, their health benefits will remain limited (WHO, 2021). Transparent communication, clear labeling, and nutrition education at the community level are therefore essential. It is also important to note that biotechnology's contributions extend beyond direct gene editing. Synthetic biology and advanced fermentation technologies allow the production of vitamins, probiotics, and functional compounds that can support both nutritional adequacy and chronic disease prevention. Taken together, these innovations show that biotechnology is not only about producing more food — it is about producing better food that promotes population health.

Fermentation Technologies and Precision Fermentation

Fermentation has been part of human diets and cultures for thousands of years. Long before the science of microbiology was understood, communities relied on fermented foods such as bread, beer, yogurt, kefir, kimchi, and miso. These products not only extended the shelf life of perishable ingredients but also enriched diets with distinctive flavors and beneficial microorganisms. What began as a method of preservation has now developed into one of the most advanced areas of modern food biotechnology.

In the 21st century, fermentation has expanded far beyond its traditional role. A new approach known as precision fermentation has emerged, in which genetically optimized microorganisms are programmed to produce highly specific proteins, enzymes, and bioactive compounds (Eastham, et al., 2024; Yildiz and Yıldız, 2025a; Yildiz et al., 2025). Unlike conventional fermentation, where outcomes are determined by natural microbial metabolism, precision fermentation allows researchers to design organisms with targeted genetic instructions. As a result, ingredients such as milk proteins, egg proteins, and complex enzymes can be produced at an industrial scale without involving animals. The sustainability benefits of this technology are profound. Casein and whey, essential components of dairy products, can now be synthesized in bioreactors without cows. Similarly, ovalbumin — the main protein in egg whites — can be produced without raising poultry (Augustin et al., 2024). These advances substantially reduce land use, water consumption, and methane emissions, while also eliminating animal welfare concerns. Precision fermentation is not limited to proteins; it is also being used to create natural flavors, colorants, and vitamins with greater consistency and lower environmental costs than conventional farming (Yildiz and Yıldız, 2025b).

Beyond environmental advantages, precision fermentation has growing applications in human health. Scientists are engineering microbes to function as "living factories" within the gut. Tailored probiotic strains may one day synthesize vitamins such as B12 or folate directly in the digestive system, or deliver therapeutic peptides that help regulate immunity and reduce inflammation (Marco et al., 2021). This represents a new frontier in functional foods, where nutrition and medicine converge. The industrial impact is already visible. The global precision fermentation market is expanding rapidly as demand for animal-free and plant-based alternatives increases (Karabulut et al., 2023). Companies are launching dairy-free cheeses, egg-free desserts, and microbial enzymes for baking and brewing. These products illustrate how biotechnology is reshaping consumer preferences by offering sustainable and health-oriented options without compromising flavor or texture. Yet, several challenges remain. Production costs, though steadily decreasing, are still higher than those of conventional farming in some cases. Technical hurdles include scaling up bioreactor systems and ensuring reliable feedstocks such as sugars. Consumer acceptance is another obstacle: while some markets welcome animal-free proteins, others remain cautious. Overcoming this hesitation will require clear labeling, transparent communication, and robust nutritional validation.

Taken together, fermentation has transitioned from a traditional cultural practice to a sophisticated biotechnological strategy. Its future lies in expanding food diversity, reducing environmental impacts, and contributing to medical nutrition. Among emerging technologies, precision fermentation stands out as one of the clearest examples of how food engineering can reimagine the future of food production.

Synthetic Biology and Novel Food Compounds

Synthetic biology represents one of the boldest and most ambitious frontiers of modern food biotechnology. While genetic engineering focuses on modifying or inserting individual genes, and fermentation relies on microbial metabolism, synthetic biology goes further: it reprograms or even constructs entirely new biological systems. The aim is not simply to enhance nature but to redesign metabolic pathways so that microorganisms can serve as "cell factories," producing compounds that otherwise require resource-intensive farming or complex chemical synthesis (da Fonseca-Pereira et al., 2022).

A clear example of this potential is the production of natural flavoring agents. Vanillin, one of the most widely used flavor compounds, illustrates the challenge. Extracting vanillin from vanilla orchids is labor-intensive, costly, and environmentally demanding, while petrochemical synthesis raises sustainability and safety concerns. Synthetic biology offers an alternative: engineered microbes programmed to produce vanillin at scale, ensuring reliable supply and reducing ecological pressure (Arun et al., 2023). Similar efforts are underway to create saffron, stevia, and other high-value compounds, helping the food and fragrance industries reduce their dependence on fragile crops. Beyond flavors, synthetic biology is being applied to nutrient biosynthesis. Essential nutrients that are difficult to obtain from plants, such as vitamin B12, can be generated using engineered microbial pathways. Similarly, amino acids like lysine and tryptophan — vital for human health — are now produced with improved efficiency and purity compared to conventional fermentation. These advances illustrate how synthetic biology can strengthen global nutrition security in a scalable way.

The technology also opens the door to functional and therapeutic foods. Engineered probiotics are a particularly promising example. Instead of passively residing in the gut, these microbes can be designed to actively release compounds that influence metabolism, immunity, or even neurological health (Mays et al., 2018). For instance, probiotics may be tailored to secrete short-chain fatty acids that support gut health or bioactive peptides with anti-inflammatory effects. In this vision, foods become active vehicles for targeted health interventions, reinforcing the idea of "food as medicine." Despite these exciting opportunities, synthetic biology raises important ethical and regulatory questions. Who owns the intellectual property rights for synthetic organisms? How should society evaluate potential risks associated with releasing or scaling up these engineered systems? (Arun et al., 2023). Furthermore, consumer acceptance may be more complex than with genetically modified crops, since products derived from synthetic biology often lack direct natural equivalents. Building trust will require transparency, rigorous safety assessments, and honest communication about uncertainties alongside potential benefits.

Ultimately, synthetic biology is more than just another tool within biotechnology. It represents a paradigm shift in how food systems can be designed. On one hand, it offers a chance to reduce environmental costs, expand nutritional quality, and integrate healthcare with diet. On the other, it forces societies to confront fundamental ethical debates about the extent to which life can or should be reprogrammed for human benefit.

Enzyme Engineering and Consumer-Oriented Solutions

Enzymes, though often unnoticed, are central to modern food production. They act as natural catalysts in countless processes, shaping the flavor, texture, and nutritional quality of foods. From tenderizing meat to enabling bread to rise, enzymes have long been indispensable in the food industry. For decades, producers relied on naturally occurring enzymes, but with the advent of enzyme engineering, scientists can now redesign these proteins to function with greater stability, efficiency, and versatility (Ameer, et al., 2020).

One of the most well-known examples is lactase, the enzyme that breaks down lactose in dairy products. Engineered variants of lactase are more stable under industrial processing conditions, making large-scale production of lactose-free milk, cheese, and yogurt feasible (Sagawa et al., 2021). This innovation has transformed the experience of individuals with lactose intolerance, expanding food choices and creating an entire market sector. Likewise, engineered amylases and proteases are now widely used in baking, where they improve dough handling, enhance flavors, and extend the shelf life of baked goods.

Enzyme engineering also contributes directly to sustainability. Many traditional industrial food processes required high heat or chemical additives, both of which increase energy use and environmental impact. By designing enzymes that work effectively at lower temperatures or within specific pH ranges, food manufacturers can cut energy costs and reduce reliance on additives. For instance, engineered lipases that perform under mild conditions can substitute for synthetic emulsifiers in margarine or baked products (Abdullahi et al., 2021). This not only lowers industrial footprints but also supports the rising demand for clean-label foods — products made with natural and recognizable ingredients rather than artificial chemicals.

Beyond industrial benefits, engineered enzymes are finding roles in health-related applications. Research on gluten-degrading enzymes shows potential to assist individuals with celiac disease by breaking down the proteins responsible for autoimmune reactions (Henningan et al., 2022). Although these enzymes cannot yet replace strict gluten-free diets, they represent an important step in medical nutrition innovation. Another example is phytase enzymes, which have been engineered to release more phosphorus from cereals and legumes, improving mineral bioavailability in populations that rely heavily on plant-based diets.

Ultimately, enzyme engineering illustrates how biotechnology directly connects science to daily life. These innovations are not only about industrial efficiency but also about meeting consumer needs — whether that involves easier digestion, longer-lasting bread, or healthier and more sustainable food options. By combining industrial performance with consumer-centered outcomes, enzymes remain one of the most versatile and impactful areas of food biotechnology.

Nanotechnology and Smart Packaging in Food Engineering

Nanotechnology has rapidly emerged as one of the most transformative tools in food science and engineering. By manipulating materials at the nanoscale, researchers can influence how nutrients behave, how foods are preserved, and even how consumers interact with packaging. Although the concept may sound futuristic, nanotechnology is already present in fortified foods, dietary supplements, and intelligent packaging systems (Biswas et al., 2022). One of its most valuable contributions is nutrient encapsulation. Many vitamins, antioxidants, and bioactive compounds are unstable when exposed to oxygen, light, or heat. Through techniques such as nanoemulsions, liposomes, and polymeric nanoparticles, these sensitive compounds can be protected until they reach the digestive tract. Encapsulation not only preserves stability but also increases bioavailability, allowing the body to absorb more of the nutrient compared to traditional formulations. For instance, encapsulated curcumin — known for its anti-inflammatory effects — demonstrates much higher absorption than standard curcumin preparations.

Packaging is another area where nanotechnology is driving major changes. Conventional packaging materials primarily act as passive barriers, but nanomaterials can add active protective functions. Nanocomposite films containing silver or zinc oxide nanoparticles show antimicrobial activity, slowing bacterial growth on food surfaces and extending shelf life (del Rosario Herrera-Rivera et al., 2024). Other nanofilms are engineered with barriers that limit oxygen or moisture transfer, which helps preserve taste, aroma, and texture. Perhaps the most striking application is smart packaging. By incorporating nanosensors, packaging can signal real-time freshness information. A package of meat, for example, might change color when microbial spoilage begins. More advanced systems even transmit digital data that consumers can scan with smartphones to track freshness, storage conditions, or even farm-to-table origins. Such technologies could significantly reduce food waste by helping consumers distinguish between spoiled and still-safe food.

Despite these advantages, nanotechnology raises important concerns. Safety is the foremost issue: what are the long-term effects of consuming nanoparticles regularly? While most current research suggests that many food-related nanomaterials are safe, regulatory agencies such as the European Food Safety Authority (EFSA) and the U.S. Food and Drug Administration (FDA) stress the importance of thorough, long-term risk assessments (He et al., 2016). Equally critical is consumer trust. Without clear labeling, public education, and transparent communication, even highly effective nanotechnological solutions may face resistance.

In conclusion, nanotechnology in food engineering demonstrates how advanced science can improve everyday products. From nutrient delivery to antimicrobial protection and intelligent freshness indicators, its applications have the potential to make foods safer, longer-lasting, and more nutritious. Moving forward, the challenge will be balancing rapid innovation with safety, regulation, and consumer acceptance to ensure that nanotechnology fulfills its promise in food systems.

3D/4D Food Printing and Personalized Nutrition

What once seemed like science fiction — the idea of printing food — is now becoming a reality through 3D food printing. At its core, this technology relies on the layer-by-layer deposition of edible "inks," which may include proteins, starches, fats, or bioactive compounds (Chao, et al., 2024). This method allows precise control over the shape, texture, and nutrient composition of food, offering possibilities far beyond novelty desserts or futuristic dining concepts. One of the most transformative potentials of 3D printing is its application in personalized nutrition. Traditional food systems are built around standardized products, yet individual nutritional needs vary according to age, health status, and lifestyle. For instance, an athlete may require protein-rich meals, older adults may need texture-modified foods for safe swallowing, and individuals with diabetes may benefit from carbohydrate-controlled diets. With 3D printing, it becomes feasible to design meals tailored to these diverse needs — not only in nutrient content but also in appearance and taste.

Clinical nutrition is one area where 3D printing could have an especially significant impact. Patients suffering from dysphagia (difficulty swallowing) often rely on bland purées that are safe but unappetizing. Using 3D printers, pureed foods can be reshaped into familiar, attractive forms while maintaining the required texture (Dankar et al., 2018). For hospitalized patients, the elderly, or those recovering from illness, this approach can improve appetite, dignity, and overall quality of life.

Looking ahead, 4D food printing adds an even more dynamic dimension. Foods produced with this method can respond to environmental cues such as heat, humidity, or pH, changing in shape, texture, or even flavor (Ahmed et al., 2021). For example, pasta that changes form when boiled could reduce packaging requirements, while snack bars designed to release flavors gradually in response to saliva could enhance the eating experience. Although still in early stages, these innovations highlight the potential of food as an interactive and adaptable product.

Sustainability is another important aspect. 3D printing has the ability to incorporate alternative proteins derived from algae, insects, or fermentation processes, which may otherwise be less appealing to consumers. By transforming these ingredients into attractive, palatable designs, the technology could make sustainable food sources more acceptable, helping shift diets toward lower environmental impact without sacrificing enjoyment.

Challenges, however, remain. Current printers are relatively slow, limiting large-scale adoption. Heat and mechanical stress during printing can degrade sensitive nutrients, raising questions about stability. Costs are also higher compared to conventional production, and widespread adoption will depend on whether the technology can deliver affordability alongside convenience. Nevertheless, research and development are moving quickly, with universities, research centers, and startups exploring everything from personalized snacks to complete meals for space travel. In short, 3D and 4D food printing represent the convergence of biotechnology, engineering, and personalized healthcare. They point toward a future where meals are not only cooked but designed — tailored to the needs of athletes, patients, older adults, and environmentally conscious consumers alike. As these technologies mature, printed foods may become one of the defining features of next-generation food systems.

Smart Food Systems, Sensors, and Food Safety

Food safety has always been central to public health, as millions of people worldwide suffer foodborne illnesses every year. Contamination often goes undetected until products have already reached consumers, and conventional safety checks—while accurate—are often slow. In today's globalized food supply chains, such delays mean contaminated products can travel thousands of kilometers before recalls are issued. To close this gap, biotechnology and digital engineering are converging to create smart food systems capable of monitoring safety in real time. One of the most significant advances in this field is the development of biosensors. These devices can identify pathogens, toxins, allergens, or spoilage markers directly in food or along production lines. Unlike traditional laboratory tests, biosensors deliver results within minutes, making rapid intervention possible. For example, sensors able to detect Salmonella or E. coli can be installed in meat processing plants to flag contaminated batches before they leave the facility (Sagawa et al., 2021).

Smart packaging represents another major innovation. Conventional packaging provides passive protection, but new technologies enable packaging to actively monitor food quality. For instance, colorimetric indicators shift color when pH changes signal spoilage. More advanced systems embed nanosensors that respond to temperature fluctuations or volatile compounds associated with microbial growth (del Rosario Herrera-Rivera et al., 2024). For consumers, this means a simple glance at a label or a scan with a smartphone could confirm freshness, thereby reducing both health risks and unnecessary food waste. These innovations are increasingly connected to digital traceability systems. Blockchain technology, originally designed for secure financial transactions, is being adapted to food supply chains. Each stage—from farm production to processing, transport, and retail—can be recorded in a tamper-proof ledger (Dong et al., 2020). In the event of contamination, companies and regulators can quickly pinpoint the source, limit recalls to affected batches, and avoid broad, costly product withdrawals.

Beyond safety, these systems also support consumer trust. Modern consumers increasingly expect transparency regarding food origins, production methods, and additives. Smart labels linked to blockchain records allow them to access such information directly, empowering more informed choices. Still, challenges persist. Implementing real-time monitoring infrastructure is costly, making it difficult for small producers to adopt. Data-sharing raises additional concerns: Who owns the data? How can privacy and fairness be ensured? On a global scale, trade complicates matters further. A sensor-verified standard accepted in one country may not be recognized elsewhere, reducing the effectiveness of these systems unless greater international harmonization is achieved.

In summary, smart food systems bring together biotechnology, sensors, and digital infrastructure to address long-standing gaps in food safety. By providing rapid detection, transparent traceability, and consumer-friendly communication, they have the potential to reduce illness, waste, and mistrust. However, widespread adoption will require solutions that are cost-effective, equitable, and globally coordinated.

Ethical, Regulatory, and Consumer Acceptance Challenges

The development of food biotechnology raises not only technical issues but also deep ethical and societal questions. While scientists and industry leaders emphasize the nutritional and environmental benefits of these technologies, consumers and policymakers often ask: Who controls them? Who gains from them? And what risks might they pose? One of the most visible concerns is consumer acceptance. Decades of studies show that attitudes toward genetically modified or genome-edited foods vary widely depending on culture, geography, and public trust in regulatory institutions (Frewer et al., 2013). In the United States, surveys suggest moderate acceptance, particularly when products are presented as environmentally beneficial or health-promoting. In contrast, skepticism remains strong in many parts of Europe, where GM foods are still associated with health risks or environmental harm. These differences are not simply about scientific evidence; they reflect broader cultural values and ideas of what qualifies as "natural" food. Another central issue involves equity and access. Critics argue that the benefits of biotechnology often favor wealthier nations or large corporations, while smallholder farmers may be left at a disadvantage. Patents and licensing fees for engineered seeds can place farmers in dependent relationships with agribusiness companies, raising concerns about corporate control and the erosion of farming autonomy (Howard, 2015). If the majority of profits accrue to a handful of multinational firms, the promise of biotechnology as a global solution may be undermined by inequitable outcomes. Regulatory frameworks face similar challenges. Many were designed for earlier generations of transgenic crops and are not well-suited for new tools such as CRISPR or base editing. In some countries, genome-edited crops without foreign DNA are regulated differently than GMOs, while others apply the same strict rules across all genetic modifications (Dederer, et al., 2019). This lack of international alignment creates uncertainty: producers face fragmented markets, and consumers may encounter inconsistent or confusing information. Transparency and labeling are therefore essential. Many scholars argue that food labels should clearly differentiate between conventional, genetically engineered, and genome-edited products. This empowers consumers to make informed decisions. Yet, labeling also carries risks: highlighting genetic modification without context can spark unnecessary fear, while the absence of labeling may erode trust. Regulators must therefore balance transparency with careful communication. Another ethical dimension involves biosafety and dual-use risks. While food biotechnology is generally associated with sustainability and nutrition, the same tools could theoretically be misapplied in ways that threaten ecosystems or concentrate control of food systems. Ensuring responsible governance requires ongoing monitoring, not only within laboratories but also through inclusive dialogue and international cooperation.

Finally, meaningful public engagement is crucial. Treating consumers as passive recipients of technology risks alienation and rejection. Participatory approaches — such as citizen forums, public consultations, and transparent policymaking — help ensure that technological development reflects not only scientific feasibility but also social values. In short, the future of biotechnology in food will be shaped as much by trust, fairness, and governance as by technical breakthroughs. Without broad social legitimacy, even the most promising innovations may face rejection. With it, however, biotechnology could be embraced not as an imposed technology but as a shared solution to pressing global challenges.

Sustainability and Environmental Impacts of Biotechnological Foods

Among the strongest arguments for applying biotechnology to food systems is its potential to improve sustainability. Today's agriculture is a major contributor to environmental degradation, responsible for roughly one-quarter of global greenhouse gas emissions. Livestock production alone is a significant source of methane release, soil depletion, and water scarcity (FAO, 2020). Against this backdrop, biotechnological approaches offer opportunities to reduce ecological pressures while maintaining, or even expanding, food supply. One immediate benefit is the development of climate-resilient crops. Plants engineered for drought or salinity tolerance can grow in marginal soils, reducing irrigation needs and limiting pressure to convert forests into farmland (Ndudzo et al., 2024). Similarly, crops modified for pest and disease resistance decrease reliance on chemical pesticides, thereby reducing soil and water contamination. Such innovations also protect pollinators and biodiversity, which have been negatively affected by conventional agricultural practices.

Biotechnology also contributes to sustainability through alternative protein production. Precision fermentation and synthetic biology enable the creation of dairy, egg, and meat substitutes without the heavy land and water requirements of livestock farming. Studies suggest that microbial protein production requires only a fraction of the resources needed for beef or poultry while cutting greenhouse gas emissions significantly (Xu et al., 2021). Broad adoption of these alternatives could free up land for reforestation or renewable energy projects, multiplying environmental benefits.

Reducing food waste is another key area where biotechnology can make a difference. Currently, nearly one-third of food produced globally is lost or discarded before it is consumed. Innovations such as nanotechnology-enabled smart packaging extend shelf life and provide freshness indicators, helping consumers distinguish between spoiled and still-edible products (del Rosario Herrera-Rivera et al., 2024). Preventing unnecessary waste not only saves food but also reduces greenhouse gas emissions from decomposing organic matter in landfills.

However, sustainability benefits are not guaranteed. Large-scale biotechnological production, particularly fermentation in bioreactors, consumes energy and requires sugar-based feedstocks. If these inputs are produced unsustainably, the ecological advantages may be reduced (Parodi et al., 2018). This makes life-cycle assessment (LCA) critical. Evaluating every stage — from raw material sourcing to waste management — is necessary to confirm that biotechnological foods truly provide a net environmental benefit. Encouragingly, many companies are turning to renewable energy and exploring circular systems where by-products are repurposed into animal feed or biofuels.

Another important dimension is resilience. Climate change brings volatility, from droughts and floods to supply chain disruptions. By diversifying the food portfolio beyond a handful of traditional crops and livestock, biotechnology can strengthen resilience. Microbial proteins, engineered cereals, and algae-based foods expand dietary options, reducing dependence on vulnerable systems and improving preparedness for global shocks.

In summary, biotechnology offers both immediate and long-term strategies for transforming food systems into more sustainable models. From conserving water and land to reducing waste and creating climate-adapted crops, its potential is considerable. Yet realizing these benefits depends on responsible design, transparent assessment, and integration with renewable energy and circular economy principles. If managed effectively, biotechnology can shift food production from being a driver of environmental harm to becoming a powerful ally in ecological recovery.

Future Directions and Conclusion

The trajectory of biotechnology in food systems points strongly toward the future. What began with selective breeding and early genetic modification has expanded into a diverse toolkit that includes genome editing, precision fermentation, synthetic biology, nanotechnology, and digital traceability systems. Each of these technologies addresses specific aspects of the food chain, yet their true power lies in convergence. The future of food production will not be shaped by one single breakthrough but by how these different tools are integrated and reinforce one another. One of the most important directions involves the creation of multi-trait, climate-resilient crops. Rather than focusing solely on yield or a single stress factor, researchers are now working to develop varieties that can withstand multiple challenges simultaneously—such as drought, heat, salinity, and pests—while also providing improved nutritional value (Li, et al., 2021). Achieving this goal requires combining CRISPR with advanced phenotyping, big data analytics, and AI-based modeling, resulting in a holistic approach to crop design.

Another key priority is aligning biotechnology more closely with public health objectives. As rates of chronic diseases such as diabetes and cardiovascular disorders continue to rise, food innovation must address prevention as well as treatment. This includes biofortified crops that tackle hidden hunger, oils with healthier lipid profiles, and reduced levels of harmful compounds such as acrylamide formed during processing (Palmer et al., 2025; Zhang et al., 2022; Augustin et al., 2024). In the longer term, precision nutrition — integrating genomic data, microbiome science, and engineered foods — may enable dietary interventions tailored to individual health needs, significantly reducing disease burdens. Sustainability and circularity will also define the next phase of development. Industrial fermentation and enzyme engineering must increasingly adopt renewable energy sources, minimize waste, and repurpose by-products into secondary goods such as animal feed or biofuels. Life-cycle assessment (LCA) should become standard practice to ensure that environmental gains are genuine and not offset by hidden costs (Parodi et al., 2018). Scaling microbial proteins, plant-based foods, and smart packaging technologies will contribute directly to global climate goals, while conserving land and water resources. At the same time, innovation must be accompanied by ethical and governance frameworks. Without transparent communication, clear labeling, and equitable access, public skepticism could slow adoption. Including farmers, consumers, and local communities in decision-making processes is essential to prevent biotechnology from being perceived as an elite-driven initiative. Instead, it should be understood as a shared tool for addressing collective challenges (Frewer et al., 2013; Howard, 2015; Dederer, et al., 2019).

Looking further ahead, it is possible to envision food systems that are deeply connected to healthcare, environmental stewardship, and digital infrastructure. Foods may be designed not only to nourish but also to deliver therapeutic compounds. Blockchain-based supply chains could guarantee transparency and safety, while precision nutrition could redefine dietary practices at both the individual and population level. In this scenario, biotechnology becomes a cornerstone of resilient societies capable of withstanding climate change, pandemics, or economic disruptions.

In conclusion, biotechnological innovations in food and food engineering are no longer speculative ideas but rapidly advancing solutions. They offer powerful pathways to address hunger, improve nutrition, reduce environmental pressures, and strengthen resilience. Yet success will depend on more than technological achievement alone. Trust, transparency, equity, and sustainability must guide their application. If these principles are upheld, biotechnology has the potential to redefine food not only as a commodity but as a tool for health, justice, and planetary well-being (Walait et al., 2024).

Healthy Product Trends in the Food Industry: Market Drivers

Rising demand for health-promoting foods is reshaping product portfolios and marketing across the global food sector. Greater health consciousness—amplified by digital information flows and social networks—has pushed prevention-focused eating into the mainstream (Prabhu et al., 2018). In response, companies are pairing deeper market analysis with new solutions such as convenient better-for-you formats, functional foods, and tailored foodservice offerings (Hanus, 2018). This shift reframes diet as a proactive lever for health optimization and disease risk reduction, helping to explain the sustained expansion of the functional category (Khan et al., 2012; Sultana et al., 2025). Within that space, functional beverages alone were projected to reach \$208.13 billion by 2024 with a 7.5% CAGR—an indicator of both commercial momentum and consumer willingness to seek specific benefits from foods and drinks (Gupta et al., 2023; Corbo et al., 2014).

Conceptually, functional foods emerged from advances linking nutrition to measurable health outcomes and now offer a practical route to support well-being and potentially lower disease risk (Siró et al., 2008). Often enriched with targeted bioactives, these products aim at defined physiological effects—improving gut function, supporting immune competence, or modulating risk factors for chronic conditions (Riezzo et al., 2005). Coupled with growing investment and a more discerning consumer base, the sector retains substantial—yet still underrealized—potential for innovation and impact (Yang et al., 2023).

Probiotic and Fermented Foods

Growing insight into the human microbiome has pushed probiotics and fermented foods to the forefront of product innovation. A broadening evidence base links gut microbial composition to digestive and immune outcomes—and, more tentatively, to aspects of mental well-being—driving consumers toward probiotic-rich options such as yogurts, kefirs, and fermented vegetables. In parallel, R&D has focused on strain selection, dose, and delivery formats that maintain viability through processing and gastrointestinal transit (Ninchan & Noidee, 2021).

Dairy has long served as the primary vehicle for probiotic delivery and remains a major growth engine (Kechagia et al., 2013). Yet lactose intolerance, milk-protein allergy, and the rise of plant-forward eating have accelerated development of non-dairy alternatives. Fermented products based on cereals, fruits, and vegetables—along with oat, almond, and coconut "yogurt" analogues—are gaining traction as they broaden access while aiming to preserve microbial functionality (Gupta & Abu-Ghannam, 2011; Kårlund et al., 2020; Demarinis et al., 2023; Francis et al., 2024).

Active research now targets the fit between probiotic strains and plant-based matrices, addressing sensory quality, stability, and nutrient interactions in both fermented and non-fermented formats (Jang et al., 2024; Pua et al., 2022; Kechagia et al., 2013). Cultural or religious restrictions and vegetarian preferences can also limit dairy's universality, reinforcing the need for alternative carriers (Setta et al., 2020; Boukid et al., 2023). To that end, encapsulation and related protection technologies are becoming central tools to shield cells from processing stress and gastric conditions, improving survival and the likelihood of benefit in diverse product types.

Functional Foods and Beverages

Rising interest in health benefits beyond basic nutrition has positioned functional foods and drinks as a key growth area in the food industry. Formulations increasingly incorporate characterized bioactives—prebiotics, probiotics, omega-3 fatty acids, and selected antioxidants—with the aim of supporting chronic-disease risk management and physiological functions, a trend reinforced by population ageing and greater awareness of diet–health links (Pateiro et al., 2021; Yildiz and Bayat, 2022; Yavruturk et al., 2024).

A central technical lever is encapsulation. Nano- and micro-scale systems enhance stability, bioavailability, and controlled release of sensitive ingredients during processing, storage, and gastrointestinal transit. Oil-in-water nanoemulsions and related nanostructures are particularly useful carriers for hydrophobic compounds (e.g., vitamins, essential oils, nutraceuticals) that otherwise show poor solubility or degrade within complex food matrices (Salvia-Trujillo et al., 2017; Pateiro et al., 2021). Because many natural bioactives are heat-labile, microencapsulation helps preserve integrity through thermal and mechanical stresses while enabling sustained delivery (Mehta et al., 2022; Rezagholizade-Shirvan et al., 2024).

Encapsulation also mitigates practical hurdles in product design—limiting unwanted reactions with other ingredients and tempering strong flavors or aromas—by enclosing a bioactive core within polymeric or lipid shells that can be tuned for targeted release (Días et al., 2015; Comunian & Fávaro-Trindade, 2016; Simões et al., 2017; Yildiz et al., 2021). Lipid-oriented methods are widely applied to both hydrophobic and hydrophilic payloads in foods and supplements, helping maintain bioactivity against environmental stressors and matrix interactions (Subramani & Ganapathyswamy, 2020; Zhu, 2017). Together, these strategies support longer shelf life and more reliable nutritional performance, enabling functional products that are both technologically robust and consumer-friendly.

Low-Sugar and Sugar-Free Options

Rising rates of obesity, type 2 diabetes, and related metabolic conditions have accelerated consumer interest in lowering added sugar. Public-health messaging and dietary guidelines reinforce this direction, prompting manufacturers to reformulate and to adopt both synthetic and natural sweeteners—such as aspartame, sucralose, stevia, and xylitol—so that products can reduce energy while preserving palatability (Senda et al., 2024). Parallel research continues to examine longer-term health implications; some studies report possible effects on gut microbiota and metabolic responses even when immediate calorie reduction is achieved (Welsh et al., 2013).

A second front targets ingredient innovation: allulose and other rare sugars, for example, provide sweetness with minimal caloric contribution and are being evaluated for broader use (Sievenpiper, 2016). Category-specific case work illustrates what is possible; the Kuwait Danish Dairy Company, for instance, reformulated ice creams with no added sugar, substantially lowering net carbohydrates while maintaining consumer-acceptable taste (Al-Ozairi et al., 2024). Clear labeling remains important here—"no added sugar" is not synonymous with "zero total sugar," as intrinsic sugars can still influence glycemic response (Al-Ozairi et al., 2024).

Finally, sugar plays multiple functional roles beyond sweetness—affecting texture, bulk, freezing point, browning, and flavor release—so successful reduction depends on systems that replace these functions without compromising eating quality (Alipoor et al., 2020; Karl et al., 2020). Ongoing progress will come from combining smarter sweetener systems with process design and honest front-of-pack communication, helping consumers make informed choices while manufacturers safeguard sensory performance.

Gluten-Free Products

Once positioned primarily for individuals with celiac disease, the gluten-free (GF) category has expanded into the mainstream as broader consumer segments pursue perceived health benefits (Intrasook et al., 2024). Greater awareness of dietary sensitivities, diagnostic advances that increase recognition of gluten-related disorders, and self-identified intolerances have all contributed to shelf growth across supermarkets and specialty retailers. This shift is also sustained by popular beliefs—sometimes independent of medical diagnosis—that GF patterns may help digestion, weight management, or inflammation control (Kroes & Walker, 2004).

Industry has responded with rapid product development, from staple grain alternatives to complex processed foods designed to meet this demand. At the same time, GF does not automatically mean "healthier": nutritional equivalence depends on formulation, fortification, and overall diet quality. As the category matures, successful offerings increasingly emphasize whole-food bases, fiber and micronutrient parity with gluten-containing counterparts, and clear communication about intended use and target consumers.

Organic and Natural Products

The shift towards organic and "natural" products has gained momentum as consumers pay more attention to the source of food, production methods, and the impact of these processes on health and the environment. This tendency; While increasing the demand for products that are free from artificial additives, pesticides and GMOs, and processed as little as possible, it also raises ethical and sustainability expectations that go beyond their nutritional content. Manufacturers meet these expectations; It responds with transparency in raw material procurement, stricter ingredient criteria, and reformulations to reduce its environmental footprint throughout the supply chain. As a result, as the organic market grows, both large firms and SMEs are trying to reinforce consumer confidence by investing in organic certification and sustainable agriculture. The "natural" claim—though more loosely defined than organic—pushes formulations to simplify through short ingredient lists and "clean label" expectations.

This transformation is also influencing microencapsulation and similar technology choices in sensitive categories such as products containing live cultures and plant-based matrices. Appropriate encapsulation strategies; It helps maintain viability and functionality by increasing the resistance of beneficial microorganisms to processing conditions and the gastrointestinal environment, even in non-dairy or plant-based systems (Agriopoulou et al., 2023; Călinoiu et al., 2019; Sever and Yildiz, 2025). Thus, the "natural" component approach goes hand in hand with technology-enabled stability solutions.

On the preservatives front, natural antimicrobials are also gaining traction. Plant-derived phenolic derivatives and essential oils, as well as bacteriocins derived from lactic acid bacteria, offer the potential to extend shelf life and improve safety by being added directly to the product or used in active packaging systems (Valdés et al., 2015; Hemmati et al., 2021; Rollers, 1995). This approach, which is in line with the expectation of a "clean label", produces alternatives without being left without an alternative to synthetic preservatives, while also making use of controlled-release techniques such as nanoencapsulation to strengthen oxidative/antimicrobial effects (Dutta et al., 2008; Taylor, 2018; Mogoşanu et al., 2016; Bahrami et al., 2020; Mahmud & Khan, 2018).

However, efficacy and safety assessments are decisive for the large-scale applicability of natural preservatives. Detection of minimum inhibitory concentrations (MIC), monitoring of possible interactions and sensory effects in different food matrices, long-term stability and effect verification under storage conditions are critical steps on the road to commercialization (Bondi et al., 2017; Tiwari et al., 2009). In short, the organic/natural product approach; When it combines the principles of transparent procurement, clean label and sustainability with scientific evidence-based process and protection strategies, it can meet consumer expectations in a permanent way.

Impact on Public Health

Industrial shifts toward healthier formulations have direct implications for population health. Broader access to products with improved nutrient profiles—and clearer on-pack guidance—can help nudge dietary patterns away from highly processed options and, over time, contribute to lower risk of obesity, type 2 diabetes, and cardiovascular disease. Framing dietary modification as a first-line intervention not only supports individual agency but may also ease pressure on health systems by reducing the incidence of lifestyle-related conditions. Within this context, probiotic-enriched foods are of interest for their potential to support gut microbial diversity and immune modulation, factors relevant to chronic disease pathways (Tyutkov et al., 2022).

The evidence base around probiotics is expanding beyond traditional digestive outcomes. Emerging studies explore links between specific strains, the gut–brain axis, and facets of mental well-being and cognition (Kim et al., 2016). There are reports of strain-dependent benefits across diverse indications—including inflammatory conditions, cardiometabolic risk markers, and infectious disease contexts—alongside antimicrobial effects and support for a balanced gut ecosystem (Baral et al., 2021; Young et al., 2021). Preclinical work suggests behavioral and mood improvements in dysbiosis models, yet translation to consistent human benefit requires cautious interpretation and further validation (Esquivel, 2021; Palamutoğlu et al., 2024). Accordingly, large, well-controlled trials remain essential to clarify efficacy, mechanisms, and who benefits most, especially for neuropsychological endpoints (Gnanavel, 2015; Sequeira et al., 2022).

A next wave of precision approaches—next-generation probiotics and postbiotics—aims to pair clearer mechanisms with better practicality. Candidate strains (e.g., selected Bacteroides spp.) are being studied for their interactions with host immunity and microbiome metabolism, with potential relevance from intestinal disorders to cardiometabolic health and oncology settings (Jan et al., 2024; Tufail & Schmitz, 2024). Postbiotics—defined preparations of inactivated cells or cellular components—offer stability and safety advantages where live organisms are impractical or contraindicated, while seeking similar functional benefits (Mosca et al., 2022). Taken together, the public-health opportunity is substantial, but progress depends on rigorous human evidence, transparent labeling, and fit-for-purpose products that balance nutritional quality with sensory acceptance and accessibility.

Impact on the Food Industry

Rising demand for health-promoting foods has reshaped industry priorities, pushing manufacturers to build portfolios anchored in functional ingredients rather than traditional line extensions. The resulting expansion of the functional segment—underpinned by greater public awareness of diet–health links—has accelerated investment in R&D, quality systems, and claims substantiation (Sbehat et al., 2022). Development pipelines now routinely explore bioactive additions, targeted fortification, and the incorporation of beneficial microbes into everyday formats, with attention to stability, sensory performance, and labeling (Agriopoulou et al., 2023).

Within this shift, probiotics have become a flagship category: global sales are projected to reach about USD 77 billion by 2025, reflecting consumer interest in gut health and broader systemic effects (Voidarou et al., 2020). Translating that interest into dependable products has, in turn, advanced microencapsulation and related delivery technologies that protect viability during processing, storage, and gastrointestinal transit, and enable more precise, strain-specific dosing across diverse matrices (Samedi & Charles, 2019).

Plant-Based Nutrition

The term "plant-based nutrition" refers to a dietary pattern that emphasizes fruits, vegetables, whole grains, legumes, nuts, and oilseeds, as well as popular plant-based alternatives such as peanut butter, while limiting or completely excluding animal products. As the popularity of plant-based eating has increased, various sub-dietary models that can be defined as plant-based have emerged. These can be described as plant-centered, plant-focused, plant-rich, or plant-forward approaches (Kent et al., 2022).

In general, a healthy plant-based dietary model aims to minimize processed foods, animal-based products, and fats while maximizing the consumption of nutrient-dense plant-based foods. Accordingly, the plant-based nutrition model encourages the consumption of abundant amounts of raw or cooked vegetables, fruits, legumes, and oilseeds (Tuso et al., 2013).

Since the beginning of human life, plants have been used for numerous purposes beneficial to humans, primarily nutrition, as well as medicinal substances, timber production, fiber, and more. Plants have been regarded as valuable sources of essential compounds and considered reservoirs of many primary and secondary metabolites with therapeutic effects. Primary metabolites (proteins, carbohydrates, fats, and nucleic acids) constitute the building blocks of human life. In addition, secondary metabolites are produced by plants as mechanisms to defend against predators and pathogens, cope with environmental stress, attract pollinators, and act as protective agents (Langyan et al., 2022).

Since the mid-twentieth century, the consumption of animal proteins (meat, eggs, dairy products, fish) has increased significantly, whereas the proportion of plant-based foods (vegetables, fruits, legumes, nuts, grains, seeds) has decreased. Today, in high-income countries, a shift from animal protein products toward plant-based foods has begun (Dagevos, 2021). Particularly in Western societies, the number of individuals adopting plant-based diets has increased. However, research shows that 98–99% of people still consume animal products. Since this percentage includes dietary models with limited consumption of animal products, the actual proportion of plant-based eaters is not precisely known, and due to the variations in the definitions of plant-based diets, it is difficult to express in exact figures (Kent et al., 2022).

The transition from meat-based dietary patterns to those containing more plant-based foods is considered beneficial for both planetary and personal health. Due to the adverse environmental and health effects of meat production and consumption, a significant reduction is required. Reducing overall meat consumption and replacing meat products with plant-based alternatives is essential. However, adopting a flexitarian dietary model, where meat is not completely eliminated but consumption is consciously reduced, can play a crucial role in such a dietary transition without the necessity of becoming fully vegetarian (Dagevos, 2021).

Plant-based dietary models are generally categorized into six subgroups: raw vegan, vegan (also referred to as strict vegetarian), lacto-vegetarian, ovo-vegetarian, lacto-ovo vegetarian, and Mediterranean-type diets (Tuso et al., 2013).

Raw Vegan Diet Model

Individuals' relationships with their diets can be influenced by various factors such as self-efficacy, health control, susceptibility to illness, social support, willingness to change, previous habits, and socio-demographic characteristics. In the raw vegan model, individuals consider cooking as a form of "death," distinguishing between living and dead foods, and prefer living foods. This belief is grounded in the fact that during cooking, some nutrients are degraded, essential enzymes are denatured, and certain pro-inflammatory compounds may be generated (Raba et al., 2019).

This model includes plant-derived foods that have not undergone thermal processing. Fresh foods, products dried at low temperatures (maximum 40°C), cold-smoked items, or fermented plant-based products are consumed. Foods containing eggs or other animal products are excluded. Additionally, processed foods and foods cooked above 48°C are not consumed (Raba et al., 2019).

Vegan Diet Model

Individuals adhering to a vegan diet exclude all animal products, particularly meat, seafood, poultry, eggs, and dairy products. No restrictions are made on fats or refined sugars (Tuso et al., 2013). Despite excluding animal products from their diets, some vegans consume honey (Koeder & Perez-Cueto, 2024). As part of the "Western" lifestyle, the vegan diet differs from the "traditional" plant-based diet commonly practiced in developing countries, where lower income and education levels are often associated with limited food availability and reduced energy intake (Richter et al., 2016).

Vegans can be further categorized into three subgroups: those who follow veganism strictly for dietary purposes, those who adopt veganism as a lifestyle and extend it to all aspects of life, and those who emphasize vegan ethics, showing a completely ethical stance against animal use (Koeder & Perez-Cueto, 2024).

Vegans are generally defined as individuals who adopt a vegan diet, yet the majority also avoid using non-food animal-derived products such as leather, fur, wool, and silk. Many vegans also refrain from using products manufactured by companies that conduct animal testing (Koeder & Perez-Cueto, 2024).

Vegans require a well-planned diet. In poorly planned vegan diets, the most common nutrient deficiencies include iron, vitamins D and B_{12}, calcium, zinc, and omega-3 fatty acids (Akbulut & Yeşilkaya, 2021).

Lacto-Vegetarian Diet Model

This model excludes eggs, meat, seafood, and poultry but, unlike the vegan diet, includes dairy products (Tuso et al., 2013).

Ovo-Vegetarian Diet Model

This model excludes meat, seafood, poultry, and dairy products but, unlike veganism, allows the consumption of eggs (Tuso et al., 2013).

Lacto-Ovo Vegetarian Diet Model

This is a combination of lacto-vegetarianism and ovo-vegetarianism. Meat, seafood, and poultry are not consumed, while eggs and dairy products are permitted (Tuso et al., 2013).

Pescetarian Diet Model

Pescetarianism refers to individuals who, by personal preference, follow a plant-based diet but, unlike vegetarians, also consume fish and other seafood. In this model, fish serves as the main source of dietary protein and also provides healthy fat intake. The healthy fats in fish are associated with a reduced risk of various cardiovascular diseases (Althubaiti, 2024).

Flexitarianism

In addition to vegetarian variations, flexitarianism is defined as a plant-based diet characterized by consciously reducing meat consumption while still allowing occasional intake of meat and meat products. Different definitions include consuming meat once a week or less, consuming it 2–3 times per week, or eating it occasionally without fully eliminating it (Bruns et al., 2023).

Plant-Based Diets and Health

Plant-based dietary models, by either excluding or minimizing animal-based sources, are associated with positive health effects. Evidence from prospective cohort studies shows that the consumption of predominantly plant-based foods such as fruits, vegetables, nuts, and whole grains significantly reduces the risk of cardiovascular diseases. The protective effects of these foods are likely due to their rich content of monounsaturated and polyunsaturated fatty acids, omega-3 fatty acids, antioxidant vitamins, minerals, phytochemicals, fiber, and plant proteins (Patel et al., 2017).

Numerous studies have investigated the effects of plant-based diets on human health. In a meta-analysis conducted by Yokoyama and colleagues, which examined the impact of plant-based diets on lipid profiles compared to omnivorous diets, plant-based eaters were reported to have lower total cholesterol (TC) and low-density lipoprotein cholesterol (LDL-C), higher high-density lipoprotein cholesterol (HDL-C), but no significant difference in triglyceride levels. Since elevated LDL-C and reduced HDL-C levels play a key role in the development of coronary artery diseases, plant-based diets may significantly help in reducing coronary artery disease risk (Yokoyama et al., 2017).

Another meta-analysis by Dinu and colleagues, which examined the effects of plant-based diets on cancer, reported that the incidence of cancer among vegetarians and vegans was significantly lower compared to omnivores. After analyzing cancer localization, the authors found a non-significant reduction in mortality rates due to breast, colorectal, prostate, and lung cancers among vegetarians compared to omnivores (Dinu et al., 2017).

A study conducted on 74 individuals with type 2 diabetes over 24 weeks investigated the effect of plant-based diets on glycemic control. It was reported that a low-calorie vegetarian diet improved insulin sensitivity more effectively than a conventional diabetes diet over the same period. The same study also found that greater visceral fat loss, along with improvements in plasma adipokine and oxidative stress markers, contributed to reduced insulin resistance. Additionally, exercise training introduced after the third month enhanced the effects of the vegetarian diet (Kahleova et al., 2011).

Beyond these outcomes, plant-based diets are also associated with reduced risk of type 2 diabetes, maintenance of body mass index, and improved blood pressure. While animal-based foods are nutrient-rich, they may lead to undesirable health consequences (Perez-Cueto, 2020).

Plant-Based Dietary Models and Sustainability

Over the past 70 years, the world population has increased by approximately 200%, rising from 2.6 billion in 1950 to 7.6 billion in 2017, and it is projected to surpass 9 billion by 2030. This rapid population growth has become a major concern for the environment, necessitating sustainable development to safeguard the needs of future generations. Without actions that promote healthy aging and sustainable development, chronic diseases and environmental impacts arising from limited resource use will intensify, creating risks of intergenerational inequality in resource utilization. In response, countries around the world have adopted the 2030 Agenda for Sustainable Development and its 17 Sustainable Development Goals (SDGs). These international goals provide a roadmap to end poverty, ensure health and well-being for all, and protect the planet (Grosso et al., 2020).

Within the framework of the 2030 Sustainable Development Goals, which have evolved into international cooperation, achieving all 17 goals is targeted regardless of a country's level of development. SDG 2, "Zero Hunger," is the most directly relevant to nutrition. This goal is defined as: "end hunger, achieve food security, improve nutrition, and promote sustainable agriculture." Globally, 150 million children under the age of five are stunted, 50 million are wasted, and 20 million newborns have low birth weight. Among adults, nearly 800 million people suffer from hunger due to poverty worldwide, while within the European Union, 43 million people are unable to consume a quality meal every other day (Grosso et al., 2020). Despite some progress in reducing global hunger, around 800 million people remain chronically undernourished in terms of energy, and 2 billion suffer from micronutrient deficiencies. At the same time, 1.9 billion people worldwide are overweight or obese (Qaim, 2017).

Sustainable diets are characterized by their low carbon footprint, protection of biodiversity and ecosystems, and respect for natural resources. A sustainable diet must also be nutritionally adequate, safe, healthy, culturally acceptable, and economically affordable (Perez-Cueto, 2020).

Animal-based foods such as meat, dairy, eggs, and fish are significant sources of protein, vitamins, and minerals vital to human health. Their consumption helps reduce nutritional deficiencies and provides considerable benefits, especially in low-income countries. One of their most important roles is converting grass and by-products that humans cannot consume into edible food, thereby enhancing dietary diversity. Nevertheless, animal-based foods have larger environmental and climatic impacts compared to plant-based alternatives. The livestock sector is a leading driver of deforestation and biodiversity loss, while also contributing substantially to greenhouse gas emissions. Moreover, overfishing for seafood production has severe negative effects on marine ecosystems (Qaim et al., 2024).

By 2050, the global population is expected to exceed 9 billion, implying a potential 70% increase in food demand. To meet this rising demand, new sources are being explored, with plant-derived foods, vitamins, phenolic compounds, and other bioactive components considered key contributors (Langyan et al., 2022). If current dietary practices persist, diet-related diseases will account for three-quarters of the global disease burden by 2050, while water scarcity, excessive land use, species loss, and greenhouse gas emissions will increase global temperatures by more than 1.5°C (Carey et al., 2023). Although animal-based foods supply less than 20% of total dietary energy, they occupy 70% of all agricultural land and 40% of arable land. Livestock production also consumes much of the world's freshwater and contributes to about two-thirds of total agricultural greenhouse gas emissions (Qaim et al., 2024). In the United States, food production uses nearly 80% of freshwater resources, with animal protein production requiring about 100 times more water than plant protein production (Pimentel & Pimentel, 2003). Additionally, the production of some plant-based foods, such as nuts, requires more water than dairy, eggs, processed meats, or fish (Clark et al., 2019).

Definitions of sustainability often encompass ecology, economy, and society, and may vary depending on context. A sustainable diet may not mean the same thing to consumers as it does to farmers or food producers. In 2010, the Food and Agriculture Organization (FAO) defined sustainable diets as "those with low environmental impacts which contribute to food and nutrition security and to healthy life for present and future generations." Such diets protect biodiversity and ecosystems, are culturally acceptable, accessible, economically fair, and affordable; are nutritionally adequate, safe, and healthy; and optimize natural and human resources (Sabaté & Soret, 2014).

Food production consumes numerous resources, making it necessary to evaluate the environmental sustainability of current practices. The most fundamental inputs of the food system include soil, water, sunlight, fossil energy, and chemicals, combined with human labor and machinery. While food is the main output, the system also generates unwanted solid, liquid, and gaseous waste. Consumer preferences shape this entire cycle, from production to processing, transportation, storage, and retail (Sabaté & Soret, 2014).

Despite existing targets to reduce animal-based food production and consumption, little progress has been achieved. Instead, production continues to rise annually in parallel with population and income growth. Over the past 60 years, global meat production has increased fivefold, milk production threefold, egg production sixfold, and fish production 4.5-fold. Asia has been identified as the region with the highest increase in animal-based food production (Qaim et al., 2024). Individuals in less developed regions tend to consume fewer animal products compared to those in developed areas. For example, the average person in North America consumes seven times more animal-based foods than the average person in Africa (Qaim et al., 2024).

The EAT-Lancet Commission has proposed strategies to feed a projected global population of 10 billion by 2050 through healthy and sustainable food systems. Their four-step plan included:

- Defining a healthy reference diet based on current research,
- Identifying planetary boundaries for six critical environmental systems and processes, including greenhouse gas emissions, cropland use, water use, nitrogen and phosphorus application, and biodiversity loss,
- Developing a global food systems modeling framework to sustain healthy diets through 2050, and
- Outlining strategies to achieve the necessary transformations to meet these goals (Fanzo et al., 2022).

The same commission also introduced the "Planetary Health Diet" model, which aims to safeguard planetary sustainability through science-based, flexible, and individually adaptable targets for daily food groups. This dietary plan emphasizes vegetables, fruits, whole grains, legumes, nuts, and healthy fats while keeping red meat, sugar, and saturated fat consumption to a minimum (Willett et al., 2019).

Plant-Based Nutrition and Environmental Impacts

Plant-based diets are considered environmentally sustainable because the production of animal-based foods consumes resources at a faster rate, requiring intensive energy, water, and land use, while also generating higher greenhouse gas emissions (Poore & Nemecek, 2018). Approximately 14.5% of global anthropogenic greenhouse gas emissions are attributed to the livestock sector (Gerber et al., 2013).

Roughly one-third of global greenhouse gas emissions originate from agriculture. The production of red meats such as beef releases significant amounts of methane and nitrous oxide. A study by Clark et al. (2020) reported that red meat production, regardless of processing method, is associated with the highest levels of environmental damage, whereas plant-based foods have relatively low carbon footprints. Similarly, a systematic review by Aleksandrowicz et al. (2016) found that plant-based dietary patterns could reduce greenhouse gas emissions by up to 70–80%.

Currently, about 38% of Earth's land area is devoted to agriculture, much of it for livestock and feed production (Foley et al., 2011). Growing demand for animal products accelerates deforestation in tropical regions, threatening ecosystems and biodiversity. This demand is considered a major driver of biodiversity loss (Godfray et al., 2018). Moreover, livestock farming increases the risk of zoonotic diseases (Jones et al., 2008). Biodiversity loss linked to animal product production threatens ecosystems across the planet. Around 60% of global agricultural production does not depend on pollination, while 35% relies on animal pollinators. For 13 crops, pollinators are indispensable; for 30 crops, production is highly dependent on pollinators; for 27 crops, moderately dependent; and for 21 crops, slightly dependent. Increasing agricultural activities jeopardize wild bee populations and the stability of pollination services at the landscape scale (Klein et al., 2007).

Animal food production also has significant impacts on water sustainability. Producing beef requires 20 times more water compared to grains. Approximately 15,400 liters of water are required to produce just one kilogram of beef (Mekonnen & Hoekstra, 2012). Livestock farming further affects aquatic ecosystems through nitrogen and phosphorus pollution

(Davidson et al., 2014). In contrast, plant-based diets reduce both water use and pollution. It has been reported that 24% of freshwater resources are used for food that is ultimately wasted, highlighting the potential contribution of food waste reduction to water sustainability (Kummu et al., 2012).

Water sustainability is also classified through green, blue, and gray water footprints. Green water refers to rainwater absorbed by plants, typically in rain-fed agricultural systems. Blue water refers to freshwater extracted from rivers, lakes, and groundwater. Corn and soy, often grown as animal feed, usually have a high blue water footprint. Gray water refers to the volume of freshwater needed to dilute pollutants such as fertilizers and chemicals from agricultural activities (Mekonnen & Hoekstra, 2012).

Soil fertility preservation is equally vital for sustainable food systems. Intensive livestock farming can deplete soil organic carbon and increase erosion risk (Lal, 2015).

If current food production practices persist, environmental impacts are projected to increase by up to 90% by 2050 (Springmann et al., 2018). Transitioning to plant-based diets could substantially reduce emissions and land use. According to the EAT-Lancet Commission report, shifting toward plant-based eating could enable Earth's resources to sustain 10 billion people (Willett et al., 2019).

Current Protein Sources

The global population is expected to surpass nine billion by 2050, and the accompanying rise in food demand suggests that reliance on conventional livestock alone will struggle to meet future protein needs (Liang et al., 2024). This outlook has accelerated work on alternative, resource-efficient protein options to support food security (Pastrana-Pastrana et al., 2024). Heightened concern over greenhouse-gas emissions and the broader environmental footprint of high animal-protein intake further underscores the need for new approaches (Malek & Umberger, 2023). Emerging sources—plant-based proteins, edible insects, fungi/mycoprotein, and algae, as well as microbe-derived and cultivated meat—are being developed to help meet demand in ways aligned with sustainability goals (Pastrana-Pastrana et al., 2024). These alternatives aim to deliver indispensable amino acids and key micronutrients at levels comparable to those in conventional animal proteins, with some reports noting advantages for specific metrics or products (Bull et al., 2022). Moving from promise to practice will require comparative assessments that integrate life-cycle impacts, nutritional quality, and consumer acceptance to enable responsible integration into food systems (Malila et al., 2024). Reflecting this trajectory, alternative proteins have become a focal point for research and innovation, propelled by growing consumer interest in foods that are both nutritious and environmentally responsible (Liu et al., 2024).

Types of Alternative Protein Sources

Plant-based options stand out within the alternative-protein landscape for their culinary flexibility and comparatively small environmental footprint, while proteins sourced from insects, filamentous fungi/mycoprotein, and algae are also progressing as efficient production platforms (Tolnay et al., 2020). To meet rising global protein needs and support future food security, these sources must be compared systematically—considering their nutrient profiles, techno-functional behavior, and life-cycle impacts—so they can be deployed at scale in appropriate contexts (Malek & Umberger, 2023; Zhang et al., 2024).

Plant-Based Proteins

Plant-based proteins derived from legumes, grains, oilseeds, and other crops are attracting growing interest as lower-impact, health-aligned alternatives to animal proteins. In the dairy arena, however, developers still grapple with reproducing the layered textural, functional, and nutritional traits of animal counterparts while sustaining consumer appeal and overall sustainability (Alehosseini et al., 2025). Meaningful progress requires looking beyond crude protein levels to the broader food matrix and nutrient bioaccessibility, and deploying strategies that raise both sensory quality and nutrient density (Pontonio & Rizzello, 2021). Central to this is a clearer understanding of protein polysaccharide interactions and how they govern mouthfeel and stability in plant-based milk analogues. Current research therefore targets the techno-functional and sensory performance of plant proteins and explores new processing routes to integrate them more effectively into cheese and yogurt formats, addressing persistent technical and sensory hurdles (Díaz-Bustamante et al., 2023). Despite notable advances, mimicking dairy's organoleptic profile remains challenging because many plant materials carry inherent off-flavors and atypical textures (Pua et al., 2022). In practice, careful ingredient selection, tailored processing, and—where appropriate—fermentation are combined to refine flavor and texture, strengthening the market competitiveness and consumer acceptance of plant-based milk alternatives (Pua et al., 2022).

In addition to plant-based milk alternatives, plant-based cheeses are also produced. The historical process, from fermented nut and seed products to modern industrial formulations, reveals the constant effort to mimic the complex properties of traditional dairy cheeses with plant-derived ingredients (Fabiszewska et al., 2024). However, the biggest challenge in the development of plant-based cheeses is mimicking the unique structure of animal proteins, which provides the textural and sensory properties of traditional dairy products (Short et al., 2021). A significant challenge in the development of plant-based cheeses is to reproduce properties such as melting, stretching, and firmness, which are often provided by the complex protein networks in dairy cheese (Grasso et al., 2023). Unlike traditional dairy cheeses, which derive their structure from casein micelles, plant-based alternatives often rely on mixtures of different ingredients such as hazelnuts and legumes to mimic these properties, which brings with it some nutritional and textural compromises (Tekin and Hayaloğlu, 2024; Kovačević et al., 2024). As a matter of fact, a significant proportion of commercially produced plant-based cheeses show low protein content, with many of them having less than 5 g of protein; this indicates a significant nutrient imbalance compared to dairy cheeses (Chuang et al., 2025). This nutritional gap often raises concerns about the overall health benefits of plant-based cheeses over their dairy counterparts, despite their environmental advantages (Mefleh et al., 2022).

The growing interest in plant-based alternatives extends beyond milk and cheese to a wider range of dairy products, including milk, yogurt, and ice cream, each presenting unique rheological and sensory challenges. Therefore, the integration of plant-based proteins into a wide range of products, such as milk, cheese, yogurt and ice cream, is critical for future research and industrial applications, not only in terms of sustainability and environmental benefits, but also in terms of the development of innovative strategies to improve sensory quality and nutritional value.

Insect-Based Proteins

Interest in eating insects has expanded as they offer a protein source with low environmental demands and high biological efficiency compared with conventional livestock (Lima et al., 2022). Many edible species provide substantial protein together with essential fatty acids, vitamins, and minerals, and their amino-acid patterns often align well with dietary requirements (Quintieri et al., 2023). While life-cycle impacts are context-dependent, the combination of nutrient density and efficient feed conversion positions insects as a credible option within sustainable protein strategies.

Translate that potential into foods people actually enjoy, current work concentrates on processing and formulation. Protein extraction and fractionation aim to yield concentrates with reliable emulsifying, gelling, and water-/oil-binding behavior and improved bioavailability (Mintah et al., 2019; Kurek et al., 2022). Notably, high-moisture extrusion of insect protein concentrates can generate fibrous meat analogues with textures and protein contents comparable to conventional meat, illustrating a feasible route to palatable products that fit familiar formats (Smetana et al., 2017; 2018).

Adoption nonetheless hinges on consumer acceptance, which remains limited in many Western settings due to unfamiliarity and negative perceptions (Shockley & Dossey, 2013). Even so, the nutritional profile and versatility of insect proteins support applications in both human foods and animal feeds (Benes et al., 2022; Rumpold & Schlüter, 2013), and international organizations have highlighted their role in food-security initiatives (Khan, 2018). The next step is scale: rearing, harvesting, and downstream processing must progress from lab- or pilot-scale to cost-effective, continuous, and quality-assured industrial operations—a shift that will require targeted research, investment, and clear regulatory pathways (Ssepuuya et al., 2017; Hawkey et al., 2020).

Cultured Meat

Cultured (cultivated) meat refers to producing edible muscle and fat by expanding animal cells in controlled bioreactor conditions, rather than raising and slaughtering livestock. By decoupling meat from conventional farming, this biotechnology can lower land and water use and reduce greenhouse-gas emissions while addressing ethical concerns about animal slaughter (Kirsch et al., 2023). It also aims to mitigate issues often associated with conventional production—such as certain food-borne risks and resource pressure—by shifting to cleaner, more efficient cell-culture systems (Arshad et al., 2017).

In practice, a small biopsy provides starter cells that proliferate in nutrient media and differentiate into muscle and adipose tissues (Kirsch et al., 2023). Although in-vitro growth of skeletal muscle and fat has been explored for more than a decade in biomedical contexts, translating this know-how to food requires culture conditions that reliably reproduce the sight, smell, texture, and flavor of meat (Post, 2012). Recent advances in tissue culture and process design suggest commercial feasibility is becoming more realistic—provided the final products can match conventional meat on color, aroma, texture, palatability, and cost (Kadim et al., 2015).

Key bottlenecks remain. Culture parameters must be tuned for efficient growth, and affordable, animal-component-free media are needed to support proliferation and differentiation at scale (Kadim et al., 2015; Benny et al., 2022). Many platforms still rely on serum or face hurdles in scaling cell expansion from lab to industrial volumes (Kashim et al., 2022; Kumar et al., 2021). Long-term cell viability, functional performance over passages, and suitable scaffolds that mimic the three-dimensional architecture of meat are active areas of research (Zheng et al., 2023). Promising directions include engineered scaffolds—such as 3D-printed prolamin structures—that can help deliver recognizable texture while controlling costs (Su et al., 2022).

If these challenges are met, cultivated meat could meaningfully shrink the environmental footprint of protein supply and advance animal welfare (Treich, 2021). Precision control over differentiation in bioreactors also opens the door to "designer" cuts with tuned nutrition and sensory profiles, along with more consistent raw material than is possible with live-animal production (Bhat & Fayaz, 2010). Reduced exposure to animal-borne diseases and fewer logistics around transport and slaughter add further advantages (Alam et al., 2023). Regulatory pathways are evolving: Singapore has authorized sales of cultivated chicken, and U.S. oversight is being built jointly by the FDA and USDA (Hallman et al., 2023). Moving from pilot runs to broad market availability will still require substantial investment, safety validation, and consumer outreach; transparent communication has already been shown to improve attitudes toward these products (Teng et al., 2021; Siddiqui et al., 2022; Chong et al., 2023; Xie et al., 2025).

Challenges and Opportunities

Shifting diets away from conventional meat is as much a cultural task as a technological one. Long-standing preferences shape expectations for taste, price, and convenience, so adoption improves when communication is specific: clear ingredient lists, transparent production steps, and evidence on safety, nutrition, and sensory quality, alongside the environmental and ethical co-benefits (Siddiqui et al., 2022). Trust is also reinforced by regulatory clarity—safety assessment pathways, labeling and nomenclature standards, and fair marketing rules that help consumers compare products on equal terms.

Economics remains pivotal. To compete with established protein sources, alternative platforms must bring costs down through scale and process efficiency—improved yields, cheaper inputs, and streamlined, continuous operations (Hamdan et al., 2017). These gains depend on manufacturing learning curves but also on supportive supply chains and investment that can carry technologies from pilot to commercial plants.

For cultivated meat specifically, several technical hurdles still limit scale. Cells need to proliferate and differentiate reliably in vitro while maintaining functionality over passages; robust, serum-free, animal-component-free media are essential yet remain costly (Zheng et al., 2023; Xie et al., 2025). At the same time, products must achieve three-dimensional structure and texture, which calls for advances in scaffolding and mass-transfer-friendly bioreactor design capable of sustaining high cell densities at competitive costs (Su et al., 2022).

Overcoming these scientific, engineering, and cost barriers—through coordinated work across cell biology, bioprocessing, materials science, sensory science, and consumer research—is key to moving from niche offerings to mainstream protein options.

Nutritional Considerations

As alternative proteins advance, careful nutritional profiling against conventional animal sources is essential. Beyond total protein, evaluation should cover amino-acid patterns (including limiting amino acids), digestibility and protein quality metrics (e.g., PDCAAS/DIAAS), micronutrient density, bioaccessibility, and antinutritional factors such as phytates or tannins. Attention to potential allergenicity and to needs of vulnerable groups (e.g., infants, older adults, pregnant individuals, and patients with specific conditions) helps ensure that novel products support complete and balanced diets and address concerns about equivalence with familiar proteins (Siddiqui et al., 2022).

Processing choices strongly influence nutritional value. Fermentation, germination, soaking, and appropriate heat or extrusion treatments can lower antinutrients and improve mineral uptake, while excessive heat or shear may degrade heat-labile vitamins or reduce available lysine. Product design should therefore aim to preserve or enhance nutrient density while meeting sensory targets, with systematic assessment of protein quality and micronutrient bioaccessibility to guide formulation and labeling (Pastrana-Pastrana et al., 2024).

Fortification and blending offer additional levers: adding nutrients that can be limiting in some plant-forward patterns (e.g., vitamin B12, iron, zinc, iodine, calcium, vitamin D) and tailoring lipid profiles can produce options that are both sustainable and nutritionally robust (Fasolin et al., 2019; Bull et al., 2022). Taken together, rigorous characterization, smart processing, and targeted fortification enable alternative proteins to contribute to nutrition security at population scale rather than introduce new gaps.

Conclusion

Plant-based dietary patterns have emerged as a crucial component of both human health and environmental sustainability. Evidence from epidemiological and clinical studies demonstrates that diets emphasizing fruits, vegetables, legumes, whole grains, nuts, and seeds are associated with reduced risks of cardiovascular diseases, type 2 diabetes, obesity, and certain cancers. In addition, the reduced reliance on animal-sourced foods significantly contributes to the mitigation of climate change, preservation of biodiversity, and more efficient use of natural resources.

Nevertheless, transitioning toward plant-based diets requires careful planning to ensure sufficient intake of critical nutrients. Flexitarian, vegetarian, and vegan models all provide varying degrees of plant-based nutrition, enabling individuals to adapt dietary changes according to their cultural, social, and personal contexts.

From a global perspective, the increasing population, limited natural resources, and urgent need for sustainable development goals highlight the importance of adopting plant-forward eating practices. The EAT-Lancet Commission and similar frameworks illustrate that planetary health diets can balance nutritional adequacy with ecological preservation.

In conclusion, plant-based nutrition is not only a dietary choice but also a multidimensional strategy to promote public health, environmental sustainability, and intergenerational equity. Widespread adoption of such models, supported by policy interventions, educational efforts, and food innovation, will be essential in achieving a healthier future for both humanity and the planet.

References

Abdul Aziz, M., Brini, F., Rouached, H., & Masmoudi, K. (2022). Genetically engineered crops for sustainably enhanced food production systems. *Frontiers in plant science,* 13, 1027828.

Abdullahi, N., Atiku, M. K., & Umar, N. B. (2021). The roles of enzyme in food processing-an overview. *Fudma Journal of Sciences*, 5(1), 157-164.

Acquah, C., Ohemeng-Boahen, G., Power, K. A., & Tosh, S. M. (2021). The Effect of Processing on Bioactive Compounds and Nutritional Qualities of Pulses in Meeting the Sustainable Development Goal 2. *Frontiers in Sustainable Food Systems,* 5.

Aghababaei, F., McClements, D. J., & Hadidi, M. (2025). Strategies to overcome nutritional and technological limitations of pulse proteins. *Future Foods,* 100624.

Agriopoulou, S., Tarapoulouzi, M., Varzakas, T., & Jafari, S. M. (2023). Application of Encapsulation Strategies for Probiotics: From Individual Loading to Co-Encapsulation. *Microorganisms,* 11(12), 2896.

Ahmed, A., Arya, S., Gupta, V., Furukawa, H., & Khosla, A. (2021). 4D printing: Fundamentals, materials, applications and challenges. *Polymer,* 228, 123926.

Akbulut, Ş., & Yeşilkaya, B. (2021). Evaluation of vegan nutrition in regards to health. *Sağlık Akademisyenleri Dergisi,* 8(2), 163-168

Alam, A. N., Kim, C. J., Kim, S. H., Kumari, S., Lee, S. Y., Hwang, Y. H., & Joo, S. T. (2023). Trends in Hybrid Cultured Meat Manufacturing Technology to Improve Sensory Characteristics. *Food Science of Animal Resources,* 44(1), 39.

Alehosseini, E., McSweeney, P. L. H., & Miao, S. (2025). Recent updates on plant protein-based dairy cheese alternatives: outlook and challenges [Review of Recent updates on plant protein-based dairy cheese alternatives: outlook and challenges]. *Critical Reviews in Food Science and Nutrition,* 1. Taylor & Francis.

Alipoor, A., Daneshi, M., & Sadeghian, A. (2020). Aroma and texture related cross-modal interactions for sugar reduction in chocolate milk. *Food Science and Technology*, 17(105), 169.

Althubaiti, H. (2024). Pescatarianism VS Vegetarianism: Which is Healthier?. *Nutrition,* 2735(5934), 2735-590X

Al-Ozairi, E., Mandani, Y., Alfaleh, G., Raj, J., Alshammari, S., & Roux, C. W. le. (2024). Reformulating ice cream to improve postprandial glucose response: an opportunity for industry to create shared value. *Frontiers in Nutrition,* 11.

Ameer, K., Das, P. R., & Eun, J. B. (2020). Advances in enzymatic applications in food industry. In *Microbial fermentation and enzyme technology* (pp. 85-97). CRC Press.

Arshad, M. S., Javed, M., Sohaib, M., Saeed, F., Imran, A., & Amjad, Z. (2017). Tissue engineering approaches to develop cultured meat from cells: A mini review [Review of Tissue engineering approaches to develop cultured meat from cells: A mini review]. *Cogent Food & Agriculture,* 3(1), 1320814. Cogent OA.

Arun, K. B., Anoopkumar, A. N., Sindhu, R., Binod, P., Aneesh, E. M., Madhavan, A., & Awasthi, M. K. (2023). Synthetic biology for sustainable food ingredients production: recent trends. *Systems Microbiology and Biomanufacturing,* 3(1), 137-149.

Augustin, M. A., Riley, M., Stockmann, R., Bennett, L., Kahl, A., Lockett, T., ... & Cobiac, L. (2016). Role of food processing in food and nutrition security. *Trends in Food Science & Technology,* 56, 115-125.

Avezum, L., Rondet, É., Mestres, C., Achir, N., Madodé, Y. E., Gibert, O., Lefèvre, C., Hémery, Y., Verdeil, J., & Rajjou, L. (2022). Improving the nutritional quality of pulses via germination. *Food Reviews International*, 39(9), 6011.

Bahrami, A., Delshadi, R., Assadpour, E., Jafari, S. M., & Williams, L. (2020). Antimicrobial-loaded nanocarriers for food packaging applications [Review of Antimicrobial-loaded nanocarriers for food packaging applications]. *Advances in Colloid and Interface Science,* 278, 102140. Elsevier BV.

Baral, K. C., Bajracharya, R., Lee, S. H., & Han, H. (2021). Advancements in the Pharmaceutical Applications of Probiotics: Dosage Forms and Formulation Technology [Review of Advancements in the Pharmaceutical Applications of Probiotics: Dosage Forms and Formulation Technology]. *International Journal of Nanomedicine,* 7535. Dove Medical Press.

Beara, I., & Marija, L. (2019). *Role of Phenols in Energy and Functional Beverages.* In Elsevier eBooks (p. 229). Elsevier BV.

Bedsaul, J., Monroy-Gomez, J., Zutphen, K. G. van, & Kraemer, K. (2023). An Introduction to Traditional and Novel Alternative Proteins for Low- and Middle-Income Countries. *Current Developments in Nutrition,* 102014.

Benes, E., Bíró, B., Fodor, M., & Gere, A. (2022). Analysis of wheat flour-insect powder mixtures based on their near infrared spectra. *Food Chemistry X,* 13, 100266.

Benny, A., Pandi, K., & Upadhyay, R. (2022). Techniques, challenges and future prospects for cell-based meat [Review of Techniques, challenges and future prospects for cell-based meat]. *Food Science and Biotechnology,* 31(10), 1225. Springer Science+Business Media.

Bessada, S. M. F., Barreira, J. C. M., & Oliveira, M. B. P. P. (2019). Pulses and food security: Dietary protein, digestibility, bioactive and functional properties. *Trends in Food Science & Technology*, 93, 53.

Bhat, H. F., & Fayaz, H. (2010). Prospectus of cultured meat—advancing meat alternatives. *Journal of Food Science and Technology,* 48(2), 125.

Biswas, R., Alam, M., Sarkar, A., Haque, M. I., Hasan, M. M., & Hoque, M. (2022). Application of nanotechnology in food: processing, preservation, packaging and safety assessment. *Heliyon,* 8(11).

Biswas, S., Zhang, D., & Shi, J. (2021). CRISPR/Cas systems: opportunities and challenges for crop breeding. *Plant Cell Reports,* 40(6), 979-998.

Bondi, M., Lauková, A., Niederhäusern, S. D., Messi, P., & Papadopoulou, C. (2017). Natural Preservatives to Improve Food Quality and Safety. *Journal of Food Quality,* 2017, 1.

Boukid, F., Hassoun, A., Zouari, A., Tülbek, M., Mefleh, M., Aït-Kaddour, A., & Castellari, M. (2023). Fermentation for Designing Innovative Plant-Based Meat and Dairy Alternatives [Review of Fermentation for Designing Innovative Plant-Based Meat and Dairy Alternatives]. *Foods,* 12(5), 1005. Multidisciplinary Digital Publishing Institute.

Boye, J. I., Zare, F., & Pletch, A. (2009). Pulse proteins: Processing, characterization, functional properties and applications in food and feed. *Food Research International*, 43(2), 414.

Bruns, A., Nebl, J., Jonas, W., Hahn, A., & Schuchardt, J. P. (2023). Nutritional status of flexitarians compared to vegans and omnivores-a cross-sectional pilot study. *BMC nutrition,* 9(1), 140

Bull, C. F., Belobrajdic, D. P., Hamzelou, S., Jones, D., Leifert, W. R., Reyes, R. P., Terefe, N. S., Williams, G., & Colgrave, M. L. (2022). How Healthy Are Non-Traditional Dietary Proteins? The Effect of Diverse Protein Foods on Biomarkers of Human Health [Review of How Healthy Are Non-Traditional Dietary Proteins? The Effect of Diverse Protein Foods on Biomarkers of Human Health]. *Foods,* 11(4), 528. Multidisciplinary Digital Publishing Institute.

Camerlengo, F., Frittelli, A., & Pagliarello, R. (2022). CRISPR towards a sustainable agriculture. *Encyclopedia,* 2(1), 538-558.

Carey, C. N., Paquette, M., Sahye-Pudaruth, S., Dadvar, A., Dinh, D., Khodabandehlou, K., ... & Jenkins, D. J. (2023). The environmental sustainability of plant-based dietary patterns: A scoping review. *The Journal of nutrition,* 153(3), 857-869.

Chalmers, J. J. (2015). Mixing, aeration and cell damage, 30+ years later: what we learned, how it affected the cell culture industry and what we would like to know more about. *Current Opinion in Chemical Engineering*, 10, 94.

Chao, C., Nam, H. K., Park, H. J., & Kim, H. W. (2024). Potentials of 3D printing in nutritional and textural customization of personalized food for elderly with dysphagia. *Applied Biological Chemistry,* 67(1), 25

Chong, M., Leung, A. K. -Y., Fernandez, T. M., & Ng, S. T. (2023). Effects of framing, nomenclature, and aversion to tampering with nature on consumer acceptance of cultivated meat in Singapore. *Journal of Environmental Psychology,* 91, 102140.

Chuang, B. F., Chen, S. Y., Lin, J., & Yen, G. (2025). Modifying pea protein by cold plasma for the development of functional vegan cheese. *Food Bioscience*, 65, 106017.

Clark, M. A., Springmann, M., Hill, J., and Tilman, D. (2019). Multiple health and environmental impacts of foods. *Proc. Natl. Acad. Sci.* 116, 23357–23362. doi: 10.1073/pnas.1906908116.

Comunian, T. A., & Fávaro-Trindade, C. S. (2016). Microencapsulation using biopolymers as an alternative to produce food enhanced with phytosterols and omega-3 fatty acids: A review [Review of Microencapsulation using biopolymers as an alternative to produce food enhanced with phytosterols and omega-3 fatty acids: A review]. *Food Hydrocolloids,* 61, 442. Elsevier BV.

Corbo, M. R., Bevilacqua, A., Petruzzi, L., Casanova, F. P., & Sinigaglia, M. (2014). Functional Beverages: The Emerging Side of Functional Foods. *Comprehensive Reviews in Food Science and Food Safety,* 13(6), 1192.

Călinoiu, L. F., Ştefănescu, B. E., Pop, I., Muntean, L. S., & Vodnar, D. C. (2019). Chitosan Coating Applications in Probiotic Microencapsulation. *Coatings,* 9(3), 194.

da Fonseca-Pereira, P., Siqueira, J. A., de Cássia Monteiro-Batista, R., Vaz, M. G. M. V., Nunes-Nesi, A., & Araújo, W. L. (2022). Using synthetic biology to improve photosynthesis for sustainable food production. *Journal of Biotechnology,* 359, 1-14.

Çabuk, B., Nosworthy, M. G., Stone, A. K., Korber, D. R., Tanaka, T., House, J. D., & Nickerson, M. T. (2018). Effect of Fermentation on the Protein Digestibility and Levels of Non-Nutritive Compounds of Pea Protein Concentrate. *Food Technology and Biotechnology,* 56(2).

Dagevos, H. (2021). Finding flexitarians: Current studies on meat eaters and meat reducers. *Trends in Food Science & Technology,* 114, 530-539.

Dankar, I., Haddarah, A., Omar, F. E., Sepulcre, F., & Pujolà, M. (2018). 3D printing technology: The new era for food customization and elaboration. *Trends in food science & technology,* 75, 231-242.

Das, A. B., Goud, V. V., & Das, C. (2019). *Phenolic Compounds as Functional Ingredients in Beverages.* In Elsevier eBooks (p. 285). Elsevier BV.

Das, R., & Biswas, S. (2016). Nutraceutical-prophylactic and Therapeutic Role of Functional Food in Health. *Journal of Nutrition & Food Sciences,* 6(4).

Davidson, N. C. (2014). How much wetland has the world lost? Long-term and recent trends in global wetland area. *Marine and Freshwater Research,* 65(10), 934–941.

Dederer, H. G., & Hamburger, D. (2019). *Regulation of genome editing in plant biotechnology.* Cham: Springer International Publishing.

del Rosario Herrera-Rivera, M., Torres-Arellanes, S. P., Cortés-Martínez, C. I., Navarro-Ibarra, D. C., Hernández-Sánchez, L., Solis-Pomar, F., ... & Román-Doval, R. (2024). Nanotechnology in food packaging materials: role and application of nanoparticles. *RSC advances,* 14(30), 21832-21858.

Demarinis, C., Montemurro, M., Torreggiani, A., Pontonio, E., Verni, M., & Rizzello, C. G. (2023). Use of Selected Lactic Acid Bacteria and Carob Flour for the Production of a High-Fibre and "Clean Label" Plant-Based Yogurt-like Product. *Microorganisms,* 11(6), 1607.

Dent, T., & Maleky, F. (2022). Pulse protein processing: The effect of processing choices and enzymatic hydrolysis on ingredient functionality [Review of Pulse protein processing: The effect of processing choices and enzymatic hydrolysis on ingredient functionality]. *Critical Reviews in Food Science and Nutrition,* 63(29), 9914. Taylor & Francis.

Dini, I. (2019). *An Overview of Functional Beverages.* In Elsevier eBooks (p. 1). Elsevier BV.

Dinu M, Abbate R, Gensini GF, Casini A, Sofi F. (2017) Vegetarian, vegan diets and multiple health outcomes: A systematic review with meta-analysis of observational studies *Crit Rev Food Sci Nutr.* 57:3640–9.

Dobson, S., & Marangoni, A. G. (2023). Methodology and development of a high-protein plant-based cheese alternative. *Current Research in Food Science,* 7, 100632.

Dong, Z., Chen, J., Chen, Y., & Shao, R. (2020, October). Food traceability system based on blockchain. In *Proceedings of the 2020 International Conference on Aviation Safety and Information Technology* (pp. 571-576).

Dutta, P. K., Tripathi, S., Mehrotra, G. K., & Dutta, J. (2008). Perspectives for chitosan based antimicrobial films in food applications. *Food Chemistry,* 114(4), 1173.

Días, M. I., Ferreira, I. C. F. R., & Barreiro, M. F. (2015). Microencapsulation of bioactives for food applications [Review of Microencapsulation of bioactives for food applications]. *Food & Function,* 6(4), 1035. Royal Society of Chemistry.

Díaz-Bustamante, M. L., Keppler, J. K., Reyes, L. H., & Álvarez, Ó. (2023). Trends and prospects in dairy protein replacement in yogurt and cheese. *Heliyon,* 9(6).

Eastham, J. L., & Leman, A. R. (2024). Precision fermentation for food proteins: ingredient innovations, bioprocess considerations, and outlook — a mini-review. *Current Opinion in Food Science,* 58, 101194.

Eastham, J. L., & Leman, A. R. (2024). Precision fermentation for food proteins: ingredient innovations, bioprocess considerations, and outlook—a mini-review. *Current Opinion in Food Science*, 58, 101194.

Esquivel, M. (2021). Probiotics for Mental Health: A Review of Recent Clinical Trials [Review of Probiotics for Mental Health: A Review of Recent Clinical Trials]. *American Journal of Lifestyle Medicine,* 16(1), 21. SAGE Publishing.

Essa, M. M., Bishir, M., Bhat, A., Chidambaram, S. B., Al-Balushi, B., Hamdan, H., Govindarajan, N., Freidland, R. P., & Qoronfleh, M. W. (2021). Functional foods and their impact on health [Review of Functional foods and their impact on health]. *Journal of Food Science and Technology,* 60(3), 820. Springer Science+Business Media.

Etter, B., Michel, F., & Siegrist, M. (2024). Which are the most promising protein sources for meat alternatives? *Food Quality and Preference,* 119, 105226.

Fabiszewska, A., Wierzchowska, K., Dębkowska, I., Śliczniak, W., Ziółkowska, M., Jasińska, K., Kobus, J., Nowak, D., & Zieniuk, B. (2024). Plant-Based Alternatives to Mold-Ripened Cheeses as an Innovation among Dairy Analogues. *Foods*, 13(14), 2305.

Fanzo, J., Rudie, C., Sigman, I., Grinspoon, S., Benton, T. G., Brown, M. E., ... & Willett, W. C. (2022). Sustainable food systems and nutrition in the 21st century: A report from the 22nd annual Harvard Nutrition Obesity Symposium. *The American Journal of Clinical Nutrition,* 115(1), 18-33.

FAO. (2020). The State of Food and Agriculture 2020: Overcoming water challenges in agriculture. Food and Agriculture Organization of the United Nations.

Fasihi, M., Jouzi, F., Tervasmäki, P., Vainikka, P., & Breyer, C. (2025). Global potential of sustainable single-cell protein based on variable renewable electricity. *Nature Communications,* 16(1).

Fasolin, L. H., Pereira, R. N., Pinheiro, A. C., Martins, J. T., Andrade, C. C. P. de, Ramos, Ó. L., & Vicente, A. A. (2019). Emergent food proteins – Towards sustainability, health and innovation [Review of Emergent food proteins – Towards sustainability, health and innovation]. *Food Research International,* 125, 108586. Elsevier BV.

Fitriarni, D., Martanto, M., & Rifkowaty, E. E. (2021). Formulation of Indonesian traditional functional drink wedang empon based on Zingiberaceae rhizomes mixed with fruits. IOP Conference Series Earth and Environmental Science, 913(1), 12028.

Foley, J. A., Ramankutty, N., Brauman, K. A., Cassidy, E. S., Gerber, J. S., Johnston, M., ... Zaks, D. P. M. (2011). Solutions for a cultivated planet. *Nature,* 478(7369), 337–342.

Francis, D. V., Dahiya, D., Gokhale, T., & Nigam, P. S. N. (2024). Sustainable packaging materials for fermented probiotic dairy or non-dairy food and beverage products: challenges and innovations [Review of Sustainable packaging materials for fermented probiotic dairy or non-dairy food and beverage products: challenges and innovations]. *AIMS Microbiology,* 10(2), 320. AIMS Press.

Frewer, L. J., van der Lans, I. A., Fischer, A. R., Reinders, M. J., Menozzi, D., Zhang, X., ... & Zimmermann, K. L. (2013). Public perceptions of agri-food applications of genetic modification–a systematic review and meta-analysis. *Trends in Food Science & Technology,* 30(2), 142-152.

Garrido-Galand, S., Asensio-Grau, A., Calvo-Lerma, J., Heredia, A., & Andrés, A. (2021). The potential of fermentation on nutritional and technological improvement of cereal and legume flours: A review [Review of The potential of fermentation on nutritional and technological improvement of cereal and legume flours: A review]. *Food Research International*, 145, 110398. Elsevier BV.

Gnanavel, S. (2015). Psychobiotics: The Latest Psychotropics. *Indian Journal of Psychological Medicine,* 37(1), 110.

Godfray, H. C. J., Aveyard, P., Garnett, T., Hall, J. W., Key, T. J., Lorimer, J., ... Jebb, S. A. (2018). Meat consumption, health, and the environment. *Science,* 361(6399), eaam5324.

Godfray, H. C. J., Beddington, J. R., Crute, I. R., Haddad, L., Lawrence, D., Muir, J. F., ... & Toulmin, C. (2010). Food security: the challenge of feeding 9 billion people. *Science,* 327(5967), 812-818.

Granato, D., Barba, F. J., Kovačević, D. B., Lorenzo, J. M., Cruz, A. G., & Putnik, P. (2020). Functional Foods: Product Development, Technological Trends, Efficacy Testing, and Safety [Review of Functional Foods: Product Development, Technological Trends, Efficacy Testing, and Safety]. *Annual Review of Food Science and Technology,* 11(1), 93. *Annual Reviews.*

Grasso, N., Bot, F., Roos, Y. H., Crowley, S. V., Arendt, E. K., & O'Mahony, J. A. (2023). Plant-Based Alternatives to Cheese Formulated Using Blends of Zein and Chickpea Protein Ingredients. *Foods,* 12(7), 1492.

Grosso, G., Mateo, A., Rangelov, N., Buzeti, T., & Birt, C. (2020). Nutrition in the context of the Sustainable Development Goals. *European journal of public health,* 30(Supplement_1), i19-i23.

Guiné, R. P. F., & Lima, M. R. (2008). Overview and Developments Regarding Functional Foods and Beverages. *Current Nutrition & Food Science,* 4(4), 298.

Guo, D., Liu, C., Zhu, H., Cheng, Y., Guo, Y., Yao, W., He, Q., & Jiang, J. (2024). Large edible mushrooms and mycelial proteins: A sustainable, nutritious protein source with health benefits and processing innovations. *Innovative Food Science & Emerging Technologies,* 103870.

Gupta, A., Sanwal, N., Bareen, M. A., Barua, S., Sharma, N., Olatunji, O. J., Nirmal, N. P., & Sahu, J. K. (2023). Trends in functional beverages: Functional ingredients, processing technologies, stability, health benefits, and consumer perspective [Review of Trends in functional beverages: Functional ingredients, processing technologies, stability, health benefits, and consumer perspective]. *Food Research International,* 170, 113046. Elsevier BV.

Gupta, S., & Abu-Ghannam, N. (2011). Probiotic Fermentation of Plant Based Products: Possibilities and Opportunities [Review of Probiotic Fermentation of Plant Based Products: Possibilities and Opportunities]. *Critical Reviews in Food Science and Nutrition,* 52(2), 183. Taylor & Francis.

Hallman, W. K., Hallman, W. K., & Hallman, E. E. (2023). Cell-based, cell-cultured, cell-cultivated, cultured, or cultivated. What is the best name for meat, poultry, and seafood made directly from the cells of animals? *Npj Science of Food,* 7(1).

Hamdan, M. N., Post, M. J., Ramli, M. A., & Mustafa, A. R. (2017). Cultured Meat in Islamic Perspective. *Journal of Religion and Health,* 57(6), 2193.

Hanus, G. (2018). Food market innovations as a response to consumer requirements: a review of litearture [Review of Food market innovations as a response to consumer requirements: a review of litearture]. *Optimum Economic Studies,* 251. University of Białystok.

Hasler, C. M. (2002). Functional Foods: Benefits, Concerns and Challenges—A Position Paper from the American Council on Science and Health. *Journal of Nutrition,* 132(12), 3772.

Hassoun, A., Cropotova, J., Trif, M., Rusu, A., Bobiş, O., Nayik, G. A., Jagdale, Y. D., Saeed, F., Afzaal, M., Mostashari, P., Khaneghah, A. M., & Regenstein, J. M. (2022). Consumer acceptance of new food trends resulting from the fourth industrial revolution technologies: A narrative review of literature and future perspectives [Review of Consumer acceptance of new food trends resulting from the fourth industrial revolution technologies: A narrative review of literature and future perspectives]. *Frontiers in Nutrition,* 9. Frontiers Media.

Hawkey, K., Lopez-Viso, C., Brameld, J. M., Parr, T., & Salter, A. M. (2020). Insects: A Potential Source of Protein and Other Nutrients for Feed and Food [Review of Insects: A Potential Source of Protein and Other Nutrients for Feed and Food]. Annual Review of Animal Biosciences, 9(1), 333. *Annual Reviews.*

He, X., & Hwang, H. M. (2016). Nanotechnology in food science: Functionality, applicability, and safety assessment. *Journal of food and drug analysis,* 24(4), 671-681.

Hemmati, F., Bahrami, A., Esfanjani, A. F., Hosseini, H., McClements, D. J., & Williams, L. (2021). Electrospun antimicrobial materials: Advanced packaging materials for food applications. *Trends in Food Science & Technology,* 111, 520.

Henchion, M., Hayes, M., Mullen, A. M., Fenelon, M. A., & Tiwari, B. K. (2017). Future Protein Supply and Demand: Strategies and Factors Influencing a Sustainable Equilibrium [Review of Future Protein Supply and Demand: Strategies and Factors Influencing a Sustainable Equilibrium]. *Foods,* 6(7), 53. Multidisciplinary Digital Publishing Institute.

Hennigan, J. N., & Lynch, M. D. (2022). The past, present, and future of enzyme-based therapies. *Drug Discovery Today,* 27(1), 117-133.

Howard, P. H. (2015). Intellectual property and consolidation in the seed industry. *Crop Science,* 55(6), 2489-2495.

Intrasook, J., Tsusaka, T. W., & Anal, A. K. (2024). Trends and current food safety regulations and policies for functional foods and beverages containing botanicals [Review of Trends and current food safety regulations and policies for functional foods and beverages containing botanicals]. *Journal of Food and Drug Analysis,* 32(2), 118. Elsevier BV.

Jan, T., Negi, R., Sharma, B., Kumar, S., Singh, S., Kumar, A., Shreaz, S., Rustagi, S., Chaudhary, N., Kaur, T., Kour, D., Sheikh, M. A., Kumar, K., Yadav, A. N., & Ahmed, N. (2024). Next generation probiotics for human health: An emerging perspective [Review of Next generation probiotics for human health: An emerging perspective]. *Heliyon,* 10(16). Elsevier BV.

Jang, H. J., Lee, N., & Paik, H. (2024). Overview of Dairy-based Products with Probiotics: Fermented or Non-fermented Milk Drink. *Food Science of Animal Resources,* 44(2), 255.

Jones, K. E., Patel, N. G., Levy, M. A., Storeygard, A., Balk, D., Gittleman, J. L., & Daszak, P. (2008). Global trends in emerging infectious diseases. *Nature,* 451(7181), 990–993.

Kadim, I. T., Mahgoub, O., Baqir, S., Faye, B., & Purchas, R. W. (2015). Cultured meat from muscle stem cells: A review of challenges and prospects [Review of Cultured meat from muscle stem cells: A review of challenges and prospects]. *Journal of Integrative Agriculture,* 14(2), 222. Elsevier BV.

Kahleova, H., Matoulek, M., Malinska, H., Oliyarnik, O., Kazdova, L., Neskudla, T., ... & Pelikanova, T. (2011). Vegetarian diet improves insulin resistance and oxidative stress markers more than conventional diet in subjects with Type 2 diabetes. *Diabetic Medicine,* 28(5), 549-559.

Karabulut, G., Kapoor, R., Yildiz, G., Kahraman, O., & Feng, H. (2023). Ultrasonication-assisted modification of plant-based proteins. In *Novel plant protein processing: Developing the foods of the future* (pp. 116-143). CRC Press.

Karl, C. M., Wendelin, M., Lutsch, D., Schleining, G., Dürrschmid, K., Ley, J. P., Krammer, G., & Lieder, B. (2020). Structure-dependent effects of sweet and sweet taste affecting compounds on their sensorial properties. *Food Chemistry X,* 7, 100100.

Kashim, M. I. A. M., Haris, A. A. A., Mutalib, S. Abd., Anuar, N., & Shahimi, S. (2022). Scientific and Islamic perspectives in relation to the Halal status of cultured meat [Review of Scientific and Islamic perspectives in relation to the Halal status of cultured meat]. *Saudi Journal of Biological Sciences,* 30(1), 103501. Elsevier BV.

Kechagia, M., Basoulis, D., Konstantopoulou, S., Dimitriadi, D., Gyftopoulou, K., Skarmoutsou, N., & Fakiri, E. M. (2013). Health Benefits of Probiotics: A Review [Review of Health Benefits of Probiotics: A Review]. *ISRN Nutrition,* 2013, 1. Hindawi Publishing Corporation.

Kent, G., Kehoe, L., Flynn, A., & Walton, J. (2022). Plant-based diets: a review of the definitions and nutritional role in the adult diet. *Proceedings of the Nutrition Society*, 81(1), 62-74

Khan, R. S., Grigor, J., Winger, R. J., & Win, A. (2012). Functional food product development – Opportunities and challenges for food manufacturers. *Trends in Food Science & Technology*, 30(1), 27.

Khan, S. (2018). Recent advances in role of insects as alternative protein source in poultry nutrition. *Journal of Applied Animal Research,* 46(1), 1144.

Kim, B., Hong, V. M., Yang, J., Hyun, H., Im, J. J., Hwang, J., Yoon, S., & Kim, J. E. (2016). A Review of Fermented Foods with Beneficial Effects on Brain and Cognitive Function [Review of A Review of Fermented Foods with Beneficial Effects on Brain and Cognitive Function]. Preventive Nutrition and Food Science, 21(4), 297. *Korean Society of Food Science and Nutrition.*

Kirsch, M., Morales-Dalmau, J., & Lavrentieva, A. (2023). Cultivated meat manufacturing: Technology, trends, and challenges [Review of Cultivated meat manufacturing: Technology, trends, and challenges]. *Engineering in Life Sciences,* 23(12). Wiley.

Klein, A. M., Vaissière, B. E., Cane, J. H., Steffan-Dewenter, I., Cunningham, S. A., Kremen, C., & Tscharntke, T. (2007). Importance of pollinators in changing landscapes for world crops. *Proceedings of the Royal Society B: Biological Sciences,* 274(1608), 303–313.

Koeder, C., & Perez-Cueto, F. J. (2024). Vegan nutrition: a preliminary guide for health professionals. *Critical reviews in food science and nutrition,* 64(3), 670-707

Kovačević, J., Bechtold, T., & Pham, T. (2024). Plant-Based Proteins and Their Modification and Processing for Vegan Cheese Production. Macromol—A Journal of Macromolecular Research, 4(1), 23.

Kroes, R., & Walker, R. (2004). Safety issues of botanicals and botanical preparations in functional foods. *Toxicology*, 198, 213.

Kumar, P., Sharma, N., Sharma, S., Mehta, N., Verma, A. K., Chemmalar, S., & Sazili, A. Q. (2021). In-vitro meat: a promising solution for sustainability of meat sector [Review of *In-vitro* meat: a promising solution for sustainability of meat sector]. Journal of Animal Science and Technology, 63(4), 693. *Korean Society of Animal Sciences and Technology.*

Kumar, Y., Basu, S., Goswami, D., Devi, M., Shivhare, U. S., & Vishwakarma, R. K. (2021). Anti-nutritional compounds in pulses: Implications and alleviation methods. *Legume Science,* 4(2).

Kummu, M., de Moel, H., Porkka, M., Siebert, S., Varis, O., & Ward, P. J. (2012). Lost food wasted resources: Global food supply chain losses and their impacts on freshwater, cropland, and fertiliser use. *Science of the Total Environment*, 438, 477–489.

Kurek, M. A., Onopiuk, A., Pogorzelska-Nowicka, E., Szpicer, A., Zalewska, M., & Półtorak, A. (2022). Novel Protein Sources for Applications in Meat Alternative Products—Insight and Challenges [Review of Novel Protein Sources for Applications in Meat-Alternative Products—Insight and Challenges]. *Foods,* 11(7), 957. Multidisciplinary Digital Publishing Institute.

Kårlund, A., Gómez-Gallego, C., Korhonen, J., Palo-oja, O.-M., El-Nezami, H., & Kolehmainen, M. (2020). Harnessing Microbes for Sustainable Development: Food Fermentation as a Tool for Improving the Nutritional Quality of Alternative Protein Sources [Review of Harnessing Microbes for Sustainable Development: Food Fermentation as a Tool for Improving the Nutritional Quality of Alternative Protein Sources]. *Nutrients*, 12(4), 1020. Multidisciplinary Digital Publishing Institute.

Lagouri, V. (2019). Introductory Chapter: Functional Foods. In *IntechOpen eBooks*. IntechOpen.

Lal, R. (2015). Restoring soil quality to mitigate soil degradation. *Sustainability,* 7(5), 5875–5895.

Langyan, S., Yadava, P., Khan, F. N., Dar, Z. A., Singh, R., & Kumar, A. (2022). Sustaining protein nutrition through plant-based foods. *Frontiers in Nutrition,* 8, 772573.

Lee, J., Kim, M. J., & Kim, C. Y. (2024). The Development of New Functional Foods and Ingredients. *Foods,* 13(19), 3038.

Li, A., Zheng, J., Han, X., Yang, S., Cheng, S., Zhao, J., ... & Lu, Y. (2023). Advances in low-lactose/lactose-free dairy products and their production. *Foods,* 12(13), 2553.

Li, Y. P., Ahmadi, F., Kariman, K., & Lackner, M. (2024). Recent advances and challenges in single cell protein (SCP) technologies for food and feed production [Review of Recent advances and challenges in single cell protein (SCP) technologies for food and feed production]. *Npj Science of Food,* 8(1). Nature Portfolio.

Liang, Z., Zhu, Y., Leonard, W., & Fang, Z. (2024). Recent advances in edible insect processing technologies. *Food Research International*, 182, 114137.

Lima, M., Costa, R., Rodrigues, I., Lameiras, J., & Botelho, G. (2022). A Narrative Review of Alternative Protein Sources: Highlights on Meat, Fish, Egg and Dairy Analogues [Review of A Narrative Review of Alternative Protein Sources: Highlights on Meat, Fish, Egg and Dairy Analogues]. *Foods,* 11(14), 2053. Multidisciplinary Digital Publishing Institute.

Liu, X., Xu, M., Wang, H., & Zhu, L. (2025). Integrating Precision Medicine and Digital Health in Personalized Weight Management: The Central Role of Nutrition. *Nutrients,* 17(16), 2695.

Liu, Y., Aimutis, W. R., & Drake, M. A. (2024). Dairy, Plant, and Novel Proteins: Scientific and Technological Aspects. *Foods,* 13(7), 1010.

Magbanua, T. O., & Ragaza, J. A. (2023). Growth and whole-body proximate composition of Oreochromis niloticus Nile tilapia fed pea meal: A systematic review and meta-analysis [Review of Growth and whole-body proximate composition of Oreochromis niloticus Nile tilapia fed pea meal: A systematic review and meta-analysis]. *Frontiers in Sustainable Food Systems,* 7. Frontiers Media.

Mahmud, J., & Khan, R. A. (2018). Characterization of Natural Antimicrobials in Food System. *Advances in Microbiology*, 8(11), 894.

Malek, L., & Umberger, W. J. (2023). Protein source matters: Understanding consumer segments with distinct preferences for alternative proteins. *Future Foods*, 7, 100220.

Maleš, I., Pedišić, S., Zorić, Z., Garofulić, I. E., Repajić, M., You, L., Vladimir-Knežević, S., Butorac, D., & Dragović-Uzelac, V. (2022). The medicinal and aromatic plants as ingredients in functional beverage production. *Journal of Functional Foods,* 96, 105210.

Malila, Y., Owolabi, I. O., Chotanaphuti, T., Sakdibhornssup, N., Elliott, C. T., Visessanguan, W., Karoonuthaisiri, N., & Petchkongkaew, A. (2024). Current challenges of alternative proteins as future foods [Review of Current challenges of alternative proteins as future foods]. *Npj Science of Food,* 8(1). Nature Portfolio.

Marco, M. L., Heeney, D., Binda, S., Cifelli, C. J., Cotter, P. D., Foligné, B., ... & Hutkins, R. (2017). Health benefits of fermented foods: microbiota and beyond. *Current opinion in biotechnology,* 44, 94-102.

Martins, N., Oliveira, M. B. P. P., & Ferreira, I. C. F. R. (2017). *Development of Functional Dairy Foods. In Reference series in phytochemistry* (p. 1). Springer International Publishing.

Martirosyan, D., & Singh, J. (2015). A new definition of functional food by FFC: what makes a new definition unique? *Functional Foods in Health and Disease,* 5(6), 209.

Mays, Z. J., & Nair, N. U. (2018). Synthetic biology in probiotic lactic acid bacteria: At the frontier of living therapeutics. *Current opinion in biotechnology*, 53, 224-231.

Mefleh, M., Pasqualone, A., Caponio, F., & Faccia, M. (2021). Legumes as basic ingredients in the production of dairy-free cheese alternatives: a review [Review of Legumes as basic ingredients in the production of dairy-free cheese alternatives: a review]. *Journal of the Science of Food and Agriculture*, 102(1), 8. Wiley.

Mefleh, M., Pasqualone, A., Caponio, F., Angelis, D. D., Natrella, G., Summo, C., & Faccia, M. (2022). Spreadable plant-based cheese analogue with dry-fractioned pea protein and inulin–olive oil emulsion-filled gel. *Journal of the Science of Food and Agriculture,* 102(12), 5478.

Mehta, N., Kumar, P., Verma, A. K., Umaraw, P., Kumar, Y., Malav, O. P., Sazili, A. Q., Domínguez, R., & Lorenzo, J. M. (2022). Microencapsulation as a Noble Technique for the Application of Bioactive Compounds in the Food Industry: A Comprehensive Review [Review of Microencapsulation as a Noble Technique for the Application of Bioactive Compounds in the Food Industry: A Comprehensive Review]. *Applied Sciences,* 12(3), 1424. Multidisciplinary Digital Publishing Institute.

Mekonnen, M. M., & Hoekstra, A. Y. (2012). A global assessment of the water footprint of farm animal products. *Ecosystems,* 15(3), 401–415.

Messina, V., Skylas, D. J., Roberts, T. H., Valtchev, P., Whiteway, C., Li, Z., Hopf, A., Dehghani, F., Quail, K., & Kaiser, B. N. (2025). Pulse Proteins: Processing, Nutrition, and Functionality in Foods [Review of Pulse Proteins: Processing, Nutrition, and Functionality in Foods]. *Foods,* 14(7), 1151. Multidisciplinary Digital Publishing Institute.

Mintah, B. K., He, R., Dabbour, M., Xiang, J., Agyekum, A. A., & Ma, H. (2019). Techno-functional attribute and antioxidative capacity of edible insect protein preparations and hydrolysates thereof: Effect of multiple mode sonochemical action. *Ultrasonics Sonochemistry,* 58, 104676.

Mogoşanu, G. D., Grumezescu, A. M., Bejenaru, C., & Bejenaru, L. E. (2016). *Natural products used for food preservation.* In Elsevier eBooks (p. 365). Elsevier BV.

Mohsin, A. Z., Norsah, E., Marzlan, A. A., Rahim, M. H. A., & Hussin, A. S. M. (2023). Exploring the applications of plant-based coagulants in cheese production: A review [Review of Exploring the applications of plant-based coagulants in cheese production: A review]. *International Dairy Journal,* 148, 105792. Elsevier BV.

Mondal, S., Soumya, N. P. P., Mini, S., & Sivan, S. K. (2021). Bioactive compounds in functional food and their role as therapeutics. *Bioactive Compounds in Health and Disease* - Online ISSN 2574-0334 Print ISSN 2769-2426, 4(3), 24.

Mosca, A., Abreu, A. T. A. y, Gwee, K., Ianiro, G., Tack, J., Ha, N. T. V., & Hill, C. (2022). The clinical evidence for postbiotics as microbial therapeutics [Review of The clinical evidence for postbiotics as microbial therapeutics]. *Gut Microbes,* 14(1). Landes Bioscience.

Muha-Ud-Din, G., Ali, F., Hameed, A., Naqvi, S. A. H., Nizamani, M. M., Jabran, M., ... & Yong, W. (2024). CRISPR/Cas9-based genome editing: A revolutionary approach for crop improvement and global food security. *Physiological and Molecular Plant Pathology,* 129, 102191.

Naraoka, Y., Mabuchi, Y., Kiuchi, M., Kumagai, K., Hisamatsu, D., Yoneyama, Y., Takebe, T., & Akazawa, C. (2024). Quality Control of Stem Cell-Based Cultured Meat According to Specific Differentiation Abilities. *Cells,* 13(2), 135.

Naumovski, N., Mellor, D., & Ranadheera, C. S. (2020). Functional Beverages, from Idea to Functionality. *Beverages,* 6(4), 71.

Ndudzo, A., Makuvise, A. S., Moyo, S., & Bobo, E. D. (2024). CRISPR-Cas9 genome editing in crop breeding for climate change resilience: Implications for smallholder farmers in Africa. *Journal of Agriculture and Food Research,* 16, 101132.

Ninchan, B., & Noidee, C. (2021). Production and prebiotic properties of oligofructans from sugarcane juice fermentation by Bacillus subtilis TISTR 001. 3 *Biotech,* 11(5).

Ohanenye, I. C., Ekezie, F.-G. C., Sarteshnizi, R. A., Boachie, R. T., Emenike, C., Sun, X., Nwachukwu, I. D., & Udenigwe, C. C. (2022). Legume Seed Protein Digestibility as Influenced by Traditional and Emerging Physical Processing Technologies [Review of Legume Seed Protein Digestibility as Influenced by Traditional and Emerging Physical Processing Technologies]. *Foods,* 11(15), 2299. Multidisciplinary Digital Publishing Institute.

Palamutoğlu, M. İ., Köse, G., & Baş, M. (2024). Probiotics and Prebiotics Affecting Mental and Gut Health. *Healthcare,* 12(5), 510.

Palmer, A. C. (2025). Golden Rice: a quarter-century of innovation, challenges, and the promise of better nutrition. *The Journal of Nutrition.*

Parodi, A., Leip, A., De Boer, I. J. M., Slegers, P. M., Ziegler, F., Temme, E. H., ... & Van Zanten, H. H. E. (2018). The potential of future foods for sustainable and healthy diets. *Nature Sustainability,* 1(12), 782-789.

Parveen, S., Jamil, A., Pasha, I., & Ahmad, F. (2022). Pulses: A Potential Source of Valuable Protein for Human Diet. In IntechOpen eBooks. *IntechOpen.*

Pastrana-Pastrana, Á. J., Rodríguez-Herrera, R., Solanilla-Duque, J. F., & Flores-Gallegos, A. C. (2024). Plant proteins, insects, edible mushrooms and algae: more sustainable alternatives to conventional animal protein. *Journal of Future Foods,* 5(3), 248.

Pateiro, M., Gómez, B., Munekata, P. E. S., Barba, F. J., Putnik, P., Kovačević, D. B., & Lorenzo, J. M. (2021). Nanoencapsulation of Promising Bioactive Compounds to Improve Their Absorption, Stability, Functionality and the Appearance of the Final Food Products [Review of Nanoencapsulation of Promising Bioactive Compounds to Improve Their Absorption, Stability, Functionality and the Appearance of the Final Food Products]. *Molecules,* 26(6), 1547. Multidisciplinary Digital Publishing Institute.

Patel, H., Chandra, S., Alexander, S., Soble, J., & Williams Sr, K. A. (2017). Plant-based nutrition: An essential component of cardiovascular disease prevention and management. *Current cardiology reports*, 19(10), 104

Pendrill, F., Persson, U. M., Godar, J., & Kastner, T. (2019). Deforestation displaced: Trade in forest-risk commodities and the prospects for a global forest transition. *Environmental Research Letters,* 14(5), 055003.

Perez-Cueto, F. J. (2020). Sustainability, health and consumer insights for plant-based food innovation. *International Journal of Food Design,* 5(1-2), 139-148.

Pimentel, D., and Pimentel, M. (2003). Sustainability of meat-based and plant-based diets and the environment. *Am. J. Clin. Nutr.* 78, 660S–663S.

Pontonio, E., & Rizzello, C. G. (2021). Milk Alternatives and Non-Dairy Fermented Products: Trends and Challenges. *Foods,* 10(2), 222.

Poore, J., & Nemecek, T. (2018). Reducing food's environmental impacts through producers and consumers. *Science,* 360(6392), 987–992.

Post, M. J. (2012). Cultured meat from stem cells: Challenges and prospects [Review of Cultured meat from stem cells: Challenges and prospects]. *Meat Science*, 92(3), 297. Elsevier BV.

Prabhu, A., Kamath, G. B., Karthik, A., & Sinha, K. C. (2018*). Influence of Farmer demographics and Supply Chain Issues on Organized Retailer Performance.* MATEC Web of Conferences, 144, 5010.

Pua, A., Tang, V. C. Y., Goh, R. M. V., Sun, J., Lassabliere, B., & Liu, S. (2022). Ingredients, Processing, and Fermentation: Addressing the Organoleptic Boundaries of Plant-Based Dairy Analogues [Review of

Ingredients, Processing, and Fermentation: Addressing the Organoleptic Boundaries of Plant-Based Dairy Analogues]. *Foods,* 11(6), 875. Multidisciplinary Digital Publishing Institute.

Qaim, M. (2017). Globalisation of agrifood systems and sustainable nutrition. *Proceedings of the Nutrition Society,* 76(1), 12–21.

Qaim, M., Barrangou, R., & Ronald, P. C. (2024). Sustainability of animal-sourced foods and plant-based alternatives. *Proceedings of the National Academy of Sciences,* 121(50), e2400495121

Quintieri, L., Nitride, C., Angelis, E. D., Lamonaca, A., Pilolli, R., Russo, F., & Monaci, L. (2023). Alternative Protein Sources and Novel Foods: Benefits, Food Applications and Safety Issues [Review of Alternative Protein Sources and Novel Foods: Benefits, Food Applications and Safety Issues]. *Nutrients*, 15(6), 1509. Multidisciplinary Digital Publishing Institute.

Raba, D. N., Iancu, T., Bordean, D. M., Adamov, T., Popa, V. M. & Pîrvulescu, L. C. (2019). Pros and Cons of Raw Vegan Diet. *Advanced Research in Life Sciences,* 3(1), 2019. 46-51.

Rad, A. H., Aghebati-Maleki, L., Kafil, H. S., Zavoshti, H. F., & Abbasi, A. (2020). Postbiotics as novel health-promoting ingredients in functional foods. *Health Promotion Perspectives,* 10(1), 3.

Raffan, S. (2020). Genome Editing for Low-Acrylamide Wheat (Doctoral dissertation, University of Bristol).

Reganold, J. P., & Wachter, J. M. (2016). Organic agriculture in the twenty-first century. *Nature Plants,* 2(2), 15221.

Rezagholizade-shirvan, A., Soltani, M., Shokri, S., Radfar, R., Arab, M., & Shamloo, E. (2024). Bioactive compound encapsulation: Characteristics, applications in food systems, and implications for human health [Review of Bioactive compound encapsulation: Characteristics, applications in food systems, and implications for human health]. *Food Chemistry X*, 24, 101953. Elsevier BV.

Ribeiro, G. O., Rodrigues, L. de A. P., Santos, T. B. S. dos, Alves, J. P. S., Oliveira, R. S., Nery, T. B. R., Barbosa, J. D. V., & Soares, M. B. P. (2023). Innovations and developments in single cell protein: Bibliometric review and patents analysis. *Frontiers in Microbiology,* 13.

Ribéreau, S., Aryee, A. N. A., Tanvier, S., Han, J., & Boye, J. I. (2017). Composition, digestibility, and functional properties of yellow pea as affected by processing. *Journal of Food Processing and Preservation,* 42(1).

Richter, M., Boeing, H., Grünewald-Funk, D., Heseker, H., Kroke, A., Leschik-Bonnet, E., ... & Watzl, B. (2016). Vegan diet. Position of the German nutrition society (DGE). *Ernahrungs umschau,* 63(04), 92-102.

Riezzo, G., Chiloiro, M., & Russo, F. (2005). Functional Foods: Salient Features and Clinical Applications [Review of Functional Foods: Salient Features and Clinical Applications]. Current Drug Targets - *Immune Endocrine & Metabolic Disorders,* 5(3), 331. Bentham Science Publishers.

Rischer, H., Szilvay, G. R., & Oksman-Caldentey, K. (2020). Cellular agriculture — industrial biotechnology for food and materials [Review of Cellular agriculture — industrial biotechnology for food and materials]. *Current Opinion in Biotechnology,* 61, 128. Elsevier BV.

Roller, S. (1995). The quest for natural antimicrobials as novel means of food preservation: Status report on a European research project. *International Biodeterioration & Biodegradation,* 36, 333.

Rumpold, B. A., & Schlüter, O. (2013). Nutritional composition and safety aspects of edible insects [Review of Nutritional composition and safety aspects of edible insects]. *Molecular Nutrition & Food Research,* 57(5), 802. Wiley.

Sabaté, J., & Soret, S. (2014). Sustainability of plant-based diets: back to the future. *The American journal of clinical nutrition,* 100, 476S-482S.

Sagawa, C. H., Thomson, G., Mermaz, B., Vernon, C., Liu, S., Jacob, Y., & Irish, V. F. (2024). An efficient multiplex approach to CRISPR/Cas9 gene editing in citrus. *Plant Methods,* 20(1), 148.

Salvia-Trujillo, L., Soliva-Fortuny, R., Rojas-Graü, M. A., McClements, D. J., & Martín-Belloso, O. (2017). Edible Nanoemulsions as Carriers of Active Ingredients: A Review [Review of Edible Nanoemulsions as Carriers of Active Ingredients: A Review]. *Annual Review of Food Science and Technology,* 8(1), 439. Annual Reviews.

Samedi, L., & Charles, A. L. (2019). Viability of 4 Probiotic Bacteria Microencapsulated with Arrowroot Starch in the Simulated Gastrointestinal Tract (GIT) and Yoghurt. *Foods,* 8(5), 175.

Sbehat, M., Mauriello, G., & Altamimi, M. (2022). Microencapsulation of Probiotics for Food Functionalization: An Update on Literature Reviews [Review of Microencapsulation of Probiotics for

Food Functionalization: An Update on Literature Reviews]. *Microorganisms,* 10(10), 1948. Multidisciplinary Digital Publishing Institute.

Semba, R. D., Ramsing, R., Rahman, N., Kraemer, K., & Bloem, M. W. (2021). Legumes as a sustainable source of protein in human diets. *Global Food Security,* 28, 100520.

Senda, M. V. G., Raposo, A., Teixeira-Lemos, E., Chaves, C., Alturki, H. A., Alsharari, Z. D., & Romão, B. (2024). Brazilian Front-of-Package Labeling: A Product Compliance Analysis 12 Months after Implementation of Regulations. *Nutrients*, 16(3), 343.

Seo, S., & Jin, Y. (2022). Next-Generation Genetic and Fermentation Technologies for Safe and Sustainable Production of Food Ingredients: Colors and Flavorings [Review of Next-Generation Genetic and Fermentation Technologies for Safe and Sustainable Production of Food Ingredients: Colors and Flavorings]. *Annual Review of Food Science and Technology,* 13(1), 463. Annual Reviews.

Sequeira, C. L. M. de, Hengstberger, C., Enck, P., & Mack, I. (2022). Effect of Probiotics on Psychiatric Symptoms and Central Nervous System Functions in Human Health and Disease: A Systematic Review and Meta-Analysis [Review of Effect of Probiotics on Psychiatric Symptoms and Central Nervous System Functions in Human Health and Disease: A Systematic Review and Meta-Analysis]. *Nutrients,* 14(3), 621. Multidisciplinary Digital Publishing Institute.

Setta, M., Matemu, A., & Mbega, E. R. (2020). Potential of probiotics from fermented cereal-based beverages in improving health of poor people in Africa [Review of Potential of probiotics from fermented cereal-based beverages in improving health of poor people in Africa]. *Journal of Food Science and Technology,* 57(11), 3935. Springer Science+Business Media.

Sever, S., & Yildiz, G. (2025). Non-thermal strategies in dairy processing: Enhancing the quality of cow, sheep, goat, and buffalo milk via sonication and thermosonication. *Journal of Raw Materials and Processing Foods*, *6*(1), 31–40.

Shang, Z., Pai, L., & Patil, S. (2024). Unveiling the dynamics of gut microbial interactions: a review of dietary impact and precision nutrition in gastrointestinal health [Review of Unveiling the dynamics of gut microbial interactions: a review of dietary impact and precision nutrition in gastrointestinal health]. *Frontiers in Nutrition,* 11. Frontiers Media.

Shockley, M., & Dossey, A. T. (2013). *Insects for Human Consumption.* In Elsevier eBooks (p. 617). Elsevier BV.

Short, E. C., Kinchla, A. J., & Nolden, A. A. (2021). Plant-Based Cheeses: A Systematic Review of Sensory Evaluation Studies and Strategies to Increase Consumer Acceptance [Review of Plant-Based Cheeses: A Systematic Review of Sensory Evaluation Studies and Strategies to Increase Consumer Acceptance]. *Foods,* 10(4), 725. Multidisciplinary Digital Publishing Institute.

Shrestha, S., Hag, L. van't, Haritos, V. S., & Dhital, S. (2022). Lentil and Mungbean protein isolates: Processing, functional properties, and potential food applications. *Food Hydrocolloids,* 135, 108142.

Siddiqui, S. A., Alvi, T., Sameen, A., Khan, S., Блинов, A. B., Nagdalian, A., Mehdizadeh, M., Adli, D. N., & Onwezen, M. C. (2022). Consumer Acceptance of Alternative Proteins: A Systematic Review of Current Alternative Protein Sources and Interventions Adapted to Increase Their Acceptability [Review of Consumer Acceptance of Alternative Proteins: A Systematic Review of Current Alternative Protein Sources and Interventions Adapted to Increase Their Acceptability]. *Sustainability,* 14(22), 15370. Multidisciplinary Digital Publishing Institute.

Siddiqui, S. A., Khan, S., Murid, M., Asif, Z., Oboturova, N., Nagdalian, A., Блинов, A. B., Ibrahim, S. A., & Jafari, S. M. (2022). Marketing Strategies for Cultured Meat: A Review [Review of Marketing Strategies for Cultured Meat: A Review]. *Applied Sciences,* 12(17), 8795. Multidisciplinary Digital Publishing Institute.

Sievenpiper, J. L. (2016). Sickeningly Sweet: Does Sugar Cause Chronic Disease? No. *Canadian Journal of Diabetes,* 40(4), 287.

Simões, L. S., Madalena, D. A., Pinheiro, A. C., Teixeira, J. A., Vicente, A. A., & Ramos, Ó. L. (2017). Micro- and nano bio-based delivery systems for food applications: *In vitro* behavior [Review of Micro- and nano bio-based delivery systems for food applications: *In vitro* behavior]. *Advances in Colloid and Interface Science,* 243, 23. Elsevier BV.

Singh, N. (2017). Pulses: an overview. *Journal of Food Science and Technology,* 54(4), 853.

Siró, I., Kápolna, E., Kápolna, B., & Lugasi, A. (2008). Functional food. Product development, marketing and consumer acceptance—A review [Review of Functional food. Product development, marketing and consumer acceptance—A review]. *Appetite,* 51(3), 456. Elsevier BV.

Smetana, S., Larki, N. A., Pernutz, C., Franke, K., Bindrich, U., Toepfl, S., & Heinz, V. (2017). Structure design of insect-based meat analogs with high-moisture extrusion. *Journal of Food Engineering,* 229, 83.

Smetana, S., Pernutz, C., Toepfl, S., Heinz, V., & Campenhout, L. V. (2018). High-moisture extrusion with insect and soy protein concentrates: cutting properties of meat analogues under insect content and barrel temperature variations. *Journal of Insects as Food and Feed,* 5(1), 29.

Springmann, M., Clark, M., Mason-D'Croz, D., Wiebe, K., Bodirsky, B. L., Lassaletta, L., ... Willett, W. (2018). Options for keeping the food system within environmental limits. *Nature,* 562(7728), 519–525.

Ssepuuya, G., Namulawa, V. T., Mbabazi, D., Mugerwa, S., Fuuna, P., Nampijja, Z., Ekesi, S., Fiaboe, K. K. M., & Nakimbugwe, D. (2017). Use of insects for fish and poultry compound feed in sub-Saharan Africa – a systematic review [Review of Use of insects for fish and poultry compound feed in sub-Saharan Africa – a systematic review]. *Journal of Insects as Food and Feed,* 3(4), 289.

Stephens, N., Silvio, L. D., Dunsford, I., Ellis, M. J., Glencross, A., & Sexton, A. (2018). Bringing cultured meat to market: Technical, socio-political, and regulatory challenges in cellular agriculture [Review of Bringing cultured meat to market: Technical, socio-political, and regulatory challenges in cellular agriculture]. *Trends in Food Science & Technology,* 78, 155. Elsevier BV.

Su, L., Jing, L., Zeng, X., Chen, T., Liu, H., Kong, Y., Wang, X., Yang, X., Fu, C., Sun, J., & Huang, D. (2022). 3D-Printed Prolamin Scaffolds for Cell-Based Meat Culture. *Advanced Materials,* 35(2).

Subramani, T., & Ganapathyswamy, H. (2020). An overview of liposomal nano-encapsulation techniques and its applications in food and nutraceutical [Review of An overview of liposomal nano-encapsulation techniques and its applications in food and nutraceutical]. *Journal of Food Science and Technology,* 57(10), 3545. Springer Science+Business Media.

Sufficiency, E., Qamar, S. A., Ferreira, L. F. R., Franco, M., Iqbal, H. M., & Bilal, M. (2022). Emerging biotechnological strategies for food waste management: A green leap towards achieving high-value products and environmental abatement. *Energy Nexus,* 6, 100077.

Sultana, H., Alakeel, K. A., Hassan, J., Mallick, S., Solaiman, Z. M., Kayesh, E., Gomasta, J., Zubayer, Md., Billah, M. M., Ozaki, Y., Alfagham, A. T., & Alamri, S. (2025). Nutrients, bioactive compounds and antinutritional properties of marigold genotypes as promising functional food. *Scientific Reports,* 15(1).

Sözer, N., Holopainen-Mantila, U., & Poutanen, K. (2016). Traditional and New Food Uses of Pulses. *Cereal Chemistry,* 94(1), 66.

Takors, R. (2011). Scale-up of microbial processes: Impacts, tools and open questions [Review of Scale-up of microbial processes: Impacts, tools and open questions]. *Journal of Biotechnology,* 160, 3. Elsevier BV.

Taylor, T. M. (2018). Natural Food Antimicrobials: Recent Trends in Their Use, Limitations, and Opportunities for Their Applications in Food Preservation. In *ACS symposium series* (p. 25). American Chemical Society.

Tekin, A., & Hayaloğlu, A. A. (2024). Overcoming the flavour and textural/rheological problems of plant-based cheese alternatives: Challenges and solution strategies. *Future Foods,* 100531.

Teng, T. S., Chin, Y. L., Chai, K. F., & Chen, W. N. (2021). Fermentation for future food systems. *EMBO Reports,* 22(5).

Tian, S., Yao, L., Zhang, Y., Rao, X., & Zhu, H. (2025). Prime Editing for Crop Improvement: A Systematic Review of Optimization Strategies and Advanced Applications. *Genes,* 16(8), 965.

Tilman, D., & Clark, M. (2014). Global diets link environmental sustainability and human health. *Nature,* 515(7528), 518–522.

Tiwari, B. K., Valdramidis, V., Donnell, C. P. O., Muthukumarappan, K., Bourke, P., & Cullen, P. J. (2009). Application of Natural Antimicrobials for Food Preservation [Review of Application of Natural Antimicrobials for Food Preservation]. *Journal of Agricultural and Food Chemistry,* 57(14), 5987. American Chemical Society.

Tolnay, A., Nath, A., & Kőris, A. (2020). Challanges of sustainable food technology. *Analecta Technica Szegedinensia,* 14(1), 118.

Treich, N. (2021). Cultured Meat: Promises and Challenges. *Environmental and Resource Economics,* 79(1), 33.

Tso, R., & Forde, C. G. (2021). Unintended consequences: Nutritional impact and potential pitfalls of switching from animal- to plant-based foods. *Nutrients,* 13(8), 2527.
Tufail, M. A., & Schmitz, R. A. (2024). Exploring the Probiotic Potential of Bacteroides spp. Within One Health Paradigm [Review of Exploring the Probiotic Potential of Bacteroides spp. Within One Health Paradigm]. *Probiotics and Antimicrobial Proteins.* Springer Science+Business Media.
Tuso, P. J., Ismail, M. H., Ha, B. P., & Bartolotto, C. (2013). Nutritional update for physicians: plant-based diets. *The Permanente Journal,* 17(2), 61.
Tyutkov, N., Zhernyakova, A., Birchenko, A. A., Eminova, E. R., Надточий, Л., & Baranenko, D. (2022). Probiotics viability in frozen food products. *Food Bioscience,* 50, 101996.
Valdés, A., Mellinas, C., Ramos, M., Burgos, N., Jiménez, A., & Garrigós, M. C. (2015). Use of herbs, spices and their bioactive compounds in active food packaging. *RSC Advances,* 5(50), 40324.
Voidarou, C., Αντωνιάδου, M., Rozos, G., Tzora, A., Skoufos, I., Varzakas, T., Lagiou, A., & Bezirtzoglou, E. (2020). Fermentative Foods: Microbiology, Biochemistry, Potential Human Health Benefits and Public Health Issues [Review of Fermentative Foods: Microbiology, Biochemistry, Potential Human Health Benefits and Public Health Issues]. *Foods,* 10(1), 69. Multidisciplinary Digital Publishing Institute.
Walait, M., Ajaz, M., Rasool, W., Irfan, M., Fatima, M., & Tariq, F. (2024). Biotechnological Innovations in Sustainable Food Production: Sustainable Food Production. DIET FACTOR (*Journal of Nutritional and Food Sciences*), 02-09.
Wang, K., Lin, X., Zhang, M., Li, Y., Luo, C., & Wu, J. (2022). Review of electrochemical biosensors for food safety detection. *Biosensors,* 12(11), 959.
Welsh, J. A., Lundeen, E. A., & Stein, A. D. (2013). The sugar-sweetened beverage wars [Review of The sugar-sweetened beverage wars]. *Current Opinion in Endocrinology Diabetes and Obesity,* 20(5), 401. Lippincott Williams & Wilkins.
WHO. (2021). Micronutrient deficiencies. World Health Organization. Retrieved from https://www.who.int/news-room/fact-sheets/detail/micronutrient-deficiencies.
Willett, W., Rockström, J., Loken, B., Springmann, M., Lang, T., Vermeulen, S., Garnett, T., Tilman, D., DeClerck, F., Wood, A., Jonell, M., Clark, M., Gordon, L. J., Fanzo, J., Hawkes, C., Zurayk, R., Rivera, J. A., De Vries, W., Sibanda, L. M., … Murray, C. J. L. (2019). Food in the Anthropocene: the EAT–Lancet Commission on healthy diets from sustainable food systems. *The Lancet,* 393(10170), 447–492.
Xavier, G. R., Jesus, E. da C., Dias, A., Coelho, M. R. R., Castro, Y., & Rumjanek, N. G. (2023). Contribution of Biofertilizers to Pulse Crops: From Single-Strain Inoculants to New Technologies Based on Microbiomes Strategies [Review of Contribution of Biofertilizers to Pulse Crops: From Single-Strain Inoculants to New Technologies Based on Microbiomes Strategies]. *Plants,* 12(4), 954. Multidisciplinary Digital Publishing Institute.
Xie, Y., Cai, L., Ding, S., Wang, C., Wang, J., Ibeogu, I. H., Li, C., & Zhou, G. (2025). An overview of recent progress in cultured meat: focusing on technology, quality properties, safety, industrialization and public acceptance [Review of An overview of recent progress in cultured meat: focusing on technology, quality properties, safety, industrialization and public acceptance]. *Journal of Nutrition.* Elsevier BV.
Xu, X., Sharma, P., Shu, S., Lin, T. S., Ciais, P., Tubiello, F. N., ... & Jain, A. K. (2021). Global greenhouse gas emissions from animal-based foods are twice those of plant-based foods. *Nature food,* 2(9), 724-732.
Yang, X., Tang, C., Zhao, Q., Jia, Y., Qin, Y., & Zhang, J. (2023). Melanin: A promising source of functional food ingredient. *Journal of Functional Foods*, 105, 105574.
Yavruturk, M., Celik, P., Kazanci, H., Demiragac, R., & Yildiz, G. (2024). Potential uses of dried apricot pulp in obtaining functional products, *Latin American Applied Research Journal,* 54(4): 525–530.
Yildiz, G., & Bayat, G. (2022). The effect of different drying methods on the production and quality characteristics of home-made kurut, a traditional dried dairy product, *Latin American Applied Research Journal,* 52(4):365-369.
Yildiz, G., Cox, G., & Moran, L. (2021). Identification of Aroma Compounds of Dried and Fresh Thyme (*Thymus Vulgaris L.)* by Gas Chromatography-Olfactometry and Gas Chromatography-Mass spectroscopy Analysis, *Latin American Applied Research Journal,* 51 (1): 51-55.
Yildiz, G., & Yıldız, G. (2025a). Effect of high-intensity ultrasound-induced acoustic cavitation on enzymatic activity and quality attributes of Granny Smith apples during storage. *Food Physics, 2:* 100058.

Yildiz, G., & Yıldız, G. (2025b). Alteration of proteins upon high-pressure processing treatment. In S. R. Falsafi, H. Rostamabadi, & N. K. Rastogi (Eds.), *Non-thermal processing of major food macromolecules* (pp. 91–117). Academic Press.

Yildiz, G., Sadeghi, H., Zhu, S., Chen, G., & Feng, H. (2025). Molecular and functional modulation of mycoprotein extract from edible *Trametes versicolor* via pH cycling and ultrasound-enhanced transglutaminase treatment. *International Journal of Biological Macromolecules, 323(1)*: 147128.

Yokoyama Y, Levin SM, Barnard ND. Association between plant-based diets and plasma lipids: A systematic review and meta-analysis *Nutr Rev.* 2017;75:683–98

Young, R. B., Marcelino, V. R., Chonwerawong, M., Gulliver, E. L., & Forster, S. C. (2021). Key Technologies for Progressing Discovery of Microbiome-Based Medicines [Review of Key Technologies for Progressing Discovery of Microbiome-Based Medicines]. *Frontiers in Microbiology*, 12. Frontiers Media.

Zhang, W., Boateng, I. D., Xu, J., & Zhang, Y. (2024). Proteins from Legumes, Cereals, and Pseudo-Cereals: Composition, Modification, Bioactivities, and Applications [Review of Proteins from Legumes, Cereals, and Pseudo-Cereals: Composition, Modification, Bioactivities, and Applications]. *Foods,* 13(13), 1974. Multidisciplinary Digital Publishing Institute.

Zheng, Y., Hu, Z., Liu, Z., Jiang, Y., Guo, R., Ding, S., & Zhou, G. (2023). The Effect of Long-Term Passage on Porcine SMCs' Function and the Improvement of TGF-β1 on Porcine SMCs' Secretory Function in Late Passage. *Foods,* 12(14), 2682.

Zhu, F. (2017). Encapsulation and delivery of food ingredients using starch based systems [Review of Encapsulation and delivery of food ingredients using starch based systems]. *Food Chemistry,* 229, 542. Elsevier BV.

Chapter 8

The Role of Alternative Foods in Health and Sustainability

Ayse Nur Elmaskaya[1,*]
and Gulperi Demir Hakli[2]
[1]Department of Gastronomy and Culinary Arts, Igdir University, Iğdır, Turkiye
[2]Department of Nutrition and Dietetics, Selcuk University, Konya, Turkiye

Abstract

This section evaluates the sustainability and health potential of alternative food sources—including plant-based products, lab-grown meat, single-cell microorganism-derived proteins, and edible insects—amid the environmental and health challenges of current food systems. Literature and comparative analyses were reviewed to assess production processes, environmental impacts (greenhouse gas emissions, water and land use), nutrient profiles, and health effects, as well as consumer perception, regulatory frameworks, and technological feasibility. Plant-based foods offer disease-preventive benefits due to their high fiber, vitamin, and phytochemical content and low environmental footprint. Cultured meat may reduce land and water use, though energy-intensive production limits environmental gains. Single-cell proteins (algae, yeast, bacteria) provide rapid, low-carbon, high-quality protein, while edible insects offer protein, essential nutrients, and efficient bioconversion, supporting circular economy objectives. Challenges include food safety, cultural acceptance, regulatory gaps, and consumer perception. Realizing the potential of alternative foods requires optimized production technologies, stronger regulatory frameworks, cultural adaptation, and increased public awareness. Interdisciplinary research and public–private partnerships will further advance their role in sustainable food systems.

Keywords: alternative foods, sustainability, health impact, environmental impact

---

* Corresponding Author's Email: aysenur.elmaskaya@igdir.edu.tr.

In: Let Food Be Your Medicine, and Medicine Be Your Food
Editors: Gulcin Yildiz and Arzu Imece
ISBN: 979-8-90134-204-6

Introduction

In recent years, the growing global population and the depletion of natural resources have emerged as some of the most pressing challenges worldwide. The world population is projected to reach 9 to 10 billion by 2050 (UN, 2019), which will increase the demand for sustainable, healthy, and nutritious food. Current food production systems are major contributors to environmental problems, including greenhouse gas emissions, deforestation, and excessive water use (Vermeulen et al., 2012). Meat production, in particular, has a significant impact on both greenhouse gas emissions and water consumption (Charles et al., 2018). Consequently, the need for alternative protein sources has become increasingly urgent.

In its report titled *"How Will the World Be Fed in 2050?,"* the Food and Agriculture Organization of the United Nations (FAO) predicts that most of the world's population growth will occur in developing countries, with the urbanization rate reaching 70% by 2050. Food demand is expected to increase by 60% over the next 40 years, in line with population growth. Consequently, annual grain production is projected to rise from 2.1 billion tons to 3 billion tons, and annual meat production from 200 million tons to 470 million tons (FAO, 2021).

Population growth, climate change, environmental degradation, and the limitations of traditional agricultural production worldwide place significant pressure on food security and nutrition. This situation has highlighted the need to develop sustainable food production systems and explore alternative food sources (FAO, 2020). Challenges in producing conventional agricultural products have increased interest in alternative protein sources and other nutrients. In particular, the high costs of animal proteins, their environmental impact, and ethical concerns underscore the necessity of seeking alternative food sources for human consumption (Godfray et al., 2010).

According to the United Nations Food System Summit and Agenda 2030 held in 2021, it is emphasized that global food systems need to be urgently restructured to achieve sustainable development goals. Food proteins play a critical role in preventing inadequate protein intake, which is the main cause of nutritional deficiencies in the modern era (Aiking and de Boer, 2020). To ensure sustainable protein production, Katz-Rosene and other researchers (2023) proposed three main meta-analysis coalitions: protein rebuilding, protein modernizing, and protein renewing. This approach aims to sustainably transform protein through raw material, technology, and production innovations and thus reshape food systems. In particular, the item "protein rebuilding" stands out as an important strategy that suggests reducing animal protein consumption and introducing alternative new proteins. This issue is based on various concerns about animal protein: insufficient supply (animal protein may not meet the protein needs of a growing population), health issues (meat consumption is associated with diseases such as coronary heart disease and obesity), environmental impacts (greenhouse gas emissions related to animal farming), ethical issues (animals are kept in small spaces and killed inhumane ways) (Katz-Rosene et al., 2023). In this context, the development of new protein sources with similar nutritional value to animal proteins has become an important research topic for researchers and the food industry.

In recent years, various alternative protein sources have emerged as potential replacements for animal proteins, including plants, insects, algae, and even bacteria or fungi. These new protein sources are generally considered healthier and more environmentally sustainable than traditional animal proteins. For instance, producing animal protein requires approximately 6 kg of plant protein through feed and animal husbandry (Aiking and de Boer, 2020). Moreover, plant proteins may offer preventive and protective effects against chronic diseases such as obesity and type II diabetes. Different plant proteins, including those from legumes, cereals, pseudocereals, and oilseeds, have been extensively studied. In contrast, proteins derived from insects, algae, bacteria, and fungi are still in the early stages of research compared to the widespread study of plant proteins. Nevertheless, these emerging protein sources are widely recognized as nutritionally valuable and environmentally friendly (Zhang et al., 2024). This chapter aims to explore the diversity of alternative food sources, their potential benefits, challenges, and future research directions. Additionally, it evaluates the role of alternative foods in the food system, emphasizing their sustainability and potential to meet the nutritional needs of populations.

Alternative Food Sources

Alternative food sources are foods that cannot be produced through conventional agriculture or whose production is limited, including edible insects, algae, legumes, mushrooms, seafood, and lab-grown meats (Van Huis et al., 2013). These foods offer significant potential for environmental sustainability, human nutrition, and economic viability. Plant-based foods reduce greenhouse gas emissions and resource use while providing fiber and phytochemicals that lower the risk of chronic diseases (Poore and Nemecek, 2018). Insects are rich in protein, vitamins, and minerals and emit 80–90% fewer greenhouse gases than traditional livestock (Van Huis, 2013). Lab-grown meats can decrease carbon footprint and land use while minimizing pathogenic risks (Post, 2012). Single-cell proteins, derived from algae, yeast, or bacteria, provide rapid, resource-efficient production and high-quality amino acids (Anupama and Ravindra, 2000). A multidisciplinary approach is needed to ensure these alternative foods are safe, accessible, and acceptable for widespread use.

Plant-Based Alternative Food Sources

Plant-based protein sources are critically important not only for human health but also for environmental sustainability. In addition to offering eco-friendly alternatives with low greenhouse gas emissions and water usage, foods such as cereals, legumes, and pseudocereals have been reported to reduce the risk of cardiovascular diseases due to their high fiber content and low saturated fat levels (Poore and Nemecek, 2018). Oilseeds and nuts support neurological health because they are rich in protein, beneficial oils, and antioxidant compounds. Notably, the production of these foods has a low carbon footprint (Mertens et al., 2022). Tubers and microbial protein sources also contribute to sustainability by providing essential amino acids and promoting efficiency in land and water use (Ritala et al., 2017). Collectively, these alternative food sources represent an innovative and sustainable solution to mitigate climate change and promote a healthier society.

Plant proteins are widely recognized as sustainable alternatives for the future. Therefore, legumes, cereals, pseudocereals, and oilseeds are examined individually, considering differences in macronutrient composition and protein classification. Legume seeds have been described by the United Nations and FAO as "nutritious seeds for a sustainable future" due to their high protein and dietary fiber content, richness in micronutrients and bioactive compounds, lower water requirements, and ability to enhance soil fertility through nitrogen fixation (Venkidasamy et al., 2019). Protein is generally the most abundant macronutrient in legumes. Major legume species include soybean [Glycine max (Linn.) Merr.], pea (Pisum sativum L.), chickpea (Cicer arietinum L.), mung bean (Vigna radiata L.), and lentil (Lens culinaris Medikus). Soy protein has been extensively studied and widely applied in the food industry due to its nutritional and functional properties, appearing in commercial products such as meat additives, plant-based meat alternatives, and dairy products (Zhang et al., 2021). Beyond soy, other legumes like peas and chickpeas have emerged as non-soy alternative protein sources with considerable potential (Zhang et al., 2024).

In addition to protein quantity, protein quality is crucial for health. High-quality protein provides sufficient essential amino acids that are easily digestible and efficiently utilized for protein synthesis. Amino acid composition is the key determinant of protein quality, and different plant-based proteins offer distinct health benefits due to their unique amino acid profiles (Kumar et al., 2022). Plant proteins are primarily derived from cereals, legumes, oilseeds, green leafy vegetables, nuts, pseudocereals, and other plant species (Zhang et al., 2024).

Although plant-based protein sources are highly diverse, they may have limitations in structural and sensory properties compared to animal-based proteins such as meat, milk, and eggs. As a result, fully replacing animal proteins with plant-based options presents certain challenges. Nevertheless, their diversity and potential benefits make them a highly valuable alternative (Ewy et al., 2022).

Cereals

Cereals are a significant source of energy and carbohydrates for humans and contribute to the diet through their dietary fiber content. Moreover, they serve as one of the primary protein sources in the human diet. While cereals are affordable, accessible, and renewable products, most are primarily used for animal production (e.g., grazing, feed, silage), and only a few undergo further processing to produce food products suitable for human consumption. Research on cereal proteins has generally been limited to whole grain utilization. However, in recent years, with the rising global demand for new and sustainable alternative protein sources (Poutanen et al., 2022), cereal proteins have attracted increased attention due to their significant content of essential amino acids (except lysine) and beneficial physiological functions for human health. For instance, cereals contain between 7% and 18% protein (on a dry matter basis). Today, proteins extracted from various cereals such as wheat (Triticum aestivum L.), maize (Zea mays L.), rice (Oryza sativa L.), and oats (Avena sativa L.) are utilized in both nutritional and applied research contexts (Zhang et al., 2024).

Cereals are among the key nutritional sources that can serve as alternatives to animal-derived foods and are of great significance in global human nutrition. Plants such as wheat, barley, maize, rice, oats, and sorghum belong to this group. Cereals are relatively rich in protein, with their primary protein sources being storage proteins. Additionally, they are good sources of fiber, vitamins, and minerals, and they primarily contribute to dietary energy through the digestion of starch. Furthermore, cereals contain polyphenols, antioxidants, and bioactive compounds such as calcium, magnesium, zinc, and iron. Among cereals, rice is widely consumed in both developed and developing countries. When the amino acid profile of rice proteins is examined, albumin is found to be rich in lysine, while globulin is abundant in sulfur-containing amino acids (Sá et al., 2020). Rice protein is also considered valuable due to its hypocholesterolemic properties (cholesterol-lowering), hypoallergenic nature (low risk of allergic reaction), mild taste, colorlessness, and high content of essential amino acids, offering significant added value in various food products. Another important cereal, wheat, contains gluten, a low-cost byproduct obtained during starch extraction from wheat. However, gluten is a component that individuals with celiac disease must avoid. In recent years, oats have been used in the production of food items such as plant-based milk alternatives and fermented products (Barbosa et al., 2022). Maize, on the other hand, is a nutritious cereal protein source widely consumed in developing countries. Corn provides a significant advantage in enhancing protein intake in human nutrition by containing high amounts of essential amino acids such as lysine compared to others (Sá et al., 2020).

Pseudocereals

The seeds of pseudocereals share functional properties and composition similar to traditional cereals such as wheat and corn, but they offer higher bioavailability due to their rich, high-quality protein content. Pseudocereal proteins are particularly abundant in lysine, an essential amino acid often limited in cereals. Additionally, these seeds are gluten-free, making them suitable for individuals with celiac disease (Zhang et al., 2023). These characteristics have increased interest in pseudocereals as a high-quality and promising alternative protein source. The main pseudocereals include quinoa (Chenopodium quinoa Willd.), amaranth (Amaranthus tricolor L.), buckwheat (Fagopyrum esculentum Moench), and chia (Salvia hispanica L.) (Zhang et al., 2024). In recent years, consumer interest in nutritious and balanced plant-based foods has driven the growing use of these pseudocereals. Although their seeds resemble true grains, they are botanically distinct, which is why they are referred to as 'pseudocereals.' Their protein amino acid profiles, high essential amino acid content, and good bioavailability make them nutritionally valuable. Furthermore, pseudocereals are rich in protein, dietary fiber, healthy unsaturated fatty acids, and essential vitamins and minerals, making them an important component of a balanced diet (López et al., 2018).

Legumes

With the increasing interest in plant-based protein sources, the amount of protein obtained from legumes is also increasing. Soybeans, which have a low glycemic index and are rich in dietary fiber, vitamins, minerals and various bioactive components, have recently attracted more attention than others. Studies report that soybeans contain all the essential proteins necessary for growth and development (Tan et al., 2023). It explains that the amino acids such as phenylalanine, methionine, threonine, valine, isoleucine, tryptophan and lysine found in soybean proteins have a composition close to the quality of animal proteins (Kudełka et al., 2021). On the other hand, some negative characteristics of soybeans are mentioned along with their high protein content. This feature, in addition to the tasteless and unpleasant aroma of soybeans, causes allergic reactions in some individuals, restricting its use (Hayaloğlu, 2024). Another legume that has not been utilized sufficiently, lentil protein, is the subject of research thanks to its essential amino acid content and its positive effects on health. Lentil proteins form three-dimensional network structures containing food components such as water, aroma, and prebiotics, in addition to their gelling feature, which is an important parameter in food products (Tang et al., 2024). Mung bean, another legume similar to soy protein, has a wider area of use due to its lower allergenic potential compared to soy protein. When the gelling capacities of soy protein and mung bean proteins are compared, it has been reported that mung bean has a better gelling feature. Thanks to this feature, mung bean protein has gained more usability in the food industry (Tojan et al., 2024). Although legumes offer high protein content, they cannot be sufficiently productive due to the indigestible carbohydrates they contain, which causes gas problems in individuals. Researchers are working to increase the consumption of legumes by reducing these undesirable effects. For example, heat treatments such as soaking, germination, frying, extrusion, baking, steaming, microwaving, boiling and autoclaving reduce the formation of gas and increase the digestibility of proteins, thus reducing the effects of antinutrients (Bolarinwa et al., 2019).

Oilseeds

Oilseeds are defined as high-quality and edible vegetable oil sources due to their oil content of more than 15%. Thanks to their functional properties, oilseeds are quite diverse in terms of nutritional and functional components such as fiber, sterols, unsaturated fatty acids, proteins and biopeptides. The second most abundant macronutrient in oilseeds is protein, and these proteins have a balanced amino acid profile, containing higher levels of leucine and valine amino acids compared to other alternative protein sources (Kotecka-Majchrzak et al., 2020). More than 200 species of oilseed crops are produced worldwide. Oilseeds have a broad definition encompassing any plant source with high oil content and include other species such as soybeans. Some of the other important oilseeds used in food production are rapeseed (Brassica napus L.), sunflower (Helianthus annuus L.), coconut (Cocos nucifera L.), olive (Canarium album Lour. Raeusch.), pumpkin (Cucurbita moschata Duch. Ex Poiret L.), peanut (Arachis hypogaea Linn.), sesame (Sesamum indicum L.) and hemp (Cannabis sativa L.). Among these oilseeds, rapeseed, pumpkin and hemp are determined as alternative protein sources in the future due to their low allergenicity, high nutritional value and processing properties (Zhang et al., 2024). Oilseeds and their by-products, pulps, are seen as an important source of protein and bioactive components. For this reason, in recent years, oilseeds have become increasingly preferred alternative protein sources in the food and nutraceutical sector. These potential sources, known by nutritionists and food manufacturers, have also attracted significant consumer interest worldwide. Processed foods, especially meat, bakery products and dairy products, are offered by being enriched with oilseed meal or oilseed protein hydrolysates for consumers with various health problems (Zhang et al., 2024). In addition, significant awareness has been raised recently about the amino acid profiles of different vegetable oilseeds and their health benefits. Proteins obtained from plants such as chia, hemp, black cumin, pumpkin seed and rapeseed have been found to have bioactive properties such as angiotensin converting enzyme (ACE) inhibitory activity, antioxidant effects, anti-inflammatory, antimicrobial, hypoglycemic and cardioprotective. However, it is known that there is still limited information on the bioactive properties of primrose, milk thistle and sesame proteins. It is reported that more studies are needed to investigate the health benefits of these alternative proteins with clinical and animal experiments. According to all these facts, oilseed proteins show great potential as functional ingredients or protein alternatives in the food industry (Kotecka-Majchrzak et al., 2020).

Nuts

Nuts, which are mostly used as snacks, have a wide area of use worldwide. In general, as the interest in plant-based protein sources increases, it is seen that nuts are also quite efficient protein sources. One of the most well-known nuts, peanuts, contains 24-36% protein on dry matter. Peanuts are quite strong in terms of essential amino acids and other food components that provide positive effects on health. Due to their high nutritional value, efficiency and low cost, there is an increasing interest in processing peanut protein with other food products. The main fractions, different amino acid profiles and molecular structures of peanut proteins distinguish them from others and provide unique physical and functional properties (Cui et al., 2023). Walnuts, which have an economically important place, also stand out with their high fat and protein content. The fat-free dry matter in walnuts contains approximately 45% protein, and this protein content provides the balanced essential amino acids desired for the body (Ma et al., 2024). Another of the most consumed nuts in the world, almonds, attract attention with an annual production of 1.2 million tons. Protein extraction from almond kernels is a very simple process, and the extraction efficiency can reach up to 70%. With its high nutritional content, easy processability and high extraction efficiency, almond protein is defined as an alternative food source to fill the protein gap (Shi et al., 2023). In addition, cashews stand out from the others with their high protein content and emulsification properties; in addition to these properties, thanks to their oil retention capacity, their use in plant-based food production is widespread (Oyeyinka et al., 2019).

Tubers

When the nutritional content and nutritional value of tubers, which are alternative food sources, are investigated, it is seen that they can be used as plant-based protein alternatives thanks to the amount of protein they contain. Potato, which is widely consumed around the world, is among the plant protein alternatives thanks to its high protein content and similar nutritional quality to egg protein (Romero et al., 2011). In addition, the fact that potato proteins do not cause allergic reactions allows their areas of use to increase even more. Potato proteins are seen to have very important functional properties in the food industry, such as gelation, foaming and emulsification. Studies have clearly reported that potato proteins have similar emulsification properties to whey powder (Tan et al., 2023).

Edible Mushrooms

Edible mushrooms are reported to be another plant-based protein source that has an important place in human nutrition due to their high protein content and nutritional value. Mushrooms require less water for their development than traditional protein sources and are quite different in that they are fed with some industrial waste. Along with their easy cultivation, the fact that their dry matter contains 19-35% protein makes them a very valuable food source. Edible mushrooms are notable for containing more protein than some vegetables and having similar protein ratios to soybeans. In addition, they also provide 8 essential amino acids required for the human body. When compared to other common plant-based protein sources such as soybeans and wheat, the allergenicity level of edible mushrooms has been found to be lower. Since food allergens are proteins that can be affected by physical, chemical and biological factors in general, allergenic effects can be reduced or eliminated with various methods. The nutritional properties of mushroom proteins play an important role in the development of plant-based food products and their inclusion in the food industry (Wang and Zhao, 2023).

Algae

In recent years, microalgae, which are notable for their high protein content and obtained from edible algae, have emerged as an important alternative protein source. When microalgae proteins are evaluated in terms of dry matter content, it has been reported that they have high protein ratios ranging from 50-80% (Muys et al., 2023). Protein production of microalgae may vary depending on their species. Species such as Spirulina platensis, Chlorella sp., Dunaliella terticola, Dunaliella salina and Aphanizomenon flos-aquae also stand out due to their high protein content and nutritional values. There are studies in the literature that present the protein amounts obtained from the dry matter of some microalgae species in detail (Koyande et al., 2019).

Green Leafy Vegetables

Green leafy plants are traditionally used in non-food areas such as feed and green manure. However, these plants have the potential to be used in food production thanks to the high protein, vitamins, minerals, dietary fiber and essential fatty acids they contain. It is reported that proteins in green leafy plants are found at significant levels in different plant species such as alfalfa, spinach, cassava, tobacco, grass, moringa, broccoli and sugar beet (Hadidi et al., 2023). For example, spinach is a good source of minerals thanks to its unique content, containing 827 mg sodium, 28.4 mg iron, 827 mg magnesium and 1036 mg calcium per 100 grams (Kumari et al., 2004). These edible green leafy plants are generally more economical and abundant sources of protein. In addition, the absorption of carbon dioxide from the atmosphere during the cultivation of these plants is a remarkable potential protein source in terms of supporting the reduction of greenhouse gas emissions.

Desired Properties of Plant-Based Protein Sources

Nowadays, to adopt and implement a sustainable diet, the tendency towards plant protein sources is increasing by reducing the use of animal protein. Essentially, plant proteins are seen as alternatives that can replace animal proteins, therefore, studies conducted to examine the functional and physicochemical properties of the derivatives of these proteins in detail and to understand their physiological effects are considered important. In addition, it is very important to investigate the integrability of these proteins into food products. For the use of plant-based protein sources in food production, important factors, such as consumer acceptance, food safety, and production costs, need to be improved. Generally, these proteins are preferred in the food industry as various functional products such as isolates, edible coatings, food stabilizers, hydrogel adhesives and bioactive peptides (Kumar et al., 2022). In order for plant-based proteins to be used more efficiently in food products, they are desired to have some basic functional properties. These properties include water retention capacity, lack of allergenic effects, good solubility, gelation ability, emulsification capacity, foam formation ability, binding ability, abundance and easy availability, high purity, low cost and high nutritional value (Hewage et al., 2022; Hadidi et al., 2023). In addition, it is reported that the characterization and quality analysis of proteins after extraction is very important. The effectiveness of protein sources is evaluated by the level at which they meet the metabolic needs of the body. A full understanding of their functional and physicochemical properties makes these proteins more widely used in the food industry and in processed food products that increase their nutritional value. This approach provides a clear guide to the use of protein for a specific purpose. For example, purified antioxidant peptides obtained from plant sources can be preferred to extend the shelf life of food products because these peptides have a slowing effect on oxidation processes in foods (Kumar et al., 2022).

Functional Components of Plant-Based Proteins

Plant-based protein sources are being developed as alternative products to replace animal-derived foods. Animal milk has long been the primary source for fermented dairy products, making it widely used in the industry. To produce plant-based fermented dairy products that are nutritionally comparable to traditional dairy, it is essential to evaluate the nutritional values of these plant protein sources relative to milk. Plant-based ingredients such as soy protein, coconut, and rice are reported to contain high carbohydrate levels (Tangyu et al., 2019). Plant proteins play important roles in food formulations. They function as gelling, thickening, and foaming agents, and are also reported to stabilize emulsions and bind water and oil. Additionally, some plant proteins exhibit biological activities, including antimicrobial and antioxidant effects. Common plant protein sources include cereals (wheat, rice, corn, barley, sorghum), legumes (soybean, kidney bean, pea, black bean, pigeon pea, peanut, cowpea), and oilseeds (sesame, rapeseed, cottonseed). Many studies also highlight the potential of legumes and oilseeds (e.g., wheat, barley, soybean, peanut, canola, flaxseed, grape seed, sesame, cottonseed, pumpkin seed, rapeseed, and sunflower), as well as by-products such as bran, oilseed meals, and cereal leaves, as valuable sources of plant protein (Kumar et al., 2022; Singh et al., 2022).

Food Applications of Plant-Based Proteins

Plant-Based Milk Alternatives

From the past to the present, cow milk has been a fundamental component of human nutrition. This animal-derived liquid is considered a balanced food, providing essential nutrients such as proteins, minerals, vitamins, fats, and lactose, which support vital bodily functions. Cow milk is widely recognized for its contribution to human health and is included as a basic element in nutritional guidelines in many countries (Haas et al., 2019).

Since 2012, however, cow's milk has become difficult to digest and absorb for some individuals (Silva et al., 2020). Factors such as lactose intolerance, cow's milk protein allergy, high saturated fat and sugar content, hormonal residues, and antibiotic use are among the primary concerns for consumers regarding milk consumption (Chalupa-Krebzdak et al., 2018). Consequently, there has been a growing search for alternatives to cow's milk, and the demand for dairy-free beverages has increased by 61% since 2012 (Reyes-Jurado et al., 2023).

Additionally, the rising popularity of vegan and vegetarian diets, environmental concerns, and ethical considerations related to cow's milk consumption have further boosted interest in plant-based milk alternatives (Jeske et al., 2017). Alongside this growing demand, research on the functional properties of these products has expanded. Plant-based milk alternatives can be categorized into five main groups: legumes, nuts, seeds, pseudocereals, and grains. While soy-based beverages are the most common alternatives in Western countries, other sources include almonds, coconut, sunflower seeds, chickpeas, lupins, hemp, sesame, quinoa, peas, and rice. The texture and flavor of these beverages vary significantly depending on the raw materials used (Sethi et al., 2016).

Plant-Based Yogurt Alternatives

Yogurt is considered a functional food due to the health benefits of lactic acid bacteria (LAB) during fermentation, including protection against gastrointestinal infections, cholesterol reduction, immune support, and intestinal microflora balance (Ayivi et al., 2020). Recently, consumer interest in fermented plant-based yogurt alternatives has grown due to their sustainability, environmental friendliness, and natural antioxidant content. Soy-based beverages fermented with LAB can produce yogurt-like products, although non-dairy yogurts may differ structurally from traditional milk-based yogurts (Grupta et al., 2022). Studies have examined the structure and rheological properties of soy- and oat-protein gels in acidic environments, yet differences in physical, chemical, and volatile flavor components before and after fermentation remain underexplored (Łopusiewicz, 2023; Wang et al., 2023).

The nutritional profile of plant-based yogurts varies with raw materials. Cereals such as oats and rice provide carbohydrates and fiber, while pseudocereals and legumes like quinoa, lupin, lentil, and soy are higher in protein. Plant proteins, however, often have low amino acid content, digestibility issues, and antinutrients. Combining cereals and legumes can complement amino acid deficiencies and improve bioavailability via bioprocessing (Dhakal et al., 2023).

A study comparing reconstituted milk, soy milk, and oat milk yogurts found that milk-based yogurt had higher titratable acidity, making it more suitable for fermentation. Soymilk yogurt exhibited the highest hardness, consistency, water-holding capacity, and stability, while oat milk yogurt showed a looser structure and lower hardness and G' values, reflecting weaker protein interactions during fermentation (Wang et al., 2023). Acceptable plant-based yogurts were produced using soy and quinoa. Incorporating quinoa affected starter culture metabolism, altering composition, water-holding capacity, cell viability, color, and rheological properties. Sensory evaluation indicated that soy–quinoa blends in 8:2 or 2:8 ratios improved nutritional value and consumer preference (Huang et al., 2022).

Plant-Based Cheese Alternatives

The plant-based cheese sector is currently one of the fastest-growing areas in the food industry. Unlike traditional cheese, which relies primarily on milk, plant-based cheese analogues are produced using diverse plant proteins such as pea, soy, lupin, potato, hazelnut, and corn. These proteins must provide essential functional properties, including emulsification, gelation, water retention, and flavor (Grossman et al., 2021).

In a study on optimizing pea protein matrices for fermented plant-based cheeses, emulsions containing 10% pea protein isolate (BPI) and varying olive oil levels (0–20%) were tested. All BPI emulsions remained physically stable after 7 hours, with a consistent bimodal particle size distribution and unchanged potency over time. Higher fat content weakened gel formation and reduced elastic modulus values, but gel hardness remained stable during 7 days of refrigerated storage (Masiá et al., 2022).

Correia et al. (2022) investigated the addition of raw and extruded sorghum flour to cream cheese. Raw sorghum flour offers greater sensory and operational advantages due to lower cost and simpler processing, highlighting the potential for innovative plant-based cream cheese with functional, nutritional, and sensory benefits.

Another study utilized legume flours (boiled and roasted yellow pea and broad beans) to produce cheese analogues. These flours proved suitable for plant-based cheese production, offering high fiber content (6–8 g/100 g product) and functional food potential. Such developments not only enhance the plant-based cheese market but also encourage legume consumption as part of a healthier diet (Ferawati et al., 2021).

Plant-Based Meat Analogue

In the production of plant-based meat analogues, plant proteins are generally preferred as the main ingredient, with cereals and legumes being the most common sources. Soybeans, in particular, are widely used by consumers and form the basis of traditional plant-based products in Asia, including soybeans, yuba, tofu, and fermented products (Lou et al., 2023).

In one study, experiments using a Couette cell demonstrated that simple shear flow combined with heat can create fibrous structural patterns in a granular mixture of plant proteins under mild processing conditions. A 7-liter Couette cell was used to produce soy-based structured meat substitutes at a constant temperature of 120°C, with optimal process conditions determined by process time and rotation speed. The result was fiber-structured products with a high anisotropy index (Krintiras et al., 2016).

Another study evaluated the fibrous structure and rheological properties of soy protein-based food gels using a freezing method to mimic animal meat texture. Soybean flour and two types of soy protein isolate powders were examined for their ability to form meat-like structures. The study reported that specific freezing conditions and isolated powder combinations could produce multilayered structures in food protein gels. Formulations with 10% soybean flour yielded a more porous and anisotropic structure, providing breaking strength comparable to real meat and forming dense layers crucial for texture. Despite its low gelation ability, soybean flour was identified as a suitable option for freezing soy protein-based gels (Sun et al., 2015).

Cell Culture-Based Protein Sources: Cultured Meat

Meat has been an important source of protein, fat, iron, and other nutrients that meet the basic dietary needs of humans for centuries. Every year, more than 60 billion animals are raised for meat production in commercial settings. Animals raised for commercial purposes in places such as chicken farms, goat farms, cattle farms, pig farms, and fish farms have been the cause of common diseases such as "Mad Cow Disease," "Bird Flu," "Scombroid," and "Foot-and-Mouth Disease." These diseases, when not properly disposed of, can spread to humans, yet these meats continue to be slaughtered, sold, and consumed. Another negative impact of livestock farming is that approximately 70% of agricultural land is used for animal feed production. This leads to the depletion of water resources, loss of biodiversity, soil erosion, and the destruction of habitats, negatively affecting the natural balance (Akshay Patil et al., 2024).

Livestock raised for the production of meat, milk, and eggs account for about 15% of global greenhouse gas emissions, a figure that, according to a report by Valin and other researchers (2014), is higher than emissions from the entire transportation sector (Valin et al., 2014).

Meat is largely composed of muscle fibers (90%); the remaining part consists of fat, connective tissue (10%), and proteins containing some blood vessels (Thavamani et al., 2020). The idea of *in vitro* multiplication of muscle fibers for the production and imitation of meat components was first put forward in the 1930s (Weele et al., 2019). Cultured meat, which is called laboratory-produced or cell-based meat, describes bio-artificial muscle tissue created by in vitro multiplication of animal muscle stem cells (myosatellite cells) for consumption (Ong et al., 2020). In this process, stem cells are grown in a nutrient-rich environment and then begin to differentiate on a three-dimensional scaffold. During this differentiation, the cells settle on the scaffold to form myotubes (small, primitive muscle fibers) (Ong et al., 2020). These myotubes are then cultured in a gel that contains a large amount of water (99%) to help the muscle fibers take shape. The most commonly used nutrient medium in meat culturing processes is a combination of fetal bovine serum and growth factors (Thavamani et al., 2020).

Some scientists argue that it is not accurate to call this product made from animal cells in vitro "meat" and that it should instead be defined as an "animal-origin food product." However, in the Turkish literature, the term "artificial meat" is most commonly used. Scientists in this field have used various other terminologies as well. The most well-known among these are synthetic meat, in vitro meat, clean meat, lab meat, test-tube meat, Petri dish meat, animal-free meat, cell-based meat, fake burger, Frankenstein meat, and factory-grown meat (Post, 2012; Bhat et al., 2019; Benny et al., 2022; Mateti et al., 2022).

Recently, the term "cultivated meat" has begun to be adopted, especially by producing companies (Good Food Institute, 2023). On the other hand, some researchers suggest that cloned or genetically modified animal meats should also be classified under the artificial meat category (Mateti et al., 2022; Lee et al., 2023).

Differences Between Animal and Plant-Based Origins in Cultured Meat

Artificial meat is a broad concept, and it is important to distinguish between animal-sourced artificial meat and plant-based meat analogs (or alternatives) that are entirely plant-derived and do not contain any animal products (Ismail et al., 2020; Rubio et al., 2020; Mateti et al., 2022; Elhalis et al., 2023; Wang and Jian, 2023; Munialo and Vriesekoop, 2023). Plant-based meat analogs are produced mainly to meet the needs of individuals who do not consume meat, while also providing a sustainable protein source for the future. These analogs closely mimic the nutritional profile of real meat and are considered a source of protein (Bryngelsson et al., 2022). In addition, plant-based meat analogs are planned to be created in a way that mimics real meat in terms of meat products' unique texture, color, taste and nutritional value (Choudhury et al., 2020). There are many factors that increase the interest in artificial meat, and one of the most important factors among these factors is the rapid increase in the world population and the inadequacy of natural resources. The United Nations announced on November 15, 2022 that the world population reached 8 billion and it was reported that it is estimated to be 8.5 billion in 2030, 9.7 billion in 2050 and 10.4 billion in 2100 (UN, 2023). The increasing global population's demand for meat brings environmental problems such as climate change, environmental pollution and the reduction of environmental damage to the agenda, as well as important reasons such as animal welfare, more efficient use of natural resources and less efficient food conversion. While current meat production requires the use of billions of animals worldwide each year, artificial meat production is seen as quite advantageous in that it only uses a limited number of animals or animal resources. This situation is explained as one of the strongest arguments of the advocates of artificial meat. The use of antibiotics in relation to conventional meat production, the risk of meat spreading diseases due to fecal contamination, foodborne diseases related to meat consumption, health problems due to pesticides, heavy metals and other environmental pollutants, and the spread of possible zoonotic diseases, which threaten public health, can also be shown as reasons for the increasing interest in artificial meat. Furthermore, religious restrictions on meat consumption and exotic meat production are among the factors influencing the adoption of artificial meat (Godfray et al., 2018; Bhat et al., 2019; Yetim and Tekiner, 2020; Balasubramanian et al., 2021; Hong et al., 2021; Melzener et al., 2021; Mateti et al., 2022).

Historical Development of Animal-Derived Cultured Meat

Information and news about artificial meat have become more frequent in recent decades, especially after Dutch scientist and physician Dr. Mark Post announced that he had produced the first artificial meatball, which attracted more attention to this topic. However, the foundations of this idea date back much earlier. For example, politician and writer Frederick Edwin Smith stated, "From now on, we will not have to go through the trouble of raising a bull to eat steak" (Ford, 2011). Additionally, Winston Churchill, influenced by Smith's views in his work "Fifty Years Hence," predicted that "In fifty years, we will no longer need to raise a whole chicken to eat a chicken breast or wing," and that the way to achieve this would be "to grow these chicken parts separately in the right environment" (Churchill, 1931). Furthermore, the first scientific evidence that cells could survive outside of organisms comes from Alexis Carrel's original work in 1912. Carrel successfully maintained embryonic chicken heart tissues in culture for 34 years, from 1912 to 1946 (Carrel, 1912).

One of the first scientific studies related to artificial meat production was a research project supported by the National Aeronautics and Space Administration (NASA). The purpose of this study was to explore the feasibility of in vitro meat production systems for long-duration space missions. In the study, the back and belly muscle tissues of adult aquarium fish (Carassius auratus) were cultured in environments containing different concentrations of fetal calf serum (Benjaminson et al., 2002). The fish meat cultured in Petri dishes was cooked with spices and olive oil and evaluated by panelists, who deemed the product "acceptable as food" (Bhat et al., 2019).

An article titled "In Vitro Cultured Meat Production," published in the Tissue Engineering journal in 2005, made artificial meat production a new topic for both scientific communities and the public (Edelman et al., 2005). It was also predicted that with artificial meat production, the ratio of saturated fatty acids to polyunsaturated fatty acids could be better controlled, the incidence of foodborne diseases would decrease, and natural resources would be used more efficiently (Edelman et al., 2005).

A major milestone in artificial meat production came on August 5, 2013, when Dutch clinical scientist Dr. Mark J. Post unveiled the tasting of the world's first artificial meat, developed over three months (O'Riordan et al., 2017). Weighing approximately 5 ounces, made with 20,000 muscle fibers, and costing $325,000, this first edible artificial meat was cooked in a public event, proving that artificial meat had become a reality (Toor, 2013). After this event, many companies and research groups accelerated their efforts to produce artificial meat on an industrial scale (Bhat et al., 2019).

There have been patent applications made by different entrepreneurs regarding the production of artificial meat since the 1990s. Dutch entrepreneur Willem Frederik Van Eelen, who has been advocating artificial meat production since the 1950s, took an important step by obtaining the first patent related to the production of artificial meat for human or animal consumption in 1999 (van Eelen et al., 1999; Jönsson, 2016; Bhat et al., 2017; Fernandes et al., 2022). A patentometric analysis conducted in 2023 revealed that more than 190 patent applications related to artificial meat were made between 1997 and 2020, and more than half of these applications were made in five countries: the United States (US), Australia, China, Canada, and Japan (Ng et al., 2023).

Main Animal-Derived Cultured Meat Production Methods

The main techniques for artificial meat production using in vitro conditions include different methods such as cell culture, tissue culture, organ printing, nanotechnology and biophotonics (Bhat et al., 2017; Hopkins and Dacey, 2008; Bhat et al., 2019; Balasubramanian et al., 2021; Mateti et al., 2022). However, some of these techniques (e.g., biophotonics) are currently only theoretically applicable but hold potential for the future; others (e.g., tissue culture) are usable for small-scale production but are not yet suitable for large-scale commercial production (Bhat et al., 2019). It is expected that nanotechnology, especially when combined with biophotonic techniques, will provide improvements at various stages of conventional meat production, and these developments will also play a significant role in artificial meat production. For example, better cell adhesion can be achieved with nanotechnology-assisted biophotonic methods (Lamri et al., 2021; Ng and Kurisawa, 2021; Hong et al., 2021).

In vitro cell and tissue culture techniques are of great importance for scientific research and experimental medicine, especially because they provide alternatives to animal experiments (Weihe, 1985). Animal-based artificial meat production primarily relies on the multiplication of animal cells in vitro. Although the concepts of cell culture and tissue culture produced in artificial environments are often used interchangeably, there are major differences between these two techniques. Cell culture is based on the preservation of cells taken from different tissues in special environments and their reproduction in these environments. Suitable environments have been developed for the cells to reproduce, and these environments generally need to contain components such as complex carbohydrates, amino acids, salts, vitamins, hormones, animal serums and various growth factors. In tissue culture, unlike cell culture, tissue parts are separated from their entirety and kept alive and grown (Yao and Asayama, 2017).

The first step in vitro cell culture and scaffold-based techniques in artificial meat production is to identify suitable cell sources that can proliferate and transform into functional skeletal muscle tissue. For this purpose, cell isolation is carried out using embryonic myoblasts obtained from the embryo of the animal selected for artificial meat production or adult skeletal muscle cells isolated by muscle biopsy (Bhat et al., 2015; Bhat et al., 2017).

One of the most critical stages in the production of artificial meat is the selection of the cell source to be used (Reiss et al., 2021). The most common method is to obtain primary cells by taking tissue biopsies from the desired region of the targeted live animal species or by using living tissues after slaughtering. Another option is to use pluripotent cell sources, such as embryonic stem cells or induced pluripotent stem cells. When primary cells are used, muscle-resident progenitor cells can be isolated from skeletal muscle tissues. When pluripotent stem cells are used, these cells must first be transformed into mesodermal cells, i.e. muscle-resident progenitor cells (Bhat et al., 2017; Reiss et al., 2021). The isolated cells are then placed on a support structure such as a scaffold or collagen mesh and multiplied in bioreactors and appropriate growth media (Bhat et al., 2015, Bhat et al., 2017, Sürek and Uzun, 2020; Benny et al., 2022). Especially with the development of large-volume bioreactors, it is becoming more possible to produce large-scale animal cell cultures (Varley and Birch, 1999).As a result of the differentiation of cells in bioreactors to form myotubes, a food product composed of myofibrils, i.e., artificial meat, is obtained (Bhat et al., 2015; Bhat et al., 2017; Sürek and Uzun, 2020).

In animal cell culture-based artificial meat production, the timing and source of the cells, as well as the specific type of cells, are of great importance. Whether the cells to be used in artificial meat production are taken from a live animal or postmortem (after slaughter), and if postmortem, how the animal was slaughtered and the conditions of slaughter, as well as the sources and production methods of other components used in cell culture, can influence the product's acceptability and marketability. For consumers with religious sensitivities, these factors play a determining role in the acceptance and commercial success of the product (Izhar Ariff Mohd Kashim et al., 2023).

The most common method in artificial meat production is to utilize primary cell sources. In this method, tissue biopsies are taken from the desired regions (skeletal muscle, bone marrow, adipose tissue, etc.) of the target animal species or postmortem tissues are used to obtain cells (Reiss et al., 2021). For this purpose, muscle satellite cells that have the ability to transform into myotubules have been produced in cell culture environments with appropriate conditions from muscle biopsies taken from various animals such as cattle, chicken, sheep and fish (Dodson et al., 1987; Yablonka-Reuveni et al., 1987; Dodson et al., 1986; Doumit and Merkel, 1992). Similarly, mesenchymal stem cells obtained from muscle biopsy and blood and other tissue samples taken from different parts of animals such as cattle, chicken and sheep have been isolated and these cells, which have the potential to transform into cell types such as adipocytes, chondrocytes and fibroblasts, have been propagated in cell cultures (Bosnakovski et al., 2005; Khatri et al., 2009; Lyahyai et al., 2012). Another option for cell culture-based artificial meat production is the use of embryonic stem cells or induced pluripotent stem cells (Post, 2012; Reiss et al., 2021; Lee et al., 2023). Embryonic stem cells, in addition to their unlimited proliferation and self-renewal due to their pluripotent properties, have been reported to attract great interest in areas such as artificial meat production, as well as medical applications such as cell therapy and regenerative medicine, due to their ability to transform into cells representing the three embryonic layers (Post, 2012; Post, 2014; Reiss et al., 2021; Lee et al., 2023). Embryonic stem cells are produced from the "inner cell mass" of embryos at the blastocyst stage, and these cells have the potential to renew themselves unlimitedly. In addition, it is stated that they have the ability to transform into all cell types. It has been reported that embryonic stem cells have been successfully isolated and cultured from meat donor animals such as cattle, chicken, sheep, fish, and pigs (Cao et al., 2013; Lavial and Pain, 2010; Vilarino et al., 2020; Holen et al., 2010; Li et al., 2003). In addition, it has been discovered that cells taken from adult tissues can be reprogrammed and transformed into pluripotent cells with embryonic properties. Reprogrammed somatic cells can be directed and differentiated into the desired cell types in vitro. It has been stated that induced pluripotent stem cells from edible animals such as cattle, chicken, sheep, fish, and pigs have also been characterized in detail (Talluri et al., 2015; Kim et al., 2017; Liu et al., 2012; Gao et al., 2014).

Single-Cell Microorganism-Based Protein Sources

The term single cell protein (STP) was first mentioned in 1968 at a meeting at the Massachusetts Institute of Technology in the USA, where scientists came together to investigate the most appropriate alternatives to commonly used terminologies such as microbial protein (Paraskevopoulou et al., 2003). Single cell proteins are produced by various microbial organisms such as algae, bacteria, fungi and yeasts. Among these, fungi and bacteria are reported to be of importance as the main producers of this single cell protein (Anupama and Ravindra, 2000). Because these organisms are known to have characteristic features with their fast growth rates and high protein content (Gao et al., 2007). In addition, some algae species cultivated in aquatic environments have been created for this purpose (Voltolina et al., 2005). Single-cell proteins are obtained by drying the cells of microorganisms with high protein content and/or isolating purified proteins. These proteins stand out due to their wide range of amino acids, low fat content, and higher protein:carbohydrate ratio compared to animal feeds, making them a valuable nutritional supplement for humans (Srividya et al., 2013). SCPs contain vitamins such as thiamine, riboflavin, pyridoxine, nicotinic acid, pantothenic acid, folic acid, biotin, cyanocobalamin, ascorbic acid, β-carotene, and α-tocopherol; essential amino acids like lysine and methionine; minerals, nucleic acids, and lipids (Suman et al., 2015).

Compared to well-known staple food sources rich in protein such as meat (21%), fish (18%), and soybeans (39%), SCP sources (30-80%) have higher protein content (Fukagawa et al., 2022). Looking at their amino acid profiles, SCP sources contain all essential amino acids such as methionine, threonine, and lysine, which cannot be synthesized by the body and must be obtained through nutrition (Khan et al., 2022; Pires et al., 2016; Saejung and Salasook, 2020; Yap et al., 2022). When examining industrial-scale produced SCP sources, it is typically observed that microorganism groups such as microalgae, bacteria, yeasts, and fungi are preferred (Haznedaroğlu, 2024).

Microorganisms Used in Single-Cell Protein Production

Some microorganisms (e.g., Chlorella vulgaris) provide an advantage in industrial production due to their possession of unsaturated fatty acids, such as Omega-3, which are important for health. These microorganisms can be used not only for SCP production but also for the production of single-cell oils based on the biorefinery model. Additionally, Limnospira species, cyanobacteria commonly known by the commercial name "Spirulina," contain essential amino acids as well as phycocyanin, a photosynthetic pigment. Phycocyanin is a water-soluble, important pigment that carries antioxidant properties similar to Vitamin C (ascorbic acid). This pigment, especially outside the nutraceutical sector, is also used as a food dye due to its natural blue color (Haznedaroğlu, 2024).

One of the strongest antioxidants in nature, the astaxanthin molecule, is obtained from the Haematococcus pluvialis microalga species. Astaxanthin, a keto-carotenoid molecule, is produced by growing the H. pluvialis species under various stress conditions (e.g., light, nutrient media). Thanks to its ability to neutralize reactive oxygen species, astaxanthin is widely used not only in the cosmetic, nutraceutical, and pharmaceutical sectors but also in the aquaculture and fishing industries (Haznedaroğlu, 2024).

The cyanobacterium species Limnospira (commonly known as Spirulina), known for its high protein content, is rich in essential amino acids that humans cannot produce and must obtain through nutrition. These amino acids include histidine, leucine, isoleucine, methionine, phenylalanine, threonine, tryptophan, and valine. The presence of amino acids such as methionine, which is found in animal protein sources but in low amounts in legumes, and leucine, which is limited in grains, makes Spirulina an important SCP source (Haznedaroğlu, 2024).

Another species that stands out as an important protein source among microalgae is Chlorella vulgaris (Sharma et al., 2012; Ursu et al., 2014). Chlorella species are not only rich in protein but also in B vitamins, especially B1, B2, B3, B6, and B12, and they contain antioxidant secondary metabolites as well as minerals like magnesium, iron, calcium, and zinc (Pratt and Johnson, 1965; Seyfabadi et al., 2011). A recent study conducted at Boğaziçi University's Microalgae Biotechnology Research and Development Unit observed that when Chlorella vulgaris was used in functional foods such as bread and cookies, more of some heat-sensitive B vitamins were released compared to those that were exposed to heat, while other vitamins remained intact despite thermal processing (Haznedaroğlu, 2023).

Production Models

Single-cell protein (SCP) production processes are carried out through (photo) bioreactors or fermenters in carefully monitored nutrient environments due to the use of microorganisms. The goal of this method is to produce biomass rich in protein from microorganisms. Special closed systems are used, designed to provide optimal conditions for growth and reproduction of the selected microorganism, taking into account its metabolic requirements. These reactor systems are optimized to allow microorganisms to grow and proliferate effectively (Raziq et al., 2020).

Single-cell protein production processes consist of two main stages: upstream and downstream operations. The main steps and components of the upstream processes are as follows (Li et al., 2024):

- *Selection and Strain/Species Optimization of Microorganisms:* Based on the industry where the single-cell protein will be used, the selection of microorganisms is made according to factors such as rapid growth, high protein content, and suitability for the targeted application.
- *Cultivation of Microorganisms:* After preparing inoculum cultures of the selected microorganism strains suitable for production, these strains are inoculated into the bioreactor system for propagation. The reactors are operated by optimizing the most suitable temperature, pH, and oxygen levels according to the metabolic needs of the microorganism (autotrophic, heterotrophic, mixotrophic, etc.). In addition, carbon sources and other macro (nitrogen, phosphorus, etc.) and micro (vitamins and minerals) nutrients necessary for microorganism growth are provided to ensure that the microorganisms proliferate with high efficiency and reach maximum cell density in the bioreactor.

After the cultivation of microorganisms, the main steps and components of the downstream processes are as follows (Li et al., 2024):

- *Harvesting:* Once the biomass in the bioreactor reaches the desired cell density and protein content, it is harvested. Harvesting methods vary depending on the microorganism species used and the characteristics of the biomass. The most commonly preferred techniques include filtration, centrifugation, sedimentation, and flotation technologies.
- *Processing:* Various processing steps are applied to separate the protein and other valuable components (such as fats or carbohydrates) from the harvested biomass. Depending on the structure of the microorganism cell wall, processes like homogenization or grinding may be required. Additionally, spray drying or lyophilization techniques may be used for biomass drying, and purification steps can be carried out based on the protein concentration obtained.
- *Formulation:* At this stage, formulation studies are carried out to make the SCP (single-cell protein) products suitable for final use. For example, SCPs can be processed as raw protein and converted into protein-rich powders, or they can be formulated into granules for functional food products or animal feed (for small/large ruminants or fish feed).

Quality Control: This stage is particularly critical for nutritional, pharmaceutical, and nutraceutical products to ensure that the final product meets safety and nutritional standards. The detection of potential contaminants during the production process is essential, and measures are taken to eliminate the risks these contaminants could pose to the product. Furthermore, the nutritional content and other important parameters of the final product are evaluated and approved through quality control tests.

Safety and Application Challenges

During single-cell protein production, various microorganisms through common nutrient media can contaminate the products with toxins, nucleic acids, allergens, pathogens, pesticide residues that are difficult to detect, and heavy metals, which could potentially harm human health. For example, certain nucleic acids (especially purines) can increase uric acid accumulation in the body, leading to health issues such as gout or kidney stones (Ritala et al., 2017). These types of issues are generally not seen in microalgae production; however, in SCP productions using other microorganisms like fungi or bacteria, these risks can be reduced by using salts, acids, or alkaline solutions to reduce ribonuclease enzymes or nucleic acids (Haznedaroğlu, 2024).

Some fungi and bacteria used in single-cell protein production can produce toxins such as mycotoxins, ochratoxins, and aflatoxins, as well as various exotoxins and endotoxins. These substances can have negative effects on human health. For example, some Fusarium species have the potential to produce mycotoxins that can damage the central nervous system (Desjardins and Proctor, 2007; Desjardins, 2006). Another example is the production of microcystins by the cyanobacterium Microcystis aeruginosa, which can proliferate in nutrient media commonly used in Spirulina production, potentially causing harmful algal blooms. These toxins can contaminate the final product if adequate care is not taken during production (Gilroy et al., 2000; Vichi et al., 2012). Other potential allergens in the nutrient media of microorganisms used in single-cell protein production, contaminants or harmful substances from the water source used, especially those originating from food waste or industrial production processes with different regulations, can increase safety risks. Therefore, controlled nutrient media, such as the use of reliable carbon sources like glucose, dextrose, or sucrose, as well as pure carbon dioxide for autotrophic microorganisms, may be preferred. Such precautions help minimize harmful substances in the production processes and ensure a high level of safety for products intended for human or animal consumption (Haznedaroğlu, 2024).

The substrates required for the growth of some microorganisms used in single-cell-based protein production can be evaluated within the framework of circular economy by providing them from wastewater systems. For example, in the production of a Candida species, a protein yield of up to 39% was obtained using industrial coffee processing wastewater containing glucose, mannose and fructose (Pillaca-Pullo et al., 2023). In a study conducted with another Candida species, 30 g dry biomass/L productivity and 12.2 g/L protein production were achieved by using wastewater obtained from potato peels and containing 5% glycerol as a carbon source (Kurcz et al., 2018). The use of wastewater and various food wastes (e.g. orange peels, potato peels, etc.) as low-cost substrates in single-cell-based protein production can reduce the environmental impacts of production systems (Haznedaroğlu, 2024). However, it should be noted that such applications require further research in terms of safety and risk management.

Insect-Based Protein Sources

Plant-based proteins may be deficient in some essential amino acids and are generally less digestible by the human body compared to animal proteins. Therefore, proteins derived from plant sources alone are not sufficient for a complete and balanced diet. Instead, a combination of various plant sources is necessary to enhance protein quality. The Food and Agriculture

Organization of the United Nations (FAO) recommends edible insects as alternative food sources due to their environmentally friendly production conditions and high nutritional value (FAO, 2013). Moreover, it is considered that eight of the 17 Sustainable Development Goals (1, 2, 3, 6, 7, 8, 9, 12) announced by the FAO in 2015 could be directly addressed by insect farming, while the remaining nine (4, 5, 10, 11, 13, 14, 15, 16, 17) could be addressed indirectly through insect farming (Lin et al., 2023; Ravi et al., 2020). Edible insects, with growing interest, provide a high-quality protein source as a cost-effective alternative to traditional animal foods. Proteins obtained from insects such as locusts, crickets, termites, and mealworms are gaining popularity as more practical and economical alternatives compared to beef and poultry meat (de Castro et al., 2018; Nissen et al., 2020).

Insects are invertebrates belonging to the Phylum Arthropoda, showing a wide biological diversity that makes up approximately 95% of the animal kingdom. Edible insects have lower environmental impacts in terms of carbon dioxide, methane, ammonia emissions, water usage, and land area requirements for producing 1 kg of protein or 1 kg of biomass compared to traditional animal protein sources. Additionally, the production cycle of insects is much shorter than other protein sources (Lange and Nakamura, 2021). For example, the development of Bombyx mori (silkworm) larvae takes 20-24 days, while producing 1 kg of beef can take as long as 243 days. B. mori can produce 92% of the beef's protein, but more than one production cycle (520% protein yield) is required to produce this amount, making beef protein production more sustainable than beef itself (Liceaga, 2022; Lin et al., 2023; Siddiqui et al., 2023). Additionally, the feed for insects is typically sourced from industrial by-products and agricultural waste, providing high feed conversion efficiency (El, 2024).

Insect farming is considered a new "mini livestock" model within the scope of sustainable nutrition solutions. Currently, most edible insect farms operating on an industrial scale are located in Europe (e.g., France and the Netherlands) and North America (e.g., Canada and the USA). These farms typically raise crickets (Acheta domesticus) and yellow mealworms (Tenebrio molitor). The farms produce their own insects to prevent diseases, thus maintaining biological balance. Moreover, since antibiotics, hormones, pesticides, or other chemicals commonly used in traditional agriculture are not utilized during insect farming, this provides significant advantages in terms of environmental and consumer health (Lin et al., 2023).

Entomophagy

Paleoanthropologists suggest that insects played a significant role in the diet of early humans, based on historical evidence. Ancient civilizations, including Greek, Roman, and Aztec cultures, clearly consumed insects. Today, approximately 2.5 billion people worldwide include insects in their diets as part of cultural practices. In more than 100 countries—such as Thailand, China, Mexico, Colombia, and regions in Africa—and among over 3,000 ethnic groups, insect consumption is traditional. Research indicates that over 1,900 insect species are considered edible, particularly in tropical regions (van Huis et al., 2013).

However, consumer acceptance remains limited in some regions, especially in Western countries, where insects and insect-like invertebrates are still viewed with skepticism as sustainable food sources (Ordoñez-Araque & Egas-Montenegro, 2021; Zielińska & Pankiewicz, 2020). To date, approximately 2,000 edible insect species have been recorded, classified into eight main orders: Blattodea (cockroaches, termites), Coleoptera (beetles),

Diptera (flies), Hemiptera (cicadas, stink bugs), Hymenoptera (bees, wasps, ants), Lepidoptera (butterflies, moths), Odonata (dragonflies), and Orthoptera (crickets, grasshoppers, locusts) (FAO, 2013).

In many countries, insects are consumed directly or used as nutritional additives, but European consumers often struggle to accept whole insects or larvae as food (Van Huis, 2013). Grinding insects into flour has emerged as an effective strategy to improve acceptability. Insects can be consumed at different life stages—egg, larva, pupa, or adult. Consumer acceptance varies depending on whether insects are presented whole or in processed forms, such as flour, powder, puree, protein isolates, or protein hydrolysates. However, terms like "insect flour" can be misleading, as they may suggest characteristics similar to traditional grain flour used in baking. Unlike wheat flour, which is primarily starch (65–70%) with 12–13% protein and 0.3% minerals/ash, insect flour is mainly composed of protein, followed by fat and fiber (El, 2024).

Nutrient Composition

Insects are rich in protein, monounsaturated and polyunsaturated fatty acids, and fiber, making them highly nutritious food sources. For example, consuming 100 g of caterpillars can meet 76% of an individual's daily protein requirement. Additionally, 3 silkworm pupae are equivalent in nutritional value to 1 chicken egg, with the composition of these pupae consisting of approximately 50% protein and 30% lipids (Bessa et al., 2020; Jantzen da Silva Lucas et al., 2020).

The nutritional value of edible insects is highly diverse. The nutrient content of the same insect species can vary depending on its developmental stage (egg, larva, pupa, adult), living conditions, and diet (Bessa et al., 2020). Moreover, protein levels can differ between species depending on growth conditions, diet variety, and harvest time (Boulos et al., 2020; Jonas-Levi and Martinez, 2017). Edible insects emerge as a potential solution for meeting global protein demands because they have higher protein content, a rich amino acid profile, and lower environmental production costs compared to other protein sources (Jantzen da Silva Lucas et al., 2020).

Protein quality plays a critical role in meeting individuals' nutritional needs. For proteins in food to be used by the body after their amino acids are digested and absorbed, it is not only necessary to have an adequate amount of protein and amino acid composition, but the digestibility and absorption rates are also of great importance (Loveday, 2023). In this context, a high-quality protein is defined as one that contains all the essential amino acids in the required amounts for the human body. Animal-derived proteins such as meat, milk, and eggs are generally considered examples of complete quality proteins. Plant-based proteins, however, are more diverse in quality; when multiple plant sources are consumed together, these combinations can approach the quality of complete proteins (Rojas Conzuelo et al., 2022). High-quality proteins are easily digestible and provide a complete amino acid profile. Insects are defined as complete quality protein sources when evaluated in terms of their essential amino acid (EAA) profiles. Although the EAA composition varies according to insect species, most insect proteins are rich in lysine and are reported to contain high levels of other EAAs such as methionine, threonine and tryptophan (Oibiokpa et al., 2018; Rumpold and Schlüter, 2013). In studies conducted, for example, although the amino acid tryptophan is limited in cricket flour, the Amino Acid Score was reported as 0.85 and in vitro protein digestibility was reported as 76.2% (Payne et al., 2016; Dong et al., 2017; Oonincx et al., 2015). The digestibility of insect proteins is lower compared to animal proteins, but higher than plant proteins. For example, in comparative studies, it has been reported that the digestibility rates of cricket and mealworm proteins are similar to beef protein, but cricket protein has better digestibility than mealworm protein (Ojha et al., 2021). In addition, it has been clearly determined that grasshopper protein is better digestible than silkworm protein (Yang et al., 2012). Although the digestibility feature of insect proteins varies depending on the species and life stage, insects are seen as an alternative to high-quality animal proteins with their EAA composition and high digestibility rates. Chitin, which forms the exoskeleton of insects, is considered an important component that can affect the digestion of proteins. Chitin is a polysaccharide located in the external structure of the insect and is generally not digestible by humans. However, it is noteworthy that bacteria in the human intestine can produce chitinolytic enzymes and that these enzymes will only be active in people who regularly consume insects. It is also reported that chitin has antioxidant, antitumor and antimicrobial properties. One of the most important approaches in the development of edible products obtained from insects is the use of this exoskeleton structure from proteins (Baiano, 2020).

Protein Composition of Selected Insect Species

Cunha and colleagues (2023) conducted a systematic review of insect consumption to understand its effects on human health. The study analyzed 1,063 sources published over the past 10 years, 9 randomized controlled trials, and 5 studies on allergenicity. The research findings revealed that insect protein consumption led to an increase in blood amino acid levels, with no significant difference when compared to other protein sources, and that the level of leucine was closer to that of milk and soy proteins rather than beef protein (Cunha et al., 2023). This suggests that insect proteins may be a suitable alternative for muscle protein synthesis. Additionally, the chitin in insect-based products slows down digestion, delaying the release of amino acids. It was observed that glucose levels did not change after insect consumption, but insulin levels were lower. Allergenicity assessments showed sensitivity and cross-reactivity between insect allergens and other known allergens, with this cross-reactivity being partially reduced by thermal processes (Cunha et al., 2023).

In addition to research on the economic advantages and environmental sustainability potential of insect consumption, challenges related to food safety and consumer acceptance need to be examined further. Moreover, while the nutritional properties of insects, their status as quality protein sources, and the positive effects of bioactive peptides derived from these proteins are advantages, potential risks such as the presence of antinutritional compounds (phytates, tannins, saponins, oxalates), pesticide residues, allergens, heavy metals, and pathogens are also significant (Rivas-Navia et al., 2023). All these factors must be thoroughly researched in terms of edible insect species.

Insect Proteins as Bioactive Peptide Sources

Bioactive peptides are small protein fragments, containing 3 to 20 amino acids, that are inactive in the main protein structures of foods. These peptides have the capacity to regulate various metabolic processes in the body. They are generally derived from animal-based foods (milk, eggs, meat, fish), plant-based sources (soy, legumes, wheat), and insects. Bioactive peptides can be formed through food processing methods (such as heat treatment, non-thermal processes, fermentation), hydrolysis by proteolytic microorganisms, or the action of proteolytic enzymes released by microorganisms or plants, which can result in changes to the protein structure or the action of enzymes during digestion. However, not all peptides exhibit biological activity. Factors affecting the biological activity of peptides include the amino acid sequence, N- and C-terminal structures, types of amino acids, the length of the peptide chain, and the charge and polarity of the amino acids. The main biological activities of peptides in the body include antioxidant, antimicrobial, antihypertensive, antidiabetic, and anti-inflammatory effects (Nongonierma et al., 2017).

The biological potential of bioactive peptides derived from edible insects can be similar to or even higher than that found in common food proteins. These peptides demonstrate antioxidant, antimicrobial, anti-inflammatory, antidiabetic, and antihypertensive properties, warranting further detailed clinical studies (Nongonierma & FitzGerald, 2017). Early studies on bioactive peptides derived from insects, such as those from the edible species Spodoptera littoralis (cotton leafworm), have shown in vitro ACE inhibitory activity (Vercruysse et al., 2005; Vercruysse et al., 2009). These peptides were then purified, and their antihypertensive effects were tested in vivo on hypertensive rats, yielding significant results. Later, several studies using enzymes such as thermolysin and alcalase for hydrolysis found that insect proteins exhibited ACE inhibitory and antioxidant properties, which were determined through simulated digestion systems and in silicon analyses. These studies also noted that while the antioxidant activity of insect peptides was lower compared to Vitamin C, it was still similar. The results suggest that insect proteins with ACE inhibitory and antioxidant activities could be used as multifunctional components in functional food products and also offer potential applications in the pharmaceutical field (Vercruysse et al., 2008; Vercruysse et al., 2010).

Tenebrio molitor larvae, after being degreased with petroleum ether, were hydrolyzed using alcalase enzymes. The highest ACE inhibitory activity was achieved at a 20% hydrolysis degree, measuring at 75% or IC50 = 0.39/mL. The hydrolysates were then fractionated, and the fractions with the strongest ACE inhibitory activity (molecular weights between 180-500 Da) were purified. One of these purified peptides, Tryptophan-Alanine-Asparagine, was evaluated for its antihypertensive effect, showing a significant reduction in systolic blood pressure in hypertensive rats (Dai et al., 2013).

Zielińska et al., (2017) studied the antioxidant activities, ion chelation capacities, and reducing powers of peptides obtained from the in vitro gastrointestinal digestion of insects such as Dubia cockroach, Madagascar hissing cockroach, cricket, superworm, and house cricket. Amphiacusta annulipes hydrolysates showed high radical activity (IC50 = 19.1 μg/mL) against 2,2-diphenyl-1-picrylhydrazyl (DPPH), Fe^{2+} ion chelation ability (58.82%), and reducing power (0.652), demonstrating strong antioxidant effects. Zophobas morio hydrolysates exhibited radical activity (IC50 = 4.6 μg/mL) against ABTS radicals. The highest Cu^{2+} chelation capacity was observed in Locusta migratoria peptides (86.05%). The study also observed high concentrations of peptides after digestion. These findings suggest that edible insect hydrolysates could be a good peptide source with antioxidant activity (Zielińska et al., 2017).

In a study on insect species such as Tenebrio molitor, Phyllophaga rugosa, Nudaurelia melanops, Oecophylla smaradigna, Locusta migratoria, and Acheta domesticus, the protein profiles and peptide releases of these species during gastrointestinal digestion were examined. Proteins extracted from insect flours were characterized using SDS-PAGE. In silico analysis, potential peptides were conducted to identify potential peptides among the proteins. The study found that these peptides exhibited antioxidant activity as well as DPP-IV (Dipeptidyl peptidase-4) and ACE inhibitory effects (Rebollo-Hernanz et al., 2021). The study emphasized the need for a multidisciplinary approach for large-scale production of bioactive peptides.

Food Applications

Insects are not only used to enhance the nutritional value of food products as protein sources, but they also have various applications in improving technological and functional properties. These properties include solubility, gelling, foaming, emulsifying, and colloidal behaviors. For protein supplementation, the production of protein powders, extracts, isolates, and hydrolysates is commonly highlighted; processes such as isoelectric precipitation and enzymatic proteolysis (using commercial enzymes such as alcalase, flavourzyme, etc.) can also be added, depending on the intended purpose of the final product (Rivas-Navia et al., 2023; Mannozzi et al., 2023). However, the quantity, composition, and functional properties of insect proteins can vary depending on the insect's rearing conditions, nutritional status, and hunger levels.

Before the protein extraction process, preparatory steps such as drying and grinding insects are crucial for improving the powder structure, as these processes increase the surface area, thereby enhancing the extraction efficiency. Drying can be performed at the beginning of the process or after protein isolation and purification (Queiroz et al., 2023). Studies on various drying methods have observed that freeze-drying does not cause significant changes in the insect's color, taste, odor, and physical properties. In experiments with mealworms (Tenebrio molitor), different drying techniques such as microwave, fluidized bed, and vacuum drying were found to reduce protein solubility, while freeze-dried samples showed more lipid oxidation. For mealworms, microwave and vacuum oven drying methods were considered more advantageous alternatives compared to traditional methods (Mishyna et al., 2021).

Insect Protein Allergen

In addition to the nutritional benefits provided by edible insects, their potential to carry allergens can also pose a risk. The World Health Organization (WHO) has recorded 239 different allergens. Allergic reactions have also been associated with insect consumption, and these reactions can include anaphylaxis. Such allergic responses are often linked to cross-reactive proteins such as tropomyosin and arginine kinase. Therefore, individuals with shellfish or dust mite allergies may also develop allergic reactions to foods containing insects like mealworms or crickets. Tropomyosin, a common allergen found in shrimp, is a protein with myofibril structure and consists of a spiral-shaped dimer with a molecular weight ranging from

35 to 38 kDa. This protein shares 75-85% similarity with other arthropods, including crickets. According to existing literature, tropomyosin is identified as the major allergen in insects (Papia et al., 2021). Tropomyosin is highly resistant to gastrointestinal proteases. During this process, immune cells in the digestive system are responsible for recognizing pathogens and enterotoxins, but they are unable to recognize these undigested polypeptide chains, triggering an allergic reaction (El, 2024).

Thermal processes can cause changes in the molecular structures of other allergenic proteins, affecting their ability to bind antibodies, while tropomyosin protein resists heat treatment and maintains its structure. This situation makes the investigation of alternative processing methods to render insects hypoallergenic increasingly important. In particular, there is a growing need to examine the effects of non-thermal food processing techniques (green technologies) such as pulsed light, high hydrostatic pressure, irradiation, cold plasma, ultrasound waves, and pulsed electric fields on protein allergenicity (Cunha et al., 2023).

Health Effects of Alternative Food Sources and Food Safety

Plant-based protein sources are noted to be associated with lower risks of cardiovascular diseases, metabolic syndrome, certain types of cancer, and premature death when compared to primary protein sources such as meat, poultry, fish, dairy products, and eggs (Lonnie et al., 2020). Microalgae have been identified as a rich source of various bioactive compounds, such as carotenoids, fatty acids, amino acids, antioxidants, and other secondary metabolites that contribute to health protection and disease prevention. These compounds have been reported to help lower blood sugar and cholesterol levels, balance blood pressure, and increase hemoglobin concentration (Köksal, 2024). Plant-based protein consumption includes not only protein but also beneficial components like dietary fibers, isoflavones, and phospholipids. Compared to animal proteins, plant proteins are associated with more health benefits, such as reducing the risk of cardiovascular diseases, type 2 diabetes, and certain cancers. For example, while milk protein is known to aid in weight management and lower triglyceride levels, soy protein isolate has been found to show a stronger lipid-lowering effect than milk protein (Shi et al., 2019). Huang and colleagues (2020) suggested that replacing animal protein with plant-based protein could significantly reduce mortality rates and the risk of cardiovascular diseases (Huang et al., 2020). Plant proteins have been reported to contribute to healthier gut flora and help reduce the potential toxicity of some human metabolites, which are commonly linked to chronic inflammation and carcinogenic metabolites (Saeed et al., 2022). Active peptides derived from plant proteins exhibit antioxidant, anti-inflammatory, and antihypertensive properties while also lowering blood cholesterol levels (Durand et al., 2021). The world-renowned nutrition expert Dr. Campbell has demonstrated that plant-based diets are beneficial in the treatment and prevention of many chronic diseases (Kim et al., 2021). Additionally, diets that include large amounts of red meat, poultry, eggs, and fish have been associated with higher BMI compared to plant-based protein-rich diets with fresh fruits and vegetables (Tuso et al., 2013). These animal foods are high in saturated fatty acids and low in fiber, which may contribute to the development of chronic diseases over time, leading to significant long-term negative health outcomes (Päivärinta et al., 2020).

Research has revealed that edible insects contain bioactive and antioxidant compounds that strengthen the immune system and reduce disease risks. Due to these properties, edible insects are used in many regions worldwide. Furthermore, their bioactive compounds are being evaluated in both clinical and non-clinical studies for their anti-inflammatory, anticancer, antimicrobial, anti-ulcer, antidiabetic, hypolipidemic, and cardioprotective effects (Roos & van Huis, 2017; Mariutti et al., 2021). In one study, the water- and fat-soluble extracts of insects were compared for antioxidant capacity to fresh orange juice and olive oil. It was found that crickets, locusts, silkworms, African caterpillars, and evening cicadas had two to three times higher antioxidant capacity than orange juice or olive oil. However, the results of these in vitro tests need to be verified in humans (van Huis, 2020). The chitin found in insects, along with short and medium-chain fatty acids, glycosaminoglycans, and antimicrobial peptides, is reported to offer various health benefits for the gastrointestinal system and resistant pathogens. Chitin, a significant fiber source, supports the natural development of gut microbiota due to its prebiotic effects. This may help prevent foodborne illnesses and digestive issues. Research has shown that harmful microorganisms like Salmonella Typhimurium, Escherichia coli, and Vibrio cholera are reduced due to chitin and its derivatives, while beneficial gut bacteria like Bifidobacteria and Lactobacillus are promoted (Imathiu, 2020). A study examining the effects of edible insects on gut microbiota found that individuals consuming 25 grams of cricket (Gryllodes sigillatus) flour daily had a 5.7-fold increase in the Bifidobacterium animalis species in their fecal samples (Stull et al., 2018). A polysaccharide called glycosaminoglycan found in Gryllus bimaculatus crickets has been reported to show significant anti-inflammatory effects in mouse models. These anti-inflammatory effects are seen as the glycosaminoglycan polysaccharide suppresses inflammatory biomarkers such as C-reactive protein (CRP), Interleukin-6 (IL-6), and rheumatoid factor in mice (Ahn et al., 2014).

Interest in edible insects worldwide is steadily increasing as an alternative food source. However, concerns have been raised regarding food safety issues such as their allergen content (the World Health Organization Allergen Naming Subcommittee has identified 239 different allergens related to arthropods), chemical hazards (especially pesticide residues), and biological risks (pathogens and mycotoxins). Furthermore, bioavailability may decrease due to antinutrient content (e.g., oxalates, phylates, tannins, saponins, and alkaloids). Additionally, contaminants found in edible insects, such as heavy metals (lead, mercury, arsenic, and cadmium), vary depending on insect species, collection stages, production methods, and feed substrates, which can pose various health risks (Lange & Nakamura, 2021; Banach et al., 2022).

The production of cultured meat is said to have both positive and negative impacts on public health and food safety. The composition of laboratory-produced meat is adjustable, allowing for the optimization of the product's taste, quality, and fatty acid profile. Additionally, risks associated with foodborne pathogens found in animal-based meats can be reduced, leading to safer and healthier meat production under highly controlled conditions. The fact that cultured meat is not derived from live animals aligns with religious and philosophical beliefs, such as those of Hinduism and vegetarianism. Furthermore, it can significantly reduce the risks of pesticide, arsenic, dioxin, and hormone exposure, as well as the frequency of foodborne diseases. Cultured meat also contributes to the prevention of non-communicable chronic diseases, such as cardiovascular diseases, diabetes, and colon cancer. However, cultured meat technology may also help reduce costs in long-duration human space missions. Controlled ecological life support systems can provide astronauts with fresh food while simultaneously using waste to produce oxygen and water (Bhat & Fayaz, 2011; Hocquette, 2016). Additionally, cultured meat minimizes disease risks from human-animal interactions (Hocquette et al., 2015). Supporters of cultured meat argue that these products will be free from microorganisms, as they are produced without animals. However, there is concern that the intensive proliferation of cells may trigger the growth of potential cancer cells. Furthermore, it has not yet been guaranteed that the chemicals used in cell culture (hormones, nutrients, etc.) are entirely safe for food consumption (Hocquette, 2016). There are also risks associated with the consumption of untested or new materials (Hopkins & Dacey, 2008). Nevertheless, cultured meat producers argue that the product contains essential nutrients and can meet both psychological and physiological needs of individuals (Post, 2012). It is believed that cultured meat production could help prevent health issues associated with meat consumption, such as cardiovascular diseases, diabetes, and colon cancer (Capper, 2011). Additionally, components like omega-3 can be added to cultured meat, making it more health-beneficial (Macintyre, 2007).

Single-cell microorganism-based protein sources may have higher protein content compared to traditional food groups like meat, milk, and fish, and contain essential components necessary for nutrition, such as essential amino acids. It has been demonstrated that adding these protein sources to foods can increase their nutritional value and provide a richer composition than either animal-based or plant-based protein alternatives. Furthermore, single-cell proteins (SCPs) have been found to offer health benefits due to their antioxidant properties and secondary metabolites that exhibit anti-inflammatory, antibacterial, and antiviral effects. These properties clearly indicate that SCPs are a promising alternative protein source (Haznedaroğlu, 2024). SCPs are a rich source of essential amino acids, vitamins (e.g., B12), minerals (e.g., iron and zinc), and unsaturated fatty acids. Therefore, they play a significant role as a supporting alternative in cases where protein deficiency is common (Anupama & Ravindra, 2000). SCPs exhibit antioxidant, anti-inflammatory, and immune-boosting effects through their secondary metabolites. These properties help in the prevention of chronic diseases, such as cardiovascular diseases and diabetes (Kurbanoglu, 2021). Specifically, SCPs derived from microalgae possess prebiotic properties that improve gut health and positively impact gut microbiota (Mæhre et al., 2018). SCP production uses far less land, water, and energy compared to traditional agriculture, presenting a sustainable alternative for ensuring food security (Matassa et al., 2016). Since SCP production occurs through controlled fermentation processes, the risk of contamination is low. However, errors in the substrates used (e.g., waste organic materials) and fermentation conditions may lead to the formation of mycotoxins and other harmful compounds. To minimize these risks, strict control and monitoring mechanisms are necessary (Ritala et al., 2017). Some microorganisms' proteins may cause allergic reactions. Therefore, comprehensive toxicological and allergenic analyses are required to ensure product safety (Anupama & Ravindra, 2000).

Conclusion

The increasing global population, depletion of environmental resources, and climate change pose serious threats to the sustainability of current food production systems. In this context, developing alternative food sources has become a strategic necessity for ensuring a sustainable and nutritious food supply. Plant-based foods, cell culture-based products, proteins from single-celled organisms, and edible insects are particularly important due to their potential to reduce environmental impacts while enhancing nutritional quality (Oh & Kim, 2025).

Plant-based foods offer environmental benefits, including lower greenhouse gas emissions and reduced land and water use, while also providing rich micronutrients and positive health effects, increasing their appeal (Leonard et al., 2024). Although cell culture-based meat can mitigate environmental burdens from conventional animal farming, it faces technical and economic challenges such as high production costs and energy consumption (Tuomisto & Mattos, 2011). Proteins from single-celled microorganisms are promising for the future due to their rapid production cycles, high protein quality, and low environmental footprint (Zhuang et al., 2024). Edible insects also present a sustainable option because of their efficient biological conversion, high nutritional value, and minimal environmental impact (Papastavropoulou et al., 2023).

The widespread adoption of these alternative food sources depends on multiple factors, including consumer perception, food safety, regulatory frameworks, and cultural acceptance. Expanding scientific knowledge of long-term health effects, standardizing production processes, and establishing economic feasibility are also critical for integrating these products into mainstream food systems.

Overall, alternative food sources have substantial potential to support sustainability and public health goals. To realize this potential, interdisciplinary collaboration, supportive public policies, and comprehensive strategies to raise consumer awareness are essential.

References

Ahn MY, Han JW, Hwang JS, Yun EY, Lee BM, 2014. Anti-inflammatory effect of glycosaminoglycan derived from Gryllus bimaculatus (a type of cricket, insect) on adjuvanttreated chronic arthritis rat model. J *Toxicol Environ Health Part A,* 77(22-24), 1332-45.

Aiking H, de Boer J, 2020. The next protein transition. *Trends in Food Science & Technology*, 105, 515-22.

Akshay Patil R, Bhavana A, Roopa Patil B, 2024. Cultured Meat: The Upcoming Meat Production having Sustainable Benefits over Conventional Meat Production: A Review. *Agricultural Reviews,* 45(1).

Anupama R, Ravindra P, 2000. Value-added food: Single-cell protein. *Biotechnology Advances,* 18(6), 459-79.

Ayivi RD, Gyawali R, Krastanov A, Aljaloud SO, Worku M, Tahergorabi R, da Silva RC, Ibrahim SA, 2020. Lactic acid bacteria: Food safety and human health applications. *Dairy,* 1(3), 202-32.

Baiano A, 2020. Edible insects: An overview on nutritional characteristics, safety, farming, production technologies, regulatory framework, and socio-economic and ethical implications. *Trends in Food Science and Technology,*100, 35–50. Elsevier Ltd.

Balasubramanian B, Liu W, Pushparaj K, Park S, 2021. The epic of in vitro meat production-A fiction into reality. *Foods,* 10(6), 1395.

Banach JL, Berg JP, Kleter G, Veen HVV, Bastiaan-Net S, Pouvreau L, Asselt ED, 2022. Alternative proteins for meat and dairy replacers: Food safety and future trends. *Critical Reviews in Food Science and Nutrition.* 63(32),11063-11080.

Barbosa JC, Rodrigues D, Almeida D, Machado D, Franco I, Inácio J, Freitas AC, Gomes, AM, 2022. The role of cereals in the industry of plant-based foods. In 20th ICC Conference: future challenges for cereal science and technology.

Bhat ZF, Kumar S, Bhat HF, 2017. In vitro meat: A future animal-free harvest. *Crit Rev Food Sci Nutr,* 57(4), 782-9.

Bhat ZF, Fayaz H, 2011. Prospectus of cultured meat—advancing meat alternatives *J Food Sci Technol* 48(2), 125–40.

Bhat ZF, Morton JD, Mason SL, Bekhit AEA, Bhat HF, 2019. Technological, Regulatory, and Ethical Aspects of In Vitro Meat: A Future Slaughter-Free Harvest. *Compr Rev Food Sci Food Saf.* 18(4), 1192-208.

Bhat ZF, Kumar S, Fayaz H, 2015. In vitro meat production: Challenges and benefits over conventional meat production. J *Integr Agric.* 14(2), 241-8.

Bhat ZF, Kumar S, Bhat HF, 2017. In vitro meat: A future animal-free harvest. *Crit Rev Food Sci Nutr,* 57(4), 782-9.

Benny A, Pandi K, Upadhyay R, 2022. Techniques, challenges and future prospects for cell-based meat. F*ood Sci Biotechnol.* 31(10), 1225-42.

Bessa L, Pieterse E, Sigge G, Hoffman LC, 2020. Insects as human food; from farm to fork. *J Sci Food Agric,* 100(14), 5017-22.

Benjaminson MA, Gilchriest JA, Lorenz M, 2002. In vitro edible muscle protein production system (MPPS): stage 1, fish. *Acta Astronaut.* 51(12), 879-89.

Bosnakovski D, Mizuno M, Kim G, Takagi S, Okumura M, Fujinaga T, 2005. Isolation and multilineage differentiation of bovine bone marrow mesenchymal stem cells. *Cell Tissue Res.* 19(2), 243-53.

Boulos S, Tännler A, Nyström L, 2020. Nitrogen-to-Protein Conversion Factors for Edible Insects on the Swiss Market: T. molitor, A. domesticus, and L. migratoria. *Front Nutr,* 7(89).

Bolarinwa IF, Al-Ezzi MFA, Carew IE, Muhammad K, 2019. Nutritional value of legumes in relation to human health: a review. *Advance Journal of Food Science and Technology*, 17, 72–85.

Bryngelsson S, Moshtaghian H, Bianchi M, Hallström E, 2022. Nutritional assessment of plant-based meat analogues on the Swedish market. *Int J Food Sci Nutr.* 73(7), 889-901.

Capper JL, 2011. The environmental impact of beef production in the United States: 1977 compared with 2007. *Journal of Animal Science,* 89(12), 4249-61.

Cao S, Wang F, Liu L, 2013. Isolation and culture of bovine embryonic stem cells. *Methods Mol Biol.* 1074, 111-23.

Carrel A, 1912. On the permenant life of tissıes outside of the organism. *J Exp Med.,* 15(5), 516-28.

Chalupa-Krebzdak S, Long CJ, Bohrer BM, 2018. Nutrient density and nutritional value of milk and plant-based milk alternatives. *Int. Dairy J.* 87. 84–92.

Charles H, Godfray J, Aveyard P, Garnett T, Hall JW, Key TJ, Jebb SA, 2018. Meat consumption, health, and the environment. *Science.* 361, 6399.

Choudhury D, Singh S, Seah JSH, Yeo DCL, Tan LP, 2020. Commercialization of plant-based meat alternatives. *Trends Plant Sci.* 25(11), 1055-8.

Churchill W, 1931. https://teachingamericanhistory.org/document/fifty-years-hence (Erişim tarihi: 24 Ekim, 2024).

Cui S, McClements DJ, Xu X, Jiao B, Zhou L, Zhou H, Xiong L, Wang Q, Sun Q, Dai, L, 2023. Peanut proteins: Extraction, modifications, and applications: A comprehensive review. *Grain & Oil Science and Technology.* 6(3), 135-47.

Cunha N, Andrade V, Ruivo P, Pinto P, 2023. Effects of Insect Consumption on Human Health: A Systematic Review of Human Studies. *Nutrients,* 15(14), 3076.

Çakır Biçer, N., Baş, D., Seçkiner, S., Kahrıman, M., & Baş, M. (2023). Evaluation of Consumers' Perceptions and Purchase Decisions Regarding Plant-Based Milk Alternatives in Turkey. *Izmir Democracy University Health Sciences Journal,* 6(1), 82–102.

Dai C, Ma H, Luo L, Yin X, 2013. Angiotensin I-converting enzyme (ACE) inhibitory peptide derived from Tenebrio molitor (L.) larva protein hydrolysate. *European Food Research and Technology.* 236(4), 681-9.

da Veiga Correia VT, D'Angelis DF, dos Santos AN, Roncheti EFS, Queiroz VAV, Figueiredo JEF, da Silva WA, Ferreina AA, Fante CA, 2022. Tannin-sorghum flours in cream cheese: Physicochemical, antioxidant and sensory characterization. *LWT.* 154, 112672.

de Castro RJS, Ohara A, dos Santos Aguilar JG, Domingues MAF, 2018. Nutritional, functional and biological properties of insect proteins: Processes for obtaining, consumption and future challenges. *Trends in food science & technology,* 76, 82-9.

Desjardins A, Proctor R, 2007. Molecular biology of Fusarium mycotoxins. *International journal of food microbiology.* 119(1-2), 47-50.

Dhakal D, Younas T, Bhusal RP, Devkota L, Henry CJ, Dhital S, 2023. Design rules of plant-based yoghurt-mimic: Formulation, functionality, sensory profile and nutritional value. *Food Hydrocolloids.* 142, 108786.

Desjardins AE, 2006. Fusarium mycotoxins: chemistry, genetics and biology. *The American Phytopathological Society*, 260.
Dodson MV, Martin EL, Brannon MA, Mathison BA, McFarland DC, 1987. Optimization of bovine satellite cell-derived myotube formation in vitro. *Tissue Cell.* 19(2), 159-66.
Dodson MV, McFarland DC, Martin EL, Brannon MA, 1986. Isolation of satellite cells from ovine skeletal muscles. *J. Tissue Cult. Methods*. 10, 233–7.
Doumit ME, Merkel RA, 1992. Conditions for isolation and culture of porcine myogenic satellite cells. *Tissue Cell.* 24(2), 253-62.
Dong X, Fu X, Dong M, Zhan S, Luo L, Sun J, Yang Y, 2017. Evaluation of the protein quality of two insect species, Tenebrio molitor and Hermetia illucens, as a potential ingredient in dog food. *Journal of the World Aquaculture Society.* 48, 727-37.
Durand E, Beaubie S, Ilic I, Fine F, Kapel R, Villeneuve P, 2021. Production and antioxidant capacity of bioactive peptides from plant biomass to counteract lipid oxidation. *Curr. Res. Food Sci.* 4, 365–97.
Edelman PD, McFarland DC, Mironov VA, Matheny JG, 2005. Commentary: *In vitro*-cultured meat production. *Tissue Eng.,* 11(5-6), 659-62.
El SN, 2024. Böcek Bazlı Protein Kaynakları, Alternatif Gıda Kaynakları, Şahin K, Ankara, *Türkiye Bilimler Akademisi Yayınları,* 144-163.
Elhalis H, See XY, Osen R, Chin XH, Chow Y, 2023. The potentials and challenges of using fermentation to improve the sensory quality of plant-based meat analogs. *Front Microbiol,* 14, 1267227.
Ewy MW, Patel A, Abdelmagid MG, Mohamed Elfadil O, Bonnes SL, Salonen BR, Hurt RT, Mundi MS, 2022. Plant-based diet: is it as good as an animal-based diet when it comes to protein? *Current nutrition reports,* 11(2), 337-46.
FAO, 2013. Edible insects. Future prospects for food and feed security, Food and Agriculture Organization of the United Nations Rome.
FAO, 2021. The future of food and agriculture alternative pathways to 2050. Food and Agriculture Organization of the United Nations Rome.
Ford BJ, 2011. Cultured meat: food for the future. *The Microscope,* 59:2, 73-81.
FAO, 2020. The state of food security and nutrition in the world 2020. Rome: Food and Agriculture Organization. Erişim tarihi, 24.10.2024.
Ferawati F, Hefni M, Östbring K, Witthöft C, 2021. The application of pulse flours in the development of plant-based cheese analogues: proximate composition, color, and texture properties. *Foods,* 10(9):2208.
Fernandes AM, Teixeira OS, Revillion JP, Souza ÂRL, 2022. Panorama and ambiguities of cultured meat: an integrative approach. *Crit Rev Food Sci Nutr.*, 62(20), 5413-23.
Fukagawa NK, McKillop K, Pehrsson PR, Moshfegh A, Harnly J, Finley J, 2022. USDA's fooddata central: what is it and why is it needed today? *The American journal of clinical nutrition,* 115(3), 619-24.
Gao L, Chi Z, Sheng J, Ni X, Wang L, 2007. Single-cell protein production from Jerusalem artichoke extract by a recently isolated marine yeast Cryptococcus aureus G7a and its nutritive analysis. *Appl. Microbiol. Biotechnol,* 77, 825–32.
Gao Y, Guo Y, Duan A, Cheng D, Zhang S, Wang H, 2014. Optimization of culture conditions for maintaining porcine induced pluripotent stem cells. *DNA Cell Biol,* 33(1), 1-11.
Gilroy DJ, Kauffman KW, Hall RA, Huang X, Chu FS, 2000. Assessing potential health risks from microcystin toxins in blue-green algae dietary supplements. *Environmental health perspectives,* 108(5), 435-9.
Good Food Institute, 2023.
Godfray HCJ, Beddington JR, Crute IR, Haddad L, Lawrence D, Muir JF, Pretty J, Robinson S, Thomas SM, Toulmin C, 2010. Food security: the challenge of feeding 9 billion people. *Science,* 327(5967), 812-8.
Godfray HCJ, Aveyard P, Garnett T, Hall JW, Key TJ, Lorimer J, Pierrehumbert RT, Scarborough P, Springmann M, Jebb SA, 2018. Meat consumption, health, and the environment. *Science,* 361(6399), eaam5324.
Grossmann L, McClements DJ, 2021. The science of plant-based foods: Approaches to create nutritious and sustainable plant-based cheese analogs. *Trends in Food Science & Technology,* 118, 207-29.
Gupta MK, Torrico DD, Ong L, Gras SL, Dunshea FR, Cottrell JJ, 2022. Plant and dairy-based yogurts: a comparison of consumer sensory acceptability linked to textural analysis. *Foods,* 11(3), 463.

Haas R, Schnepps A, Pichler A, Meixner O, Cow milk versus plant-based milk substitutes: a comparison of product image and motivational structure of consumption. *Sustainability,* 2019, 11(18), 1–25.

Hadidi M, Aghababaei F, McClements DJ, 2023. Enhanced alkaline extraction techniques for isolating and modifying plant-based proteins. *Food Hydrocolloids,* 145(March).

Hayaloğlu AA, 2024, Bitki bazlı protein kaynakları, Alternatif Gıda Kaynakları, Şahin K, Ankara, *Türkiye Bilimler Akademisi Yayınları,* 80-104.

Haznedaroğlu BZ, 2023. Chlorella vulgaris türü mikroalglerde B vitamini içeriklerinin uzun süreli pişirme koşulunda değişimi. *Akademik Gıda,* 21(2), 167-73.

Haznedaroğlu BZ, 2024. Tek hücre-mikroorganizma bazlı protein kaynakları. Alternatif Gıda Kaynakları. *Türkiye Bilimler Akademisi Yayınları,* Şahin K, Ankara, Doi: 10.53478/TUBA. 978-625-6110-00-7.

Hewage A, Olatunde OO, Nimalaratne C, Malalgoda M, Aluko RE, Bandara N, 2022. Novel extraction technologies for developing plant protein ingredients with improved functionality. *Trends in Food Science & Technology,* 129, 492-511.

Hopkins PD, Dacey A, 2008. Vegetarian meat: could technology save animals and satisfy meat eaters? *J Agric Environ Ethics,* 21(6), 579-96.

Hong TK, Shin DM, Choi J, Do JT, Han SG, 2021. Current issues and technical advances in cultured meat production: a review. *Food Sci Anim Resour.*, 41(3), 355- 72,

Holen E, Kausland A, Skjærven K, 2010. Embryonic stem cells isolated from Atlantic cod (Gadus morhua) and the developmental expression of a stage-specific transcription factor ac-Pou2. *Fish Physiol. Biochem.* 36 (4), 1029–39.

Hocquette JF, 2016. Is in vitro meat the solution for the future?. *Meat science,* 120, 167- 76.

Huang K, Liu Y, Zhang Y, Cao H, Luo DK, Yi C, Guan X, 2022. Formulation of plant-based yoghurt from soybean and quinoa and evaluation of physicochemical, rheological, sensory and functional properties. *Food Bioscience,* 49, 101831.

Huang JQ, Liao LM, Weinstein SJ, Sinha R, Graubard BI, Albanes D, 2020. Association between plant and animal protein intake and overall and cause-specific mortality. *JAMA Inter. Med.* 180, 1173–84.

Huis AV, Itterbeeck JV, Klunder H, Mertens E, Halloran A, Muir G, Vantomme P, 2013. *Edible insects: future prospects for food and feed security.*

Ismail I, Hwang YH, Joo ST, 2020. Meat analog as future food: a review. *J Anim Sci Technol.* 62(2), 111-20.

Imathiu S, 2020. Benefits and food safety concerns associated with consumption of edible insects. *NFS Journal,* 18, 1-11.

Izhar Ariff Mohd Kashim M, Abdul Haris AA, Abd Mutalib S, Anuar N, Shahimi S, 2023. Scientific and Islamic perspectives in relation to the Halal status of cultured meat. *Saudi J Biol Sci.,* 30(1), 103501.

Jantzen da Silva Lucas A, Menegon de Oliveira L, da Rocha M, Prentice C, 2020. Edible insects: An alternative of nutritional, functional and bioactive compounds. *Food Chem*, 311, 126022.

Jeske S, Zannini E, Arendt EK, Evaluation of physicochemical and glycaemic properties of commercial plant-based milk substitutes. *Plant Foods Hum. Nutr.* 2017, 72(1), 26–33.

Jonas-Levi A, Martinez JJI, 2017. The high level of protein content reported in insects for food and feed is overestimated. *Journal of Food Composition and Analysis,* 62, 184-188.

Jönsson E, 2016. Benevolent technotopias and hitherto unimaginable meats: Tracing the promises of in vitro meat. *Soc Stud Sci.* 46(5), 725-48.

Katz-Rosene R, Heffernan A, Arora A, 2023. Protein pluralism and food systems transition: A review of sustainable protein meta-narratives. *World Development,* 161, 106121.

Ketelings L, Benerink E, Havermans RC, Kremers SPJ, de Boer A, 2023. Fake meat or meat with benefits? How Dutch consumers perceive health and nutritional value of plantbased meat alternatives. *Appetite,* 188.

Khan MKI, Asif M, Razzaq ZU, Nazir A, Maan AA, 2022. Sustainable food industrial waste management through single cell protein production and characterization of protein enriched bread. *Food Bioscience,* 46, 101406.

Kim YM, Park YH, Lim JM, Jung H, Han JY, 2017. Technical note: Induction of pluripotent stem cell-like cells from chicken feather follicle cells. *J Anim Sci.* 95(8), 3479- 86.

Kim J, Kim H, Giovannucci EL, 2021. Plant-based diet quality and the risk of total and disease-specific mortality, A population-based prospective study. *Clin. Nutr.* 40, 5718–25.

Krintiras GA, Diaz JG, Van Der Goot AJ, Stankiewicz AI, Stefanidis GD, 2016. On the use of the Couette Cell technology for large scale production of textured soy-based meat replacers. *Journal of Food Engineering,* 169, 205-13.

Kotecka-Majchrzak K, Sumara A, Fornal E, Montowska M, 2020. Oilseed proteins–Properties and application as a food ingredient. *Trends in Food Science & Technology,* 106, 160-70.

Koyande AK, Chew KW, Rambabu K, Tao Y, Chu DT, Show PL, 2019. Microalgae: A potential alternative to health supplementation for humans. *Food Science and Human Wellness,* 8(1), 16-24.

Köksal E, 2024. Alternatif protein kaynaklı beslenmenin toplum sağlığı açısından yararları ve riskleri, Alternatif gıda kaynakları, Şahin K, 1. Baskı, Ankara, *Türkiye Bilimler Akademisi Yayınları,* 165-82.

Kumari M, Gupta S, Lakshmi AJ, Prakash J, 2004. Iron bioavailability in green leafy vegetables cooked in different utensils. *Food chemistry,* 86(2), 217-22.

Kumar M, Tomar M, Potkule J, Punia S, Dhakane-Lad J, Singh S, Dhumal S, Pradhan PC, Bhushan B, Anitha T, Alail O, Alhariri A, Amaroicz R, Kennedy JF, 2022. Functional characterization of plant-based protein to determine its quality for food applications. *Food Hydrocolloids,* 123, 106986.

Kurcz A, Błażejak S, Kot AM, Bzducha-Wróbel A, Kieliszek M, 2018. Application of industrial wastes for the production of microbial single-cell protein by fodder yeast candida utilis. *Waste and Biomass Valorization,* 9(1), 57-64.

Kumar M, Tomar M, Punia S, Dhakane-Lad J, Dhumal S, Changan S, Senapathy M, Berwal MK, Sampathrajan V, Sayed AAS, Chandran D, Pandiselvam R, Rais N, Mahato DK, Udikeri SS, Satankar V, Anitha T, Reetu, Radha, Singh S, Amarowicz R, Kennedy JF, 2022. Plant-based proteins and their multifaceted industrial applications. *LWT,* 154, 112620.

Kudełka W, Kowalska M, Popis M, 2021. Quality of soybean products in terms of essential amino acids composition. *Molecules,* 26(16), 5071.

Khatri M, O'Brien TD, Sharma JM, 2009. Isolation and differentiation of chicken mesenchymal stem cells from bone marrow. *Stem Cells Dev.* 18(10), 1485-92.

Kurbanoglu S, 2021. Single-cell protein as a potential source for functional food production. *Critical Reviews in Food Science and Nutrition.*

Lange KW, Nakamura Y, 2021. Edible insects as future food: chances and challenges. In *Journal of Future Foods,* 1(1), 38–46. *Beijing Academy of Food Sciences.*

Lamri M, Bhattacharya T, Boukid F, Chentir I, Dib AL, Das D, Djenane D, Gagaoua M, 2021. Nanotechnology as a processing and packaging tool to improve meat quality and safety. *Foods,* 10(11), 2633.

Lavial F, Pain B, 2010. Chicken embryonic stem cells as a non-mammalian embryonic stem cell model. *Dev. Growth Differ.*, 52(1), 101–14.

Lee DK, Kim M, Jeong J, Lee YS, Yoon JW, An MJ, Jung HY, Kim CH, Ahn Y, Choi KH, Jo C, Lee CK, 2023. Unlocking the potential of stem cells: Their crucial role in the production of cultivated meat. *Curr Res Food Sci.,* 7, 100551.

Leonard, U. M., Leydon, C. L., Arranz, E., & Kiely, M. E. (2024). Impact of consuming an environmentally protective diet on micronutrients: a systematic literature review. *The American Journal of Clinical Nutrition.*

Lin X, Wang F, Lu Y, Wang J, Chen J, Yu Y, Tao X, Xiao Y, Peng Y, 2023. A review on edible insects in China: Nutritional supply, environmental benefits, and potential applications. In *Current Research in Food Science* Elsevier B.V.

Liceaga AM, 2022.Chapter Four - Edible insects, a valuable protein source from ancient to modern times, Editor(s): Jianping Wu, Advances in Food and Nutrition Research, *Academic Press,* 101,129-52,

Liu J, Balehosur D, Murray B, Kelly JM, Sumer H, Verma PJ, 2012. Generation and characterization of reprogrammed sheep induced pluripotent stem cells. *Theriogenology.* 77(2), 338-46.

Li M, Zhang D, Hou Y, Jiao L, Zheng X, Wang WH, 2003. Isolation and culture of embryonic stem cells from porcine blastocysts. *Mol Reprod Dev.,* 65(4), 429-34.

Li YP, Ahmadi F, Kariman K, Lackner M, 2024. Recent advances and challenges in single cell protein (SCP) technologies for food and feed production. *NPJ Science of Food,* 8(1), 66.

Li H, Van Loo EJ, Trijp HCM van, Chen J, Bai J, 2023. Will cultured meat be served on Chinese tables? A study of consumer attitudes and intentions about cultured meat in China. *Meat Science,* 197, 109081.

López DN, Galante M, Robson M, Boeris V, Spelzini D, 2018. Amaranth, quinoa and chia protein isolates: Physicochemical and structural properties. *International journal of biological macromolecules,* 109, 152-9.

Loveday SM, 2023. Protein digestion and absorption: The influence of food processing. *Nutrition research reviews*, 36(2), 544-59.

Lou J, Ou M, Teng Y, Guo Y, Pan D, Yang H, Wu Z, 2023. Comparative metabolomics analysis of the fermented plant-based meat analogue of soybean proteins. *Food Bioscience,* 56, 103143.

Łopusiewicz Ł, 2023. Comparison of homemade and commercial plant-based drinks (almond, oat, soy) fermented with yogurt starter culture for fresh consumption. *Fermentation,* 10(1), 35.

Lonnie M, Laurie I, Myers M, Horgan G, Russell WR, Johnstone AM, 2020. Exploring health-promoting attributes of plant proteins as a functional ingredient for the food sector: a systematic review of human interventional studies. *Nutrients,* 12(8), 2291.

Lyahyai J, Mediano DR, Ranera B, Sanz A, Remacha AR, Bolea R, Zaragoza P, Rodellar C, Martn-Burriel I, 2012. Isolation and characterization of ovine mesenchymal stem cells derived from peripheral blood. *BMC Vet Res.*, 8, 169.

Ma X, Huang C, Zheng C, Wang W, Ying H, Liu C, 2024. Effect of oil extraction methods on walnut oil quality characteristics and the functional properties of walnut protein isolate. *Food Chemistry,* 438, 138052.

Matassa S, Boon N, Verstraete W, Pikaar I, 2016. Microbial protein: Sustainable food production. *Current Opinion in Biotechnology,* 38, 102-9.

Mannozzi C, Foligni R, Mozzon M, Aquilanti L, Cesaro C, Isidoro N, Osimani A, 2023. Nonthermal technologies affecting techno-functional properties of edible insect-derived proteins, lipids, and chitin: a literature review. *Innovative Food Science & Emerging Technologies,* 103453.

Mateti T, Laha A, Shenoy P, 2022. Artificial meat industry: production methodology, challenges, and future. *JOM,* 74(9), 3428–44.

Masiá C, Jensen PE, Petersen IL, Buldo P, 2022. Design of a functional pea protein matrix for fermented plant-based cheese. *Foods,* 11(2), 178.

Macintyre B, 2007. Test-tube meat science's next leap. *Weekend Australian*, 20(29), 16.

Mæhre HK, Dalheim L, Edvinsen GK, Elvevoll EO, Jensen IJ, 2018. Protein determination—Method matters. *Foods,* 7(1), 5.

Maehle N, Skjeret F, 2022. Microalgae-based food: Purchase intentions and willingness to pay. *Future Foods,* 6, 100205.

Mariutti LRB, Rebelo KS, Bisconsin-Junior A, de Morais JS, Magnani M, Maldonade IR, Madeira NR, Tiengo A, Maróstica MR Jr, Cazarin CBB, 2021. The use of alternative food sources to improve health and guarantee access and food intake. *Food research international* (Ottawa, Ont.), 149, 110709.

Mellor C, Embling R, Neilson L, Randall T, Wakeham C, Lee MD, Wilkinson LL, 2022. Consumer Knowledge and Acceptance of "Algae" as a Protein Alternative: A UKBased Qualitative Study. *Foods,* 11(12), 1703.

Mertens E, Van't Veer P, Jager L, 2022. Environmental and health benefits of plant-based diets: A review. *Trends in Food Science & Technology*, 120, 234–49.

Melzener L, Verzijden KE, Buijs AJ, Post MJ, Flack JE, 2021. Cultured beef: from small biopsy to substantial quantity. *J Sci Food Agric.*, 101(1), 7-14.

Mishyna M, Keppler JK, Chen J, 2021. Techno-functional properties of edible insect proteins and effects of processing. *Current Opinion in Colloid & Interface Science,* 56, 101508.

Muys M, Cámara SJG, Derese S, Spiller M, Verliefde A, Vlaeminck SE, 2023. Dissolution rate and growth performance reveal struvite as a sustainable nutrient source to produce a diverse set of microbial protein. *Science of the Total Environment,* 866, 161172.

Munialo CD, Vriesekoop F, 2023. Plant-based foods as meat and fat substitutes. *Food Sci Nutr.* 11(9), 4898-911.

Ng ET, Singh S, Yap WS, Tay SH, Choudhury D, 2023. Cultured meat - a patentometric analysis. *Crit Rev Food Sci Nutr.*, 63(16), 2738-48.

Ng S, Kurisawa M, 2021. Integrating biomaterials and food biopolymers for cultured meat production. *Acta Biomater.* 124, 108-29.

Nissen L, Samaei SP, Babini E, Gianotti A, 2020. Gluten free sourdough bread enriched with cricket flour for protein fortification: Antioxidant improvement and volatilome characterization. *Food Chem,* 333, 127410.

Nongonierma AB, FitzGerald RJ, 2017. Unlocking the biological potential of proteins from edible insects through enzymatic hydrolysis: A review, *Innovative Food Science & Emerging Technologies,* Volume 43, 239-52.

Oh, Y. N., & Kim, H. Y. (2025). Exploring sustainable future protein sources. *Food Science of Animal Resources,* 45(1), 81.

Oibiokpa FI, Akanya HO, Jigam AA, Saidu AN, Egwim EC, 2018. Protein quality of four indigenous edible insect species in Nigeria. *Food Science and Human Wellness,* 7(2), 175–83.

Ojha S, El-Din Bekhit A, Grune T, Schlüter OK, 2021. Bioavailability of nutrients from edible insects, *Current Opinion in Food Science,* 41, 240-8.

Ong S, Choudhury D, Naing MW, 2020. Cell-based meat: current ambiguities with nomenclature. *Trends Food Sci Technol.* 102:223–31.

Oonincx DG, de Boer IJ, Smetana S, 2015. An exploration on greenhouse gas and ammonia production by insect species suitable for animal or human consumption. *PLOS ONE.* 10: e0144601.

Ordoñez-Araque R, Egas-Montenegro E, 2021. Edible insects: A food alternative for the sustainable development of the planet. *Int J Gastron Food Sci,* 23, 100304.

O'Riordan K, Fotopoulou A, Stephens N, 2017. The first bite: Imaginaries, promotional publics and the laboratory grown burger. *Public Underst Sci.*, 26(2), 148-63.

Oyeyinka AT, Odukoya JO, Adebayo YS, 2019. Nutritional composition and consumer acceptability of cheese analog from soy and cashew nut milk. *Journal of Food Processing and Preservation,* 43(12), 1-6.

Papastavropoulou, K., Xiao, J., & Proestos, C. (2023). Edible insects: Tendency or necessity (a review). *eFood,* 4(1), e58.

Papia F, Bellia C, Uasuf CG, 2021. Tropomyosin: A panallergen that causes a worldwide allergic problem. *In Allergy & Asthma Proceedings,* 42(5).

Paraskevopoulou A, Athanasiadis I, Kanellaki M, Bekatorou A, Blekas G, Kiosseoglou V, 2003. Functional properties of single cell protein produced by kefir microflora. *Food Res. Int.* 36, 431–8.

Payne CL, Scarborough P, Rayner M, Nonaka K, Burgess G, 2016. Health benefits and risks of cricket consumption. *Journal of Insects as Food and Feed* 2: 93-105.

Päivärinta E, Itkonen ST, Pellinen T, Lehtovirta M, Erkkola M, Pajari AM, 2020. Replacing animal-based proteins with plant-based proteins changes the composition of a whole Nordic diet-a randomised clinical trial in healthy Finnish adults. *Nutrients.* 12(4).

Pillaca-Pullo OS, Lopes AM, Rodriguez-Portilla LM, Estela-Escalante W, 2023. Optimizing medium composition with wastewater from Coffea arabica processing to produce single-cell protein using Candida sorboxylosa. *Journal of Chemical Technology & Biotechnology,* 98(1), 106-16.

Pires JF, Ferreira GMR, Reis KC, Schwan RF, Silva CF, 2016. Mixed yeasts inocula for simultaneous production of SCP and treatment of vinasse to reduce soil and fresh water pollution. *Journal of Environmental Management,* 182, 455- 63.

Poutanen KS, Kårlund AO, Gómez-Gallego C, Johansson DP, Scheers NM, Marklinder IM, Eriksen AK, Silventoinen PC, Nordlund E, Sozer N, Hanhineva K, Kolehmainen M, Landberg R, 2022. Grains–a major source of sustainable protein for health. *Nutrition reviews,* 80(6), 1648-63.

Powell RL, Dodson MV, Cloud JG, 1989. Cultivation and differentiation of satellite cells from skeletal muscle of the rainbow troutSalmo gairdneri. *J. Exp. Zool.,* 250, 333–8.

Post MJ, 2012. Cultured meat from stem cells: challenges and prospects. *Meat Sci.,* 92(3), 297-301.

Post MJ, 2014. An alternative animal protein source: cultured beef. *Ann N Y Acad Sci.*, 1328, 29-33.

Poore J, Nemecek T, 2018. Reducing food's environmental impacts through producers and consumers. *Science,* 360(6392), 987–92.

Pratt R, Johnson E, 1965. Production of thiamine, riboflavin, folic acid, and biotin by Chlorella vulgaris and Chlorella pyrenoidosa. *Journal of Pharmaceutical Sciences,* 54(6), 871-4.

Queiroz LS, Silva NFN, Jessen F, Mohammadifar MA, Stephani R, de Carvalho AF, Perrone ÍT, Casanova F, 2023. Edible insect as an alternative protein source: A review on the chemistry and functionalities of proteins under different processing methods. *Heliyon,* 9(4).

Raziq A, Lateef M, Ullah A, Ullah H, Khan W, 2020. Single cell protein (SCP) production and potential substrates: A comprehensive review. *Pure and applied biology,* 9(3), 1743-54.

Ravi HK, Degrou A, Costil J, Trespeuch C, Chemat F, Vian MA, 2020. Larvae mediated valorization of industrial, agriculture and food wastes: biorefinery concept through bioconversion, processes, procedures, and products. *Process,* 8(7), 857.

Ravindra P, 2000. Value-added food: Single cell protein. *Biotechnology advances,* 18(6), 459-79.

Rebollo-Hernanz M, Cañas S, Martin-Cabrejas M, Aguilera Y, 2021. *In Silico* Identification of Bioactive Proteins and Peptides in Insects for Human Consumption. First Canadian Peptide and Protein Community Virtual Symposium.

Reiss J, Robertson S, Suzuki M, 2021. Cell sources for cultivated meat: Applications and considerations throughout the production workflow. *Int J Mol Sci.,* 22(14), 7513.

Reyes-Jurado F, Soto-Reyes N, Dávila-Rodríguez M, Lorenzo-Leal AC, Jiménez-Munguía MT, Mani-López E, López-Malo A, 2023. Plant-based milk alternatives: Types, processes, benefits, and characteristics. *Food Reviews International,* 39(4), 2320-51.

Rivas-Navia DM, Dueñas-Rivadeneira AA, Dueñas-Rivadeneira JP, Aransiola SA, Maddela NR, Prasad R, 2023. Bioactive compounds of insects for food use: Potentialities and risks. *Journal of Agriculture and Food Research,* 14.

Ritala A, Häkkinen ST, Toivari M, Wiebe MG, 2017. Single cell protein—stateof-the-art, industrial landscape and patents 2001–2016. *Frontiers in microbiology,* 8, 2009.

Romero A, Beaumal V, David-Briand E, Cordobes F, Guerrero A, Anton M, 2011. Interfacial and oil/water emulsions characterization of potato protein isolates. *Journal of Agricultural and Food Chemistry,* 59(17), 9466-74.

Romero A, Beaumal V, David-Briand E, Cordobes F, Guerrero A, Anton M, 2011. Interfacial and oil/water emulsions characterization of potato protein isolates. *Journal of Agricultural and Food Chemistry,* 59(17), 9466-74.

Roos N, van Huis A, 2017. Consuming insects: are there health benefits? *Journal of Insects as Food and Feed,* 3(4), 225–9.

Rojas Conzuelo Z, Robyr R, Kopf-Bolanz KA, 2022. Optimization of protein quality of plant-based foods through digitalized product development. *Frontiers in Nutrition,* 9, 902565.

Rumpold BA, Schlüter OK, 2013. Nutritional composition and safety aspects of edible insects. *Mol Nutr Food Res,* 57(5), 802-23.

Rumpold BA, Schlüter OK, 2013. Insect and the future of food production. *Food Research International,* 54(1), 303-9.

Rubio NR, Xiang N, Kaplan DL, 2020. Plant-based and cell-based approaches to meat production. *Nat Commun.,* 11(1), 6276.

Saejung C, Salasook P, 2020. Recycling of sugar industry wastewater for single-cell protein production with supplemental carotenoids. *Environmental Technology,* 41(1), 59-70.

Saeed M, Shoaib A, Kandimalla R, Javed S, Almatroudi A, Gupta R, Aqil F, 2022. Microbe-based therapies for colorectal cancer: Advantages and limitations. *Semin. Cancer Biol.* 86, 652–65.

Sá AGA, Moreno YMF, Carciofi BAM, 2020. Plant proteins as high-quality nutritional source for human diet. *Trends in Food Science & Technology,* 97, 170-84.

Seyfabadi J, Ramezanpour Z, Amini Khoeyi Z, 2011. Protein, fatty acid, and pigment content of Chlorella vulgaris under different light regimes. *Journal of applied phycology,* 23, 721-6.

Sethi S, Tyagi SK, Anurag RK, 2016. Plant-based milk alternatives an emerging segment of functional beverages: a review. *Journal of food science and technology,* 53, 3408-23.

Shi T, Cao J, Cao J, Zhu F, Cao F, Su E, 2023. Almond (Amygdalus communis L.) kernel protein: A review on the extraction, functional properties and nutritional value. *Food Research International,* 167, 112721.

Shi W, Hou T, Guo DJ, He H, 2019. Evaluation of hypolipidemic peptide (Val-Phe-Val-Arg-Asn) virtual screened from chickpea peptides by pharmacophore model in high-fat diet-induced obese rat. *J. Funct. Foods.* 54, 136–45.

Sharma R, Singh GP, Sharma V, 2012. Effects of culture conditions on growth and biochemical profile of Chlorella vulgaris. *Journal of Plant Pathology and Microbiology,* 3(5), 1-6.

Silva ARA, Silva MMN, Ribeiro BD, 2020. Review: Health Issues and Technological Aspects of Plant-based Alternative Milk. *Food Res. Int.* 131, 108972.

Siddiqui SA, Osei-Owusu J, Yunusa BM, Rahayu T, Fernando I, Shah MA, Centoducati G, 2023. Prospects of edible insects as sustainable protein for food and feed–a review. *Journal of Insects as Food and Feed,* 1, 1-27.

Singh R, Langyan S, Sangwan S, Rohtagi B, Khandelwal A, Shrivastava M, 2022. Protein for human consumption from oilseed cakes: a review. *Frontiers in Sustainable Food Systems,* 6, 856401.

Srividya AR, Vishnuvarthan VJ, Murugappan M, Dahake PG, 2013. Single cell protein- a review. *Int. J. Pharm. Res. Sch.* 2, 472–85.

Stull VJ, Finer E, Bergmans RS, Febvre HP, Longhurst C, Manter DK, Jonathan AP, Weir, T. L. 2018. Impact of edible cricket consumption on gut microbiota in healthy adults, a double-blind, randomized crossover trial. *Sci Rep,* 8(1), 10762.

Sun J, Peng Z, Yan L, Fuh JY, Hong GS, 2015. 3D food printing—An innovative way of mass customization in food fabrication. *International Journal of Bioprinting,* 1(1), 27-38.

Suman G, Nupur M, Anuradha S, Pradeep B, 2015. Single cell protein production: A review. *Int. J. Curr. Microbiol. Appl. Sci.* 4, 251–62.

Sürek E, Uzun P, 2020. *Geleceğin alternatif protein kaynağı:* Yapay Et. Akademik Gıda, 18(2), 209-16.

Tan ST, Tan SS, Tan CX, 2023. Soy protein, bioactive peptides, and isoflavones: a review of their safety and health benefits. *PharmaNutrition,* 100352.

Tangyu M, Muller J, Bolten CJ, Wittmann C, 2019. Fermentation of plant-based milk alternatives for improved flavour and nutritional value. *Applied microbiology and biotechnology*, 103, 9263-9275.

Tang Q, Roos YH, Miao S, 2024. Comparative studies of structural and thermal gelation behaviours of soy, lentil and whey protein: A pH-dependency evaluation. *Food Hydrocolloids,* 146, 109240.

Tan Y, Wannasin D, McClements DJ, 2023. Utilization of potato protein fractions to form oil-in-water nanoemulsions: Impact of pH, salt, and heat on their stability. *Food Hydrocolloids,* 137, 108356.

Talluri TR, Kumar D, Glage S, Garrels W, Ivics Z, Debowski K, Behr R, Niemann H, Kues WA, 2015. Derivation and characterization of bovine induced pluripotent stem cells by transposon-mediated reprogramming. *Cell Reprogram.,* 17(2), 131-40.

Thavamani A, Sferra TJ, Sankararaman S, 2020. Meet the meat alternatives: The value of alternative protein sources. *Current nutrition reports*, 9, 346-55.

Tojan S, Kaur L, Singh J, 2024. Hybrid Paneer: Influence of mung bean protein isolate (Vigna radiata L.) on the texture, microstructure, and in vitro gastro-small intestinal digestion. *Food Chemistry,* 434, 137434.

Toor A, 2013. *World's first lab-grown burger unveiled at public tasting.* https://www. theverge.com/ 2013/8/5/4589744/cultured-beef-burger-public-tasting-markpost-sergey-brin.

Tuomisto, H. L., & Teixeira de Mattos, M. J. (2011). Environmental impacts of cultured meat production. *Environmental science & technology*, 45(14), 6117-6123.

Tuso PJ, Ismail MH, Ha BP, Bartolotto C, 2013. Nutritional update for physicians: plant-based diets. *Perm J.*17(2):61–6.

United Nations. Report of the Secretary-General on the Work of the Organization (A/74/1, 74th Session); United Nations: New York, NY, USA, 2019; Available online: https://www.un.org/annualreport/2019/ files/2019/09/Annual-report-SG-2019-EN Complete-Web.pdf (October 24, 2024).

UN, 2023. Available time, November 01, 2024. Available online: https://www.un.org/en/global-issues/population.

Ursu AV, Marcati A, Sayd T, Sante-Lhoutellier V, Djelveh G, Michaud P, 2014. Extraction, fractionation and functional properties of proteins from the microalgae Chlorella vulgaris. *Bioresource technology,* 157, 134-9.

Valin H, Sands RD, Van der Mensbrugghe D, Nelson GC, Ahammad H, Blanc E, Bodirsky B, Fuimori S, Hasegawa T, Havlik P, Heyhoe E, Kyle P, Mason-D'Croz D, Paltsev S, Rolinski S, Tabeau A, van Mejil H, von Lampe M, Willenbockel D, 2014. The future of food demand: understanding differences in global economic models. *Agricultural economics,* 45(1), 51-67.

Varley J, Birch J, 1999. Reactor design for large scale suspension animal cell culture. *Cytotechnology,* 29(3), 177-205.

van Huis A, 2020. Nutrition and health of edible insects. *Current Opinion in Clinical Nutrition and Metabolic Care,* 23(3) 228–31.

Van der Stricht H, Profeta A, Hung Y, & Verbeke W, 2023. Consumers' willingnessto-buy pasta with microalgae proteins – Which label can promote sales? *Food Quality and Preference,* 110, 104948.

Vanga SK, Vijaya R. 2018. How Well Do Plant Based Alternatives Fare Nutritionally Compared to Cow's Milk? *Journal of Food Science and Technology,* 55(1), 10–20.

van Huis A, Van Itterbeeck J, Klunder H, Mertens E, Halloran A, Muir G, Vantomme P, 2013. Edible insects: future prospects for food and feed security.171.

Van Huis A, 2013. Potential of insects as food and feed in assuring food security. *Annual Review of Entomology,* 58, 563–83.

Van Eelen WF, Van Kooten WJ, Westerhof W, 1999. Industrial scale production of meat from in vitro cell cultures. World Intellectual Property Organization.

van der Weele C, Feindt P, van der Goot AJ, van Mierlo B, van Boekel M. 2019 Meat alternatives: an integrative comparison. *Trends Food Sci Technol.* 88. 505–12.

Vermeulen SJ, Campbell BM, Ingram JS, 2012. Climate change and food systems. *Annu. Rev. Environ. Resour.* 2012, 37, 195–222.

Venkidasamy B, Selvaraj D, Nile AS, Ramalingam S, Kai G, Nile SH, 2019. Indian pulses: A review on nutritional, functional and biochemical properties with future perspectives. *Trends in Food Science & Technology*, 88, 228-42.

Vercruysse L, Smagghe G, Herregods G, Van Camp J, 2005. ACE inhibitory activity in enzymatic hydrolysates of insect protein. *J. Agric. Food Chem.* 53, 5207–11.

Vercruysse L, Smagghe G, Beckers T, Van Camp J, 2009. Antioxidative and ACE inhibitory activities in enzymatic hydrolysates of the cotton leafworm, Spodoptera littoralis. *Food Chemistry,* 114(1), 38-43.

Vercruysse L, Smagghe G, Herregods G, Van Camp J, 2005. ACE inhibitory activity in enzymatic hydrolysates of insect protein. *J. Agric. Food Chem.* 53, 5207–11.

Vercruysse L, Smagghe G, Matsui T, Van Camp J, 2008. Purification and identification of an angiotensin I converting enzyme (ACE) inhibitory peptide from the gastrointestinal hydrolysate of the cotton leafworm, Spodoptera littoralis. *Process Biochemistry,* 43(8), 900-4.

Vercruysse L, Smagghe G, Beckers T, Van Camp J, 2009. Antioxidative and ACE inhibitory activities in enzymatic hydrolysates of the cotton leafworm, Spodoptera littoralis. *Food Chemistry,* 114(1), 38-43.

Vercruysse L, Van Camp J, Morel N, Rougé P, Herregods G, Smagghe G, 2010. Ala-Val-Phe and Val-Phe: ACE inhibitory peptides derived from insect protein with antihypertensive activity in spontaneously hypertensive rats. *Peptides,* 31(3), 482-8.

Vercruysse L, Van Camp J, Morel N, Rougé P, Herregods G, Smagghe G, 2010. Ala-Val-Phe and Val-Phe: ACE inhibitory peptides derived from insect protein with antihypertensive activity in spontaneously hypertensive rats. *Peptides,* 31(3), 482-8.

Vilarino M, Alba Soto D, Soledad Bogliotti Y, Yu L, Zhang Y, Wang C, Paulson E, Zhong C, Jin M, Carlos Izpisua Belmonte J, Wu J, Juan Ross P, 2020. Derivation of sheep embryonic stem cells under optimized conditions. *Reproduction,* 160(5), 761-72.

Vichi S, Lavorini P, Funari E, Scardala S, Testai E, 2012. Contamination by Microcystis and microcystins of blue–green algae food supplements (BGAS) on the italian market and possible risk for the exposed population. *Food and Chemical Toxicology,* 50(12), 4493-9.

Voltolina D, Gómez-Villa H, Correa G, 2005. Nitrogen removal and recycling by Scenedesmus obliquus in semicontinuous cultures using artificial wastewater and a simulated light and temperature cycle. *Bioresour. Technol.* 96, 359–62.

Wang M, Zhao R. 2023. A review on nutritional advantages of edible mushrooms and its industrialization development situation in protein meat analogues. *Journal of Future Foods,* 3(1). 1–7.

Wang X, Kong X, Zhang C, Hua Y, Chen Y, Li X, 2023. Comparison of physicochemical properties and volatile flavor compounds of plant-based yoghurt and dairy yoghurt. *Food Research International,* 164, 112375.

Wang G, Plaster MT, Bai Y, Liu C, 2023. Consumers' experiences and preferences for plant-based meat food: Evidence from a choice experiment in four cities of China. *Journal of Integrative Agriculture,* 22(1), 306–19.

Wang Y, Jian C, 2023. Novel plant-based meat alternatives: Implications and opportunities for consumer nutrition and health. *Adv Food Nutr Res.,* 106, 241-74.

Weihe WH, 1985. Use and misuse of an imprecise concept: alternative methods in animal experiments. *Lab Anim.,* 19(1), 19-26.

Yap SM, Lan JCW, Kee PE, Ng HS, Yim HS, 2022. Enhancement of protein production using synthetic brewery wastewater by Haematococcus pluvialis. *Journal of Biotechnology,* 350, 1-10.

Yablonka-Reuveni Z, Quinn LS, Nameroff M, 1987. Isolation and clonal analysis of satellite cells from chicken pectoralis muscle. *Dev Biol.,* 119(1), 252-9.

Yang S, Li Q, Zeng Q, Zhang J, Yu Z, Liu Z, 2012. Conversion of solid organic wastes into oil via Boettcherisca peregrine (Diptera: Sarcophagidae) larvae and optimization of parameters for biodiesel production. *PLOS ONE.*7: e45940.

Yao T, Asayama Y, 2017. Animal-cell culture media: History, characteristics, and current issues. *Reprod Med Biol.,* 16(2), 99-117.

Yetim H, Tekiner IH, 2020. Alternatif protein kaynaklarından yapay et üretimi kavramına eleştirel bir bakış. *Helal ve Etik Araşt. Derg.,* 2(2), 85-100.

Zhang X, Shi J, Fu Y, Zhang T, Jiang L, Sui X, 2023. Structural, nutritional, and functional properties of amaranth protein and its application in the food industry: A review. *Sustainable Food Proteins,* 1(1), 45-55.

Zhang X, Zhang T, Zhao Y, Jiang L, Sui X, 2024). Structural, extraction and safety aspects of novel alternative proteins from different sources. *Food Chemistry,* 436, 137712.

Zhang T, Dou W, Zhang X, Zhao Y, Zhang Y, Jiang L, Sui X, 2021. The development history and recent updates on soy protein-based meat alternatives. *Trends in Food Science & Technology,* 109, 702-10.

Zhuang, Z., Wan, G., Lu, X., Xie, L., Yu, T., & Tang, H. (2024). Metabolic engineering for single-cell protein production from renewable feedstocks and its applications. *Advanced Biotechnology*, 2(4), 35.

Zielińska E, Karaś M, Jakubczyk A, 2017. Antioxidant activity of predigested protein obtained from a range of farmed edible insects. I*nternational Journal of Food Science & Technology,* 52(2), 306-12.

Zielińska E, Pankiewicz U, 2020. Nutritional, physiochemical, and antioxidative characteristics of shortcake biscuits enriched with tenebrio molitor flour. *Molecules,* 25(23), 5629.

Chapter 9

Nutrition and Psychological Health: The Gut-Brain Axis

Hakan Toguc*

Department of Nutrition and Dietetics, Inonu University, Malatya, Turkiye

Abstract

This study addresses the bidirectional relationship between gut microbiota and mental health from the perspective of the 'gut brain axis' and reveals the determining role of nutrition in psychological health with a multidimensional approach. Gut microbiota is defined as a microbial ecosystem that plays an active role in the holistic functioning of not only the digestive system but also the immune, endocrine and nervous systems. While the microbiota composition is shaped by environmental and genetic factors from birth, nutrition is one of the most effective determinants in the modulation of this composition. Disturbances in microbiota balance (dysbiosis) are associated with many neuropsychiatric disorders, especially depression and anxiety. In this context, the microbiota - gut - brain axis stands out as a key communication network in the sustainability of brain health and psychological functions. Diet can have direct and indirect effects on the central nervous system by influencing the microbiota's neurotransmitter production, intestinal permeability, inflammation levels and immune responses. Fiber-rich diets, fermented foods, probiotic and prebiotic consumption support brain health by increasing the production of short-chain fatty acids (especially butyrate). At the same time, microbial synthesis of neuroactive compounds such as serotonin, dopamine and Gamma aminobutyric acid (GABA) plays a decisive role in mood and cognitive functions. Micronutrients such as omega-3 fatty acids, polyphenols, vitamin D and B group vitamins also show neuroprotective effects by reducing neuroinflammation. In this study, the effects of various dietary models (Western-style diet, traditional diets, low FODMAP diet (Fermentable Oligosaccharides, Disaccharides, Monosaccharides and Polyols)), especially the Mediterranean diet, on microbiota composition and mental health were compared. The Mediterranean diet has a strong therapeutic potential in reducing the risk of depression and anxiety thanks to its anti-inflammatory effects and microbiota-friendly content. In contrast, Western diets, which are high in sugar, saturated fat and processed foods, may have negative effects on mental health. Furthermore, the dependence of serotonin production on microbial tryptophan metabolism further strengthens the role of microbiota health in psychiatric disorders. Specialized nutritional approaches such as low FODMAP diets have also been reported to alleviate symptoms of anxiety and depression, especially in conditions such as irritable bowel syndrome. In conclusion, this study reveals that the interaction between the gut microbiota and the brain can be regulated through nutrition and

In: Let Food Be Your Medicine, and Medicine Be Your Food
Editors: Gulcin Yildiz and Arzu Imece
ISBN: 979-8-90134-204-6

that this regulation has therapeutic potential for psychological well-being. The role of the microbiota in neurotransmitter synthesis, its interaction with the immune system and its regulatory effects on the hypothalamic - pituitary - adrenal (HPA) axis allow the development of new therapeutic approaches. These findings necessitate a re-evaluation of the effects of nutrition not only on physical but also on mental health and indicate that nutrition-based psychiatric interventions may find a wider place in clinical practice in the future.

Keywords: gut microbiota, gut-brain axis, nutritional therapies, omega-3 fatty acids, prebiotics, probiotics

Introduction

Gut microbiota refers to the trillions of microorganisms living in the human gastrointestinal tract. These microorganisms are an extremely complex community consisting of bacteria, viruses, fungi and archaea and are dense enough to exceed the number of human cells (Wójcik et al., 2025). The microbiota begins to develop from birth and is shaped depending on the individual's age, diet, environment, genetic makeup and drug use. The gut microbiota contributes not only to the digestive system, but also to metabolic processes, the immune system, hormonal balances and nervous system functions. When the microbial balance is disturbed, a condition called 'dysbiosis' occurs, which has been associated with the development of various chronic diseases, autoimmune disorders and, as recently demonstrated, psychiatric disorders (Emanowicz et al., 2024). Microbiota is a dynamic structure that is not limited to the digestive system; it regulates many physiological processes such as metabolic balance, immune system responses and neurological functions (Wójcik et al., 2025). The microbiota structure, which begins to take shape at birth, is determined by many factors such as mode of delivery (vaginal delivery or caesarean section), diet (breast milk or formula), environmental exposures and the genetic structure of the individual (Pushpanathan et al., 2019). These early environmental and genetic interactions lay the foundation for the microbial diversity that the individual will have in later life and ensure the continuity of the microbiota composition. A healthy microbiota plays critical roles in maintaining the homeostasis of the host organism, providing defense against external pathogens, training the immune system and maintaining energy balance (Popkes et al., 2020). Beyond its role in digestion, the gut microbiota is also involved in the synthesis of certain vitamins (e.g., vitamins B and K) and the production of short-chain fatty acids (e.g., butyrate). These biological products are involved in maintaining the integrity of the intestinal epithelium and regulating inflammatory responses. The microbiota also plays an active role in the maintenance of immune tolerance and the generation of an appropriate immune response to antigens entering the intestinal lumen. However, disruption of microbial balance is recognized as an important risk factor in the development of various metabolic and inflammatory diseases. Strong associations have been established between changes in the gut microbiota and obesity, type 2 diabetes, inflammatory bowel diseases (Crohn's disease and ulcerative colitis), allergic diseases and even some neuropsychiatric disorders (e.g., depression and anxiety) (Nuriel-Ohayon et al., 2016). This situation reveals that microbiota has an effect not only at the local level but also at the systemic level. In particular, recent studies have highlighted the concept of 'microbiota-gut-brain axis', which defines the bidirectional communication network between the gut microbiota and the central nervous system. Through this axis, it is suggested that gut flora may affect neurotransmitter production and thus play a decisive role in mood, stress response and cognitive functions (Neuman and Koren, 2016). In this context, modulation of microbiota composition is considered a potential target not only in the treatment of gastrointestinal or metabolic diseases, but also in the prevention or amelioration of psychiatric disorders.

Numerous studies in modern neuroscience and gastroenterology have revealed that mental health is closely related not only to biochemical processes in the central nervous system, but also to the functioning of the gastrointestinal tract. In this context, the physiological communication network called the 'gut-brain axis' (GBA) has become a central concept in understanding the interplay between mental health and gut health (Wójcik et al., 2025). The GIT refers to a complex communication system that provides bidirectional signaling between the central nervous system (CNS) and the gastrointestinal (GI) tract. This axis operates mainly through three main structures: the enteric nervous system (the neural network of the gut), the vagus nerve (the main neural pathway between the brain and the gut), and various metabolites synthesized by the microbiota (Bonaz et al., 2018). Signals transmitted through these components can directly influence both the functions of the gastrointestinal tract and neurological responses in the brain.

The enteric nervous system is also referred to as the 'second brain' because it has the capacity to coordinate reflexes independent of the central nervous system and plays a mediating role in communication between the gut and the brain. The gut microbiota is considered a critical component on this axis. Microorganisms in the gut not only regulate digestion but also play a direct role in neurotransmitter synthesis (e.g., serotonin, gamma-aminobutyric acid (GABA), dopamine), production of gut hormones (e.g., ghrelin and peptide YY) and regulation of systemic immune responses (Romijn et al., 2008). This bidirectional mechanism of communication and regulation is key not only in maintaining normal psychological functioning but also in understanding the mechanisms of psychiatric and neurological disorders. Recent findings reveal that changes in the composition of gut microbiota (dysbiosis) are linked to various neuropsychiatric conditions such as depression, anxiety, autism spectrum disorder, Parkinson's disease and Alzheimer's disease (Nafady et al., 2022). In particular, depression and anxiety disorders have been associated with the reduction of certain species in the microbiota or the predominance of harmful species. These imbalances can lead to increased inflammatory responses and suppression of neuroplasticity-related molecules such as brain-derived neurotrophic factor (BDNF) (Appleton, 2018). This process both supports the development of mood disorders and contributes to the emergence of treatment-resistant clinical pictures.

The complex and dynamic interaction between nutrition and microbiota is a multidisciplinary research area that profoundly affects the physiological and psychological dimensions of human health. The gut microbiota is an ecosystem that is greatly influenced by the host's diet and also undertakes critical functions such as metabolism of nutrients, modulation of the immune system and regulation of neurochemical processes. Recent studies have revealed that the composition and metabolic activities of the microbiota play a decisive role, especially on brain health and cognitive functions. In this context, the concept of 'gut-brain axis' has become an important paradigm bridging neuroscience, gastroenterology and nutrition sciences. The effect of diet on microbiota composition is closely related to the consumption of fermented foods, fibers, polyphenols and other bioactive components. For example, short-chain fatty acids (SCFAs) produced by fermentation of dietary fibers support neuronal plasticity by reducing neuroinflammation and thus contribute to the maintenance of cognitive functions (Bienenstock et al., 2015; Pan et al., 2025). Similarly, regular consumption of probiotics and prebiotics modulates stress response by increasing microbial diversity and may alleviate symptoms of mood disorders such as depression (Madireddy and Madireddy, 2019). These findings highlight the therapeutic potential of traditional dietary patterns on the microbiota and suggest that certain strains of microorganisms, particularly so-called psychobiotics, show promise in the management of neuropsychiatric disorders (Clapp et al., 2017).

The clinical importance of microbiota-brain interaction is not limited to a theoretical framework, but is now recognized as an important factor in the pathogenesis of various pathologies such as depression, anxiety, autism spectrum disorders and neurodegenerative diseases. For example, meta-analyses examining the relationship between microbial dysbiosis and psychiatric disorders have revealed that loss of diversity in gut flora is significantly correlated with these disorders (Li et al., 2020). Furthermore, it has been emphasized that the gut microbiota affects central nervous system functions through mechanisms such as neurotransmitter synthesis, immunoregulation and intestinal barrier integrity, thus playing a critical role in the development of neuropsychiatric disorders (Góralczyk-Bińkowska et al., 2022). Recent studies indicate that microbiota modulation can be used as an adjuvant approach in psychiatric treatments. In particular, regular intake of certain bacterial species (e.g. Lactobacillus and Bifidobacterium) has been reported to have potential effects in reducing symptoms of depression and anxiety (Cheng et al., 2019). In addition, it is suggested that microbiota reconstruction through dietary interventions may offer new therapeutic strategies in slowing neurodegenerative processes and promoting psychological well-being (Czarnik et al., 2024).

The Connection Between Nutrition and Mental Health

Nutrition, as well as genetic predispositions, environmental stimuli and psychosocial dynamics, has a decisive influence on the sustainability of mental health. While traditional psychiatry approaches generally explain mental disorders with neurochemical imbalances, genetic factors or environmental stressors, it is seen that this paradigm has expanded in recent years and the role of diet on brain functions and mood has gained more importance. In line with these developments, an interdisciplinary field called 'nutritional psychiatry' has emerged and has become a comprehensive scientific field that examines the relationships between mental health and dietary components at the biochemical, neurological and behavioral levels (Julius, 2025). Research in this field also examines the effects of micronutrients such as omega-3 fatty acids, B vitamins and magnesium on mood (Julius, 2025). Studies have found that the gut microbiota, along with nutrients, interact with the central nervous system through the immune system, neural and hormonal pathways, and this axis has focused on its role in the pathogenesis of depression, anxiety and other psychiatric disorders (Du et al., 2020). Thus, with nutritional psychiatry, nutrition, mental health and microbiota are examined together, nutritional effects on mental health are understood in more depth, and the integration of this field into treatment approaches offers new opportunities in the prevention and management of psychiatric disorders (Julius, 2025; Du et al., 2020).

The relationship between nutrition and mental state has a bidirectional structure. While the psychological state of individuals can have an impact on what they eat, how often and how much they consume, the quality of the food consumed can directly affect brain chemistry, neurotransmitter balance and thus mood. Polivy et al. (2005) emphasized that this bidirectional relationship is a critical mechanism that needs to be understood in order to promote the adoption of healthy eating habits. For example, individuals who are depressed or stressed are more likely to turn to unhealthy foods, binge eating behavior and irregular eating habits. On the contrary, following a balanced and adequate diet creates a positive mood, reduces stress levels and supports general psychological well-being (Polivy et al., 2005). There are numerous studies showing that long-term healthy eating patterns not only provide temporary relief but also have lasting positive effects on an individual's quality of life and psychological resilience (Janitra et al., 2024). These findings suggest that high-quality dietary habits reduce the risk of common mental health problems such as depression, anxiety and stress. In particular, strong correlations have been reported between dietary patterns with anti-inflammatory properties, such as the Mediterranean diet, and higher life satisfaction, increased emotional balance and cognitive performance (Allcock et al., 2024). Not only nutritional quality but also physiological conditions caused by nutrition are associated with mental health. At this point, nutrition-based conditions such as obesity and malnutrition should be considered important risk factors.

Obesity is defined as abnormal or excessive fat accumulation in the body to the extent that it impairs health and is generally accepted as a body mass index (BMI) ≥ 30 kg/m^2 (Alam and Agrawal, 2016). The relationship between obesity and mental health is defined as a mutual and reinforcing cycle. Studies have shown that depression and anxiety symptoms are more common in obese individuals; however, these disorders may increase the risk of obesity (Fan et al., 2024; Pan et al., 2012). In addition, some antidepressant and antipsychotic drugs may cause weight gain, further complicating the relationship between obesity and psychiatric disorders (Karoń and Kozakowski, 2023). Malnutrition is the deterioration of body composition and physiological functions due to inadequate or unbalanced intake of energy and nutrients (Poggiano et al., 2017). Malnutrition has negative effects on mental health. In particular, deficiencies of micronutrients that play a critical role in neurological functions such as B group vitamins, vitamin D, iron, zinc and magnesium have been associated with symptoms such as depression, anxiety, impaired concentration and cognitive decline. It is noteworthy that vitamin D deficiency has a bidirectional relationship with both mental disorders and obesity. These findings reveal the necessity of not only adequate but also balanced and qualified nutrition for optimal mental health (Mikkelsen et al., 2016; Dickerman et al., 2011).

In recent years, the effects of nutrition on mental health have started to be evaluated especially from the perspective of gut microbiota. In this interaction through the gut-brain axis, the transformative effect of diet on microbial composition plays a central role. The quality of the food consumed shapes the diversity and metabolic products of the gut microbiota and thus the biochemical communication process with the central nervous system (Roy, 2024). Some nutrients, especially B vitamins, zinc and magnesium, have been reported to alleviate symptoms of depression and anxiety by supporting neurotransmitter synthesis (Noah et al., 2021). These micronutrients contribute to mood improvement by supporting biochemical pathways involved in the production of happiness hormones such as serotonin and dopamine. At the same time, a healthy diet strengthens the immune system, reduces inflammation levels and regulates stress responses, reinforcing its effect on mental well-being (Janitra et al., 2024).

Many studies have shown that B vitamins, especially B6, B9 and B12, are associated with depression. These vitamins contribute to mood regulation by taking part in the synthesis of neurotransmitters such as serotonin, dopamine and noradrenaline (Mikkelsen et al., 2016). In a cohort study of more than 5,000 older adults conducted by The Trinity Ulster Department of Agriculture (TUDA) on the island of Ireland between 2008 and 2012, it was found that the risk of depression was significantly increased in individuals with low levels of B6, B9 and riboflavin (Moore et al., 2019). In a recent study using the Mendelian randomization method, the association of B vitamins and homocysteine with mental disorders was examined, and it was suggested that serum B12 levels may increase the risk of anxiety disorders and bipolar disorder. This finding indicates that B vitamins may be both protective and potentially risk factors in psychiatric disorders and emphasizes the importance of considering individual biological differences (Hu et al., 2024). Inadequate intake of B vitamins, especially during pregnancy and early life, has been associated with mental health problems in children and adolescents. In an experimental study on animal models, it was found that B6, B9 and B12 deficiencies in mothers led to epigenetic changes in offspring and these changes were associated with depressive and social behaviors (Xu et al., 2021). The role of B vitamins in the prevention of cognitive decline in elderly individuals is particularly important. In a randomized controlled trial, a significant difference was found in the preservation of visual-spatial skills within two years in individuals receiving B6, B9, B12 and riboflavin supplements (Moore et al., 2018). B vitamins play multifaceted roles at the biochemical, epigenetic and neurotransmitter level in the protection of mental health and prevention of disorders. Current findings support the monitoring of B vitamin levels and supplementation if necessary, especially in elderly individuals and malnourished groups. However, more randomized controlled trials are needed, taking into account the dose to be used, individual differences and interactions.

Zinc is the second most abundant trace element in the human body and plays an important role in many basic physiological processes such as cellular growth, immune functions, wound healing and healthy functioning of the neurological system by taking part in the structure of more than 300 enzymes. Especially in the nervous system, zinc is involved as a co-factor in communication between neurons and affects synaptic plasticity (Butola et al., 2020). Zinc deficiency has been associated with various neuropsychiatric disorders such as depression, anxiety, cognitive decline and behavioral disorders. Zinc, which is found in high concentration in the hippocampal region of the brain, regulates synaptic activity associated with learning and memory. Zinc deficiency may exacerbate depressive symptoms by causing neuroinflammation and synaptic degeneration in this region (Sun et al., 2022). Zinc deficiency during pregnancy may adversely affect the neurodevelopment of not only the mother but also the fetus, leading to congenital nervous system anomalies. Animal models show that maternal zinc deficiency is associated with neuroinflammation and increased anxiety-like behavior in offspring (Sauer and Grabrucker, 2019). There is increasing evidence that zinc supplementation can be considered as a complementary approach in the treatment of psychiatric disorders. Especially in cases of depression resistant to antidepressant treatment, zinc supplementation may increase treatment response. Zinc may have a direct effect on mood by showing regulatory effects on serotonin and N - methyl D - aspartate (NMDA) receptors (Butola et al., 2020). Zinc deficiency in advanced ages is considered a risk factor in the emergence of cognitive disorders. Zinc plays an important role in maintaining synaptic structure and reducing oxidative stress in aging neurons. Maintaining zinc levels in cognitive decline and neurodegenerative diseases may be critical for maintaining neural functions (Sun et al., 2022).

Magnesium is the second most common intracellular cation in the human body and acts as a cofactor of more than 300 enzymes, including those involved in nervous system functions. It plays an active role in processes such as energy metabolism, neuronal transmission, cell membrane stability and neurotransmitter release (Kumar et al., 2024). Recent studies have revealed that magnesium deficiency is associated with various neuropsychiatric disorders, especially depression and anxiety. Depression is one of the most common mental disorders worldwide, and neuroinflammation and oxidative stress play a role in its pathogenesis as well as neurotransmitter imbalances. Magnesium limits glutamate overactivation by inhibiting NMDA receptors; this effect contributes to the prevention of neuronal damage associated with depression (Botturi et al., 2020). In a large-scale National Health and Nutrition Examination Survey (NHANES) study conducted in 2025, it was reported that daily magnesium intake below 366 mg was significantly associated with the risk of depression (Huang et al., 2025). The regulatory effects of magnesium on the nervous system are not limited to depression. In a randomized controlled trial in particularly stressed individuals, the combination of magnesium and vitamin B6 was shown to significantly reduce anxiety and depression scores (Noah et al., 2021). It was also found that magnesium supplementation led to improvement in objective stress indicators such as heart rate variability (HRV) (Wienecke and Nolden, 2016). The effects of magnesium on cognitive health are particularly important in the elderly population. A systematic review in 2024 found that both low and high levels of serum magnesium levels exhibited a U-shaped relationship with dementia risk, meaning that both extreme levels may increase the risk of dementia, but intermediate levels of magnesium reduce this risk. The optimum serum magnesium level is recommended to be approximately 0.85 mmol/L (Chen et al., 2024). Magnesium is an indispensable microelement for neuropsychiatric functions. This mineral, which can play a role in many mental health problems such as depression, anxiety, stress and cognitive disorders, is considered as both a preventive and complementary treatment tool. However, there is a need for more randomized controlled trials in which factors such as individual differences, dosage, bioavailability and magnesium form are taken into account.

The Mediterranean diet is considered a prominent nutritional model in this regard. This diet, which is rich in foods such as olive oil, vegetables, fruits, fish and whole grains, has been shown to positively affect mental health by both supporting the gut microbiota and reducing systemic inflammation (Petkoska and Trajkovska-Broach, 2021). However, metabolites produced by the microbiota also play an important role in mental health. In particular, short-chain fatty acids (SCFAs) - butyrate, acetate and propionate - reduce neuroinflammation in the gut and support brain function (Bienenstock et al., 2015). Functional foods such as probiotics and prebiotics have been shown to modulate these processes, increase stress tolerance via the gut-brain axis and alleviate symptoms of mood disorders (Madireddy and Madireddy, 2019). The issue of how diet modifies the intestinal flora is attracting more and more scientific interest. In particular, diets high in fiber contribute to the improvement of cognitive functions by enriching the gut microbiota (Tuplin et al., 2022). One of the most comprehensive findings on the effects of fiber consumption on mental health is the study conducted in Iranian adults, which showed that high fiber consumption significantly reduced the risk of depression, anxiety and psychological distress. Individuals in the highest fiber consumption group had a 33% and 29% lower risk of anxiety and psychological distress, respectively (Saghafian et al., 2021). Similarly, in the Whitehall II cohort study, it was found that the risk of common mental disorders was lower in individuals with higher fiber intake during a five-year follow-up period. This finding supports that fiber may have a protective effect on common mental disorders such as depression and anxiety (Knüppel et al., 2018). Fermented foods and probiotics are among the important food groups that mediate psychological recovery by supporting traditional microflora (Dahiya and Nigam, 2022). In addition, some functional food components such as polyphenols and omega-3 fatty acids have been found to directly improve both gut microbiota and brain health (Zinkow et al., 2024). In addition, significant improvements in stress responses, anxiety levels and cognitive performance have been observed as a result of supporting the microbiota with diet (Provensi et al., 2019).

Finally, some probiotic species (e.g., lactobacillus, bifidobacterium, enterococcus), so-called 'psychobiotics', have been shown to alleviate symptoms of depression and support mental health by positively affecting brain function (Clapp et al., 2017). These bacteria act through the gut-brain axis, which manages the communication between the nervous system and the immune, endocrine and metabolic systems (Mörkl et al., 2020). Psychobiotics can modulate the production of neurotransmitters such as serotonin, dopamine and GABA; they can alleviate the stress response by reducing cortisol levels (Misra and Mohanty, 2019). In addition, psychobiotics are reported to reduce systemic inflammation by regulating intestinal permeability and thus provide a protective effect on neural processes. These microorganisms can be taken naturally with fermented foods or used in supplement forms (Casertano et al., 2021). One of the most researched areas of psychobiotics is depression, anxiety and stress disorders. In a systematic review, findings from 51 randomized controlled trials revealed that psychobiotics are particularly effective in reducing symptoms of depression (Mosquera et al., 2024). This effect is explained by various mechanisms such as increased neurotransmitter production, regulation of the hypothalamic - pituitary - adrenal (HPA) axis and decreased inflammation. Andreeva et al. (2022) reported that psychobiotics have the potential to be an alternative to classical antidepressants and can provide positive effects on cognitive processes such as learning, memory and emotion regulation (Andreeva et al., 2022). Similarly, the study by Kamal et al. (2024) shows that psychobiotics have strong effects on anxiety and depression at preclinical and clinical levels and that these effects are mediated by short-chain fatty acids, neurohormones and immunomodulators (Kamal et al., 2024). It is also emphasized that psychobiotics have the potential to alleviate the symptoms of neurodegenerative and neurodevelopmental diseases such as Alzheimer's disease, autism spectrum disorder and Parkinson's disease (Cheng et al., 2019). The effects of psychobiotics on cognitive functions are being investigated, especially in conditions such as age-related cognitive decline and mild cognitive impairment (MCI). In a randomized controlled clinical trial conducted by Aljumaah et al. (2022), improvement in cognitive performance was observed in individuals receiving Lactobacillus rhamnosus GG supplementation. This effect was associated with the reduction of some microbial species such as Prevotella and Dehalobacterium (Aljumaah et al., 2022). In another comprehensive review, psychobiotic supplements were reported to provide significant improvements in functions such as information processing speed, attention and memory in both healthy individuals and patients with cognitive impairment (Lv et al., 2021). These findings suggest that psychobiotics may be effective not only in mood regulation but also in maintaining and improving cognitive abilities. These effects provided by prebiotic and probiotic compounds open new horizons in the fields of complementary and preventive intervention in modern psychiatry.

Microbiota-Neurotransmitter Relationship

Although the gut microbiota has been associated only with digestive functions for many years, it is now recognized that this microbial ecosystem interacts with much wider physiological systems. In particular, the microbiota plays a central role in maintaining homeostasis through its integrated communication network with the nervous system, immune system and endocrine system. This multifaceted interaction is shaped through a series of neurobiological processes, particularly the gut-brain axis, and the effects of microbial components and products on neural signaling are increasingly being investigated. This microbial ecosystem is in constant interaction with the central nervous system (CNS) through the bioactive metabolites it produces and the signaling pathways it modulates (Bienenstock et al., 2015). This bidirectional communication network, defined as the gut-brain axis, basically operates through four main mechanisms: (1) neurotransmitter synthesis and modulation, (2) immune-mediated signaling, (3) endocrine pathways and (4) neural transmission via the vagus nerve (Pan et al., 2025). The influence of the microbiota on neurotransmitter dynamics is particularly evident in the homeostasis of critical neural messengers such as serotonin, gamma-aminobutyric acid and dopamine. The synthesis of serotonin in the intestinal mucosa, which harbors 90% of total body stores, is directly linked to microbial tryptophan metabolism. Bacterial species (e.g. Akkermansia muciniphila), while promoting the utilization of tryptophan as a serotonin precursor, can also trigger the accumulation of neurotoxic metabolites by directing them to the kynurenine pathway (Pan et al., 2025, O'Mahony et al., 2015). Disruption of this balance is associated with depressive disorders and neurodegenerative processes. The integrated communication network of the gut microbiota with the nervous system, immune system and endocrine system is shown in Figure 1.

The gut microbiota plays a critical role in the regulation of serotonin production. Enterochromaffin cells in the intestine are the main source of serotonin production. However, serotonin production and metabolism can be significantly altered by microorganisms. In particular, microorganisms in the gut modulate serotonin levels by affecting the metabolism of tryptophan required for serotonin production (Stasi et al., 2019). Changes in the gut microbiota can lead to the development of metabolic and neurological diseases associated with serotonin. Serotonin plays a key role in the regulation of brain functions and emotional states. However, imbalances in serotonin levels are closely associated with depression, anxiety and other psychiatric disorders (Jenkins et al., 2016). The gut microbiota plays an important role in the pathophysiology of these mental health conditions by regulating serotonin production and signaling. Imbalances in the gut microbiota disrupt serotonin signaling pathways, leading to exacerbation of conditions such as depression and anxiety (Marino Maia et al., 2024). The interaction between serotonin and gut microbiota builds a bridge between the digestive system and neurotransmitters in the brain. This interaction has a direct impact on mental health by influencing mood, sleep patterns, stress responses and cognitive functions. It has also been shown that microbial metabolites, especially short-chain fatty acids, can influence emotional and cognitive behaviors by regulating serotonin levels (Buey et al., 2023). The interaction between gut microbiota and serotonin offers an important strategy in treatment approaches. Interventions such as probiotics and dietary changes show promise in the treatment of psychiatric disorders by regulating serotonin lvels. In particular, there are increasing studies on the ability of probiotics to regulate the gut microbiota, increase serotonin production, and thus alleviate symptoms of depression (Akram et al., 2023). In addition, personalized treatment approaches make it possible to apply personalized probiotic treatments in the treatment of serotonin-related disorders by analyzing the microbiome profile (Potter et al., 2023). The relationship between serotonin and gut microbiota stands out as an important research area in the management of mental health and neurological disorders. Gut microbiota has a direct impact on mental health by regulating serotonin production and function. Understanding these relationships may lead to the development of new therapeutic approaches in the treatment of psychiatric disorders.

The gut microbiota is not only an ecosystem serving the digestive system, but also a multifaceted regulatory organ integrated with the endocrine, immune and nervous systems. The gut-brain axis allows microorganisms in the gut to communicate with the brain, and this communication is modulated by various biochemical processes. The most important of these processes include neurotransmitter production by microorganisms, interaction through the immune system, endocrine pathways and metabolic pathways. The gut microbiota can directly influence the production of neurotransmitters such as serotonin, dopamine and gamma-aminobutyric acid (GABA). Thus, some bacteria present in the gut (such as Akkerman Sia municipia) can increase serotonin levels by affecting tryptophan metabolism (Pan et al., 2025). In addition, SCFAs regulate inflammatory responses of the brain and maintain neurotransmitter balance (Peirce and Alviña, 2019).

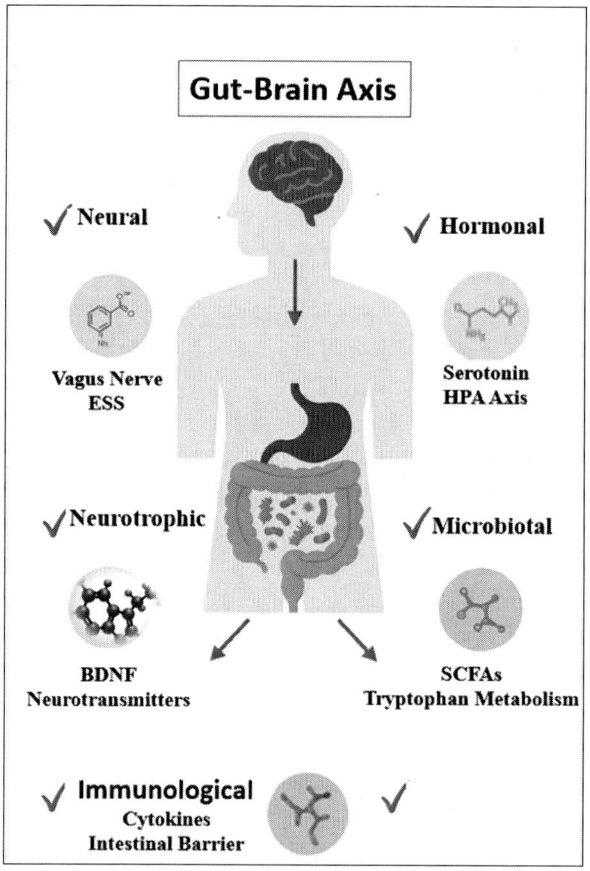

Figure 1. Gut-Brain Axis.

The microbiota communicates between the nervous system and the enteric nervous system (ENS) both via the vagus nerve and through metabolites that enter the circulation. Approximately 90% of serotonin is synthesized in enteroendocrine cells in the intestinal mucosa; tryptophan metabolism plays a key role in this synthesis process and the microbiota influences this process in two ways: (1) increasing or decreasing the absorption of tryptophan, a serotonin precursor, and (2) directing tryptophan to non-serotonin pathways (e.g., the kynurenine pathway) (O'Mahony et al., 2015). Furthermore, stress responses and the hypothalamic-pituitary-adrenal axis may be influenced by changes in the gut microbiota, with profound effects on mental health (Forssten et al., 2022).

In recent years, studies on the microbiota-brain axis have revealed that the gut microbiota has important effects on the central nervous system. Neurotransmitters such as GABA have been shown to play a mediating role in this two-way communication. GABA acts as a depressant neurotransmitter in the brain and is known to be associated with many mental illnesses such as anxiety and depression. New studies show that GABA is produced not only in the nervous system but also in the gut and that microbial balance can affect this production. Some microorganisms living in the intestine can synthesize GABA from glutamate and secrete this neuroactive compound into the intestinal environment. In particular, bacteria such as Bifidobacterium adolescents (Duranti et al., 2020), Bacteroides spp. (Strandwitz et al., 2018), and Levi lactobacillus brevis (Banerjee et al., 2021) have been shown to have the capacity to produce GABA. These bacteria produce GABA using the enzyme glutamate decarboxylase (GAD). Studies suggest that these GABA-producing bacteria may increase GABA levels when introduced into the intestinal tract with probiotic applications and thus affect central nervous system functions (Braga et al., 2024). Direct relationships have begun to be established between microbiota-derived GABA production and mental disorders. GABA imbalances have been associated with many neuropsychiatric conditions such as depression, anxiety disorder, autism spectrum disorder and Alzheimer's disease (Conn et al., 2024). In a study, transcriptome analyses revealed that Bacteroides, Parabacteroides and Escherichia species actively express GABA production pathways in healthy individuals. In patients with major depressive disorder, Bacteroides abundance in faces was found to be negatively correlated with depression-related brain activity (Strandwitz et al., 2018). Similarly, it has been reported that intestinal GABA levels are increased in individuals with autism, which may be linked to the overrepresentation of bacteria such as Escherichia coli (Wang et al., 2025). GABA has been shown to be not only a neurotransmitter but also a molecule that maintains the barrier integrity of the gut and modulates immune system responses (Conn et al., 2024). Thanks to these properties, increasing intestinal GABA levels may have beneficial effects on both intestinal and brain health. In a study on rats, it was shown that oral administration of GABA-producing probiotics reduced anxiety and depression-like behaviors (Lozano et al., 2024).

The gut microbiota produces numerous neuroactive compounds that affect the central nervous system as well as the peripheral nervous system. The synthesis of dopamine is a complex process metabolized by the gut microbiota and carried out by the enzymatic activity of microorganisms. Species such as Bacteroides, Lactobacillus and Bifidobacterium in the gut microbiota have a significant effect on the biosynthesis and metabolism of dopamine (Hamamah et al., 2022). It has also been reported that imbalances in the gut microbiota can affect dopamine levels, which can lead to the development of neurological diseases (González-Arancibia et al., 2019). The relationship between gut microbiota and dopamine also plays an important role in the pathophysiology of mental diseases. Especially in dopaminergic disorders such as Parkinson's disease, the effects of gut microbiota become evident. The gut microbiota of Parkinson's patients show significant changes compared to healthy individuals, which is associated with dopamine production and function (Liu et al., 2021). Similarly, diseases such as depression and anxiety have been associated with disruption of the gut microbiota, which has been linked to imbalances in dopamine production (Duan et al., 2022). There is strong evidence that imbalances in the gut microbiota can lead to dopamine-related disorders and cause the development of psychiatric disorders. For example, in conditions such as stress and post-traumatic stress disorder (PTSD), it has been observed that changes in the gut microbiota can increase the severity of diseases by disrupting dopamine function (Laudani et al., 2022). The production of dopamine is not limited to the central nervous system; significant levels of dopamine are also found in the intestines. It has been observed that intestinal microbiota contribute to the biosynthesis of dopamine and this process is regulated by the enzymatic activities of microorganisms. Gut bacteria produce dopamine by processing l-tyrosine, the starting material of dopamine. However, the degradation and metabolism of dopamine is also modulated by the gut microbiota. This suggests that imbalances in dopamine levels may play a critical role in the development of neurological diseases such as Parkinson's disease and depression. The interaction between dopamine and gut microbiota plays a central role in brain function and the pathophysiology of mental disorders (Liu et al., 2021). The gut microbiota may influence the development of neurological and psychiatric diseases by modulating dopamine production. Therefore, early detection of alterations in the gut microbiota and development of therapeutic strategies may provide an important advance in the treatment of dopaminergic diseases.

The relationship between the gut microbiota and the hypothalamic-pituitary-adrenal axis has been the focus of an increasing number of studies in recent years, providing a new paradigm for the microbial basis of stress physiology. The HPA axis is central to the neuroendocrine responses of the organism to stress and exerts wide-ranging effects on both the central nervous system and peripheral systems through the secretion of glucocorticoid hormones (cortisol in humans) (Rusch et al., 2023). The gut microbiota plays an important role in both the development of the HPA axis and its responsiveness. Studies in germ-free animal models have revealed that these animals show increased HPA activation and elevated glucocorticoid levels in the face of stress (Vagnerová et al., 2019). Similarly, structural disruption of the gut microbiota can lead to changes in gene expression levels of HPA axis components such as corticotropin-releasing hormone (CRH) and glucocorticoid receptor (Wu et al., 2020). Microbiota-HPA axis interaction operates not only through endocrine but also immune, neural and metabolic pathways. Microbial metabolites (e.g., short-chain fatty acids) can transmit signals to the brain via the vagus nerve, and these signals can modulate the tone of the HPA axis (Sudo, 2016). In addition, the circadian rhythms of the microbiota may also affect glucocorticoid release and alter stress sensitivity during the day (Tofani et al., 2024). The link between these biological interactions and mental health is quite strong. Chronic activation of the HPA axis plays a critical role in the pathophysiology of psychiatric disorders such as depression and anxiety. The bidirectional relationship between the HPA axis and the microbiota-gut-brain axis suggests that the microbiota may be a determining factor in the development of these disorders (Bertollo et al., 2025; Lima-Ojeda et al., 2020). Decreased microbial diversity may affect neurotransmitter production through increased intestinal permeability and inflammation, which may predispose to mood disorders. The clinical consequences of these microbial effects are more clearly understood, especially considering that serotonin production occurs largely in the intestine (Frankiensztajn et al., 2020). The interaction between the gut microbiota and the HPA axis has a direct effect on the stress response of the organism and plays an important role in the maintenance or deterioration of mental health. Interventions on this axis (probiotics, dietary changes, etc.) may offer new approaches in the treatment of psychiatric disorders in the future.

The gut microbiota also communicates with the brain via the immune system. The microbiota plays a critical role in the regulation of both innate and adaptive immune responses. Commensal bacteria support the development of regulatory T cells (Tregs) in the intestinal mucosa, creating an anti-inflammatory environment. At the same time, some bacterial species stimulate the immune system and promote the production of proinflammatory cytokines (Ghanei et al., 2019). This bidirectional interaction can influence intestinal permeability and the recruitment of immune cells into the systemic circulation, leading to the generation of immune signals to the central nervous system. The effects of gut microbiota on the immune system reach the brain through various signaling pathways. In particular, microbial products such as cytokines, lipopolysaccharides (LPS) and microbiota-derived SCFAs can directly affect both the blood-brain barrier and neural inflammation (Park and Im, 2022). Proinflammatory cytokines (IL-6, TNF-α) can increase microglia activation by triggering peripheral inflammation and thus initiate neuroinflammatory processes (Dinan and Cryan, 2017). In studies, disruptions in the gut microbiota have been associated with immune-mediated brain disorders. For example, in conditions such as multiple sclerosis (MS), autism spectrum disorder (ASD) and depression, both immune system dysfunction and microbiota changes have been observed (Yadav et al., 2023; Zhu et al., 2020).

The gut microbiota is a complex and dynamic ecosystem that exerts significant effects on various aspects of human health. Recent human and animal studies have revealed that the microbiota is not only limited to digestive and immune functions, but may also have an impact on neuropsychiatric health through bidirectional communication with the central nervous system. In particular, some microbial species have been found to be associated with psychological well-being. For example, the presence of short-chain fatty acid-producing bacteria such as Faecalibacterium and Coprococcus is positively correlated with individuals' quality of life scores. In addition, these species have been associated with the production of 3,4-dihydroxyphenylacetic acid, a metabolite of dopamine. It was found that these bacteria were significantly reduced in individuals diagnosed with depression (Valles-Colomer et al., 2019). Similarly, in a study conducted by Duan et al. (2022) on prison populations, it was observed that Bacteroides and Prevotella species were dominant in individuals with high depression and anxiety scores, whereas SCFA-producing species were significantly reduced. These findings suggest a strong link between microbial composition and psychological state (Duan et al., 2022)).

The basis of the effects of microbiota on the central nervous system is the regulation of neurotransmitter levels. Studies have shown that probiotic species such as Lactobacillus and Bifidobacterium can synthesize neurotransmitters such as GABA, serotonin, dopamine and norepinephrine. These compounds play important roles in basic psychological processes such as mood, motivation and cognitive functions (Strandwitz et al., 2018). In experiments on germ-free animal models, it was shown that serotonin and dopamine levels were severely reduced in mice lacking microbiota, and these levels were increased again with probiotic supplementation. This suggests that microbiota-neurotransmitter interaction may be one of the basic physiological mechanisms of psychiatric disorders (Liu et al., 2016). Animal studies show that microbiota is effective not only in neurotransmitter levels but also in behavioral outcomes. It has been reported that anxiety, depression and even symptoms similar to Parkinson's disease can be triggered with the change of microbiota (Qu et al., 2024). These findings suggest that microbial diversity may be determinant not only for gut health but also for neuropsychiatric conditions. However, how these findings obtained from animal models will be reflected to humans is still under investigation. Lynch and Hsiao (2023) stated that the effects of microbiota-neuron interaction on neurodevelopment and behavior are strong, but emphasized that advanced mechanistic studies are needed to reveal causality in humans (Lynch and Hsiao, 2023). However, human-based studies do not evaluate the microbiota-psychology relationship only unidirectionally. Kazemi and Djafarian (2021) suggested that depression can alter the gut microbiota and that this alteration may again exacerbate depression symptoms by affecting inflammatory responses and neurotransmitter production. This bidirectional relationship is further complicated by the fact that stress affects the microbiota via the hypothalamo-pituitary-adrenal axis. When all these studies are brought together, the effects of gut microbiota on neurotransmitter synthesis and psychiatric health are clearly demonstrated. Therefore, interventions targeting the microbiota - for example, probiotic supplementation or dietary changes - may have an important place in the prevention and treatment of psychiatric diseases in the future (Kazemi and Djafarian, 2021).

Nutritional Therapies for Depression and Anxiety

Depression is a mood disorder that negatively affects an individual's mood, thought structure and behavior and can be common and recurrent. Clinical depression is usually characterized by symptoms such as hopelessness, loss of interest, decreased energy, difficulty in concentration and a feeling of worthlessness. Biological mechanisms such as imbalances in monoamine levels, dysregulation of the hypothalamo-pituitary-adrenal axis and neuroinflammation play a role in its pathophysiology (Botturi et al., 2020). Major depression is one of the most common psychiatric disorders, affecting approximately 280 million people worldwide (WHO, 2023). Approximately 1 in 20 people worldwide have major depressive disorder; this rate is approximately 1.5 times higher in women than in men. According to a meta-analysis, the prevalence of depression increased up to 25% during the COVID-19 pandemic, with a significant increase observed especially in young adults (Sønderskov et al., 2020). Anxiety is a psychiatric disorder characterized by excessive and disproportionate fear, worry and bodily arousal in the absence of a real threat. Generalized anxiety disorder is characterized by constant and uncontrollable anxiety thoughts that disrupt the daily functioning of the individual. On the neurobiological basis, amygdala hyperactivation, impairments in neurotransmitter systems such as serotonin and GABA, and stress-related HPA axis dysregulations come to the fore (Hu et al., 2024). Anxiety disorders are among the most common mental health disorders affecting approximately 301 million people worldwide (WHO, 2023). The lifetime prevalence of anxiety disorders in the general population varies between 10-20%; this rate is significantly higher in women than in men (Baxter et al., 2013). Depression and anxiety disorders are among the most common mental health problems worldwide and seriously affect long-term quality of life due to their tendency to become chronic. In recent years, the idea that these disorders can be managed not only with pharmacological treatments but also with lifestyle adjustments has been accepted. In this context, 'nutritional therapies' stand out as an important field in supporting mental well-being, especially through the microbiota-brain axis (Julius, 2025). Many nutritional models such as traditional diets, especially the Mediterranean diet, western-style diet, low FODMAP (Fermentable Oligosaccharides, Disaccharides, Monosaccharides and Polyols) diet have been examined for the relationship between nutritional therapies and mental health.

The Mediterranean diet is a diet enriched with foods such as olive oil, vegetables, fruits, fish and whole grains, which may be effective in reducing inflammation, improving brain health and supporting neurological function. Research shows that the Mediterranean diet may help reduce the risk of depression and anxiety. The fact that the Mediterranean diet improves gut microbiota highlights this diet as a potential therapy for psychological health (Sadeghi et al., 2021; Martins et al., 2021). A meta-analysis reported that greater adherence to the Mediterranean diet was associated with a 30 per cent lower risk of depression (Psaltopoulou et al., 2013). IIn a study conducted in Iran, researchers examined the relationship between three different dietary patterns—the Mediterranean diet (rich in fruits, vegetables, and low-fat dairy products), the Western-style diet (high in processed foods, butter, and sweets), and the Iranian traditional diet (characterized by refined grains, red meat, poultry, and legumes)—and the prevalence of depression and anxiety. The study concluded that individuals adhering to a Mediterranean dietary pattern had a lower risk of depression and anxiety, whereas adherence to the Iranian traditional diet may be associated with an increased risk of these mental health conditions (Firth et al., 2019).

Traditional diets are diets shaped by historical and cultural contexts and based on natural, unprocessed foods. In recent years, studies have been conducted on the effects of these diets not only on physical health but also on mental health. Some of these diets have been found to have a protective effect, while others may pose a risk. In a study conducted in Chinese adolescents, it was determined that individuals who followed a 'traditional' diet model (rice, vegetables, low-fat animal products) had a lower risk of depression and anxiety (Weng et al., 2011). On the other hand, high adherence to the 'Iranian traditional diet' (refined grains, red meat, poultry and legumes) was associated with an increased risk of depression and anxiety (Zakizadeh et al., 2019; Firth et al., 2019). In a randomized controlled trial conducted by Canheta et al. (2020) with the Traditional Brazilian diet (DieTBra) (which includes fresh vegetables, fruits, legumes, rice and small amounts of meat), the traditional Brazilian diet was found to significantly reduce symptoms of depression and anxiety. In the study, up to 50% reduction in depression symptoms was observed after 12 weeks of intervention (Canheta et al., 2020). According to the systematic review of Silva et al. (2019) on Japanese and Indian traditional cuisines, it was emphasized that Japanese and Indian traditional diets may have protective effects on prenatal and postnatal depression and anxiety, but these findings are supported by a limited number of studies (Silva et al., 2019). These findings suggest that the effects of traditional dietary patterns on mental health in different cultural contexts may be heterogeneous, and structural differences of dietary components may play a decisive role on psychological well-being. In particular, it was emphasized that traditional diets with an emphasis on vegetables and fruits may have positive effects on depression and anxiety symptoms, but since the structure of the traditional diet varies from culture to culture, the effect of each diet may not be in the same direction. Some traditional diets (e.g., those containing excessive refined grains and red meat) may have negative consequences on psychological health.

Western diet is a diet that typically includes highly processed foods, saturated fats, refined carbohydrates, sugary drinks, red and processed meat, low fiber content, low fruit and vegetable consumption and low amounts of omega-3 fatty acids (Jacka et al., 2010). The Western-style diet has become dominant in many developed countries, especially in the United States of America, and has spread to developing countries with globalization. This prevalence has increased even more with the food industry's promotion of cheap and high-energy products. As a result, this diet type has increased the incidence of chronic diseases such as obesity and diabetes (Gibson-Smith et al., 2018). The effects of dietary habits on mental health have become one of the important research topics in recent years. The low nutrient density and inflammatory effects of the Western-style diet have been associated with cognitive dysfunctions, depression and anxiety. In addition, it has been reported that sugary and processed foods consumed frequently in this diet may cause neuroinflammation with negative effects on brain functions (Francis et al., 2022). In addition, in a study conducted with 1,046 women aged 20-93 years, the relationship between western-style diets and depression and anxiety was examined and it was found that there was a positive correlation between western-style diet consumption and anxiety and depression rates (Jacka et al., 2010). In addition, in a study conducted on university students, it was observed that as depression and anxiety symptoms increased, total calorie intake decreased and sugar consumption increased. Thus, it has been observed that the Western-style diet model increases depression and anxiety levels, and the tendency towards this diet model increases with the increase in depression (Keck et al., 2020). Francis et al. found that Western-style diet consumption led to a decrease in kynurenic acid levels, which are associated with depression and have neuroprotective properties. It has been reported that this may lead to impaired regulation of the glutamate system and an increase in depressive symptoms (Francis et al., 2022). Gibson-Smith et al. (2018) reported that individuals diagnosed with depression and anxiety generally had lower diet quality, and this showed a dose-response relationship with symptom severity (Gibson-Smith et al., 2018). In a study conducted by Kamali et al. (2016) with 1634 adults, it was shown that Western-style diet consumption was positively associated with anxiety and stress levels (Kamali et al., 2016). The Western-style diet stands out as an important risk factor in the increase in mental health problems such as depression and anxiety in modern societies due to both its contextual characteristics and its widespread use.

The FODMAP diet is a dietary pattern based on limiting short-chain carbohydrates and sugar alcohols (fermentable oligosaccharides, disaccharides, monosaccharides and polyols). Because these components are poorly absorbed in the small intestine, they ferment in the colon and can cause functional gastrointestinal disorders such as irritable bowel syndrome (IBS). The main components of the diet include fructose (apples, honey), lactose (dairy products), fructus (wheat, onions), galactooligosaccharides (legumes) and polyols (sorbitol, mannitol) (Ribichini et al., 2024). The FODMAP diet has been found to be effective in reducing symptoms such as abdominal pain, bloating, gas and diarrhea, especially in IBS patients. The FODMAP diet provides a decrease in gastrointestinal symptom severity and an increase in quality of life in a short time (Kortlever et al., 2019). In recent years, the effects of the FODMAP diet on not only physical but also psychological health have been the subject of research. In one study, the low FODMAP diet was found to significantly reduce anxiety scores (Piacentino et al., 2016). In a prospective study, Kortlever et al. found that a low FODMAP diet improved both gastrointestinal symptoms and depression and anxiety levels in patients with irritable bowel syndrome with a 6-month follow-up. These improvements were reported to be accompanied by an increase in quality of life, a decrease in fatigue, and an increase in general happiness levels (Kortlever et al., 2019). In older adults, low FODMAP diet was reported to significantly reduce anxiety scores and provide clinical improvement in depression scores, although not statistically significant (O'Brien et al., 2020). However, similar results are not obtained in every study. Chan et al. (2021) reported that a low FODMAP diet reduced gastrointestinal symptoms in IBS patients participating in group training but had no significant effect on depression and anxiety (Chan et al., 2021). In a systematic review, it was stated that the FODMAP diet may have positive effects on depression and anxiety, but the available evidence is heterogeneous and limited (Aranburu et al., 2021). This contradiction is seen due to the variables that may be caused by the flexibility in diet application, the difference in the groups applied, and environmental factors. This situation is explained through the psychological effects of diet and the gut-brain axis in connection with this. Dietary FODMAP restriction has been interpreted as reducing intestinal permeability and inflammation by reducing dysbiosis in the gut microbiota, which may indirectly have a positive effect on mental health (Ribichini et al., 2024). There are also studies examining the relationship between nutrients, their components, and mental health outcomes such as depression and anxiety. Omega-3 fatty acid is an anti-inflammatory agent rich in eicosatetraenoic acid (EPA) and docosahexaenoic acid (DHA). In a comprehensive meta-analysis on the antidepressant effects of omega-3, Hallahan et al. (2016) included 35 randomized controlled trials in their study and found that EPA in particular reduced symptom severity when used as an adjuvant to pharmacotherapy in major depressive disorder (Hallahan et al., 2016). In a double-blind randomized controlled double-blind study on the neuroprotective effects of curcumin, significant improvement in Hamilton Depression Rating Scale scores was observed in patients with major depressive disorder taking 500 mg curcumin supplements daily for 8 weeks compared to the placebo group (Sanmukhani et al., 2014). The meta-analysis of Anglin et al. (2013), which examined the relationship between 25-hydroxyvitamin D [25(OH)D] levels and depression risk, revealed that the risk of depression increased 2.14-fold in individuals with serum 25(OH)D levels <20 ng/ml (Anglin et al., 2013). In a cohort study on the relationship between homocysteine metabolism and mental health, it was shown that each 5 µmol/L increase in plasma homocysteine levels increased the risk of depression by 42% and this effect was more pronounced in individuals with vitamin B12 deficiency (Petridou et al., 2016).

Omega-3, Prebiotics, Probiotics and Cognitive Functions

Cognitive functions are shaped by complex neurological processes that change throughout life. Skills such as memory, attention, decision-making and learning can be affected by ageing, stress, inflammation and microbial dysbiosis. Omega-3 fatty acids and prebiotics have emerged as promising interventions for both prevention and improvement of these processes (Horning and Davis, 2012; Kang et al., 2024; Zhang et al., 2023).

Omega-3 fatty acids are polyunsaturated fatty acids that are essential for human health and cannot be synthesized by the body. The most well-known types are alpha-linolenic acid (ALA), eicosatetraenoic acid (EPA), and docosahexaenoic acid (DHA). ALA is of plant origin and found in flaxseeds, walnuts and chia seeds, while EPA and DHA are mainly derived from seafood, especially oily fish (White, 2015). DHA is a type of omega-3 found in high concentrations, especially in brain tissue, and is critical for neurological functions such as synaptic plasticity, neurogenesis and integrity of cell membranes. EPA is mostly known for its anti-inflammatory effects (Hashimoto, 2014). Omega-3 polyunsaturated fatty acids (PUFA), especially EPA and DHA, are among the essential nutrients for brain development and function. In recent years, new and important research has been conducted on the effects of omega-3 fatty acids on cognitive functions through the gut-brain axis. Omega-3 fatty acids take place in brain cell membranes and support synaptic plasticity, neurotransmitter release and neuronal growth (Zinkow et al., 2024). The effect of omega-3 fatty acids on cognitive functions has been examined in many studies conducted in the last decade. In one study, it was reported that adequate intake of DHA played an important role in the prevention of cognitive decline due to aging (Robinson et al., 2010), while neutral results were obtained in another study (Zolot, 2015). Increased serum omega-3 levels, especially DHA and EPA, were positively associated with processing speed and non-verbal memory performance in older adults (Duchaine et al., 2022). In addition, studies in Alzheimer's patients have observed that omega-3 supplementation improves cognitive performance in individuals with adequate vitamin B12 status (Jernerén et al., 2019). Omega-3 fatty acids, especially DHA and EPA, have been found to increase hippocampal neurogenesis and improve short-term memory in adult rodents. Especially in 'fat-1' transgenic mice enriched with fatty acids, increased neurogenesis and memory performance were observed in animals with high endogenous omega-3 levels (Dyall, 2011). Omega-3 deficiency in early life has been found to increase depressive behaviors and social interaction disorders (Robertson et al., 2017).

The effects of omega-3 on microbiota have been investigated with increasing interest in recent years. Omega-3 fatty acids, especially DHA and EPA, have been reported to regulate the inflammatory response by increasing the diversity of gut microbiota (Kerman et al., 2023). In addition, omega-3 supplementation causes an increase in beneficial intestinal bacteria (Bifidobacterium and Lactobacillus) and a decrease in harmful species (Costantini et al., 2017; Fu et al., 2021). Bifidobacterium increase was positively correlated with increased DHA intake (Horigome et al., 2019). Omega-3 positively affects the gut microbiota, regulating the immune system and reducing inflammation levels (Zorgetto-Pinheiro et al., 2022). Short-chain fatty acids produced by the gut microbiota, especially butyrate, may help improve cognitive function (Nguyen et al., 2024). Increased butyrate production with omega-3 supplementation has been associated with reduced inflammation in the brain and preservation of gut-brain barrier integrity (Fan et al., 2024). In Alzheimer's disease, omega-3 is thought to slow cognitive decline by balancing the microbiota (La Rosa et al., 2018). However, taking into account the different results of clinical studies, it was emphasized that the benefits may be limited, especially in patients with advanced Alzheimer's disease (Kerman et al., 2023). Omega-3-rich fats such as walnut oil have been found to improve cognitive function through microbiota change (Kang et al., 2024). Omega-3 has been shown to reduce intestinal permeability that increases with aging and alleviate cognitive loss through microbiota (Qiu et al., 2024). In cases where obesity and depression are seen together, omega-3 has been found to increase cognitive and psychological health with its microbiota regulatory role (Chang et al., 2025).

Cognitive health is shaped not only by medical treatment, but also by the contribution of microbial interactions favored by prebiotics. Prebiotics are indigestible carbohydrates and soluble fibers such as fructooligosaccharides (FOS), galactooligosaccharides (GOS), inulin, lactulose, resistant starch, beta-glucans and pectin. By supporting the growth of beneficial bacteria (especially Bifidobacterium and Lactobacillus) in the intestines, it positively affects the microbiota composition and may have effects on different mental problems (Slavin, 2013). In a study, classical prebiotics such as FOS and GOS were found to reduce neuroinflammation and improve cognitive functions in Alzheimer's disease model. In particular, GOS increased the proportion of Lactobacillus, reduced amyloid beta accumulation and increased 5-hydroxytryptamine (5-HT) levels (Zhang et al., 2023). Similarly, PDX-GOS (a combination of polydextrose and GOS) diet given to young rats enhanced short-term memory and increased the proportion of bacteria producing short-chain fatty acids such as Allobaculum. These findings suggest that GOS may play a potential role in improving brain health by regulating intestinal microflora and may contribute to the prevention of neurological diseases. (Waworuntu et al., 2014). In a study in rats, administration of a High Prebiotic Fiber (HPF) diet reduced cognitive impairment due to age-related hypertension, increased the level of Lactobacillus and prevented hippocampal damage by reducing blood-brain barrier permeability (Nist et al., 2023). Another study showed that prebiotics provide anxiolytic effects by increasing the expression of brain neurotropic factor (BDNF) and N-methyl-d-aspartate receptor in rats. Prebiotics such as FOS and GOS were found to increase hippocampal BDNF and NR1 subunits, and GOS intake was found to have an effect on d-serine and NR2A. These findings suggest that prebiotics may improve brain health through gut hormones and may be used as a potential treatment for neuropsychiatric disorders (Savignac, 2013). In Alzheimer models, prebiotics (especially FOS and GOS) have been shown to contribute to brain function by reducing intestinal inflammation (Kang and Zivkovic, 2021). In another study, the effect of GOS and FOS type insulins given for 6 weeks was investigated in Alzheimer's model mice. GOS supplementation was associated with an increase in Lactobacillus, decreased neuroinflammation, and improved cognitive functions (especially memory tests) (Zhang et al., 2023). A systematic review analyzed the effect levels of prebiotics (especially inulin) in human intervention studies. Some of the 14 studies reported that long-term (28 days+) inulin supplementation improved verbal episodic memory and mood (Desmedt et al., 2018).

β-glucans are soluble dietary fibers composed of glucose monomers and polysaccharides linked together by β-glycosidic bonds. They may vary structurally depending on their source: β-(1→3)/(1→4) bonds are usually present in cereals (e.g. oats and barley), whereas β-(1→3)/(1→6) bonds are common in β-glucans of fungal and yeast origin. These structural differences significantly determine their functional properties and biological effects (Singh and Bhardwaj, 2023). β-glucans cannot be digested directly by humans; therefore, they reach the large intestine where they are fermented by microbiota. Bacteroides and Bifidobacterium species, in particular, are among the primary microorganisms responsible for breaking down these complex carbohydrates. (Fernandez-Julia et al., 2021). Among prebiotics, β-glucan is thought to be associated with mental health as well as anti-inflammatory effects. In an animal study, mice obese with a high-fat diet were given beta-glucan and at the end of 15 weeks, it was observed that beta-glucan improved cognitive function and improved microglia activity, brain inflammation and synapse structures (Shi et al, 2020). In another study, it was reported that three different types of beta-glucans obtained from mushrooms, worms and oat hulls increased cognitive function and improved the gut-brain axis in mice (Hu et al., 2022).

Probiotics are living microorganisms that provide health benefits to the host when taken in adequate amounts. They usually consist of Lactobacillus, Bifidobacterium and some Saccharomyces species. They settle in the digestive tract and prevent the growth of pathogenic bacteria, regulate the immune system and maintain the integrity of the intestinal barrier (Renukadevi et al., 2024). Probiotics affect not only gastrointestinal health but also mental health by establishing bidirectional communication with the central nervous system through the microbiota-brain axis (Madireddy and Madireddy, 2019). These microorganisms act through suppression of inflammation, regulation of intestinal permeability and contribution to neurotransmitter synthesis (Matiş et al., 2023). In addition, some probiotic species have been shown to contribute indirectly to brain health by modulating the immune system and reducing oxidative stress (Yang et al., 2023). With these properties, probiotics are increasingly being investigated as functional foods and psychoneurobiotic agents.

Various animal and human studies have shown that probiotics can affect neurotransmitter production and release via the gut-brain axis. Neurotransmitters such as serotonin, GABA, dopamine and acetylcholine play an important role in this process (Grover et al., 2019). It has been reported that some Lactobacillus species show anxiolytic effect by increasing GABA production and may also affect tryptophan metabolism involved in serotonin production (Matiş et al., 2023). Renukadevi et al. (2024) showed that probiotics stabilize neurotransmitter levels by suppressing neuroinflammation, which positively affects cognitive performance. Bifidobacterium and Lactobacillus species are particularly important in regulating dopamine and glutamate levels (Renukadevi et al., 2024). The study by Mukilan et al. (2024) revealed that probiotics contribute to synaptic plasticity by increasing the passage of neurotransmitter precursors (e.g., tryptophan) across the blood–brain barrier. In conclusion, probiotics may modulate behavioral and cognitive outcomes by influencing neurotransmitter biosynthesis, making them potential psychoneurobiotic agents.

The effect of probiotics on cognitive functions has been extensively investigated in both preclinical and clinical studies. In the meta-analysis of Liu et al. (2023), probiotic supplementation was shown to significantly improve cognitive performance, especially in Alzheimer's patients and individuals with mild cognitive impairment (Liu et al., 2023). The systematic review of Zhu et al. (2021) reported that probiotic supplementation was more effective especially in individuals with mild cognitive impairment (MCI), while its effect was limited in advanced cases diagnosed with Alzheimer's disease (Zhu et al., 2021). Sánchez de Lara-Sánchez and Sánchez-Pérez (2022), in their review of studies of high methodological quality, emphasized that probiotics may slow progression, especially in early cognitive impairment (Sánchez de Lara-Sánchez and Sánchez-Pérez, 2022). Fei et al. (2023) reported that in a study conducted in elderly individuals, not only cognitive functions but also sleep quality improved with 12 weeks of probiotic supplementation (Fei et al., 2023). In a randomized controlled trial conducted by Ruiz-González et al. (2024), it was reported that probiotic intake provided significant improvements in attention, memory and executive functions in healthy elderly individuals (Ruiz-González et al., 2024).

Conclusion

This study evaluated the effects of nutrition on psychological health in the context of the microbiota-gut-brain axis with a multidisciplinary approach and comprehensively demonstrated the decisive role of this axis in the pathophysiology of neuropsychiatric disorders. The findings show that the gut microbiota is not a structure limited to the gastrointestinal system, but is the central component of a complex biological network that interacts with the nervous, immune and endocrine systems. In this context, it is understood that the gut microbiota plays an active role in many neurobiological processes such as neurotransmitter synthesis, immune modulation, protection of intestinal barrier integrity and regulation of stress response through the HPA axis.

The study showed that the modulatory effects of dietary components - especially prebiotics, probiotics, fibers, short-chain fatty acids, omega-3 fatty acids and micronutrients (B vitamins, zinc, magnesium, etc.) - on gut microbiota are indirectly reflected in central nervous system functions. It is emphasized that these components support the production of neuroactive substances such as serotonin, dopamine and GABA by increasing microbial diversity and thus may have a protective or therapeutic effect on the pathogenesis of neuropsychiatric conditions such as depression, anxiety and cognitive disorders.

In particular, it was found that microbiota-friendly dietary models such as the Mediterranean diet have positive effects on mental health with its anti-inflammatory potential, metabolic benefits and microbial diversity supportive structure; on the other hand, high sugar, saturated fat and processed foods common in the Western-type diet may increase neuroinflammation and psychiatric symptoms by triggering microbial dysbiosis, similarly, low FODMAP diet may increase the capacity to cope with stress. In addition, the findings that the mood-regulating effects of certain probiotic strains, called psychobiotics, are mediated by modulation of serotonin and GABA levels and that these microorganisms may have an antidepressant-like effect were supported in this study in accordance with the literature. The regulatory effects of microbiota on neurotransmitter synthesis pathways, its interaction with the HPA axis that governs the immune system and stress physiology, and the systemic signals it generates through intestinal permeability provide the scientific basis to justify the integration of nutrition-based interventions into clinical practice.

In conclusion, this study systematically analyzed the effects of nutrition on neuropsychiatric health in the light of current literature on the functioning of the microbiota-brain axis and suggested that nutrition can be not only supportive but also a transformative therapeutic element in the context of mental health. In this context, the proposed nutrition-based approaches have paved the way for the development of holistic and individual-specific intervention strategies in the clinical practice of modern psychiatry and brought the importance of interdisciplinary integration back to the agenda.

References

Akram N, Faisal Z, Irfan R, et al. Exploring the serotonin-probiotics-gut health axis: A review of current evidence and potential mechanisms. *Food Sci Nutr* (2023) 12(2):694-706.

Alam I, Agrawal S. Introduction to Obesity. In: Agrawal S, editor. *Obesity, Bariatric and Metabolic Surgery*. Cham: Springer; (2016).

Aljumaah MR, Bhatia U, Roach J, et al. The gut microbiome, mild cognitive impairment, and probiotics: A randomized clinical trial in middle-aged and older adults. *Clin Nutr* (Edinb) (2022) 41(11):2565-2576.

Allcock L, Mantzioris E, Villani A. Adherence to a Mediterranean Diet Is Inversely Associated with Anxiety and Stress but Not Depression: A Cross-Sectional Analysis of Community-Dwelling Older Australians. *Nutrients* (2024) 16(3):366.

Andreeva I, Tolpygo AV, Andreev V, et al. Psychobiotics: A new way in psychopharmacology, or how do bacteria manage our brain? *Clin Microbial Antimicrobe Chemotherapy* (2022) 24(2):108-133.

Anglin RES, Samaan Z, Walter SD, McDonald SD. Vitamin D deficiency and depression in adults: systematic review and meta-analysis. *Br J Psychiatry* (2013) 202:100-107.

Appleton J. The Gut-Brain Axis: Influence of Microbiota on Mood and Mental Health. *Integr Med* (Encinitas) (2018) 17(4):28-32.

Aranburu E, Matias S, Simón E, et al. Gluten and FODMAPs Relationship with Mental Disorders: Systematic Review. *Nutrients* (2021) 13(6):1894.

Banerjee S, Poore M, Gerdes S, et al. Transcriptomics reveal different metabolic strategies for acid resistance and gamma-aminobutyric acid (GABA) production in select Levilactobacillus brevis strains. *Microbe Cell Fact* (2021) 20(1):173.

Baxter AJ, Scott KM, Vos T, Whiteford HA. Global prevalence of anxiety disorders: a systematic review and meta-regression. *Psychol Med* (2013) 43(5):897-910.

Bertollo AG, Santos CF, Bagatini MD, Ignácio ZM. Hypothalamus-pituitary-adrenal and gut-brain axes in biological interaction pathway of the depression. *Front Neurosci* (2025) 19:1541075.

Bienenstock J, Kunze W, Forsythe P. Microbiota and the gut-brain axis. *Nutr Rev* (2015) 73 Suppl 1:28-31.

Bonaz B, Bazin T, Pellissier S. The Vagus Nerve at the Interface of the Microbiota-Gut-Brain Axis. F*ront Neurosci* (2018) 12:49.

Botturi A, Ciappolino V, Delvecchio G, et al. The Role and the Effect of Magnesium in Mental Disorders: A Systematic Review. *Nutrients* (2020) 12(6):1661.

Braga JD, Thongngam M, Kumrungsee T. Gamma-aminobutyric acid as a potential postbiotic mediator in the gut-brain axis. *NPJ Sci Food* (2024) 8(1):16.

Buey B, Forcén A, Grasa L, et al. Gut Microbiota-Derived Short-Chain Fatty Acids: Novel Regulators of Intestinal Serotonin Transporter. *Life (Basel)* (2023) 13(5):1085.

Butola LK, Ambad R, Kacchwa K. Zinc and psychiatric disorders: A review. *Int J Res Pharm Sci* (2020) 11 Suppl 4:1505-1514.

Canheta ABS, Santos ASEAC, Souza JD, Silveira EA. Traditional Brazilian diet and extra virgin olive oil reduce symptoms of anxiety and depression in individuals with severe obesity: Randomized clinical trial. *Clin Nutr* (Edinb) (2021) 40(2):404-411.

Casertano M, Fogliano V, Ercolini D. Psychobiotics, gut microbiota and fermented foods can help preserving mental health. *Food Res Int* (2022) 152:110892.

Chan MMH, Zarate-Lopez N, Martin L. Group education on the low FODMAP diet improves gastrointestinal symptoms but neither anxiety or depression in irritable bowel syndrome. *J Hum Nutr Diet* (2022) 35(3):425-434.

Chang JPC, Wu SK, Capuron L, Su KP. Omega-3 Polyunsaturated Fatty Acids: A Clinical Perspective on Gut Microbiota, Major Depressive Disorder, and Obesity. *Psychiatr Ann* (2025) 55(2):e34-e39.

Chen F, Wang J, Cheng Y, et al. Magnesium and Cognitive Health in Adults: A Systematic Review and Meta-Analysis. *Adv Nutr* (2024) 15(8):100272.

Cheng LH, Liu YW, Wu CC, et al. Psychobiotics in mental health, neurodegenerative and neurodevelopmental disorders. *J Food Drug Anal* (2019) 27(3):632-648.

Clapp M, Aurora N, Herrera L, et al. Gut microbiota's effect on mental health: The gut-brain axis. *Clin Pract* (2017) 7(4):987.

Conn KA, Borsom EM, Cope EK. Implications of microbe-derived ɣ-aminobutyric acid (GABA) in gut and brain barrier integrity and GABAergic signaling in Alzheimer's disease. *Gut Microbes* (2024) 16(1):2371950.

Costantini L, Molinari R, Farinon B, Merendino N. Impact of Omega-3 Fatty Acids on the Gut Microbiota. *Int J Mol Sci* (2017) 18(12):2645.

Czarnik W, Fularski P, Gajewska A, et al. The Role of Intestinal Microbiota and Diet as Modulating Factors in the Course of Alzheimer's and Parkinson's Diseases. *Nutrients* (2024) 16(2):308.

Dahiya D, Nigam PS. Clinical Potential of Microbial Strains, Used in Fermentation for Probiotic Food, Beverages and in Synbiotic Supplements, as Psychobiotics for Cognitive Treatment through Gut-Brain Signaling. *Microorganisms* (2022) 10(9):1687.

Desmedt O, Broers VJV, Zamariola G, et al. Effects of prebiotics on affect and cognition in human intervention studies. *Nutr Rev* (2019) 77(2):81-95.

Dickerman B, Liu J. Do the Micronutrients Zinc and Magnesium Play a Role in Adult Depression? *Top Clin Nutr* (2011) 26(3):257-267.

Dinan TG, Cryan JF. The Microbiome-Gut-Brain Axis in Health and Disease. *Gastroenterol Clin North Am* (2017) 46(1):77-89.
Du Y, Gao XR, Peng L, Ge JF. Crosstalk between the microbiota-gut-brain axis and depression. *Heliyon* (2020) 6(6):e04097.
Duan Y, Wu X, Yang Y, et al. Marked shifts in gut microbial structure and neurotransmitter metabolism in fresh inmates revealed a close link between gut microbiota and mental health: A case-controlled study. *Int J Clin Health Psychol* (2022) 22(3):100323.
Duchaine CS, Fiocco AJ, Carmichael PH, et al. Serum ω-3 Fatty Acids and Cognitive Domains in Community-Dwelling Older Adults from the NuAge Study: Exploring the Associations with Other Fatty Acids and Sex. *J Nutr* (2022) 152(9):2117-2124.
Duranti S, Ruiz L, Lugli G, et al. Bifidobacterium adolescentis as a key member of the human gut microbiota in the production of GABA. *Sci Rep* (2020) 10:14112.
Dyall S. The role of omega-3 fatty acids in adult hippocampal neurogenesis. *Oléagineux Corps Gras Lipides* (2011) 18:242-245.
Emanowicz P, Średnicka P, Wójcicki M, Zapaśnik A. Gut Microbiota as the Key to Mental and Physical Health. *Food Biotechnol Agric Sci* (2024) 78:32-36.
Fan C, Xu J, Tong H, Fang Y, et al. Gut-brain communication mediates the impact of dietary lipids on cognitive capacity. *Food Funct* (2024) 15(4):1803-1824.
Fan Y, Wang D, Wang W, et al. Depression, anxiety, and development of obesity from childhood to young adulthood in China. *J Psychosom Res* (2024) 185:111867.
Fei Y, Wang R, Lu J, et al. Probiotic intervention benefits multiple neural behaviors in older adults with mild cognitive impairment. *Geriatr Nurs* (New York) (2023) 51:167-175.
Fernandez-Julia PJ, Munoz-Munoz J, van Sinderen D. A comprehensive review on the impact of β-glucan metabolism by Bacteroides and Bifidobacterium species as members of the gut microbiota. *Int J Biol Macromol* (2021) 181:877-889.
Firth J, Marx W, Dash S, et al. The Effects of Dietary Improvement on Symptoms of Depression and Anxiety: A Meta-Analysis of Randomized Controlled Trials. *Psychosom Med* (2019) 81(3):265-280.
Forssten SD, Ouwehand AC, Griffin SM, Patterson E. One Giant Leap from Mouse to Man: The Microbiota-Gut-Brain Axis in Mood Disorders and Translational Challenges Moving towards Human Clinical Trials. *Nutrients* (2022) 14(3):568.
Francis HM, Stevenson RJ, Tan LSY et al. Kynurenic acid as a biochemical factor underlying the association between Western-style diet and depression: A cross-sectional study. *Front Nutr* (2022) 9:945538.
Frankiensztajn LM, Elliott E, Koren O. The microbiota and the hypothalamus-pituitary-adrenocortical (HPA) axis, implications for anxiety and stress disorders. *Curr Opin Neurobiol* (2020) 62:76-82.
Fu Y, Wang Y, Gao H, et al. Associations among Dietary Omega-3 Polyunsaturated Fatty Acids, the Gut Microbiota, and Intestinal Immunity. *Mediators Inflamm* (2021) 2021:8879227.
Ghanei N, Saghazadeh A, Rezaei N. Gut Microbiome and Immunity. In: Mahmoudi M, Rezaei N, editors. *Nutrition and Immunity*. Cham: Springer; (2019). p. 10.
Gibson-Smith D, Bot M, Brouwer IA, et al. Diet quality in persons with and without depressive and anxiety disorders. *J Psychiatr Res* (2018) 106:1-7.
González-Arancibia C, Urrutia-Piñones J, Illanes-González J, et al. Do your gut microbes affect your brain dopamine? *Psychopharmacology* (2019) 236(5):1611-1622.
Góralczyk-Bińkowska A, Szmajda-Krygier D, Kozłowska E. The Microbiota-Gut-Brain Axis in Psychiatric Disorders. *Int J Mol Sci* (2022) 23(19):11245.
Grover S, Patil A, Kaur A, Garg G. Probiotics: A Potential Immunotherapeutic Approach for the Treatment of Schizophrenia. *J Pharm Bioallied Sci* (2019) 11(4):321-327.
Hallahan B, Ryan T, Hibbeln JR, et al. Efficacy of omega-3 highly unsaturated fatty acids in the treatment of depression. *Br J Psychiatry* (2016) 209(3):192-201.
Hamamah S, Aghazarian A, Nazaryan A, et al. Role of Microbiota-Gut-Brain Axis in Regulating Dopaminergic Signaling. *Biomedicines* (2022) 10(2):436.
Hashimoto M. Nihon rinsho. *Jpn J Clin Med* (2014) 72(4):648-656.
Horigome A, Okubo R, Hamazaki K, et al. Association between blood omega-3 polyunsaturated fatty acids and the gut microbiota among breast cancer survivors. *Benef Microbes* (2019) 10(7):751-758.

Horning S, Davis H. Encyclopedia of human behavior. Aging and cognition. 2nd ed. San Diego: Academic Press; (2012). p. 44-52.

Hu Y, Yu M, Wang Y, et al. Exploring the Association between Serum B Vitamins, Homocysteine and Mental Disorders: Insights from Mendelian Randomization. *Nutrients* (2024) 16(13):1986.

Huang F, Wu X. Brain Neurotransmitter Modulation by Gut Microbiota in Anxiety and Depression. *Front Cell Dev Biol* (2021) 9:649103.

Huang Y, Ruan S, Yang Y, et al. Impact of dietary magnesium intake on depression risk in American adults: a cross-sectional study of the National Health and Nutrition Examination Survey 2005-2020. *Front Nutr* (2025) 12:1484344.

Jacka FN, Pasco JA, Mykletun A, et al. Association of Western and traditional diets with depression and anxiety in women. *Am J Psychiatry* (2010) 167(3):305-311.

Janitra FE, Chen R, Lin HC, et al. Efficacy of resilience-related psychological interventions in patients with long-term diseases: A meta-analysis of randomized controlled trials. *Int J Ment Health Nurs* (2024) 33(5):1388-1406.

Jenkins TA, Nguyen JC, Polglaze KE, Bertrand PP. Influence of Tryptophan and Serotonin on Mood and Cognition with a Possible Role of the Gut-Brain Axis. *Nutrients* (2016) 8(1):56.

Jernerén F, Cederholm T, Refsum H, et al. Homocysteine Status Modifies the Treatment Effect of Omega-3 Fatty Acids on Cognition in a Randomized Clinical Trial in Mild to Moderate Alzheimer's Disease: The OmegAD Study. *J Alzheimers Dis* (2019) 69(1):189-197.

Julius S. Nutritional Psychiatry: The Impact of Diet on Mental Health. *Newport Int J Res Med Sci* (2025) 6(1):9-13.

Kamal N, Saharan BS, Duhan JS, et al. Exploring the promise of psychobiotics: Bridging gut microbiota and mental health for a flourishing society. *Med Microecol* (2024) 23(1):100118.

Kamali M, Dastsouz F, Sadeghi F, et al. Associations between Western and Mediterranean-type dietary patterns and anxiety and stress. *Acta Aliment* (2016) 45(3):398-405.

Kang JW, Zivkovic AM. The Potential Utility of Prebiotics to Modulate Alzheimer's Disease: A Review of the Evidence. *Microorganisms* (2021) 9(11):2310.

Kang T, Zheng J, Jiang C, et al. Amelioration of walnut, peony seed and camellia seed oils against D-galactose-induced cognitive impairment in mice by regulating gut microbiota. *Food Funct* (2024) [Epub ahead of print].

Karoń K, Kozakowski J. Obesity and psychiatric disorders. *Wiedza Med* (2023) 5(1):1-7.

Kazemi A, Djafarian K. Gut microbiota and depression. In: Preedy VR, Watson RR, editors. *Neurological Modulation of Sleep.* Academic Press; (2021). p.463-472.

Keck MM, Vivier H, Cassisi JE, et al. Examining the Role of Anxiety and Depression in Dietary Choices among College Students. *Nutrients* (2020) 12(7):2061.

Kerman BE, Self W, Yassine HN. Can the gut microbiome inform the effects of omega-3 fatty acid supplementation trials on cognition? *Curr Opin Clin Nutr Metab Care* (2024) 27(2):116-124.

Knüppel A, Shipley M, Brunner E. Dietary fibre intake and common mental disorder: Prospective findings from the Whitehall II study. *Proc Nutr Soc* (2018) 77(OCE4):E126-E160.

Kortlever TL, Ten Bokkel Huinink S, Offereins M, et al. Low-FODMAP Diet Is Associated With Improved Quality of Life in IBS Patients-A Prospective Observational Study. *Nutr Clin Pract* (2019) 34(4):623-630.

Kumar A, Mehan S, Tiwari A, et al. Magnesium (Mg^{2+}): Essential Mineral for Neuronal Health: From Cellular Biochemistry to Cognitive Health and Behavior Regulation. *Curr Pharm Des* (2024) 30(39):3074-3107.

Laudani S, Torrisi SA, Alboni S, et al. Gut microbiota alterations promote traumatic stress susceptibility associated with p-cresol-induced dopaminergic dysfunctions. *Brain Behav Immun* (2023) 107:385-396.

Li J, Ma Y, Bao Z, et al. Clostridiales are predominant microbes that mediate psychiatric disorders. *J Psychiatr Res* (2020) 130:48-56.

Lima-Ojeda JM, Rupprecht R, Baghai TC. Darmflora und Depression: Pathophysiologie der Depression: Hypothalamus-Hypophysen-Nebennierenrinden- und Mikrobiota-Darm-Hirn-Achse [Gut microbiota and depression: Pathophysiology of depression: hypothalamic-pituitary-adrenal axis and microbiota-gut-brain axis]. Der Nervenarzt (2020) 91(12):1108-1114.

Liu CZ, Feng XY, Liu S, et al. Synthesis and Metabolism of Gut Dopamine. In: Zhu JX, editor. *Dopamine in the Gut*. Singapore: Springer; (2021). p. 2.

Liu N, Yang D, Sun J, Li Y. Probiotic supplements are effective in people with cognitive impairment: a meta-analysis of randomized controlled trials. *Nutr Rev* (2023) 81(9):1091-1104.

Lozano J, Fabius S, Fernández-Ciganda S, et al. Beneficial effect of GABA-producing Lactiplantibacillus strain LPB145 isolated from cheese starters evaluated in anxiety- and depression-like behaviours in rats. *Benef Microbes* (2024) 15(5):465-479.

Lv T, Ye M, Luo F, et al. Probiotics treatment improves cognitive impairment in patients and animals: A systematic review and meta-analysis. *Neurosci Biobehav Rev* (2021) 120:159-172.

Lynch JB, Hsiao EY. Toward understanding links between the microbiome and neurotransmitters. Ann N Y *Acad Sci* (2023) 1524(1):10-16.

Madireddy S, Madireddy S. Most Effective Combination of Nutraceuticals for Improved Memory and Cognitive Performance in the House Cricket, Acheta domesticus. *Nutrients* (2021) 13(2):362.

Marino Maia J, Solon IG, Soriano RN, Branco LG. Exploring the links between pro-inflammatory diets, gut dysbiosis, serotonin, and their implications for psychiatric and neurodegenerative disorders. *Brain Behav Immun Integr* (2024) 8(7):100097.

Martins LB, Braga Tibães JR, Sanches M, et al. Nutrition-based interventions for mood disorders. *Expert Rev Neurother* (2021) 21(3):303-315.

Matiş L, Alexandru BA, Fodor R, et al. Effect of Probiotic Therapy on Neuropsychiatric Manifestations in Children with Multiple Neurotransmitter Disorders: A Study. *Biomedicines* (2023) 11(10):2643.

Mikkelsen K, Stojanovska L, Apostolopoulos V. The Effects of Vitamin B in Depression. *Curr Med Chem* (2016) 23(38):4317-4337.

Misra S, Mohanty D. Psychobiotics: A new approach for treating mental illness? *Crit Rev Food Sci Nutr* (2019) 59(8):1230-1236.

Moore K, Hughes CF, Hoey L, et al. B-vitamins in Relation to Depression in Older Adults Over 60 Years of Age: The Trinity Ulster Department of Agriculture (TUDA) Cohort Study. *J Am Med Dir Assoc* (2019) 20(5):551-557.e1.

Mosquera FEC, Lizcano Martinez S, Liscano Y. Effectiveness of Psychobiotics in the Treatment of Psychiatric and Cognitive Disorders: A Systematic Review of Randomized Clinical Trials. *Nutrients* (2024) 16(9):1352

Mörkl S, Butler MI, Cichini F, et al. Psychobiotics. In: *The Oxford Handbook of the Microbiome-Gut-Brain Axis*. (2020).

Mukilan M, Adithya R, Pruthivi S. Role of Probiotic Microorganisms in the Brain Plasticity Development. *J Exp Biol Agric Sci* (2024) 12(3):354-365.

Nafady MH, Sayed ZS, Abdelkawy DA, et al. The Effect of Gut Microbe Dysbiosis on the Pathogenesis of Alzheimer's Disease (AD) and Related Conditions. *Curr Alzheimer Res* (2022) 19(4):274-284.

Nguyen HD, Vu GH, Kim WK. Aspects of Molecular Medicine. Aspects Mol Med (2024) 4:100047.

Nist KM, CapriglioneA, Wainford RD. Prebiotic Dietary Fiber Supplementation Attenuates Gut Microbiota Dysbiosis And Age-dependent Hypertension-driven Cognitive Impairment. *Hypertension* (2023) 80 Suppl_1.

Noah L, Dye L, Bois De Fer B, et al. Effect of magnesium and vitamin B6 supplementation on mental health and quality of life in stressed healthy adults: Post-hoc analysis of a randomized controlled trial. *Stress Health* (2021) 37(5):1000-1009.

Nuriel-Ohayon M, Neuman H, Koren O. Microbial Changes during Pregnancy, Birth, and Infancy. *Front Microbiol* (2016) 7:1031.

O'Brien L, Skidmore P, Wall C, et al. A Low FODMAP Diet Is Nutritionally Adequate and Therapeutically Efficacious in Community Dwelling Older Adults with Chronic Diarrhea. *Nutrients* (2020) 12(10):3002

O'Mahony SM, Clarke G, Borre YE, et al. Serotonin, tryptophan metabolism and the brain-gut-microbiome axis. *Behav Brain Res* (2015) 277:32-48.

Pan A, Sun Q, Czernichow S, et al. Bidirectional association between depression and obesity in middle-aged and older women. *Int J Obes* (Lond) (2012) 36(4):595-602.

Park JC, Im SH. The gut-immune-brain axis in neurodevelopment and neurological disorders. *Microbiome Res Rep* (2022) 1(4):23.

Peirce JM, Alviña K. The role of inflammation and the gut microbiome in depression and anxiety. *J Neurosci Res* (2019) 97(10):1223-1241.

Trajkovska Petkoska A, Trajkovska-Broach A. Mediterranean Diet—A Healthy Dietary Pattern and Lifestyle for Strong Immunity. In: Agarwal P, Nieto JJ, Ruzhansky M, Torres DFM, editors. Analysis of Infectious Disease Problems (Covid-19) and Their Global Impact. Singapore: Springer; (2021). p. 13.

Petridou ET, Kousoulis AA, Michelakos T, et al. Folate and B12 serum levels in association with depression in the aged: a systematic review and meta-analysis. *Aging Ment Health* (2016) 20(9):965-973.

Piacentino D, Rossi S, Piretta L, et al. Low-FODMAP diet improves psychopathology in IBS. *Eur Psychiatry* (2016) 33:S395.

Poggiano M, Ciarla S, Gnerre P, et al. The management of the patient with malnutrition: from evidence to clinical practice. *Ital J Med* (2017) 11:134-150.

Polivy J, Coleman J, Herman CP. The effect of deprivation on food cravings and eating behavior in restrained and unrestrained eaters. *Int J Eat Disord* (2005) 38(4):301-309.

Popkes M, Valenzano D. Microbiota–host interactions shape ageing dynamics. *Philos Trans R Soc Lond B Biol Sci* (2020) 375.

Potter K, Gayle EJ, Deb S. Effect of gut microbiome on serotonin metabolism: a personalized treatment approach. *Naunyn Schmiedebergs Arch Pharmacol* (2024) 397(5):2589-2602.

Provensi G, Schmidt SD, Boehme M, et al. Preventing adolescent stress-induced cognitive and microbiome changes by diet. *Proc Natl Acad Sci USA* (2019) 116(19):9644-9651.

Psaltopoulou T, Sergentanis TN, Panagiotakos DB, et al. Mediterranean diet, stroke, cognitive impairment, and depression: A meta-analysis. *Ann Neurol* (2013) 74(4):580-591.

Pushpanathan P, Mathew GS, Selvarajan S, et al. Gut microbiota and its mysteries. *Indian J Med Microbiol* (2019) 37(2):268-277.

Qu S, Yu Z, Zhou Y, et al. Gut microbiota modulates neurotransmitter and gut-brain signaling. *Microbiol Res* (2024) 287:127858.

Renukadevi J, Velmurugan R, Nandhinidevi G, et al. Neuroprotective role of probiotics in cognitive decline. *J Appl Pharm Res* (2024) 12(5):1-15.

Ribichini E, Scalese G, Mocci C, Severi C. Gut–Brain Axis and Psychopathology: Exploring the Impact of Diet with a Focus on the Low-FODMAP Approach. *Nutrients* (2024) 16(20):3515.

Robertson RC, Seira Oriach C, Murphy K, et al. Omega-3 polyunsaturated fatty acids critically regulate behavior and gut microbiota development in adolescence and adulthood. *Brain Behav Immun* (2017) 59:21-37.

Romijn JA, Corssmit EP, Havekes LM, Pijl H. Gut-brain axis. *Curr Opin Clin Nutr Metab Care* (2008) 11(4):518-521.

Roy N. Diet, Gut Physiology, and the Microbiota-Gut-Brain Axis. *Proc Nutr Soc* (2024) 83(OCE1):E13

Ruiz-Gonzalez C, Cardona D, Rueda-Ruzafa L, et al. Cognitive and Emotional Effect of a Multi-species Probiotic Containing Lactobacillus rhamnosus and Bifidobacterium lactis in Healthy Older Adults: A Double-Blind Randomized Placebo-Controlled Crossover Trial. *Probiotics Antimicrob Proteins* (2024) [Epub ahead of print].

Rusch JA, Layden BT, Dugas LR. Signaling cognition: the gut microbiota and hypothalamic-pituitary-adrenal axis. *Front Endocrinol* (Lausanne) (2023) 14:1130689.

Sadeghi O, Keshteli AH, Afshar H, et al. Adherence to Mediterranean dietary pattern is inversely associated with depression, anxiety and psychological distress. *Nutr Neurosci* (2021) 24(4):248-259.

Saghafian F, Sharif N, Saneei P, et al. Consumption of Dietary Fiber in Relation to Psychological Disorders in Adults. *Front Psychiatry* (2021) 12:587468.

Sánchez-de-Lara-Sánchez S, Sánchez-Pérez AM. Probiotics Treatment Can Improve Cognition in Patients with Mild Cognitive Impairment: A Systematic Review. *J Alzheimers Dis* (2022) 89(4):1173-1191.

Sanmukhani J, Satodia V, Trivedi J, et al. Efficacy and safety of curcumin in major depressive disorder: a randomized controlled trial. *Phytother Res* (2014) 28(4):579-585.

Sauer AK, Grabrucker AM. Zinc Deficiency During Pregnancy Leads to Altered Microbiome and Elevated Inflammatory Markers in Mice. *Front Neurosci* (2019) 13:1295.

Savignac HM, Corona G, Mills H, et al. Prebiotic feeding elevates central brain derived neurotrophic factor, N-methyl-D-aspartate receptor subunits and D-serine. *Neurochem Int* (2013) 63(8):756-764.

Shi H, Yu Y, Lin D, et al. β-glucan attenuates cognitive impairment via the gut-brain axis in diet-induced obese mice. *Microbiome* (2020) 8(1):143.

Silva DFO, Cobucci RN, Gonçalves AK, Lima SCVC. Systematic review of the association between dietary patterns and perinatal anxiety and depression. *BMC Pregnancy Childbirth* (2019) 19(1):212.

Singh RP, Bhardwaj A. β-glucans: a potential source for maintaining gut microbiota and the immune system. *Front Nutr* (2023) 10:1143682.

Slavin J. Fiber and prebiotics: mechanisms and health benefits. *Nutrients* (2013) 5(4):1417-1435.

Sønderskov KM, Dinesen PT, Santini ZI, Østergaard SD. The depressive state of Denmark during the COVID-19 pandemic. Acta Neuropsychiatr (2020) 32(4):226-228.

Stasi C, Sadalla S, Milani S. The Relationship Between the Serotonin Metabolism, Gut-Microbiota and the Gut-Brain Axis. *Curr Drug Metab* (2019) 20(8):646-655.

Strandwitz P, Kim KH, Terekhova D, et al. GABA-modulating bacteria of the human gut microbiota. N*at Microbiol* (2019) 4(3):396-403.

Sudo N. The Hypothalamic-Pituitary-Adrenal Axis and Gut Microbiota: A Target for Dietary Intervention? In: The Gut-Brain Axis Dietary, Probiotic, and Prebiotic Interventions on the Microbiota. Elsevier Inc.; (2016). p.293-304.

Sun R, Wang J, Feng J, Cao B. Zinc in Cognitive Impairment and Aging. *Biomolecules* (2022) 12(7):1000

Tofani GSS, Leigh SJ, Gheorghe CE, et al. Gut microbiota regulates stress responsivity via the circadian system. *Cell Metab* (2025) 37(1):138-153.e5.

Noye Tuplin EW, Alukic E, Lowry Deet al. Dietary fiber combinations to mitigate the metabolic, microbial, and cognitive imbalances resulting from diet-induced obesity in rats. *FASEB J* (2022) 36(5):e22269.

Vagnerová K, Vodička M, Hermanová P, et al. Interactions Between Gut Microbiota and Acute Restraint Stress in Peripheral Structures of the Hypothalamic-Pituitary-Adrenal Axis and the Intestine of Male Mice. *Front Immunol* (2019) 10:2655.

Valles-Colomer M, Falony G, Darzi Y, et al. The neuroactive potential of the human gut microbiota in quality of life and depression. *Nat Microbiol* (2019) 4(4):623-632.

Wang D, Jiang Y, Jiang J, et al. Gut microbial GABA imbalance emerges as a metabolic signature in mild autism spectrum disorder linked to overrepresented Escherichia. *Cell Rep Med* (2025) 6(1):101919.

Waworuntu RV, Chichlowski M, Cox S, Berg BM. Dietary prebiotics modulate fecal microbiota and impact cognitive functions in developing rodents (LB376). *FASEB J* (2014) 28 Suppl 1:LB376.

Weng TT, Hao JH, Qian QW, et al. Is there any relationship between dietary patterns and depression and anxiety in Chinese adolescents? *Public Health Nutr* (2012) 15(4):673-682.

White N. Omega-3 Fatty Acids and Cognitive Decline. *Am J Lifestyle Med* (2015) 9(4):275-277.

Wienecke E, Nolden C. Langzeit-HRV-Analyse zeigt Stressreduktion durch Magnesiumzufuhr [Long-term HRV analysis shows stress reduction by magnesium intake]. *MMW Fortschr Med* (2016) 158 Suppl 6:12-16.

Wójcik ZM, Kamińska-Omasta K, Omasta B, et al. Gut microbiota and its impact on brain function and mental health. *Qual Sport* (2025) 38:58196.

World Health Organization. Depression. https://www.who.int/news-room/fact-sheets/detail/depression (Accessed 15.05.2025).

Wu Q, Xu Z, Song S, et al. Gut microbiota modulates stress-induced hypertension through the HPA axis. *Brain Res Bull* (2020) 162:49-58.

Xu P, Pang D, Zhou J, et al. Behavioral changes and brain epigenetic alterations induced by maternal deficiencies of B vitamins in a mouse model. *Psychopharmacology* (2021) 238(4):1213-1222.

Yadav SK, Ito K, Dhib-Jalbut S. Interaction of the Gut Microbiome and Immunity in Multiple Sclerosis: Impact of Diet and Immune Therapy. *Int J Mol Sci* (2023) 24(19):14756.

Yang XQ, Zhao Y, Xue L, et al. Probiotics Improve Cognitive Impairment by Decreasing Bacteria-Related Pattern Recognition Receptor-Mediated Inflammation in the Gut-Brain Axis of Mice. *J Integr Neurosci* (2023) 22(4):92.

Zakizadeh E, Saraf-Bank S, Haghighatdoost F, et al. Associations between dietary patterns and depression and anxiety in middle-aged adults. *Adv Hum Biol* (2019) 9(3):228-235.

Zhang S, Lv S, Li Y, et al. Prebiotics modulate the microbiota-gut-brain axis and ameliorate cognitive impairment in APP/PS1 mice. *Eur J Nutr* (2023) 62(7):2991-3007.

Zhu G, Zhao J, Zhang H, et al. Probiotics for Mild Cognitive Impairment and Alzheimer's Disease: A Systematic Review and Meta-Analysis. *Foods (Basel)* (2021) 10(7):1672.

Zhu S, Jiang Y, Xu K, et al. The progress of gut microbiome research related to brain disorders. *J Neuroinflammation* (2020) 17:25.

Zinkow A, Grodzicki W, Czerwińska M, Dziendzikowska K. Molecular Mechanisms Linking Omega-3 Fatty Acids and the Gut-Brain Axis. *Molecules (Basel)* (2024) 30(1):71.

Zolot J. Omega-3 Fatty Acids and Antioxidants: No Effect on Cognitive Function. *Am J Nurs* (2015) 115(12):68.

Zorgetto-Pinheiro VA, Machate DJ, Figueiredo PS, et al. Omega-3 Fatty Acids and Balanced Gut Microbiota on Chronic Inflammatory Diseases: A Close Look at Ulcerative Colitis and Rheumatoid Arthritis Pathogenesis. *J Med Food* (2022) 25(4):341-354.

Chapter 10

From Traditional Wisdom to Scientific Evidence: The Integration of Modern Nutrition Science and Cultural Practice

Tuba Erkaya Kotan*
Department of Food Technology, Ataturk University, Erzurum, Turkiye

Abstract

Since prehistoric times, humanity has struggled to survive and thrive in its environment, with nutrition always central to this endeavor. Beyond its biological necessity, nutrition constitutes a fundamental element of social identity, cultural memory, and historical continuity, while also serving as a scientific discipline examining the interactions between food and living organisms. Traditional dietary practices shaped by cultural and ecological adaptations spanning centuries, offer a holistic and sustainable approach that emphasizes the seasonal, regional, and spiritual dimensions of food. Modern nutritional science, which emerged in the early 20th century as a response to global conflicts and food insecurity, has often overlooked the holistic wisdom embedded in cultural food practices. However, recent research highlights the value of traditional food knowledge for contemporary nutrition. Empirical studies show that incorporating local and fermented foods into modern diets increases nutrient intake, supports gut and immune health, and reduces the risk of chronic disease. Fermented foods, such as tarhana and yogurt in Turkish cuisine and miso and natto in Japanese cuisine, provide health-promoting compounds, including probiotics, bioactive peptides, vitamin K2, nattokinase, and isoflavones. This chapter aims to bridge the gap between traditional culinary wisdom and modern nutritional science. By examining the scientifically validated benefits of culturally-specific nutritional practices, this chapter proposes an integrative framework to guide public health policies and personalized nutrition strategies. In doing so, the ethical, sociological, and epistemological dimensions of this integration will be explored, including questions regarding the value of knowledge, cultural respect, and the role of traditional practices in sustainable nutritional models. Overall, traditional nutrition is more than a biological necessity; it is an expression of cultural identity and ritual practices. Modern nutritional science, with its reductionist and biochemical approach, benefits from integrating this knowledge, as demonstrated by the health benefits of fermented and plant-based foods, such as yogurt, tarhana, turmeric, and

* Corresponding Author's Email: terkaya@atauni.edu.tr.

In: Let Food Be Your Medicine, and Medicine Be Your Food
Editors: Gulcin Yildiz and Arzu Imece
ISBN: 979-8-90134-204-6

olive oil. This chapter also addresses the persistence of nutritional myths and emphasizes the importance of critical thinking and cultural sensitivity in combating misinformation. Lastly, this chapter examines the ethical and ecological challenges of globalizing traditional foods and their effects on local ecosystems, food security, and biocultural sustainability.

Keywords: traditional nutrition, fermented foods, cultural food practices

Introduction

Nutrition is a biological need and a fundamental life activity necessary for individuals to maintain their physiological, psychological, and social development. Furthermore, nutrition is a scientific discipline concerned with the study of nutrients and their interactions with food and living organisms. Nutrition encompasses the study of the effects of food components on health, growth, reproduction, and the prevention or progression of disease (Ünsal, 2019). Food choices, preparation methods, consumption timing, and culinary rituals all carry cultural codes that reflect a community's way of life, belief systems, and their relationship with the environment. In this context, traditional dietary practices serve not only as elements of daily living, but also as vehicles of knowledge transmission, experience, and healing.

Nutritional knowledge that has been conserved and transferred through generations and shaped by cultural and ecological adaptation over many centuries offers us a holistic and sustainable way of eating that is still relevant to our modern health problems (Suarez and Adibi, 2025). Traditional nutritional wisdom encompasses practices based on centuries of observation, experimentation, and interaction with nature. Often passed down by mouth, this knowledge is shaped by factors, such as age, sex, season, health condition, and environmental circumstance. From Anatolia to India and from China to Latin America, people have used food not only to nourish the body, but also to maintain balance, seek protection, and promote healing. These approaches are today referred to as *folk nutrition*, *ethno-nutrition*, or *indigenous knowledge systems*.

Every civilization around the world has developed its own unique culinary culture, shaped by their use of locally available plant-based and animal-based food resources. Traditional cuisines evolve over centuries, becoming a fundamental part of cultural identity. Local communities possess extensive knowledge regarding the identification, harvesting, and preparation of food sources specific to their ecosystems. Moreover, these communities often have a profound awareness of the biological effects of these foods on the human body, along with diverse perspectives on the importance of food, nutrition, and their effects on health (Payyappallimana and Venkatasubramanian, 2015). It is evident that traditional food wisdom is closely connected with the cultural, spiritual, and ecological lives of these communities. These practices go beyond just cooking; they are woven into ceremonies and daily rituals that convey values, foster identity, and promote a sense of belonging. For example, in Türkiye, the preparation of aşure—a traditional dessert composed of grains, legumes, and dried fruits—embodies a confluence of religious, cultural, and historical influences. Commonly prepared during the Islamic month of Muharram, aşure is deeply rooted in local traditions and agricultural practices. The preparation and communal sharing of aşure symbolizes themes of gratitude, remembrance, and social unity, reflecting the dish's broader function as a vehicle for intergenerational connection and cultural continuity (Polat, 2019). Similarly, in Argentina, locro—a hearty stew typically consumed during national celebrations—reveals the synthesis of Indigenous and colonial foodways, with its ingredients and preparation rituals tied to national identity and seasonal rhythms. Additionally, the tradition of having a sharing mate in Argentina transcends nutritional value, serving as a cherished social ritual and an expression of hospitality that bridges generations (Albala, 2011).

Conversely, from the early 20th century onward, the development of nutritional science became increasingly intertwined with sociopolitical transformations and public health imperatives. In response to the challenges posed by global conflicts, poverty, and widespread food insecurity, governmental and institutional bodies sought to standardize and coordinate national efforts aimed at meeting basic nutritional needs (Ridgway et al., 2019). As a result, food composition tables and dietary guidelines were primarily formulated to combat undernutrition, particularly in the Global South and among lower-income populations in industrialized nations, such as the USA and the UK. The discovery of vitamins in the 1920s had a considerable effect on this paradigm. Nutritional instruments categorized foods according to their chemical composition, focusing on macronutrients and micronutrients, such as energy, protein, carbohydrates, fat, minerals, and vitamins. These instruments played a crucial role in wartime nutrition strategies and post-war recovery plans (Mozaffarian, 2016; Cannon and Leitzmann, 2022).

From the mid-20th century, modern nutritional science emerged as a discipline rooted in biochemical, molecular, and clinical methodologies. Concepts, such as caloric value, glycemic index, antioxidant capacity, and nutrient density were developed within this scientific framework (Mozaffarian, 2016). However, despite its analytical rigor, modern nutrition has often adopted a reductionist perspective—termed "nutritionism" by some scholars—prioritizing quantifiable biochemical components instead of holistic, culturally grounded knowledge systems. This epistemological bias has contributed to the marginalization of traditional dietary wisdoms that emphasize seasonal, regional, and spiritual dimensions of food and health (Nisbett et al., 2022). In addition to addressing dietary inadequacies, food policies were shaped by wartime needs, including population resilience and national survival. Food rationing systems, notably in the UK during both World Wars, were instrumental in maintaining public health. Consequently, food guides published by national and international bodies, including the League of Nations and the UN, reflected a functional and survival-oriented approach to nutrition, emphasizing efficiency, adequacy, and public welfare.

In recent years, traditional food knowledge has gained renewed scholarly attention not only as a cultural heritage, but also as a resource for nutritional science. Empirical studies have shown that integrating indigenous foods into modern diets can lead to improved nutrient intake and health outcomes. For example, research among tribal communities in India showed that the inclusion of local wild and fermented foods contributed significantly to daily calcium and iron needs (Ghosh-Jerath et al., 2016, Kapoor et al., 2022). Similarly, studies in Latin America and the Himalayas emphasized the resilience and nutritional value of traditional food systems in addressing food insecurity and promoting gut and immune health (Ojha et al., 2022; Lopes et al., 2024;). These findings support a more holistic understanding of health that associates biochemical efficacy with cultural and ecological relevance. Beyond enhancing individual wellbeing, this integration promises to foster cultural sustainability, preserve biodiversity, and stimulate local economies.

For instance, traditional fermented foods in Turkish cuisine, including tarhana, yogurt, and şalgam, are increasingly recognized for their functional benefits. Research has shown that these foods are characterized by the presence of live lactic acid bacteria, bioactive peptides, and antioxidants, which reportedly promote digestive processes, stimulate the growth of beneficial microbes in the gut, and contribute to the prevention of chronic diseases (Gok, 2023; Deveci et al., 2023). In the context of Indian Ayurvedic cuisine, herbal spice blends, such as masala, have been shown to function as nutraceuticals, thereby displaying both preventive and therapeutic potential (Iyer et al., 2009; Patil et al., 2021). A comparable example can be found in Japanese fermented soy products, such as miso and natto, which have become the focus of modern scientific research. The bioactive compounds present in these products, including nattokinase, vitamin K2, and isoflavones, are associated with a range of benefits, from improved gut health to reduced mortality risk (Chan et al., 2021; Nagata et al., 2017; Weng et al., 2017).

This chapter aims to combine traditional culinary wisdom and modern nutritional science. First, by examining culture-specific nutritional practices through a scientific lens, this chapter aims to uncover the hidden potential of these traditional approaches. This process not only informs public health policies, but also enables the generation of more personalized and effective dietary strategies for individuals. Furthermore, myths and misconceptions common in traditional nutrition that lack a scientific basis are addressed. The goal is to adopt a respectful approach to cultural beliefs while disseminating accurate information. Lastly, this chapter examines the ethical, sociological, and epistemological dimensions of this integration process, raising important questions about who produces information and how it is disseminated, and how this process can be conducted without negative consequences, such as cultural appropriation or distortion. As such, this chapter aims to contribute to the more equitable, ethical, and sustainable sharing of nutritional information.

The Cultural Origins of Therapeutic Foods: Ancient Wisdom throughout History

Throughout history, food has served as sustenance and as a cornerstone of healing, protection, and cultural identity in many societies. From ancient Mesopotamia to the Far East and from the Mediterranean to Indigenous American societies, various cultures have developed distinct yet convergent practices of using food as medicine. These practices were shaped by environmental factors, spiritual beliefs, and empirical observation long before the emergence of modern scientific methods (Fabricant & Farnsworth, 2001; Jyoti, and Yadav, 2025). The relationship between nutrition and health has been recognized in various traditional medical systems for a considerable length of time. Across diverse cultural contexts, food has been regarded as a source of nourishment and as a medium for healing, balancing, and maintaining overall wellbeing (Kuhnlein et al., 2009). The prevailing wisdom regarding food has been cultivated over the course of centuries and is deeply rooted in the ecological, social, and cultural context of each community. These dietary practices, which are frequently transmitted traditionally by mouth or ritualized, represent an empirical understanding of human physiology and the environment. Despite the absence of written documentation, traditional diets and therapeutic foods are rooted in a comprehensive experiential knowledge system that has influenced the health of numerous civilizations (Willet et al., 2019).

The intellectual origins of Western dietary concepts can be traced back to ancient Egypt, where medical practice was an extensive and well documented discipline. The Ebers Papyrus, dating back to approximately 1550 BCE, represents the most comprehensive surviving record of this tradition. This medical text details a wide array of remedies that blended practical knowledge with magical and spiritual components, reflecting a holistic view of health (Nunn, 2002; David, 2013). The Egyptians held a theory of internal "channels" that carried air, water, and blood throughout the body, drawing a direct analogy to the flow of the River Nile. It was thought that when these channels became blocked, illness would arise, and physicians would prescribe substances, such as laxatives to clear the obstructions (Nunn, 2002). This approach constituted a fundamental, linear model of health predicated on a straightforward cause-and-effect relationship between a blockage and its cure. This fundamental concept underwent a substantial evolution with the emergence of Greek medicine. The Greek physician Hippocrates of Kos (c. 460–375 BCE) who was regarded as the father of medicine and traditionally credited as the author of the Hippocratic Oath, is often attributed with the aphorism: "Let food be thy medicine and medicine be thy food" (Touwaide and Appetiti, 2015). The Hippocratic–Galenic tradition systematized the concept of bodily balance into the theory of the four humors, which were fluids that regulated the human body, namely: blood, phlegm, yellow bile, and black bile (Barea, 2011). Each humor was associated with two qualities: blood was hot and moist, yellow bile was hot and dry, black bile was cold and dry, and phlegm was cold and wet (Hankinson, 2016). Galen of Pergamon, the renowned physician to Roman emperors, built upon the work of Hippocrates to create a comprehensive system of diagnostics and dietetics. He categorized foods according to their elemental properties and the influence these exerted on the humors. It was believed that humors were the equilibrium of the body, and illness was the result of an imbalance (Bendick, 2002).

In the Anatolian region and during the Islamic golden age, Greco–Arabic medical knowledge was synthesized with local dietary and therapeutic traditions. Avicenna (Ibn Sina, 980–1037 CE), emphasized the primacy of diet in maintaining health and treating disease. Avicenna wrote around 450 works on different scientific topics, especially medicine, and 240 of these have survived to the present day. He also produced illustrations of the human body that were included in his writings. Avicenna's most important medical work, The Canon of Medicine, with its clear chapters and detailed content, had a major effect on medical studies in the Islamic world and Europe for hundreds of years. This work was taught at important universities, including the universities of Montpellier and Louvain, which highlights how important Avicenna's work was for medical science (Çömlek, 2025).

In traditional medicine, food is not viewed solely as a means of survival; rather, it is considered a powerful tool in the prevention and treatment of disease. The complex relationship between nutrition and health in various medical traditions, such as Ayurveda, traditional Chinese medicine (TCM), and Unani in India, is characterized by a focus on balance, harmony, and individual health needs (Lad, 2002; Kaptchuk, 2000). The philosophies of these traditional concepts often center on the idea that consuming appropriate food at the optimal time and in the right way can restore harmony in the body and mind. For example, Ayurveda, India's ancient health system, emphasizes personalized dietary practices based on an individual's constitution, or dosha (Lad, 2002). This approach posits that nutrition plays a crucial role in the primary prevention of disease and the promotion of optimal health. The classical Ayurvedic text known as Charaka Samhita, states: "In the absence of a proper diet, the effectiveness of medicines disappears." This excerpt highlights the fundamental function of nutrition in maintaining body health and correcting imbalances.

In TCM, food is viewed as being integral to the balance of yin and yang, whereby it is believed that each type of food possesses distinct energy properties that can assist in restoring the body's equilibrium (Kaptchuk, 1986). TCM does not merely advocate specific foods based on their nutritional value; it also extols their ability to harmonize the body's internal systems, recommending cooling foods in excessive heat and warming foods in cold conditions.

The Unani system of medicine, which is rooted in the Greco–Arabic tradition, categorizes foods based on their humoral properties (i.e., hot, cold, moist, and dry) with the aim of achieving equilibrium between these qualities within the body The prescription of Unani diets is determined by an individual's temperament (i.e., mizaj) and is believed to possess therapeutic properties that can treat a variety of physical and mental health problems (Itrat et al., 2014).

Another traditional diet known as the traditional Japanese diet, or Washoku, is a culturally important dietary practice recognized by UNESCO as intangible cultural heritage, reflecting Japan's deep respect for food, seasonality, and balance (UNESCO, 2013). This diet promotes the consumption of fresh, seasonal, and locally sourced ingredients, such as rice as a staple food, abundant vegetables and fruits, fish and other seafood rich in omega-3 fatty acids, soy products (e.g., miso, natto, tofu, and edamame), seaweed, green tea, and various fermented foods (Kuriyama, 2006; Shimazu et al., 2007). Modern research shows that this diet is associated with longevity, reduced risk of chronic diseases (e.g., cardiovascular disease, type 2 diabetes, and some cancers), improved gut health via nutrient-rich fermented foods (e.g., natto), weight management, and bone health (Nagata et al., 2017; ResearchGate, n.d.). By integrating traditional cultural practices with evidence-based nutrition, Washoku offers a sustainable, holistic, and health-promoting dietary model that serves as both a cultural treasure and a global example of balanced nutrition (Poddar, 2024).

These traditions developed from centuries of careful observation and ecological adaptation, offer a comprehensive approach to diet and health that closely aligns with modern concepts of holistic medicine (Sofi et al., 2010). Building evidence and attention for these practices suggest that ancient wisdom regarding food and nutrition is increasingly recognized for the potential roles in the prevention and treatment of chronic diseases (Willet et al., 2019).

The Role of Traditional Wisdom in Modern Nutrition

Traditional wisdom in nutrition refers to the cumulative body of knowledge, practices, and beliefs developed by societies over generations with lived experience, cultural adaptation, and empirical observation (Table 1). Unlike formal scientific disciplines, traditional knowledge is often transmitted by mouth and is embedded in rituals, daily life, and collective memory (Suarez and Adibi, 2025). Although lacking systematic documentation, these practices are grounded in a deep understanding of the natural environment, human physiology, and the sociocultural fabric in which they emerge. Traditional nutritional wisdom has played a central role in shaping community health, cultural identity, and ecological balance across continents and centuries. These systems developed from lived experience and environmental adaptation continue to be a crucial source of insight into sustainable, health-enhancing dietary patterns.

Table 1. Traditional dietary practices: cultural origins, therapeutic foods, and modern scientific relevance

Traditional nutrition	Geographic or cultural origin	Core philosophy or approach	Emphasized foods	Dietary characteristics	Alignment with modern science	References
Mediterranean diet	Mediterranean basin (e.g., Greece, Italy, Spain, etc.)	Holistic lifestyle, slow eating, social connections, and seasonal and local foods	Olive oil (primary fat), plenty of vegetables and fruits, whole grains, legumes, nuts, herbs, fish and seafood (moderate), poultry (low), red meat (very low), and wine (moderate)	Avoid processed foods, drink plenty of water, physical activity, and enjoy meals mindfully	Cardiovascular health, anti-inflammatory effects, longevity, and reduced risk of type 2 diabetes (e.g., polyphenols, monounsaturated fats, fiber)	Estruch et al., 2013; Scarmeas et al., 2006; Guasch-Ferré and Willett, 2021; and Burt, 2021
Ayurvedic nutrition	India	Balance of individual constitution (Dosha: Vata, Pitta, and Kapha) and Agni (digestive fire), and six tastes (ie, sweet, sour, salty, bitter, pungent, and astringent)	Personalized based on Dosha, seasonal and local foods, spices, legumes, grains, vegetables, fruits, ghee, and dairy	Personalized diet, meal timing, food energy, digestion focused, and food combinations	Gut health (Agni, microbiome), inflammation control, metabolic balance, and emphasis on personalized nutrition (compatible with nutrigenomics)	Lad, 2002; Jude et al., 2022; and Janardhanan and Madhu, 2025
Traditional Chinese medicine nutrition	China	Yin–yang balance between body and nature, five elements theory, and energy properties of foods (e.g., warming, cooling, moistening, and drying)	Seasonal and whole foods, grains, vegetables, fruits, legumes, fermented foods (soy products), herbs, green tea, and seaweed	Food selection based on energetic properties, balancing meals, organ support via meridian theory, and slow and mindful eating	Gut health, inflammation control, metabolic balance, and antioxidant and cardioprotective effects (bioactive compounds: isoflavones, and catechins)	Kaptchuk, 1986; and Matos et al., 2021
Traditional Japanese diet (Washoku)	Japan	Seasonality, freshness, variety, balance, umami taste, and portion control	Rice (staple), fish and seafood, vegetables (including pickles), soy products (e.g., miso, natto, and tofu), seaweed, and green tea	Minimal processed foods, low fat, high fiber, abundant fermented foods, and aesthetic presentation	Cardiovascular health, longevity, gut microbiota support (fermented foods), antioxidants (green tea and seaweed), and lower chronic disease risk	Shimazu et al., 2007; and Poddar, 2024

From the energy principles of traditional Chinese and Ayurvedic nutrition to the seasonal and communal ethics of the Mediterranean diet, such practices have long viewed food as more than mere nourishment; they also recognized the potential for food to heal, connect people, and promote harmony with nature (Trichopoulou et al., 2014; Sofi et al., 2014; Davis et al., 2015).

In recent decades, there has been growing recognition within the scientific community of the potential health benefits embedded in such traditional knowledge. The increasing prevalence of non-communicable diseases (NCDs) in modern societies has prompted scholars and practitioners alike to revisit traditional dietary models for potential solutions. For instance, the Mediterranean diet is globally recognized for its cardioprotective effects and is promoted as a model for sustainable and healthy eating (Estruch et al., 2018). In a similar manner, Ayurvedic principles in India and TCM promote dietary balance and seasonal adaptation, elements that align closely with modern preventive health paradigms (Jude et al., 2022). Numerous studies have shown that many traditional food practices, including fermentation, herbal medicine, and plant-based diets, contribute considerably to nutritional adequacy, gut health, and chronic disease prevention when evaluated with modern scientific methodologies (Kuhnlein et al., 2009; Mustafa, 2025). For instance, traditional fermented foods, such as *kimchi*, *miso*, and *tarhana* contain bioactive compounds and have probiotic properties that support digestive and immune health (Bilgiçli 2006; Marco et al., 2017; Tamang et al., 2020; Moon et al., 2023). Moreover, WHO has acknowledged the relevance of traditional knowledge systems in promoting public health, particularly in regions where access to conventional health care is limited (WHO, 2013). Traditional, food-based practices are not only contextually appropriate and economically accessible, but they also have considerable cultural resonance, making them crucial tools for addressing both undernutrition and overnutrition on a global scale (Kapoor et al., 2022).

Among traditional dietary models, the Mediterranean diet has garnered the most extensive scientific attention in recent decades (Verotta, 2015). Strong evidence from longitudinal studies and meta-analyses have shown that greater adherence to a Mediterranean-style diet is associated with reduced cardiovascular disease risk, delayed cognitive decline, and increased lifespan (Guasch-Ferré and Willett, 2021). The Mediterranean diet has long been recognized for its multifaceted health benefits, including cardiovascular protection, antioxidant capacity, glycemic control, cancer prevention, and positive modulation of gut microbiota (Sofi et al., 2014; Davis et al., 2015). These effects are largely attributed to the diet's rich composition of plant-based foods, healthy fats (e.g., olive oil), moderate fish consumption, and low intake of red meat and processed foods (Estruch et al., 2018). Characterized by a high intake of monounsaturated fats from olive oil, fiber, polyphenols, vegetables, and oily fish, the Mediterranean-style diet exerts its effects via a wide array of biological mechanisms, including lipid modulation, antioxidant activity, inflammation control, and hormonal regulation. Numerous studies have established that the Mediterranean-style diet reduces cardiovascular risk by improving lipid profiles and enhancing reverse cholesterol transport via the upregulation of high-density lipoprotein functionality and the reduction of cholesteryl ester transfer protein activity (Sofi et al., 2014; Martínez-González et al., 2019; Estruch et al., 2018). The antioxidant-rich foods in the Mediterranean-style diet, such as olive oil, citrus fruits, and leafy greens, also contribute to decreased oxidative stress and lower circulating levels of pro-inflammatory cytokines, including IL-6 and TNF-α, partly via the inhibition of NF-κB and MAPK signaling pathways (Casas et al., 2014). Notably, polyphenolic compounds, such as oleuropein and hydroxytyrosol have shown anticancer potential by inducing apoptosis, inhibiting angiogenesis, and regulating tumor metabolism across several cancer models (Granados-Principal et al., 2010). Furthermore, the Mediterranean-style diet improves insulin sensitivity and glycemic control, attributed to the activation of ERK/MAPK pathways in pancreatic β cells, leading to increased insulin secretion and reduced HbA1c levels in both clinical and preclinical trials (De Bock et al., 2013). Emerging evidence also suggests that the diet modulates gut microbiota composition and short-chain fatty acid production, thereby enhancing satiety hormone secretion (e.g., GLP-1 and PYY) and reducing systemic inflammation (Barber et al., 2023). In addition, bioactive components of the Mediterranean-style diet, such as oleocanthal, induce autophagy, a crucial cellular process involved in neuroprotection and longevity (Scarmeas et al., 2006; Singh et al., 2014; Trichopoulou et al., 2014; Martínez-González et al., 2019). Importantly, hydroxytyrosol has shown antiviral activity against HIV-1, influenza A, and even SARS-CoV-2 by inhibiting viral entry and replication (Yamada et al., 2009; Giordano et al., 2023). Collectively, these findings affirm that the Mediterranean diet is a cultural dietary pattern and a clinically validated therapeutic strategy supported by a growing body of nutrigenomic and metabolomic evidence. These findings not only highlight the preventive potential of traditional dietary models, but also provide a scientifically validated foundation for incorporating culturally rooted food practices into contemporary nutrition guidelines. Also, this dietary regime is extensively promoted as a universal model to address the global burden of NCDs. However, the global applicability of the Mediterranean diet requires scrutiny. While the diet originates from a culturally and geographically specific context, it is now recommended broadly for populations with distinct genetic, epigenetic, and culinary histories. Globally, non-Western populations account for over 75% of the population (UN, 2014), but most Mediterranean diet recommendations are based on Western cohorts. Such an approach raises concerns regarding cultural appropriateness, food accessibility, and bio-individual response to dietary interventions (Trajkovska Petkoska et al., 2025).

Evidence suggests that a variety of cultures worldwide have independently developed nutritionally beneficial dietary systems, often shaped by the availability of local resources and deeply rooted culinary traditions. It is evident that traditional East Asian and African diets, which are distinguishable by a high consumption of plant-based foods, fermented products, and minimal processing, have exhibited considerable potential in promoting health and preventing NCDs. In this context, an important question is posed of whether the health benefits of such traditional diets are universally applicable across diverse ethnic and cultural populations (Gherasim et al., 2020). Recent studies indicate that individual responses to dietary patterns can have substantial inter-individual variation, attributable to genetic factors and cultural dietary influences. For instance, despite the long-standing histories and reported health potential of many traditional dietary systems, such as those from East Asia and Africa, they remain underrepresented in mainstream scientific literature (Kuhnlein et al., 2009; Willett et al., 2019). The global availability and marketing of processed foods combined with universal dietary recommendations aimed at preventing chronic diseases often ignore genetic, cultural, and local dietary contexts. Universal adoption of diets, such as Mediterranean, Japanese, or Scandinavian diets, can overlook the benefits of local foods and conflict with personalized nutrition principles (Roman et al., 2024). As the field of nutrition science evolves toward increased sensitivity and inclusiveness, the need to incorporate Indigenous and non-Western dietary patterns into research frameworks is increasing. It is imperative that these traditions are recognized globally and incorporated into the discourse of evidence-based, health-promoting dietary guidelines, which will facilitate the development of more equitable and culturally appropriate dietary recommendations (Burt, 2021). As global education progresses toward inclusiveness and healing, it is essential to emphasize and acknowledge the long-standing, nurturing potential of food in ancient medical practices. These systems, shaped by centuries of ecological adaptation, production inquiry, and clinical observation, offer valuable frameworks for exploring the interplay between nutrition and health beyond traditional Western paradigms (Roman et al., 2024).

Ayurveda, India's natural health tradition, is based on a complex theory of personalized nutrition, lifestyle, and medical guidelines. This system, aligned with the concept of personalized health and medicine, has yielded strong evidence supporting the genetic basis of Ayurvedic body types according to research by molecular biologists. Ayurveda has always placed major emphasis on prevention and nutrition. One of the foundational texts of Ayurveda, the Charaka Samhita, states: "Without proper nutrition, medicine is useless; with proper nutrition, medicine is unnecessary." Ayurveda also provides guidance on the most appropriate foods and exercises for different body types, seasons, and times of day (Verotta et al., 2015). The concept known today as a healthy lifestyle has existed in Ayurveda for thousands of years in the form of dinacharya (daily routines) and ritucharya (seasonal routines). The purpose of this dietary pattern is to maintain inner harmony and prevent disease (Lad, 2002). Recent research suggests that Ayurvedic nutritional principles could be compatible with contemporary approaches to gut health, inflammation control, and metabolic balance (Janardhanan and Madhu, 2025).

Similarly, TCM places considerable emphasis on the concept of harmony between the body and nature, which is achieved with dietary therapy guided by the principles of yin and yang, the five elements, and meridian theory. The classification of foods is not solely based on their nutrient content; rather, it is also informed by their energetic properties, which manifest as the generation of specific sensations: heat, cold, dampness, or dryness. It is widely accepted that there are therapeutic properties related to these sensations, and as such, they are prescribed to address imbalances in the body (Kaptchuk, 1986; Matos et al., 2021). Research has increasingly shown that diets comprising TCMs, with high concentrations of soy, green tea, seaweed, and fermented vegetables, can contribute to reduced incidences of chronic diseases, including cardiovascular disorders and specific types of cancer (Nagata et al., 2017; Hao et al., 2017; Kuriyama et al., 2006).

Like Ayurveda, Unani medicine is a natural medical system with extensive clinical literature and a robust theory of disease origins, offering nutritional and health guidelines tailored to individual types. In the Unani system, which evolved from Greco–Arabic medicine and flourished during the golden age of Islam, nutrition is considered a crucial factor in maintaining good health. Similar to Ayurveda and TCM, Unani nutrition is tailored to an individual's temperament, which reflects the balance of the four humors: blood, phlegm, yellow bile, and black bile. Correcting imbalances in these humors is achieved by the implementation of a specific dietary regimen, and this practice has attracted considerable research interest in the applicability of Unani to contemporary metabolic disorders (Itrat et al., 2014). Despite the empirical richness of these traditional dietary systems, they have historically been marginalized in global nutrition research. As the field evolves toward precision nutrition, the integration of culturally diverse dietary frameworks could facilitate the development of more equitable and personalized guidelines, particularly for communities underserved by Western-centric models.

Traditional and Modern: Evidence-Based Functional Foods

As discussed in section 2, in recent years, findings from modern nutritional science have validated and incorporated aspects of traditional dietary practices into contemporary health models. Research has shown that fermented foods, herbal preparations, and plant-based diets that are widely used in numerous ancient traditions, provide probiotic benefits, anti-inflammatory effects, and contain bioactive compounds that support immunity, gut health, and chronic disease prevention (Sofi et al., 2014; Willet et al., 2019). These findings highlight the enduring value of traditional dietary systems and their relevance to contemporary global health challenges. However, despite this growing awareness, a substantial amount of traditional culinary knowledge remains to be elucidated. Increasing evidence regarding the therapeutic effects of traditional diets reinforces the need to integrate this ancient wisdom into modern health practices. Thus, more inclusive and culturally appropriate approaches to nutrition can be promoted by respecting diverse cultural traditions while simultaneously addressing the global burden of NCDs (Kuhnlein et al., 2009; Willet et al., 2019). In this context, the concept of functional foods has emerged to connect traditional dietary habits with targeted health promotion (Fekete et al., 2025).

Scientifically Supported Traditional Foods

Traditional foods have long been recognized for their therapeutic potential across cultures, and modern science is increasingly validating these claims. Modern nutritional science uses biochemical, molecular, and clinical methodologies to elucidate the mechanisms by which traditional foods exert their physiological effects. These methodologies involve identifying specific bioactive compounds and conducting controlled clinical trials and analyzing their effects on health, growth, and the prevention or progression of disease. Modern clinical studies increasingly confirm the therapeutic claims of traditional foods, bridging ancient wisdom with contemporary evidence-based medicine. Studies suggest that many traditional staple foods are rich in compounds that confer considerable health benefits, which are often consistent with current understandings of gut health, inflammation control, metabolic balance, and chronic disease prevention (Fekete et al., 2025) (Figure 1).

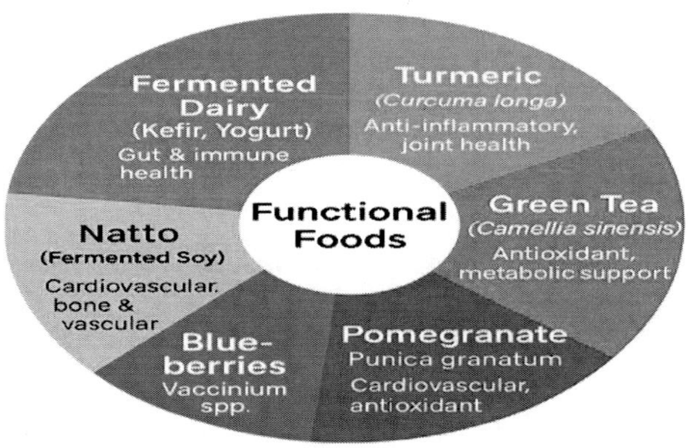

Figure 1. Functional foods and primary health benefits.

For example, turmeric (*Curcuma longa*), a fundamental component of Ayurvedic medicine, is rich in the bioactive compound curcumin, which has been widely recognized for its anti-inflammatory, antioxidant, and anticancer properties. Clinical evidence from multiple randomized controlled trials indicates that curcumin can effectively reduce markers of chronic inflammation, such as C-reactive protein, and improve clinical outcomes in patients with osteoarthritis (Hewlings & Kalman, 2017; Panahi et al., 2016). These findings highlight the therapeutic potential of turmeric as both a dietary and medicinal agent, reinforcing its traditional use in promoting joint health and mitigating inflammation-related conditions.

Fermented foods, especially dairy products, such as kefir and yogurt, have been a staple of the human diet for millennia and are widely recognized as functional foods due to their high probiotic content. It has been shown that these probiotics, which include *Lactobacillus* and *Bifidobacterium* species, contribute to the modulation of gut microbiota composition, the enhancement of digestive processes, and the regulation of immune responses (Marco et al., 2017; Sanders et al., 2019). Similarly, traditional fermented soy products, including miso and natto, are consumed for their probiotic benefits and for bioactive compounds that confer additional health advantages. Natto, for instance, contains nattokinase, which supports cardiovascular health and facilitates clot dissolution; vitamin K2 (menaquinone-7), which contributes to bone and arterial health; and isoflavones that are associated with improved vascular function and reduced mortality risk (Nagata et al., 2017; Weng et al., 2017). Therefore, these findings emphasize the therapeutic potential of fermented foods in maintaining gastrointestinal, cardiovascular, and metabolic health, thereby integrating traditional dietary practices with contemporary scientific validation.

In addition to fermented foods, other traditional functional foods, such as green tea (*Camellia sinensis*), pomegranate (*Punica granatum*), and blueberries (*Vaccinium* spp.) have been extensively studied for their health-promoting properties. Green tea is abundant in catechins, particularly epigallocatechin gallate (E.G., CG), which exhibits strong antioxidant and anti-inflammatory effects, contributing to cardiovascular health, improved metabolic function, and potential cancer prevention (Kuriyama et al., 2006). Pomegranate, a fruit rich in polyphenols and anthocyanins, has been shown to lower blood pressure, reduce oxidative stress, and provide anti-inflammatory benefits, thereby supporting cardiovascular function and overall vitality (Mohammadi et al., 2024). Blueberries are abundant in anthocyanins and other polyphenols, are recognized for their potent antioxidant capacity and have been linked to improved cognitive performance, eye health, and mitigation of oxidative damage (Miller et al., 2018; Dodd et al., 2019). In addition to the consideration of individual foods, there is a growing recognition of the efficacy of holistic nutritional principles established in traditional systems. For instance, Ayurveda's emphasis on personalized nutrition, grounded in the doshas and the yin–yang and five element food energy framework in TCM, aligns closely with contemporary scientific understandings of the gut–brain axis, chronic inflammation, and metabolic disorders (Payyappallimana and Venkatasubramanian, 2015; Suarez and Adibi, 2025). As indicated by the findings of these studies, there is a fundamental truth to be found in the traditional dietary wisdom that is often shown to be in alignment with quantifiable physiological outcomes. With the integration of ethnobotanical expertise and contemporary clinical research methodologies, health-care professionals can formulate bespoke nutritional strategies that incorporate culturally resonant foods, thereby achieving both culturally relevant health benefits and improved clinical outcomes. Collectively, these findings underscore the immense value of integrating ancient nutritional wisdom with contemporary research, thereby establishing a foundation for more equitable, personalized, and culturally informed dietary guidelines that support global health. With the systematic examination of bioactive compounds in these foods, researchers can establish a link between centuries of empirical knowledge and molecular mechanisms, which demonstrates that traditional dietary practices are not merely cultural habits, but are also interventions with measurable health effects (Jayasinghe et al., 2025).

The Contemporary Concept of Food as Medicine

The concept of food as medicine, often attributed to Hippocrates' famous aphorism: "Let food be thy medicine and medicine be thy food," has evolved from an ancient philosophical principle into a rapidly growing field of scientific research. The growing popularity of this paradigm reflects a broader shift in public and scientific discourse toward viewing food not only as a source of energy and essential nutrients, but also as a potential tool for disease prevention and therapeutic intervention (Fekete et al., 2025). This perspective is increasingly important as health-care systems face increasing pressure to manage chronic diseases while the public expresses increasing interest in preventive strategies and self-management approaches to maintain health. Modern nutritional science increasingly recognizes that foods contain bioactive compounds that can modulate physiological pathways, influence disease prevention, and complement therapeutic interventions. In contemporary practice, this concept extends beyond providing only macronutrients and micronutrients to encompass functional foods and nutraceuticals that confer specific health benefits. As mentioned earlier, curcumin in turmeric; probiotics in fermented milk and soy products; and polyphenols in green tea, pomegranates, and blueberries have been extensively studied for their anti-inflammatory, antioxidative, cardiometabolic, and neuroprotective properties (Hewlings and Kalman, 2017; Marco et al., 2017). These bioactive compounds exemplify how dietary components can actively contribute to maintaining homeostasis and preventing or managing chronic diseases, reinforcing the role of food as fuel and as an active regulator of health.

The concept of food as medicine is experiencing a renaissance in contemporary nutritional science, integrating both traditional wisdom and modern clinical evidence. Modern frameworks are moving beyond pharmacological supplementation to emphasize the preventive and therapeutic potential of diet and highlighting the role of evidence-based functional foods in personalized nutrition plans, chronic disease prevention strategies, and overall wellness programs. Personalized nutrition recognizes that individual variability in genetics, gut microbiota composition, lifestyle, and cultural dietary practices influence responses to nutritional interventions (Ordovas et al., 2018; Zeevi et al., 2015). This perspective closely aligns with traditional systems, such as Ayurveda and Unani. Advances in nutritional genomics, metabolomics, and microbiome research have clarified how functional foods and diets regulate gene expression, metabolic pathways, and inflammatory processes. These findings provide a solid scientific basis for the preventive and therapeutic roles of nutrition (Corella and Ordovás, 2014; Holscher, 2017; Wang et al., 2018; Scalbert et al., 2005). However, the integration of functional foods into mainstream dietary habits is limited by several factors, including a poor public understanding of scientific evidence, inconsistent regulation of health claims, socioeconomic barriers to access, and a lack of standardized definitions and criteria for effectiveness.

Table 2. Traditional foods and their health benefits: a scientific overview

Traditional food	Bioactive compounds	Traditional use	Modern clinical evidence	References
Turmeric (Curcuma longa)	Curcumin and curcuminoids	Anti-inflammatory, digestive health, wound healing, and pain relief in Ayurveda	Reduces C-reactive protein levels (an inflammatory marker); improves symptoms in patients with osteoarthritis by reducing pain and improving joint function; and some studies suggest a potential role in cancer prevention and treatment, although more research is needed.	Hewlings and Kalman, 2017; and Goel et al., 2008
Ginger (Zingiber officinale)	Gingerols and shogaols	Nausea relief, digestive issues, and cold and flu remedy	Highly effective for relieving nausea and vomiting related to pregnancy, chemotherapy, and surgery; and shows anti-inflammatory and analgesic effects, reducing pain and swelling in conditions, such as rheumatoid arthritis and osteoarthritis.	Mao et al., 2019; and Shahrajabian et al., 2019
Garlic (Allium sativum)	Allicin and organosulfur compounds	Antimicrobial agent, treatment for hypertension, and digestive aid	Strong evidence for its role in cardiovascular health (e.g., lowers blood pressure, regulates cholesterol levels, and prevents platelet aggregation or clotting); and its antimicrobial properties have been scientifically validated against various bacteria, fungi, and viruses.	Sagdic and Tornuk, 2012; Ahmad et al., 2019; and Bhatwalkar et al., 2021
Extra virgin olive oil	Phenolic compounds (e.g., oleuropein and hydroxytyrosol) and oleic acid	Heart health, skin care, and anti-inflammatory in Mediterranean cultures	The study and other clinical trials confirm its role in reducing the risk of cardiovascular disease by lowering oxidative stress and inflammation; and its phenolic compounds act as powerful antioxidants, protecting cells from damage.	Servili et al., 2013; Estruch et al., 2018; and Jimenez et al., 2020
Blueberries (Vaccinium spp.)	Anthocyanins, flavonoids, and polyphenols	Digestive aid, vision health, and treatment of urinary tract infections	Studies show they can slow age-related cognitive decline and improve memory in older adults; and their high antioxidant content supports eye health and can reduce eye strain.	Camire, 2002; and Krikorian et al., 2010
Kefir	Probiotics (e.g., Lactobacillus and Bifidobacterium), peptides, and polysaccharides	Gut health, immune system support, and general vitality	Improves gut microbiota composition, aids lactose digestion, and can reduce symptoms of irritable bowel syndrome; and unique probiotic strains contribute to immune modulation.	Marco et al., 2017; and Apalowo et al., 2024
Green tea (Camellia sinensis)	Catechins (E.G., CG), L-theanine, and flavonoids	Detoxification, mental alertness and weight management	Epigallocatechin gallate (E.G., CG), a key catechin, is a potent antioxidant that protects against cellular damage; L-theanine promotes a state of calm alertness; and clinical trials link regular consumption to improved cardiovascular health and a reduced risk of specific cancers.	Kuriyama, 2008; and Kochman et al., 2020
Pomegranate (Punica granatum)	Polyphenols, anthocyanins, and tannins	Cardiovascular health, vitality, and traditional anti-inflammatory	Lowers blood pressure and reduces oxidative stress, directly benefiting heart health; and its compounds have shown promising anti-inflammatory and anti-cancer properties in various studies.	Aviram et al., 2013; and Dogara et al., 2024
Honey	Antibacterial agents and phenolic compounds	Digestive aid, relieves constipation, improves gut health, and enhances nutrient absorption	Antibacterial, antioxidant, improves gut microbiota, and enhances nutrient absorption	Jaśkiewicz et al., 2025; Mohan et al., 2017; and Almasaudi et al., 2017

This discrepancy between traditional dietary practices and evidence-based nutritional interventions poses a major challenge for public health strategies (Neufeld et al., 2023). Consequently, modern approaches to food as medicine do not ignore traditional knowledge but rather reframe it within evidence-based, personalized strategies. By understanding bioactive compounds, validating their effects with rigorous clinical research and translating these findings into personalized dietary recommendations, the ancient principle of using food for health is preserved, enhanced, and firmly integrated into contemporary health practices. Table 2 shows traditional foods and their health benefits.

Myths and Misconceptions in Traditional Nutrition

Traditional dietary practices are deeply intertwined with culturally rooted beliefs, proverbs, and societal norms. Many of these practices have been passed down by mouth across generations and persist, although they often lack empirical validation. Myths, fallacies, and misconceptions have always been present in almost all societies throughout history (Qureshi and Shaikh, 2006). Such beliefs often manifest in prescriptive dietary guidelines; for example, "yogurt and fish should not be eaten together," "fruit should be avoided at night," or "vinegar helps with weight loss." While these beliefs reflect historical attempts to understand and manage health via nutrition, contemporary nutritional science offers a more nuanced understanding that often challenges or expands upon these claims. These misconceptions can influence eating behaviors, dietary preferences, and societal perceptions of health, highlighting the importance of integrating cultural context into evidence-based guidance (Qureshi and Shaikh, 2006; Kuhnlein et al., 2009).

One of these common myths involves the consumption of dairy products and fish together. In some Asian and Middle Eastern cultures, consuming these foods together is traditionally believed to cause skin conditions or gastrointestinal upset. However, scientific research indicates that there is no inherent chemical incompatibility between fish and dairy proteins under normal dietary conditions. Observed adverse effects are generally attributable to individual lactose intolerance, fish allergies, or pre-existing digestive sensitivities rather than the combination of these foods (Zaitlin et al., 2013). This example shows the strength of cultural beliefs that persist despite scientific evidence. Therefore, raising public awareness to disseminate accurate information while respecting cultural values is crucial. Similarly, warnings to avoid fruit in the evening are common, often justified by the idea that the sugar contained in fruit can disrupt sleep or contribute to weight gain. While excessive caloric intake at any time can affect metabolic outcomes, current evidence suggests that moderate fruit consumption, even at night, is metabolically neutral for healthy individuals. Fruits provide essential micronutrients, fiber, and phytonutrients that support digestive health, cardiovascular function, and antioxidant defenses (Slavin & Lloyd, 2012). Misinterpretation of such guidelines can lead to unnecessary dietary restrictions that can potentially reduce the intake of beneficial nutrients.

Vinegar is another example of the intersection of traditional beliefs and contemporary trends. Historically, vinegar has been associated with weight loss and improved digestion. Modern studies suggest that the main bioactive ingredient of vinegar, acetic acid, can modestly influence satiety and glycemic responses; however, their effect on long-term weight loss is limited and should not replace comprehensive diet and lifestyle strategies (Chen et al., 2016).

The belief that carrots dramatically improve eyesight is a common myth. While carrots are rich in beta-carotene, a precursor of vitamin A that is essential for eye health, they do not provide super vision, nor are they a miraculous cure for night blindness. This myth originated from the British promotion of carrot consumption during World War II to conceal radar technology (Snodgrass, 2004). Nevertheless, scientific literature suggests that adequate beta-carotene intake contributes to eye health, supporting normal night vision and helping to reduce the risk of age-related eye conditions, such as macular degeneration and cataracts (Anandhi et al., 2024).

The traditional concern that egg consumption substantially raises cholesterol levels has also been scientifically reevaluated. The effect of dietary cholesterol on blood cholesterol levels is less than the effect of saturated and trans fats for most healthy individuals. Current evidence suggests that consuming one egg per day does not increase the risk of cardiovascular disease in healthy individuals (Fernandez and Calle, 2010).

Beyond traditional beliefs, modern dietary myths, such as detox regimens, alkaline diets, superfood claims, the idea that gluten-free diets (GFDs) are healthier for everyone, and the notion that supplements can replace a healthy diet continue to perpetuate misconceptions. For example, detox diets often claim to remove toxins from the body; however, organs, such as the liver, kidneys, lungs, and skin continuously and efficiently perform detoxification (Klein and Kiat, 2015). There is limited empirical evidence supporting short-term detox interventions, and extreme regimens can even pose health risks. Although alkaline diets claim to alter systemic pH by food intake, they have minimal effects on blood pH regulation; however, they could offer additional nutritional advantages, primarily by promoting higher consumption of fruits and vegetables (Mohammed et al., 2023). The term superfood is often used as a marketing tool to exaggerate the health benefits of a single food, ignoring factors, such as overall diet, nutrient absorption, and lifestyle. Ignoring such factors can obscure the fact that while the nutritional value of superfoods is undeniable, excessive consumption can lead to adverse effects. Therefore, it is important to be aware of each food's characteristics and consume them in moderation (Fernández-Ríos et al., 2022).

The belief that a GFD is healthier for everyone is a common modern myth. While gluten can be harmful to individuals with celiac disease or non-celiac gluten sensitivity, the health benefits of a GFD for healthy individuals remain unproven. Despite this lack of validation, the popularity of GFDs is rising among people without gluten-related disorders, with up to 7% of adults in some countries adopting the diet due to perceived benefits, such as weight loss, increased energy, or improved skin. However, these claims are largely unsupported by scientific evidence, and unnecessary gluten avoidance can lead to nutritional deficiencies, particularly in fiber and other essential nutrients. Therefore, while GFDs are essential for those with gluten-related disorders, promoting gluten avoidance among the general population is not justified and could have unintended negative consequences (Sabença et al., 2021).

Lastly, the idea that supplements can replace a healthy diet is also misleading. Diet trends and supplements have become a major part of modern health culture. Supplements are designed to address deficiencies or increase specific nutrient intake, but they cannot provide the complex synergistic matrix of nutrients and fiber found in whole foods; therefore, they are not a substitute for a balanced diet. This industry thrives on promises of rapid weight loss and improved health, yet many of these claims lack solid scientific evidence. This lack of evidence exposes consumers to misinformation, unrealistic expectations, and potential health risks, which emphasizes the need for consumers to be critical, health experts to provide evidence-based guidance, and regulatory bodies to take stricter action against false advertising. To protect public health, collaborative research is crucial to investigate the long-term effects of these products and inform public policy (Khubchandani and Batra, 2024).

The spread of nutrition myths is further amplified by social media, online platforms, and digital information ecosystems. Viral narratives, anecdotal confirmations, and oversimplified messages often overshadow the slow and methodical communication of scientific evidence, creating an environment where misinformation can thrive. This disinformation, especially when combined with confirmation bias or selective attention to appealing narratives, complicates efforts to educate the public and can reinforce existing misconceptions (Suarez-Lledo et al., 2021). Combating these nutrition myths and misinformation requires a multifaceted approach. Health professionals and nutritionists should provide evidence-based and culturally sensitive guidance, considering both the scientific validity and cultural significance of traditional practices. Increasing nutritional literacy is crucial for individuals to critically evaluate nutrition claims, identify reliable sources, and make informed choices. Furthermore, combining traditional knowledge with modern scientific methods, such as randomized controlled trials, nutrigenomics (ie, food–gene interaction), and metabolomics provides a more holistic understanding of the nutrition–health relationship (Guasch-Ferré and Willett, 2021). This integration helps distinguish culturally relevant dietary recommendations from unproven claims, improving both public awareness and clinical practice. Ultimately, whether stemming from traditional practices or modern trends, nutrition myths profoundly influence peoples' perceptions of health and eating behaviors. By combining traditional wisdom with sound scientific evidence, nutritionists can provide both accurate and respectful guidance. This approach supports informed decisions, promotes balanced and sustainable eating habits, and contributes to better long-term health in societies (Kuhnlein et al., 2009).

Ethical Values and Sustainability in Traditional Nutrition

The intersection of traditional nutritional beliefs and modern nutritional science raises important questions regarding ethical values and biocultural sustainability (Poddar, 2024). On one hand, respecting cultural dietary norms and understanding their underlying social, historical, and religious context is essential. Traditional knowledge embodies practical wisdom acquired over generations, and completely rejecting this knowledge in favor of modern science can undermine cultural heritage (Suarez and Adibi, 2025; Kuhnlein et al., 2009). This lack of understanding creates a responsibility to use language sensitive to local beliefs and traditions when conveying scientific facts. On the other hand, the evidence provided by modern nutritional science plays a crucial role in improving individual and societal health. Therefore, disseminating evidence-based information is an ethical obligation (Fekete et al., 2025). For example, scientific findings that a traditional diet high in saturated fat increases the risk of heart disease necessitate a sensitive educational approach to changing these cultural practices (Aremu et al., 2025). The goal is to recommend healthier alternatives that is supported by scientific data, without ignoring cultural values. At the same time, the global integration of food systems poses new threats to biocultural sustainability. The use of terms, such as superfood, for marketing purposes can suddenly increase demand for local products in certain regions, which can both put pressure on local ecosystems and reduce the availability of these products to local communities. For example, global demand for popular products, such as quinoa, has increased prices in its native South America, making it difficult for local communities to access their traditional foods. Such trends highlight the need for a broader ethical framework that considers how global nutritional recommendations affect local food security and biodiversity (Johns and Sthapit, 2004). Ultimately, knowledge transfer in the field of food and nutrition is a multifaceted process that encompasses not only the presentation of scientific facts, but also complex ethical dimensions, such as cultural respect, social justice, and ecological sustainability. When evaluating traditional dietary approaches, it is essential to adopt a holistic approach that integrates scientific evidence with the cultural and environmental context.

Conclusion

The relationship between traditional nutritional wisdom and modern nutritional science is a complex process that requires mutual respect and integration, rather than a one-way interaction. Traditional practices should be viewed as cultural heritage and as a resource offering valuable insights to modern science on topics, such as preventing chronic diseases, improving gut health, and ensuring nutritional adequacy. However, this integration should be undertaken with caution; myths lacking scientific basis in traditional beliefs should be disentangled, and the negative effects of global markets on local food systems should not be overlooked. Food, both as a component of cultural identity and a subject of scientific research, demonstrates the need for a holistic, ethical, and sustainable approach to public health policy development. Consequently, the combined efforts of these two disciplines has the potential to guide humanity toward healthier and more conscious eating patterns while respecting cultural norms.

References

Ahmad, P., Alvi, S. S., & Khan, M. S. (2019). Functioning of organosulfur compounds from garlic (allium sativum linn) in targeting risk factor-mediated atherosclerosis: A cross talk between alternative and modern medicine. In *Natural Bio-active Compounds: Volume 1: Production and Applications* (pp. 561-585). Singapore: Springer Singapore.

Albala, K. (Ed.). (2011). *Food Cultures of the World Encyclopedia:[4 Volumes]*. Bloomsbury Publishing USA.

Almasaudi, S. B., Al-Nahari, A. A., El Sayed, M., Barbour, E., Al Muhayawi, S. M., Al-Jaouni, S., ... & Harakeh, S. (2017). Antimicrobial effect of different types of honey on Staphylococcus aureus. *Saudi journal of biological sciences*, *24*(6), 1255-1261.

Anandhi, E., Shams, R., Dash, K. K., Bhasin, J. K., Pandey, V. K., & Tripathi, A. (2024). Extraction and food enrichment applications of black carrot phytocompounds: A review. *Applied Food Research*, *4*(1), 100420.

Apalowo, O. E., Adegoye, G. A., Mbogori, T., Kandiah, J., & Obuotor, T. M. (2024). Nutritional characteristics, health impact, and applications of kefir. *Foods*, *13*(7), 1026.

Aremu, S. O., Akute, B., Aremu, D. O., Zando, C., Aremu, E. D., Nwachukwu, O. J., ... & Aremu, M. O. (2025). Dietary strategies for preventing and managing obesity through evidence-based nutritional interventions. *Discover Public Health*, *22*(1), 1-26.

Aviram, M., & Rosenblat, M. (2013). Pomegranate for your cardiovascular health. *Rambam Maimonides Medical Journal*, *4*(2), e0013.

Barber, T. M., Kabisch, S., Pfeiffer, A. F., & Weickert, M. O. (2023). The effects of the Mediterranean diet on health and gut microbiota. *Nutrients*, *15*(9), 2150.

Barea, F. S. (2011). The practical application of psychobiological theory of the four humors in the Jesuit colleges of the Modern Age: a model of guidance for allocating government positions. *Procedia-Social and Behavioral Sciences*, *30*, 2335-2340.

Bendick, J. (2002). *Galen: And the Gateway to Medicine*. Bethlehem Books.

Bhatwalkar, S. B., Mondal, R., Krishna, S. B. N., Adam, J. K., Govender, P., & Anupam, R. (2021). Antibacterial properties of organosulfur compounds of garlic (Allium sativum). *Frontiers in microbiology*, *12*, 613077.

Bilgiçli, N., Elgun, A., & Türker, S. (2006). Effects of various phytase sources on phytic acid content, mineral extractability and protein digestibility of tarhana. *Food chemistry*, *98*(2), 329-337.

Burt, K. (2021). The whiteness of the Mediterranean Diet: A historical, sociopolitical, and dietary analysis using Critical Race Theory. *Journal of Critical Dietetics*, *5*(2), 41-52.

Camire, M. E. (2002). Phytochemicals in the Vaccinium family: bilberries, blueberries, and cranberries. In *Phytochemicals in nutrition and health* (pp. 35-56). CRC Press.

Cannon, G., & Leitzmann, C. (2022). Food and nutrition science: the new paradigm. *Asia Pacific journal of clinical nutrition*, *31*(1), 1-15.

Casas, R., Sacanella, E., & Estruch, R. (2014). The immune protective effect of the Mediterranean diet against chronic low-grade inflammatory diseases. *Endocrine, Metabolic & Immune Disorders-Drug Targets (Formerly Current Drug Targets-Immune, Endocrine & Metabolic Disorders)*, *14*(4), 245-254.

Chan, E. W. C., Wong, S. K., Kezuka, M., Oshiro, N., & Chan, H. T. (2021). Natto and miso: An overview on their preparation, bioactive components and health-promoting effects. *Food Res*, *5*(3), 446-452.

Chen, H., Chen, T., Giudici, P., & Chen, F. (2016). Vinegar functions on health: Constituents, sources, and formation mechanisms. *Comprehensive Reviews in Food Science and Food Safety*, *15*(6), 1124-1138.

Çömlek, F. Ö. (2025). Avicenna (Ibn Sina; 980-1037): Father of early medicine. *Journal of Contemporary Medicine*, *15*(2), 104-108.

Corella, D., & Ordovas, J. M. (2014). How does the Mediterranean diet promote cardiovascular health? Current progress toward molecular mechanisms: Gene-diet interactions at the genomic, transcriptomic, and epigenomic levels provide novel insights into new mechanisms. *Bioessays*, *36*(5), 526-537.

David, R. (2013). Ancient Egyptian medicine: the contribution of twenty-first century science. *Bulletin of the John Rylands Library*, *89*(Ancient Medical and Healing Systems: Their Legacy to Western Medicine), 157-180.

Davis, C., Bryan, J., Hodgson, J., & Murphy, K. (2015). Definition of the Mediterranean diet: a literature review. *Nutrients*, *7*(11), 9139-9153.

De Bock, M., Derraik, J. G., Brennan, C. M., Biggs, J. B., Morgan, P. E., Hodgkinson, S. C., ... & Cutfield, W. S. (2013). Olive (Olea europaea L.) leaf polyphenols improve insulin sensitivity in middle-aged overweight men: a randomized, placebo-controlled, crossover trial. *PloS one*, *8*(3), e57622.

Deveci, G., Çelik, E., Ağagündüz, D., Bartkiene, E., Rocha, J. M. F., & Özogul, F. (2023). Certain fermented foods and their possible health effects with a focus on bioactive compounds and microorganisms. *Fermentation*, *9*(11), 923.

Dodd, G. F., Williams, C. M., Butler, L. T., & Spencer, J. P. (2019). Acute effects of flavonoid-rich blueberry on cognitive and vascular function in healthy older adults. *Nutrition and Healthy Aging*, *5*(2), 119-132.

Dogara, A. M., Hama, H. A., & Ozdemir, D. (2024). Update on the potential of Punica granatum L. traditional uses and pharmacological uses: A review. *Advances in Pharmacological and Pharmaceutical Sciences*, *2024*(1), 6523809.

Estruch, R., Ros, E., Salas-Salvadó, J., Covas, M. I., Corella, D., Arós, F., ... & Martínez-González, M. A. (2018). Primary prevention of cardiovascular disease with a Mediterranean diet supplemented with extra-virgin olive oil or nuts. *New England journal of medicine*, *378*(25), e34.

Fabricant, D. S., & Farnsworth, N. R. (2001). The value of plants used in traditional medicine for drug discovery. *Environmental Health Perspectives*, *109* (1), 69-75.

Fekete, M., Lehoczki, A., Kryczyk-Poprawa, A., Zábó, V., Varga, J. T., Bálint, M., ... & Varga, P. (2025). Functional Foods in Modern Nutrition Science: Mechanisms, Evidence, and Public Health Implications. *Nutrients*, *17*(13), 2153.

Fernández-Ríos, A., Laso, J., Hoehn, D., Amo-Setién, F. J., Abajas-Bustillo, R., Ortego, C., ... & Margallo, M. (2022). A critical review of superfoods from a holistic nutritional and environmental approach. *Journal of Cleaner Production*, *379*, 134491.

Fernandez, M. L., & Calle, M. (2010). Revisiting dietary cholesterol recommendations: does the evidence support a limit of 300 mg/d?. *Current atherosclerosis reports*, *12*(6), 377-383.

Gherasim, A., Arhire, L. I., Niţă, O., Popa, A. D., Graur, M., & Mihalache, L. (2020). The relationship between lifestyle components and dietary patterns. *Proceedings of the Nutrition Society*, *79*(3), 311-323.

Ghosh-Jerath, S., Singh, A., Magsumbol, M. S., Lyngdoh, T., Kamboj, P., & Goldberg, G. (2016). Contribution of indigenous foods towards nutrient intakes and nutritional status of women in the Santhal tribal community of Jharkhand, India. *Public health nutrition*, *19*(12), 2256-2267.

Giordano, D., Facchiano, A., & Carbone, V. (2023). Food plant secondary metabolites antiviral activity and their possible roles in SARS-CoV-2 treatment: an overview. *Molecules*, *28*(6), 2470.

Goel, A., Kunnumakkara, A. B., & Aggarwal, B. B. (2008). Curcumin as "Curcumin": from kitchen to clinic. *Biochemical pharmacology*, *75*(4), 787-809.

Gok, I. (2023). Functional potential of several Turkish fermented traditional foods: Biotic properties, bioactive compounds, and health benefits. *Food Reviews International*, *39*(5), 2568-2593.

Granados-Principal, S., Quiles, J. L., Ramirez-Tortosa, C. L., Sanchez-Rovira, P., & Ramirez-Tortosa, M. C. (2010). Hydroxytyrosol: from laboratory investigations to future clinical trials. *Nutrition reviews*, *68*(4), 191-206.

Guasch-Ferré, M., & Willett, W. C. (2021). The Mediterranean diet and health: a comprehensive overview. *Journal of internal medicine*, *290*(3), 549-566.

Hankinson, R. J. (2016). Humors and Humoral theory. In *The Routledge History of Disease* (pp. 21-37). Routledge.

Hao, P., Jiang, F., Cheng, J., Ma, L., Zhang, Y., & Zhao, Y. (2017). Traditional Chinese medicine for cardiovascular disease: evidence and potential mechanisms. *Journal of the American College of Cardiology*, *69*(24), 2952-2966.

Hewlings, S. J., & Kalman, D. S. (2017). Curcumin: A review of its effects on human health. *Foods*, *6*(10), 92.

Holscher, H. D. (2017). Dietary fiber and prebiotics and the gastrointestinal microbiota. *Gut microbes*, *8*(2), 172-184.

Itrat, M., Akhtar, Y., Nasir, F., Khan, T. N., Jamal, M. A., & Nikhat, S. (2023). Conceptual Understanding of Preventive Strategies for Metabolic Disorders in Unani Medicine: A Narrative Review. *Alternative Therapies in Health and Medicine*, *29*(8), 233-239.

Iyer, A., Panchal, S., Poudyal, H., & Brown, L. (2009). Potential health benefits of Indian spices in the symptoms of the metabolic syndrome: a review. *Indian journal of biochemistry & biophysics*, *46*(6), 467-481.

Janardhanan, R., & Madhu, P. M. (2025). Human Gut Microbiome in Health and Disease-An Ayurvedic Perspective. *International Journal of Ayurveda and Pharma Research*, 108-111.

Jaśkiewicz, K., Szczęsna, T., & Jachuła, J. (2025). How Phenolic Compounds Profile and Antioxidant Activity Depend on Botanical Origin of Honey—A Case of Polish Varietal Honeys. *Molecules*, *30*(2), 360.

Jayasinghe, S., Byrne, N. M., & Hills, A. P. (2025). Cultural influences on dietary choices. *Progress in Cardiovascular Diseases*.

Jimenez-Lopez, C., Carpena, M., Lourenço-Lopes, C., Gallardo-Gomez, M., Lorenzo, J. M., Barba, F. J., ... & Simal-Gandara, J. (2020). Bioactive compounds and quality of extra virgin olive oil. *Foods*, *9*(8), 1014.

Johns, T., & Sthapit, B. R. (2004). Biocultural diversity in the sustainability of developing-country food systems. *Food and nutrition bulletin*, *25*(2), 143-155.

Jude, S., Varma AC, K., Kuttappan, S., & Amalraj, A. (2022). Ayurvedic and Traditional Systems of Medicine in Clinical Trials: An Overview.

Jyoti, B., & Yadav, P. (2025). Medicinal Plants and Ethnobotany: A Nexus of Traditional Knowledge and Modern Science. *The PLANTA Research Book Series*, *6*(1), 1838-1843.

Kapoor, R., Sabharwal, M., & Ghosh-Jerath, S. (2022). Indigenous foods of India: A comprehensive narrative review of nutritive values, antinutrient content and mineral bioavailability of traditional foods consumed by indigenous communities of India. *Frontiers in sustainable food systems*, *6*, 696228.

Kaptchuk, T. J. (1986). The web that has no weaver: Understanding Chinese medicine.

Kennedy, S. J., Ryan, L., Clegg, M. E. The Effects of a Functional Food Breakfast on Gluco-Regulation, Cognitive Performance, Mood, and Satiety in Adults. *Nutrients* **2020**, *12*, 2974.

Khubchandani, J., & Batra, K. (2024). Diet fads and supplements: Navigating the allure, risks, and reality. *Journal of Medicine, Surgery, and Public Health*, 100168.

Klein, A. V., & Kiat, H. (2015). Detox diets for toxin elimination and weight management: a critical review of the evidence. *Journal of human nutrition and dietetics*, *28*(6), 675-686.

Kochman, J., Jakubczyk, K., Antoniewicz, J., Mruk, H., & Janda, K. (2020). Health benefits and chemical composition of matcha green tea: A review. *Molecules*, *26*(1), 85.

Krikorian, R., Shidler, M. D., Nash, T. A., Kalt, W., Vinqvist-Tymchuk, M. R., Shukitt-Hale, B., & Joseph, J. A. (2010). Blueberry supplementation improves memory in older adults. *Journal of agricultural and food chemistry*, *58*(7), 3996-4000.

Kuhnlein, H. V., Erasmus, B., & Spigelski, D. (2009). *Indigenous Peoples' food systems: The many dimensions of culture, diversity and environment for nutrition and health* (pp. x+-339).

Kuriyama, S., Shimazu, T., Ohmori, K., Kikuchi, N., Nakaya, N., Nishino, Y., ... & Tsuji, I. (2006). Green tea consumption and mortality due to cardiovascular disease, cancer, and all causes in Japan: the Ohsaki study. *Jama*, *296*(10), 1255-1265.

Lad, V. (2002). *Ayurveda: The science of self-healing: A practical guide*. Motilal Banarsidass Publishing House.

Lopes, C. V. A., de Sousa Alves Neri, J. L., Hunter, J., Ronto, R., & Mihrshahi, S. (2024). Interventions and Programs Using Native Foods to Promote Health: A Scoping Review. *Nutrients*, *16*(23), 4222.

Marco, M. L., Heeney, D., Binda, S., Cifelli, C. J., Cotter, P. D., Foligné, B., ... & Hutkins, R. (2017). Health benefits of fermented foods: microbiota and beyond. *Current opinion in biotechnology*, *44*, 94-102.

Martínez-González, M. A., Gea, A., & Ruiz-Canela, M. (2019). The Mediterranean diet and cardiovascular health: A critical review. *Circulation research*, *124*(5), 779-798.

Matos, L. C., Machado, J. P., Monteiro, F. J., & Greten, H. J. (2021). Understanding traditional Chinese medicine therapeutics: an overview of the basics and clinical applications. In *Healthcare* (Vol. 9, No. 3, p. 257). MDPI.

Miller, M. G., Hamilton, D. A., Joseph, J. A., & Shukitt-Hale, B. (2018). Dietary blueberry improves cognition among older adults in a randomized, double-blind, placebo-controlled trial. *European journal of nutrition*, *57*(3), 1169-1180.

Mohammadi, N., Farrell, M., O'Sullivan, L., Langan, A., Franchin, M., Azevedo, L., & Granato, D. (2024). Effectiveness of anthocyanin-containing foods and nutraceuticals in mitigating oxidative stress, inflammation, and cardiovascular health-related biomarkers: a systematic review of animal and human interventions. *Food & Function*, *15*(7), 3274-3299.

Mohammed, F. H., Al-Saily, H. M., & Awadh, E. F. A. (2023). Using Alkaline Diet for Controlling Blood Acidosis. *Journal of University of Babylon for Pure and Applied Sciences*, 312-321.

Mohan, A., Quek, S. Y., Gutierrez-Maddox, N., Gao, Y., & Shu, Q. (2017). Effect of honey in improving the gut microbial balance. *Food Quality and Safety*, *1*(2), 107-115.

Moon, H. J., Oh, S. H., Park, K. B., & Cha, Y. S. (2023). Kimchi and Leuconostoc mesenteroides DRC 1506 alleviate dextran sulfate sodium (DSS)-induced colitis via attenuating inflammatory responses. *Foods*, *12*(3), 584.

Mozaffarian, D. (2016). Dietary and policy priorities for cardiovascular disease, diabetes, and obesity: a comprehensive review. *Circulation*, *133*(2), 187-225.

Mustafa, Y. F. (2025). Coumarins in Traditional Medicine: Bridging Ancient Wisdom and Scientific Progress. *Phytomedicine Plus*, 100857.

Nagata, C., Wada, K., Tamura, T., Konishi, K., Goto, Y., Koda, S., ... & Nakamura, K. (2017). Dietary soy and natto intake and cardiovascular disease mortality in Japanese adults: the Takayama study. *The American journal of clinical nutrition*, *105*(2), 426-431.

Neufeld, L. M., Ho, E., Obeid, R., Tzoulis, C., Green, M., Huber, L. G., ... & Griffiths, J. C. (2023). Advancing nutrition science to meet evolving global health needs. *European Journal of Nutrition*, *62*(Suppl 1), 1-16.

Nunn, J. F. (2002). *Ancient Egyptian Medicine*. University of Oklahoma Press.

Ojha, S. N., Anand, A., Sundriyal, R. C., & Arya, D. (2022). Traditional dietary knowledge of a marginal hill community in the Central Himalaya: implications for food, nutrition, and medicinal security. *Frontiers in pharmacology, 12*, 789360.

Ordovas, J. M., Ferguson, L. R., Tai, E. S., & Mathers, J. C. (2018). Personalized nutrition and health. *BMJ, 361*.

Panahi, Y., Alishiri, G. H., Parvin, S., & Sahebkar, A. (2016). Mitigation of systemic oxidative stress by curcuminoids in osteoarthritis: results of a randomized controlled trial. *Journal of dietary supplements, 13*(2), 209-220.

Patil, S. Z., & Gupta, A. D. (2021). Role of Indian spices as nutraceutical and therapeutical agents. *Local food and community empowerment through tourism, 124*.

Payyappallimana, U., & Venkatasubramanian, P. (2015). Principles of Ayurveda for food, nutrition, and health. In B. Venkatraja (Ed.), *Connecting Indian Wisdom and Western Science: Plant Usage for Nutrition and Health* (pp. 15–32). India: NIPH.

Poddar, A. K. (2024). Nutrition in the New Era: Bridging Cultural Traditions and Modern Health Science in Diet Choices. *Food Studies: An Interdisciplinary Journal, 14*(2).

Polat, K. (2019). Aşure geleneğinin tarihsel arka planı ve Osmanlı kültür dünyasına yansımaları. *Akademik Dil ve Edebiyat Dergisi, 3*(4), 457-474.

Qureshi, N., & Shaikh, B. T. (2006). Myths, fallacies and misconceptions: applying social marketing for promoting appropriate health seeking behavior in Pakistan. *Anthropology & medicine, 13*(2), 131-139.

Ridgway, E., Baker, P., Woods, J., & Lawrence, M. (2019). Historical developments and paradigm shifts in public health nutrition science, guidance and policy actions: a narrative review. *Nutrients, 11*(3), 531.

Roman, S., Campos-Medina, L., & Leal-Mercado, L. (2024). Personalized nutrition: the end of the one-diet-fits-all era. *Frontiers in Nutrition, 11*, 1370595.

Sabença, C., Ribeiro, M., Sousa, T. D., Poeta, P., Bagulho, A. S., & Igrejas, G. (2021). Wheat/gluten-related disorders and gluten-free diet misconceptions: a review. *Foods, 10*(8), 1765.

Sagdic, O., & Tornuk, F. (2012). Antimicrobial properties of organosulfur compounds. In *Dietary phytochemicals and microbes* (pp. 127-156). Dordrecht: Springer Netherlands.

Scalbert, A., Manach, C., Morand, C., Rémésy, C., & Jiménez, L. (2005). Dietary polyphenols and the prevention of diseases. *Critical reviews in food science and nutrition, 45*(4), 287-306.

Scarmeas, N., Stern, Y., Tang, M. X., Mayeux, R., & Luchsinger, J. A. (2006). Mediterranean diet and risk for Alzheimer's disease. *Annals of Neurology: Official Journal of the American Neurological Association and the Child Neurology Society, 59*(6), 912-921.

Servili, M., Sordini, B., Esposto, S., Urbani, S., Veneziani, G., Maio, I. D., ... & Taticchi, A. (2013). Biological activities of phenolic compounds of extra virgin olive oil. *Antioxidants, 3*(1), 1-23.

Shahrajabian, M. H., Sun, W., & Cheng, Q. (2019). Clinical aspects and health benefits of ginger (Zingiber officinale) in both traditional Chinese medicine and modern industry. *Acta agriculturae scandinavica, section b—Soil & Plant Science, 69*(6), 546-556.

Shimazu, T., Kuriyama, S., Hozawa, A., Ohmori, K., Sato, Y., Nakaya, N., ... & Tsuji, I. (2007). Dietary patterns and cardiovascular disease mortality in Japan: a prospective cohort study. *International journal of epidemiology, 36*(3), 600-609.

Singh, B., Parsaik, A. K., Mielke, M. M., Erwin, P. J., Knopman, D. S., Petersen, R. C., & Roberts, R. O. (2014). Association of mediterranean diet with mild cognitive impairment and Alzheimer's disease: a systematic review and meta-analysis. *Journal of Alzheimer's disease, 39*(2), 271-282.

Slavin, J. L., & Lloyd, B. (2012). Health benefits of fruits and vegetables. *Advances in nutrition, 3*(4), 506-516.

Snodgrass, M. E. (2004). *Encyclopedia of kitchen history*. Routledge.

Sofi, F., Abbate, R., Gensini, G. F., & Casini, A. (2010). Accruing evidence on benefits of adherence to the Mediterranean diet on health: an updated systematic review and meta-analysis. *The American journal of clinical nutrition, 92*(5), 1189-1196.

Sofi, F., Macchi, C., Abbate, R., Gensini, G. F., & Casini, A. (2014). Mediterranean diet and health status: an updated meta-analysis and a proposal for a literature-based adherence score. *Public health nutrition, 17*(12), 2769-2782.

Suarez-Lledo, V., & Alvarez-Galvez, J. (2021). Prevalence of health misinformation on social media: systematic review. *Journal of medical Internet research*, *23*(1), e17187.

Suarez, C., & Adibi, S. (2025). Integrating Traditional Nutritional Wisdom into Digital Nutrition Platforms: Toward Culturally Adaptive and Inclusive Health Technologies. *Nutrients*, *17*(12), 1978.

Tamang, J. P., Watanabe, K., & Holzapfel, W. H. (2016). Diversity of microorganisms in global fermented foods and beverages. *Frontiers in microbiology*, *7*, 377.

Touwaide, A., & Appetiti, E. (2015). Food and medicines in the Mediterranean tradition. A systematic analysis of the earliest extant body of textual evidence. Journal of ethnopharmacology, 167, 11-29.

Trajkovska Petkoska, A., Ognenoska, V., & Trajkovska-Broach, A. (2025). Mediterranean Diet: From Ancient Traditions to Modern Science—A Sustainable Way Towards Better Health, Wellness, Longevity, and Personalized Nutrition. *Sustainability*, *17*(9), 4187.

Trichopoulou, A., Martínez-González, M. A., Tong, T. Y., Forouhi, N. G., Khandelwal, S., Prabhakaran, D., ... & de Lorgeril, M. (2014). Definitions and potential health benefits of the Mediterranean diet: views from experts around the world. *BMC medicine*, *12*(1), 112.

UNESCO. (2013). Washoku, traditional dietary cultures of the Japanese, notably for the celebration of New Year. UNESCO Intangible Cultural Heritage.

United Nations. (2014). *World Population Prospects: The 2014 Revision*. Department of Economic and Social Affairs.

Ünsal, A. (2019). Beslenmenin önemi ve temel besin öğeleri. *Kırşehir Ahi Evran Üniversitesi Sağlık Bilimleri Dergisi*, *2*(3), 1-10.

Verotta, L., Macchi, M. P., & Venkatasubramanian, P. (Eds.). (2015). *Connecting Indian wisdom and western science: plant usage for nutrition and health*. CRC Press.

Wang, M., Chen, Y., Wang, Y., Li, Y., Zhang, X., Zheng, H., ... & Liao, Q. (2018). Beneficial changes of gut microbiota and metabolism in weaned rats with Lactobacillus acidophilus NCFM and Bifidobacterium lactis Bi-07 supplementation. *Journal of Functional Foods*, *48*, 252-265.

Weng, Y., Yao, J., Sparks, S., & Wang, K. Y. (2017). Nattokinase: an oral antithrombotic agent for the prevention of cardiovascular disease. *International journal of molecular sciences*, *18*(3), 523.

Willett, W., Rockström, J., Loken, B., Springmann, M., Lang, T., Vermeulen, S., ... & Murray, C. J. (2019). Food in the Anthropocene: the EAT–Lancet Commission on healthy diets from sustainable food systems. *The lancet*, *393*(10170), 447-492.

World Health Organization. (2013). *WHO traditional medicine strategy: 2014–2023*. WHO.

Yamada, K., Ogawa, H., Hara, A., Yoshida, Y., Yonezawa, Y., Karibe, K., ... & Imai, K. (2009). Mechanism of the antiviral effect of hydroxytyrosol on influenza virus appears to involve morphological change of the virus. *Antiviral research*, *83*(1), 35-44.

Zaitlin, P., Dwyer, J., & Gleason, G. R. (2013). Mistaken beliefs and the facts about milk and dairy foods. *Nutrition Today*, *48*(3), 135-143.

Zeevi, D., Korem, T., Zmora, N., Israeli, D., Rothschild, D., Weinberger, A., ... & Segal, E. (2015). Personalized nutrition by prediction of glycemic responses. *Cell*, *163*(5), 1079-1094.

Chapter 11

When Water Leaves, Nutrients Stay: Exploring the Effects of Drying on Food Quality and Nutritional Value

Rabia Kaya Celik
and Gulcin Yildiz*
Department of Food Engineering, Igdir University, Iğdır, Turkiye

Abstract

Drying is one of the most ancient and widely applied methods for preserving food, yet it remains a critical area of study due to its complex effects on food quality and nutritional value. This chapter provides a comprehensive examination of the principles, mechanisms, and outcomes of various drying technologies, ranging from traditional sun and convective drying to advanced methods such as freeze-drying, vacuum drying, microwave-assisted drying, and emerging hybrid techniques. The impact of water removal on the retention of macronutrients, vitamins, minerals, antioxidants, and other bioactive compounds is explored, with particular attention to the delicate balance between extending shelf life and preserving the nutritional and sensory attributes of foods. The chapter discusses how different drying parameters—including temperature, time, humidity, air velocity, and pre-treatments—can influence chemical stability, enzymatic activity, and the structural integrity of food matrices. It highlights the importance of optimizing these factors to minimize nutrient loss, prevent oxidative degradation, maintain color, texture, and flavor, and support the bioavailability of key compounds. Special attention is given to bioactive-rich foods such as fruits, vegetables, herbs, and functional ingredients, where the preservation of phytochemicals, polyphenols, carotenoids, and vitamins is crucial for supporting human health. Emerging technologies such as infrared, osmotic, and ultrasonic-assisted drying are presented as innovative strategies for enhancing energy efficiency while maintaining or even improving the nutritional and functional properties of dried foods. The chapter also addresses sustainability considerations, including the reduction of food waste, energy consumption, and environmental impacts associated with conventional drying processes. By integrating current scientific knowledge with practical applications, this chapter offers valuable insights for food scientists, nutritionists, and industry professionals seeking to develop high-quality, nutrient-dense, and shelf-stable dried food products. It

* Corresponding Author's Email: gulcin.yildiz@igdir.edu.tr.

In: Let Food Be Your Medicine, and Medicine Be Your Food
Editors: Gulcin Yildiz and Arzu Imece
ISBN: 979-8-90134-204-6

underscores that, while moisture removal is essential for food preservation, careful selection and control of drying methods are key to ensuring that nutrients stay even as water leaves.

Keywords: food drying, nutrient retention, bioactive compounds, drying technologies

Introduction

The preservation of food has been a central concern for human societies since the dawn of civilization. Before the advent of modern refrigeration and freezing technologies, drying was one of the most widely employed methods for extending the shelf life of perishable foods (Lisboa et al., 2024). From sun-dried fruits in ancient Egypt to air-dried grains and herbs in traditional Asian and Mediterranean diets, the removal of water has long been recognized as a simple yet effective means of inhibiting microbial growth and slowing enzymatic and chemical reactions that cause spoilage. In contemporary times, food drying has evolved from an art rooted in tradition to a scientifically driven process, integrating advanced engineering, food chemistry, and nutritional science to optimize both safety and quality (Bourdoux et al., 2016).

Water constitutes a fundamental medium for microbial activity, chemical reactions, and enzymatic processes within foods. By reducing water activity through drying, microbial growth is significantly inhibited, and many degradative reactions are slowed (Al-Sharify et al., 2025). However, the process of removing water from food is not merely a physical change; it is accompanied by a complex series of chemical and structural transformations that can directly affect the retention of nutrients, flavor compounds, and bioactive molecules (Xie et al., 2023). Nutrients such as vitamins, minerals, polyphenols, and carotenoids vary widely in their stability under heat, oxygen exposure, and dehydration stress. As a result, the challenge in food drying lies not only in achieving adequate preservation but also in retaining the intrinsic nutritional and sensory qualities that define the food's value (Sablani, 2006).

In recent decades, the global demand for dried foods has increased significantly. This rise is driven by a combination of factors, including changing lifestyles, urbanization, and the growing emphasis on convenient, shelf-stable, and nutritious products (Fanzo et al., 2022). Consumers increasingly seek minimally processed foods that provide health benefits without relying on synthetic additives or preservatives. Concurrently, food industries face pressure to reduce waste and develop sustainable practices. Drying technologies play a crucial role in addressing these challenges by enabling the transformation of perishable raw materials into stable products with extended shelf life, reduced weight for transportation efficiency, and concentrated flavors and nutrients (Yildiz and Izli, 2019a; Radojčin et al., 2021). The science of drying encompasses a wide range of methods, each with unique impacts on the structural, chemical, and nutritional properties of foods. Traditional sun drying remains widely used because it requires little equipment and low energy input, although it is limited by slow drying rates and a higher risk of contamination Hot-air drying, a common industrial method, accelerates moisture removal but can lead to thermal degradation of heat-sensitive nutrients (Calín-Sánchez et al., 2020). Novel approaches such as freeze-drying, vacuum drying, microwave-assisted drying, and infrared-assisted drying have emerged to minimize nutrient loss while maintaining texture, color, and bioactive compound integrity. Freeze-drying, in particular, is renowned for its ability to preserve heat-sensitive vitamins, antioxidants, and flavors by sublimating ice directly into vapor under low pressure. However, such techniques can be energy-intensive and costly, necessitating careful consideration of economic and environmental factors in large-scale applications (Moutia et al., 2024). Understanding the effects of drying on food quality requires an appreciation of the chemical reactions that occur during moisture removal. Thermal exposure can trigger Maillard reactions, caramelization, oxidation of lipids, and degradation of thermolabile vitamins such as vitamin C and certain B-complex vitamins (Yildiz, 2022). The rate and extent of these reactions depend on temperature, humidity, air velocity, and food matrix characteristics, highlighting the need for tailored drying strategies for different food types (Santander et al., 2025). Furthermore, structural changes induced by drying, including shrinkage, pore formation, and modification of cellular walls, can influence water reabsorption, texture, and sensory perception, which are critical determinants of consumer acceptance (Yildiz and Izli, 2019b; İzli et al., 2022; Mahfoud et al., 2025).

Nutritional quality is a key factor in evaluating dried foods, especially as interest in functional foods, nutraceuticals, and health-promoting diets grows. Bioactive compounds such as polyphenols, carotenoids, flavonoids, and other antioxidants not only enhance nutritional value but also provide protective effects against chronic diseases, including cardiovascular disorders, diabetes, and cancer (Hajzer et al., 2025). Preserving these compounds during drying is therefore essential in both research and industrial processing. For instance, carotenoids in tomatoes and carrots are prone to oxidative degradation under high-temperature drying, while polyphenols in berries may undergo structural changes that reduce their antioxidant activity (İzli and Yildiz, 2021; Santos et al., 2025). Emerging approaches—such as protective coatings, encapsulation, and optimized drying protocols—seek to maintain these nutrients without compromising safety or shelf life (Mari et al., 2025).

Beyond nutrient retention, sensory qualities—flavor, aroma, color, and texture—are crucial for consumer acceptance. Drying can intensify flavors but may also cause the loss of volatile compounds and undesirable textural changes, such as excessive hardness or brittleness (Gómez et al., 2020). Innovative approaches, including hybrid drying techniques that combine microwave, vacuum, or infrared energy, are being developed to preserve both nutritional and sensory attributes. These advancements highlight the intersection of food chemistry, engineering, and sensory science in creating drying processes that align with modern consumer expectations (Nwankwo et al., 2023).

Sustainability and environmental considerations further highlight the importance of drying technologies in modern food systems. Reducing food waste, improving energy efficiency, and minimizing the carbon footprint of food production are critical industry goals. By transforming seasonal or perishable raw materials into stable, nutrient-dense products, drying enhances resource efficiency and supports long-term food security. Moreover, lightweight dried foods lower transportation energy costs and increase global access to nutrient-rich foods, especially in regions where fresh produce is scarce or seasonally unavailable (Corigliano et al., 2025).

In conclusion, the study of drying and its effects on food quality and nutritional value represents a convergence of traditional practices and modern scientific innovation. By elucidating the mechanisms through which water removal impacts nutrient retention, bioactive compound stability, and sensory characteristics, researchers and industry professionals can develop optimized drying strategies that meet both consumer demand and sustainability goals. This chapter aims to provide a comprehensive exploration of these dynamics, examining the interactions between drying methods, food composition, and nutritional outcomes, and highlighting emerging technologies that promise to enhance the quality, safety, and functionality of dried foods in the twenty-first century.

Types of Drying Technologies and Their Effects on Nutrients

Food drying technologies have evolved substantially from traditional methods such as sun and air drying to advanced techniques like freeze-drying, vacuum drying, and hybrid methods combining multiple energy sources. Each drying method exerts distinct effects on the chemical composition, structural integrity, and nutritional quality of foods, making the selection of the appropriate technique critical for preserving both safety and health benefits. Understanding these technologies and their impact on nutrients is essential for food scientists, industry practitioners, and researchers aiming to optimize the quality of dried products (Nainggolan et al., 2024).

Sun Drying

Sun drying is the oldest and most widely practiced method of food preservation. It involves exposing foods to direct sunlight, relying on ambient heat and airflow to remove moisture. This method is energy-efficient and low-cost, making it suitable for small-scale production, particularly in developing regions. Sun drying is commonly applied to fruits, vegetables, grains, and herbs (Fernandes and Tavares, 2024).

Despite its simplicity, sun drying has limitations. Prolonged exposure to heat and UV radiation can degrade thermolabile nutrients such as vitamin C, certain B vitamins, and carotenoids. Additionally, sun drying is susceptible to microbial contamination, dust, and inconsistent drying rates due to weather variability. Nevertheless, with proper pre-treatment (e.g., blanching, washing) and controlled drying surfaces, sun drying can retain significant levels of macronutrients and minerals, while concentrating flavors and sugars (Rana et al., 2024).

Hot-Air Drying

Hot-air, or convective, drying removes water by circulating heated air over the food product. This method is widely used in industrial settings due to its scalability, reliability, and relatively fast drying rates, making it suitable for grains, fruits, vegetables, and processed foods such as powders and snacks (Berk et al., 2023). The primary advantage of hot-air drying is the controlled environment, which reduces microbial spoilage and enables standardized production. However, prolonged exposure to elevated temperatures can lead to nutrient losses, particularly of heat-sensitive vitamins such as vitamin C and folate, and may trigger Maillard reactions that affect color, flavor, and protein functionality. Therefore, optimizing temperature, air velocity, and drying time is essential to maximize nutrient retention while achieving the desired moisture content (Yildiz, 2021).

Freeze-Drying (Lyophilization)

Freeze-drying, or lyophilization, is a low-temperature dehydration method that preserves foods by sublimating ice directly into vapor under vacuum conditions. This technique is renowned for retaining heat-sensitive nutrients, antioxidants, volatile flavor compounds, and structural integrity. Freeze-dried products often exhibit excellent rehydration properties, making them ideal for instant foods, fruits, vegetables, and nutraceuticals (Coşkun et al., 2024).

Nutritionally, freeze-drying is superior to most thermal drying methods, as it minimizes oxidative damage and preserves delicate bioactive compounds such as polyphenols, flavonoids, and carotenoids. However, the high cost and energy requirements associated with freeze-drying limit its widespread industrial use. Despite this, it remains the method of choice when maximal nutrient and sensory retention is desired, such as in high-value functional foods, probiotics, and specialized dietary products (Yildiz et al., 2025; Zhao et al., 2025).

Vacuum Drying

Vacuum drying combines reduced pressure with moderate heat to facilitate moisture removal at lower temperatures. By lowering the boiling point of water, vacuum drying enables faster dehydration while minimizing thermal degradation of sensitive nutrients. This method is particularly suitable for fruits, herbs, and pharmaceutical-grade plant extracts.

Studies indicate that vacuum drying preserves higher levels of vitamin C, phenolic compounds, and carotenoids compared to conventional hot-air drying. The reduced oxygen exposure also limits oxidative reactions, maintaining both color and flavor. However, equipment costs and processing complexity can be higher than for standard convective drying, which may restrict its application in low-resource settings (Mansour et al., 2025).

Microwave-Assisted and Infrared-Assisted Drying

Microwave-assisted and infrared-assisted drying are examples of novel energy-efficient technologies designed to accelerate drying while preserving quality. Microwave energy penetrates food matrices, causing rapid internal heating and moisture evaporation, whereas infrared radiation primarily heats the surface, inducing quick water loss and browning reactions (Puente-Díaz et al., 2020).

These innovative drying methods offer notable advantages, including reduced drying time and lower energy consumption. Nutrient retention is generally higher than with conventional hot-air drying, particularly for antioxidants and heat-sensitive vitamins. However, uneven heating and localized hotspots in microwave drying can cause partial nutrient degradation or structural damage, making careful process optimization essential. Combining these techniques with vacuum or hot-air systems often yields synergistic benefits, enhancing both drying efficiency and nutrient preservation (Petković et al., 2025).

Nutrient-Specific Effects Across Drying Methods

The impact of drying on nutrients varies widely depending on the nutrient type and drying method. Water-soluble vitamins (e.g., vitamin C, B-complex) are highly sensitive to heat, oxygen, and prolonged exposure, often showing the greatest losses during conventional thermal drying. Fat-soluble vitamins (e.g., A, D, E, and K) are relatively stable but can undergo oxidative degradation under high-temperature conditions. Minerals such as calcium, potassium, and magnesium generally remain stable during drying, though their bioavailability may be influenced by structural changes in the food matrix (Santos et al., 2025).

Bioactive compounds, including polyphenols, carotenoids, and flavonoids, show variable stability. While some compounds may degrade during thermal processing, others become more bioavailable due to structural breakdown of plant cell walls, enhancing antioxidant activity. Thus, drying can both diminish and enhance the nutritional quality of foods, depending on the interplay between method, temperature, and food composition (Yildiz, 2022).

In summary, a diverse array of drying technologies exists, each with distinct advantages, limitations, and impacts on nutrient retention. Optimizing these methods requires a nuanced understanding of food chemistry, process engineering, and nutritional science. By carefully selecting and tailoring drying processes, food producers can achieve the dual objectives of extending shelf life and preserving the health-promoting properties of foods, contributing to both consumer satisfaction and sustainable food systems.

Nutritional Changes During Drying: Vitamins, Minerals, and Bioactive Compounds

Drying is a critical food preservation method that significantly affects the nutritional profile of foods. The extent of these changes depends on factors such as the drying technology, temperature, duration, food type, and pre-treatment processes. Understanding how vitamins, minerals, and bioactive compounds respond to dehydration is essential for optimizing both nutritional quality and functional properties of dried foods.

Vitamins

Vitamins, particularly water-soluble types like vitamin C and B-complex vitamins, are highly sensitive to heat, oxygen, and light. During conventional hot-air drying, vitamin C can degrade rapidly due to oxidative reactions, often resulting in losses of 50–70% in fruits such as oranges, strawberries, and kiwifruit. B vitamins, including thiamine and folate, also suffer from thermal degradation, although they tend to be slightly more stable than vitamin C (ElGamal et al., 2023).

Fat-soluble vitamins (A, D, E, and K) are generally more resistant to thermal treatments but can still undergo oxidative degradation, particularly in lipid-rich foods. Vitamin A, in the form of beta-carotene, is sensitive to both heat and light, and substantial losses can occur during prolonged drying. Conversely, some carotenoids may become more bioavailable after mild heating due to the breakdown of plant cell walls, which facilitates their release during digestion (Feszterová et al., 2023).

Advanced drying techniques such as freeze-drying or vacuum drying minimize thermal and oxidative stress, preserving higher concentrations of heat-sensitive vitamins. Microwave-assisted and infrared-assisted drying can also help retain vitamins when carefully optimized, as rapid drying reduces exposure time. Pre-treatments like blanching, osmotic dehydration, or antioxidant addition can further protect vitamin content during dehydration (Yavruturk et al., 2024).

Minerals

Minerals such as calcium, potassium, magnesium, iron, and zinc are generally stable under most drying conditions, as they are not prone to degradation from heat or oxidation like vitamins. However, their bioavailability can be affected by structural changes in the food matrix. For example, high-temperature drying may alter mineral crystallinity or promote interactions with other food components, potentially influencing absorption (Bhardwaj et al., 2024).

Drying also concentrates minerals due to water removal. For instance, potassium content in dried fruits appears several times higher on a per-gram basis than in fresh produce. This concentration reflects a reduction in water weight rather than a true increase in mineral content (Moutia et al., 2024).

Bioactive Compounds

Bioactive compounds, such as polyphenols, flavonoids, carotenoids, anthocyanins, and glucosinolates, contribute to the health-promoting properties of foods, including antioxidant, anti-inflammatory, and cardioprotective effects. Their stability during drying varies significantly depending on chemical structure, food matrix, and processing conditions (Barel and Yildiz, 2024).

Thermal drying often reduces the content of highly labile compounds like anthocyanins, which are sensitive to heat and pH changes. Conversely, certain polyphenols may increase in extractable concentration due to cell wall breakdown, enhancing antioxidant activity. For example, studies on tomato and carrot drying show that mild heat treatments can improve total phenolic content while excessive high temperatures lead to degradation (İzli et al., 2022).

Freeze-drying is particularly effective at preserving bioactive compounds, as it minimizes heat exposure and oxidative stress. Vacuum and microwave-assisted drying also show promising results, although care must be taken to avoid uneven heating that can cause localized degradation (Yildiz and Bayat, 2022). Enzymatic activity in fresh produce, such as polyphenol oxidases, can further influence the stability of bioactive compounds during drying. Pre-treatments like blanching, chemical antioxidants, or controlled atmosphere drying can reduce enzymatic degradation and enhance bioactive retention (Cemal and Yildiz, 2022).

Interactions Between Nutrients During Drying

Drying can induce complex interactions among nutrients. For example, Maillard reactions between reducing sugars and amino acids during thermal drying may reduce the availability of certain amino acids while generating flavor and color compounds. Oxidation of lipids can compromise fat-soluble vitamin stability, and the breakdown of cellular structures can either release or degrade bioactive molecules. Understanding these interactions allows food scientists to optimize drying conditions to balance nutrient retention, sensory quality, and shelf life (Zhang, and Liu, 2024; El Hosry et al., 2025).

Strategies to Preserve Nutritional Quality

The preservation of nutritional quality during drying is a critical concern in food processing, as vitamins, minerals, and bioactive compounds are highly sensitive to environmental and processing conditions. One of the most effective ways to safeguard these nutrients is through the careful selection of drying technology. Traditional methods such as hot-air drying often expose foods to prolonged heat, resulting in significant nutrient degradation, particularly of heat-labile vitamins like vitamin C and B-complex vitamins (Santos et al., 2025). In contrast, advanced methods such as freeze-drying, vacuum drying, and optimized microwave-assisted drying provide gentler processing environments that reduce thermal damage and oxidative stress, thereby maintaining much of the food's original nutritional profile. Freeze-drying, for example, removes water at low temperatures through sublimation, which protects both structural integrity and delicate bioactive compounds, while vacuum and microwave-assisted approaches allow for faster moisture removal under reduced oxygen conditions, minimizing nutrient losses. Thus, the choice of technology serves as the foundation for ensuring nutritional retention in dried foods (Roohinejad et al., 2025).

Beyond the drying technology itself, the use of pre-treatments provides an additional layer of protection against nutrient loss. Many degradative processes during drying are driven by enzymes or oxidative reactions, which can be mitigated through simple interventions. For example, blanching inactivates enzymes responsible for undesirable color changes and vitamin degradation, while osmotic dehydration partially removes water before drying, reducing heat exposure. Similarly, dipping in antioxidant solutions creates a protective barrier against oxidative damage, particularly for phenolic compounds and other sensitive bioactives. Incorporating such pre-treatments into the processing chain can significantly improve the nutritional integrity of dried products, enhancing the retention of essential nutrients and functional components (Jia et al., 2024).

Equally important is the control of drying conditions throughout the process. Temperature, humidity, and drying time act as interdependent factors that influence nutrient stability, structural quality, and overall product acceptability. High temperatures and prolonged drying times may accelerate the breakdown of vitamins and phytochemicals, while low humidity and rapid drying can induce case hardening, which traps moisture and promotes localized spoilage. Therefore, establishing optimized drying parameters is crucial for achieving the dual goals of efficient dehydration and nutrient conservation. Fine-tuning these variables allows processors to strike a balance between energy efficiency and product quality, ensuring that sensitive compounds are preserved without compromising safety or shelf life (Hu et al., 2025).

Even after drying, nutrient stability is highly dependent on packaging and storage conditions. Vitamins and bioactive compounds remain vulnerable to environmental stressors such as oxygen, light, and residual moisture, which can continue to drive degradation. Using appropriate packaging materials—such as vacuum-sealed pouches, oxygen-impermeable films, or light-protective laminates—along with controlled storage conditions helps to preserve nutritional value over time. Poor storage not only accelerates nutrient loss but can also negate the benefits of advanced drying methods and pre-treatments. Therefore, integrating protective packaging strategies is crucial for delivering high-quality dried foods to consumers (Postružnik et al., 2024; Divya et al., 2025).

In conclusion, drying is a double-edged sword in food processing. While it extends shelf life, reduces food waste, and can concentrate certain nutrients, making dried foods an efficient source of energy and micronutrients (Coelho et al., 2022), it can also degrade heat- and oxidation-sensitive compounds if not carefully managed (Versino et al., 2023). Addressing this challenge requires a holistic approach that integrates advanced drying technologies, effective pre-treatments, precise process control, and appropriate storage solutions. By tailoring strategies to nutrient-specific sensitivities, producers can create dried products that maintain their health-promoting qualities and functional properties. This integrated approach not only meets consumer demands for nutritious, safe, and shelf-stable foods but also supports broader goals of sustainability and food security.

Impact of Drying on Protein, Carbohydrate, and Lipid Quality

Drying not only influences the stability of vitamins and minerals but also has profound effects on the structure, functionality, and nutritional quality of macronutrients—proteins, carbohydrates, and lipids (İzli, and Yildiz, 2021). These impacts are highly dependent on the drying technology applied, the temperature and duration of the process, and the intrinsic properties of the food matri (Yildiz, 2022). Because macronutrients contribute not only to nutritional value but also to functional and sensory characteristics, understanding how drying alters them is critical for the development of high-quality dried foods that balance stability with consumer acceptability.

Proteins are among the most sensitive macronutrients to the stresses of drying, particularly heat and oxidation. Exposure to elevated temperatures during processes such as hot-air drying can cause denaturation and aggregation, leading to a loss of tertiary and quaternary structure. This structural disruption decreases solubility, digestibility, and bioavailability, ultimately reducing the nutritional value of protein-rich foods (Neji et al., 2022). The formation of Maillard reaction products, such as lysinoalanine, further compromises protein quality, as observed in milk powder production where excessive heat impairs both functional and nutritional attributes. However, pre-treatment strategies, such as blanching, play a protective role by inactivating proteolytic enzymes that would otherwise degrade proteins during drying (Nielsen et al., 2022; Bolchini et al., 2025). More gentle techniques, including freeze-drying and vacuum drying, help preserve protein integrity by maintaining solubility, emulsification, and foaming capacity. In plant-based foods like legumes and cereals, careful control of drying conditions not only sustains functional properties but also reduces the loss of essential amino acids, which are especially vulnerable to heat-induced damage. Thus, the preservation of protein quality relies on the combination of appropriate pre-treatments, optimized drying parameters, and advanced drying methods (Esteghlal et al., 2019).

Carbohydrates exhibit diverse responses to drying due to differences in molecular structure and water-binding capacity. Simple sugars, such as glucose and fructose, are particularly reactive under high-temperature conditions, where they may undergo caramelization or participate in Maillard reactions with amino acids (El Hosry et al., 2025). These changes can lead to desirable color and flavor development, enhancing the sensory quality of dried products, but they also decrease the availability of free sugars and essential amino acids, potentially lowering nutritional value (Qi et al., 2025). More complex carbohydrates, such as starches, are equally influenced by drying. Under certain conditions, starch granules can undergo gelatinization or retrogradation, altering digestibility, water absorption, and textural properties. Dietary fibers are generally more resistant to heat but may still be affected, with soluble fibers undergoing partial degradation under extreme processing. The use of freeze-drying offers significant advantages in this context, as it minimizes thermal stress and preserves the structural and functional integrity of starches and fibers. Additionally, pre-treatments like osmotic dehydration help retain carbohydrate composition in fruits and vegetables by preventing excessive leaching of sugars, thereby maintaining both nutritional and sensory quality (Prosapio, and Lopez-Quiroga, 2020; Aravindakshan et al., 2021).

Lipids, perhaps more than any other macronutrient, are vulnerable to oxidative degradation during drying. This is particularly problematic in foods with high levels of unsaturated fatty acids, such as nuts, seeds, and certain fruits, where lipid oxidation leads to rancidity, off-flavors, and the loss of essential fatty acids and fat-soluble vitamins. The risk of lipid degradation is exacerbated by prolonged exposure to heat and oxygen, making high-temperature drying methods less suitable for lipid-rich foods. Pre-treatment strategies that inactivate endogenous enzymes such as lipases and lipoxygenases are vital, as these enzymes catalyze lipid hydrolysis and oxidation, accelerating nutrient losses. Furthermore, the presence of natural antioxidants, such as tocopherols, or the incorporation of added antioxidants, offers an additional line of defense against oxidative deterioration. Advanced techniques such as freeze-drying and microwave-assisted vacuum drying are particularly effective at preserving lipid quality because they reduce exposure to oxygen and limit thermal stress. As a result, these methods help maintain both the nutritional value and the sensory integrity of lipid-rich dried foods (Chemat et al., 2023; Dragoev, 2024; Alagbe et al., 2025).

Taken together, the effects of drying on proteins, carbohydrates, and lipids highlight the dual challenge of stabilizing food while preserving its nutritional and functional qualities. High temperatures and oxidative conditions can compromise protein digestibility, carbohydrate availability, and lipid stability. However, careful process optimization and the use of innovative drying methods can mitigate these drawbacks. By tailoring drying approaches to the specific macronutrient sensitivities of each food matrix, producers can enhance the nutritional value, functionality, and consumer appeal of dried products, ensuring their role as both shelf-stable and health-promoting foods.

Interactions Among Macronutrients

Drying not only influences individual macronutrients but also induces complex interactions among proteins, carbohydrates, and lipids that ultimately shape the nutritional, functional, and sensory quality of dried foods. One of the most prominent examples is the Maillard reaction, which occurs between reducing sugars and amino acids. While this non-enzymatic browning reaction is often associated with desirable color and flavor development, it can also decrease the availability of essential amino acids and certain sugars, thereby reducing overall digestibility (Duda-Seiman et al., 2025). Lipid-protein interactions represent another important set of changes that may alter emulsifying properties and modify textural characteristics. These interactions, together with carbohydrate-protein cross-linking, can influence solubility, water-binding capacity, and even digestibility (Wijegunawardhana et al., 2024). Such changes highlight that macronutrients do not behave independently during drying; rather, they undergo synergistic or antagonistic modifications that can be either beneficial or detrimental depending on the process conditions (Schefer et al., 2021). For this reason, understanding these interactions is essential for food technologists aiming to balance nutrient preservation with sensory and shelf-life improvements.

The preservation of macronutrient quality during drying requires a combination of strategies that minimize degradation while maximizing functional retention. Advanced drying technologies such as freeze-drying, vacuum drying, and microwave-assisted drying under controlled conditions offer significant advantages by reducing thermal and oxidative stress, thereby limiting denaturation, oxidation, and caramelization (Moutia et al., 2024). Complementary pre-treatment methods, including blanching, antioxidant incorporation, and enzyme inactivation, provide additional protection by reducing enzymatic activity and oxidative damage to proteins, lipids, and carbohydrates (Adetoro et al., 2021). Process control is equally important; careful regulation of temperature, humidity, and drying time ensures efficient moisture removal without excessive structural or nutritional losses (Monsour et al., 2025). Finally, post-drying preservation strategies, such as appropriate packaging and storage, are critical to safeguarding macronutrient integrity. Protection from oxygen, light, and residual moisture slows degradation processes that may otherwise continue even after drying is complete (Zhang et al., 2024).

In conclusion, drying exerts a profound influence on the quality of proteins, carbohydrates, and lipids, not only through direct effects on each nutrient but also through their complex interactions during processing. While high-temperature and uncontrolled drying conditions can compromise both nutritional value and functionality, the application of advanced technologies, optimized parameters, and protective strategies makes it possible to retain macronutrient quality to a significant degree. A thorough understanding of these effects and interactions allows the production of dried foods that maintain health benefits, desirable functional attributes, and extended shelf life. This knowledge is particularly relevant in addressing modern consumer demands for food products that are convenient, shelf-stable, and at the same time nutritionally valuable.

Sensory and Functional Effects of Drying

Drying affects not only the nutritional profile of foods but also their sensory characteristics and functional properties, both of which are crucial for consumer acceptance and overall product quality (Yildiz and Izli, 2019a). Visual cues such as color, alongside flavor, aroma, and texture, strongly influence first impressions and purchasing decisions, while functional attributes—including water-holding capacity, solubility, and rehydration behavior—determine how dried foods perform in culinary and industrial applications. The extent of these changes is influenced by the drying technique used, the intensity and duration of heat exposure, and the implementation of protective strategies during processing. Therefore, sensory and functional aspects must be considered alongside nutritional factors to produce high-quality, commercially viable dried products (Yildiz and Izli, 2019b).

Color is often the first quality attribute noticed by consumers, yet it is highly sensitive to drying conditions. High temperatures can trigger browning reactions, such as Maillard reactions and caramelization, particularly in carbohydrate-rich foods like fruits, cereals, and dairy. While moderate browning can enhance visual appeal and suggest richer flavors, excessive darkening is typically viewed as over-processing and may coincide with nutrient loss.

Techniques like freeze-drying and vacuum drying minimize these reactions by limiting heat exposure and preserving natural coloration. Pre-treatments such as blanching or antioxidant application further stabilize pigments and prevent enzymatic browning, helping maintain color throughout processing and storage (Yildiz, 2022).

Flavor and aroma are equally affected during drying. Heat-sensitive volatile compounds in fruits, vegetables, and herbs can be lost, diminishing freshness and authentic taste (Izli et al., 2018). Simultaneously, Maillard reaction products or lipid oxidation may create new flavors, which can be desirable in some foods, such as roasted coffee or dried figs, enhancing sensory complexity (Shahidi and Hossain, 2022). However, in more delicate foods, these reactions may produce off-flavors that reduce acceptability. Drying methods that limit exposure to heat and oxygen, such as freeze-drying or vacuum-assisted drying, are effective at retaining natural aroma compounds and ensuring a more authentic flavor profile (Gao et al., 2024).

Texture and mouthfeel are also heavily influenced by drying, as water removal alters the food's structural matrix. Rapid moisture loss at high temperatures may cause hardening, shrinkage, or case-hardening, where a firm outer layer traps internal moisture, resulting in tough, uneven textures. These structural changes can negatively affect rehydration properties and consumer satisfaction (Yang et al., 2024). Freeze-drying, in contrast, preserves cellular integrity by sublimating water at low temperature and pressure, producing lightweight, porous foods that are crisp when dry and rehydrate quickly and uniformly (Grassino et al., 2024). Controlling drying rates and temperature gradients carefully helps prevent structural collapse, maintaining consistent, appealing textures for both ready-to-eat products and foods intended for reconstitution (Mahfoud et al., 2025).

Overall, the sensory and functional consequences of drying highlight the delicate balance between processing efficiency, nutrient preservation, and consumer expectations. Understanding how different drying techniques and conditions impact color, flavor, aroma, and texture allows food technologists to design processes that maintain desirable attributes while minimizing defects. Integrating sensory and functional considerations with nutritional quality is essential for producing dried foods that are not only shelf-stable and nutritious but also appealing and versatile for a wide range of consumers.

Rehydration and Functional Properties

Rehydration and functional performance are essential factors when assessing dried foods, as they impact both consumer experience and industrial applications. Functional properties—including water absorption, solubility, emulsification, and gelation—depend not only on the food's composition but also on the drying method and the extent of macronutrient changes during processing (Yildiz et al., 2025). Products intended for rehydration, such as powdered milk, instant soups, and dried fruits, must retain sufficient water-holding capacity and structural integrity to approximate their original texture upon reconstitution (Yildiz, 2022). Freeze-drying and vacuum drying are considered particularly effective because they maintain cellular structure and porosity, enhancing water uptake and minimizing collapse (Bhatta et al., 2020). In contrast, hot-air drying often causes surface hardening or shrinkage, limiting water penetration and resulting in uneven or incomplete rehydration.

Drying also affects the functional roles of proteins, starches, and dietary fibers. Protein denaturation can reduce solubility and emulsifying capacity, while starch gelatinization or retrogradation can alter viscosity and gel strength. Heat-induced degradation of soluble fibers can diminish water-binding ability, affecting the texture and mouthfeel of rehydrated products. These interactions emphasize the complex relationship between drying conditions and functional properties, highlighting the need for careful process optimization to produce high-quality dried foods (Yıldız et al., 2021).

Maintaining sensory and functional quality in dried foods requires a holistic approach that spans the entire processing chain. Optimizing drying parameters such as temperature, airflow, and time is essential to balance moisture removal with the preservation of desirable attributes. Excessive heat or prolonged exposure often leads to nutrient degradation, color loss, off-flavor development, and undesirable textural changes, whereas carefully tailored conditions reduce such risks (Abreu et al., 2025). Pre-treatment methods further enhance product stability and quality; blanching inactivates enzymes responsible for discoloration and flavor deterioration, osmotic dehydration protects texture and prevents nutrient leaching, and antioxidant incorporation slows oxidative damage to pigments and lipids (Adetoro et al., 2025). Advances in drying technologies offer additional opportunities to safeguard sensory and functional properties. Freeze-drying and vacuum drying minimize thermal and oxidative stress, microwave-assisted drying shortens processing time while retaining volatile compounds, and infrared drying improves energy efficiency while reducing nutrient loss. Post-drying strategies are equally important: packaging and storage conditions that minimize exposure to moisture, oxygen, and light help preserve color vibrancy, flavor stability, and textural integrity throughout the product's shelf life (Divya et al., 2025).

In summary, the sensory and functional implications of drying are inseparable from product quality, consumer acceptance, and application versatility. The ability of dried foods to rehydrate effectively, retain natural sensory attributes, and maintain key functional properties determines their success in both household consumption and industrial food formulations. By integrating advanced drying technologies, optimizing process parameters, and applying suitable pre- and post-treatment strategies, it is possible to produce dried foods that deliver not only nutritional value but also appealing sensory characteristics and reliable functional performance. This comprehensive approach ensures that dried foods remain both enjoyable to consume and adaptable to a wide range of culinary and industrial applications, meeting the growing demand for convenient, stable, and high-quality food products.

Retention of Vitamins and Minerals During Drying

Vitamins and minerals are indispensable micronutrients that play critical roles in supporting immune health, bone development, vision, enzymatic reactions, and overall metabolic balance. Because they are not synthesized in sufficient quantities by the human body, their presence in the diet is vital, and food processing methods must therefore be carefully managed to preserve them (Razzaque, and Wimalawansa, 2025). Drying, one of the most widely used preservation techniques, has the potential to both concentrate and degrade nutrients, depending on the method, temperature, oxygen exposure, and duration of processing (Nowacka et al., 2023).

Understanding the mechanisms that govern nutrient stability during drying is essential for producing dried foods that retain their nutritional value and continue to support public health.

The stability of vitamins during drying is influenced by their chemical structure and sensitivity to heat, oxygen, and light. Water-soluble vitamins, particularly vitamin C and members of the B-complex group, are the most vulnerable to degradation under thermal stress. Vitamin C, for example, is easily oxidized in the presence of heat and oxygen, leading to substantial losses in hot-air dried fruits and vegetables (Yildiz, 2022). Similarly, thiamine (vitamin B1) and folate are prone to partial breakdown, which can alter the nutritional profile of dried products. Fat-soluble vitamins such as A, D, E, and K are generally more resilient but are not entirely immune to damage; prolonged or intense heat exposure can cause structural changes, while compounds like beta-carotene may undergo isomerization or oxidative degradation when exposed to light and air (Kaźmierczak-Barańska et al., 2025). Advanced drying methods such as freeze-drying and vacuum drying are therefore often preferred, as they impose less thermal stress and limit oxidative reactions, making them more suitable for preserving sensitive vitamins (Paparella et al., 2025).

In contrast, minerals demonstrate greater resilience during drying because they are not destroyed by heat or oxidation. However, mineral content can still be affected by processing conditions, particularly during pre-treatments like blanching or osmotic dehydration (Kulczyński et al., 2025). Elements such as potassium, magnesium, calcium, and iron can leach into water or brines used in these steps, leading to measurable losses prior to the actual drying phase. The extent of these losses depends largely on the duration of water contact and the solubility of the minerals (Feng et al., 2025). Therefore, minimizing pre-treatment times and reducing direct contact with water are important considerations for maintaining the mineral composition of dried foods.

Several factors collectively determine the degree of micronutrient retention in dried foods. Temperature and time are particularly influential, as higher temperatures accelerate the degradation of heat-labile vitamins, while extended drying times further increase cumulative losses (Yildiz, 2022). Oxygen exposure plays an equally important role, with oxidative reactions being a major cause of vitamin breakdown—hence the effectiveness of vacuum or inert gas drying methods (Moon et al., 2020). Moisture content and removal rate also influence nutrient stability, since rapid water removal can limit enzymatic activity that otherwise contributes to vitamin destruction. Furthermore, the food matrix itself can act as a protective environment; acidic foods may stabilize vitamin C, while complex matrices may physically shield vitamins and minerals from degradation (Santos et al., 2025).

To enhance the retention of vitamins and minerals, researchers and industry practitioners have developed a range of strategies. Pre-treatments with antioxidants such as ascorbic acid, citric acid, or natural polyphenols are effective in slowing oxidative damage and stabilizing sensitive compounds (De Rossi et al., 2025). Gentle drying techniques—particularly freeze-drying, vacuum drying, and microwave-assisted drying—are increasingly favored because they preserve not only micronutrients but also bioactive phytochemicals that contribute to the functional properties of foods (Yildiz et al., 2025). Encapsulation technologies represent another promising approach, as microencapsulation can protect vitamins and minerals from degradation during both processing and subsequent storage, while also improving bioavailability upon consumption. Finally, optimized storage conditions, including protection from light, heat, and excess humidity, are essential for maintaining nutrient integrity over time, ensuring that the nutritional benefits of dried foods are preserved until they reach the consumer (Yildiz et al., 2018).

Implications for Functional Foods and Public Health

Preserving vitamins and minerals during drying is not only a matter of maintaining food quality but also a key determinant of the role dried foods can play in public health nutrition. Functional foods, fortified products, and dietary supplements rely heavily on the retention of micronutrients to fulfill their intended purpose of improving health outcomes beyond basic nutrition (Moutia et al., 2025). When drying technologies are optimized to conserve heat-sensitive vitamins and minimize mineral losses, manufacturers are able to produce nutritionally dense foods that address widespread micronutrient deficiencies. This is particularly important in regions where access to fresh produce is limited by seasonal variability, geographic constraints, or economic factors, as dried foods can serve as reliable, shelf-stable alternatives that help bridge nutritional gaps (Fekete et al., 2025). In this context, dried fruits, vegetables, herbs, and even fortified cereal products can act as valuable carriers of essential micronutrients, supporting both preventive health strategies and therapeutic dietary interventions.

Beyond their role in combating deficiencies, dried foods with high micronutrient retention contribute meaningfully to daily dietary requirements and therefore support long-term wellness (Li et al., 2025). Adequate intake of vitamins and minerals is strongly linked to reduced risk of chronic diseases, improved immune response, and better cognitive and physical performance (Razzaque, and Wimalawansa, 2025). By incorporating advanced drying methods such as freeze-drying, vacuum drying, or microwave-assisted drying, it becomes possible to align food processing practices with public health goals, ensuring that consumers receive products that are both safe and nutritionally effective (Coşkun et al., 2024). Moreover, maintaining nutrient integrity in dried foods enhances their functional potential, allowing them to serve as vehicles for delivering bioactive compounds, antioxidants, and other health-promoting ingredients.

In conclusion, the retention of vitamins and minerals during drying reflects a delicate balance between processing efficiency and nutrient stability. While traditional methods may prioritize cost-effectiveness and simplicity, modern approaches increasingly focus on sustainability and health outcomes. The integration of advanced drying technologies, antioxidant pre-treatments, and optimized storage practices offers a pathway to producing dried foods that maintain their micronutrient integrity and functional properties. Such advancements hold significant promise for strengthening food systems, supporting public health initiatives, and ensuring that functional foods deliver measurable nutritional and health benefits across diverse populations.

Effects of Drying on Bioactive Compounds and Phytochemicals

Bioactive compounds, including polyphenols, flavonoids, carotenoids, phenolic acids, and other phytochemicals, are major contributors to the health-promoting properties of fruits, vegetables, and herbs (Yildiz, 2022). They exhibit antioxidant, anti-inflammatory, antimicrobial, and anticancer activities, making them essential for functional foods and nutraceutical development (Yildiz et al., 2025). However, these compounds are highly sensitive to environmental and processing conditions, and drying—though crucial for preservation—can significantly affect their concentration and bioactivity. Factors such as temperature, oxygen, light, and moisture critically influence their stability. Understanding degradation mechanisms and implementing strategies to minimize losses are therefore vital for preserving the functional and nutritional value of dried products.

Polyphenols and flavonoids, which represent one of the most studied groups of plant secondary metabolites, are particularly vulnerable to degradation during drying. High heat, extended drying times, and oxidative conditions often lead to reduced total phenolic content and diminished antioxidant capacity, especially in berries, leafy vegetables, and herbs. (ElGamal et al., 2023). Flavonoids, a subclass of polyphenols, exhibit variable stability depending on their molecular form; glycosylated flavonoids are generally more resistant to heat, while aglycone forms degrade more readily. To mitigate these losses, gentle drying approaches such as vacuum-assisted or low-temperature drying are preferable. Additionally, pre-treatments like steam blanching or immersion in antioxidant solutions can inactivate enzymes such as polyphenol oxidase, which otherwise accelerate phenolic degradation. These approaches help retain the bioactive potential of foods while improving storage stability (Ciupei et al., 2024).

Carotenoids, including beta-carotene, lutein, and lycopene, are another class of bioactive compounds highly affected by drying conditions. Beyond their role in providing vivid natural colors, carotenoids are valued for their provitamin A activity and antioxidant effects. Unfortunately, they are prone to oxidative degradation, isomerization, and color fading when exposed to high temperatures, oxygen, or light (Yildiz and Bayat, 2022). Traditional methods such as sun-drying and hot-air drying often result in substantial carotenoid losses, leading not only to reduced nutritional quality but also to diminished sensory appeal. Advanced drying technologies such as freeze-drying and vacuum drying minimize thermal and oxidative stress, allowing for better retention of carotenoids and ensuring that both the functional and aesthetic qualities of foods are preserved (Yildiz, 2022).

Phenolic acids, including chlorogenic acid and caffeic acid, represent another important category of bioactives with strong antioxidant activity and roles in chronic disease prevention (Yildiz, 2021). These compounds, however, are susceptible to thermal degradation, enzymatic oxidation, and polymerization under harsh drying conditions. Modern methods such as microwave-assisted drying and high-pressure drying have demonstrated superior retention of phenolic acids compared to conventional hot-air drying, largely due to their ability to shorten processing times and reduce oxidative stress (Yildiz and Izli, 2019b; Yavruturk et al., 2024). Other bioactive compounds, such as anthocyanins, glucosinolates, and saponins, exhibit similar sensitivities, with their stability shaped by both intrinsic factors—such as molecular structure, binding within the food matrix, and water activity—and extrinsic processing parameters like temperature and exposure to oxygen (Izli et al., 2018; Enaru et al., 2021; Generalić Mekinić, and Šimat, 2025).

To protect these valuable compounds, various strategies have been implemented. Low-temperature and vacuum drying methods can greatly reduce thermal degradation, while freeze-drying is particularly effective at preserving structural integrity, antioxidant activity, and bioactive compounds, especially in fruits, herbs, and delicate plant tissues. Microwave-assisted drying offers a faster alternative that minimizes heat stress, helping retain phenolics and carotenoids (Yildiz and Izli, 2019a). Pre-treatment techniques are also essential, with antioxidant dips in citric acid, ascorbic acid, or natural extracts forming protective barriers against oxidative losses. Additionally, encapsulation and coating methods are increasingly used to stabilize bioactives during drying and storage, while enabling controlled release during digestion (Baştürk et al., 2021). Collectively, these strategies underscore the importance of combining advanced drying technologies with complementary protective measures to maximize phytochemical retention in dried foods.

Pre-Treatment Strategies to Preserve Nutritional and Functional Qualities

Pre-treatment techniques are essential for preserving the nutritional and functional quality of foods prior to drying. They aim to minimize nutrient losses, inhibit enzymatic browning, prevent microbial growth, and enhance overall product quality (Ladika et al., 2024). By carefully selecting and optimizing pre-treatment methods, food processors can improve the efficacy of subsequent drying steps, ensuring that the final product retains both health-promoting compounds and desirable sensory attributes. These approaches can be physical, chemical, or a combination of methods, depending on the type of food, its composition, and the intended final characteristics (Sławińska, and Olas, 2023).

Blanching is one of the most widely used pre-treatment techniques, involving brief exposure of foods to hot water or steam. This step serves multiple purposes: it inactivates enzymes such as polyphenol oxidase and peroxidase, which are responsible for nutrient degradation and enzymatic browning; it softens plant tissues, improving texture and allowing for uniform dehydration; and, when carefully controlled, it helps retain nutrients such as vitamins and antioxidants by reducing the time needed for subsequent drying. Although some water-soluble vitamins like vitamin C may be partially lost during blanching, optimizing temperature and duration can minimize these effects, making blanching a valuable step for both quality and efficiency (Demircan, and Velioglu, 2024).

Osmotic dehydration is an effective pre-treatment in which foods are immersed in hypertonic solutions, such as sugar or salt, before drying. This process reduces the water content of the food prior to thermal treatment, thereby shortening drying time and limiting heat exposure. Additionally, osmotic dehydration improves sensory qualities by enhancing sweetness and flavor retention and provides a protective environment that reduces oxidation and thermal degradation of sensitive nutrients during subsequent drying. Chemical pre-treatments, including antioxidant dips using ascorbic acid, citric acid, or calcium salts, further help preserve color, texture, and nutrient content by mitigating oxidative reactions and browning. Although sulfite treatments are highly efficient at maintaining color, their use is constrained by regulatory limits and potential health risks, emphasizing the importance of developing safer alternatives (Mari et al., 2024).

Edible coatings and films provide a physical barrier around the food surface, offering protection against moisture loss, nutrient degradation, and textural changes during drying. These coatings, often based on polysaccharides, proteins, or lipids, not only improve structural stability but can also be enriched with functional ingredients such as antioxidants, probiotics, or other bioactive compounds, thereby enhancing the health-promoting properties of dried foods. Physical pre-treatments, including pulsed electric fields (PEF), ultrasound, and microwave-assisted methods, further improve drying efficiency while preserving nutritional quality. Ultrasound facilitates cell disruption, enhancing mass transfer and reducing drying time; PEF induces electroporation, which allows faster moisture removal and protects heat-sensitive compounds; and microwave pre-treatment uniformly heats the food, inactivates enzymes, and shortens drying duration, helping to maintain both vitamin content and antioxidant activity (Kocira et al., 2021).

Synergistic combinations of these pre-treatment strategies often yield superior results compared to individual methods. For example, blanching followed by osmotic dehydration or the application of an edible coating can simultaneously improve nutrient retention, color, and texture. The optimal combination depends on the specific food matrix, target nutrient preservation, and desired final product characteristics. By integrating physical, chemical, and biological pre-treatment techniques, it is possible to maximize the nutritional and functional quality of dried foods while enhancing processing efficiency and consumer appeal (Mari et al., 2025).

Conclusion

Drying remains a cornerstone of food preservation, offering solutions for extending shelf life, reducing post-harvest losses, and ensuring year-round availability of fruits, vegetables, and other perishable foods. However, conventional drying methods often compromise nutritional quality, leading to significant losses of heat-sensitive vitamins, antioxidants, and bioactive compounds. As global demand for nutrient-rich, functional, and convenient foods increases, it becomes essential to adopt innovative drying technologies that prioritize both efficiency and nutrient retention.

Emerging methods such as freeze drying, vacuum drying, microwave-assisted drying, spray drying, and infrared-based techniques demonstrate significant potential in preserving the nutritional and functional properties of foods. Hybrid technologies, which combine the advantages of multiple approaches, offer even greater control over drying kinetics, moisture removal, and bioactive compound preservation. By minimizing thermal and oxidative stress, these methods help maintain vitamins, polyphenols, carotenoids, and other health-promoting compounds, ensuring that dried foods deliver both flavor and functionality.

Beyond nutrient preservation, modern drying innovations contribute to sustainability by reducing energy consumption, minimizing food waste, and supporting environmentally friendly production practices. The integration of encapsulation, precision control, and process optimization further enhances product quality, shelf stability, and consumer acceptance.

Ultimately, the convergence of food science, biotechnology, and engineering in drying technologies provides an opportunity to transform the way foods are processed and consumed. By leveraging these innovations, the food industry can create dried products that not only meet the growing global demand for convenience and longevity but also support human health, dietary diversity, and environmental sustainability. As research continues and technologies evolve, the ability to "lock in" nutrients while removing water will remain central to developing foods that nourish, delight, and sustain.

References

Abreu, D. J. M. d., Lorenço, M. S., Machado, G. G. L., Silva, J. M., Azevedo, E. C. d., & Carvalho, E. E. N. (2025). Influence of Drying Methods on the Post-Harvest Quality of Coffee: Effects on Physicochemical, Sensory, and Microbiological Composition. *Foods*, *14*(9), 1463.

Adetoro, A. O., Opara, U. L., & Fawole, O. A. (2021). Effect of Blanching on Enzyme Inactivation, Physicochemical Attributes and Antioxidant Capacity of Hot-Air Dried Pomegranate (*Punica granatum* L.) Arils (cv. Wonderful). *Processes*, *9*(1), 25.

Alagbe, G., Urbanova, K., & Alagbe, O. (2025). Overview of Fatty Acids and Volatiles in Selected Nuts: Their Composition and Analysis. *Processes*, *13*(8), 2444.

Al-Sharify, Z. T., Al-Najjar, S. Z., Naser, Z. A., Alsherfy, Z. A. I., & Onyeaka, H. (2025). The Impact of Fluid Flow on Microbial Growth and Distribution in Food Processing Systems. *Foods*, *14*(3), 401.

Aravindakshan, S., Nguyen, T. H. A., Kyomugasho, C., Buvé, C., Dewettinck, K., Van Loey, A., & Hendrickx, M. E. (2021). The Impact of Drying and Rehydration on the Structural Properties and Quality Attributes of Pre-Cooked Dried Beans. *Foods*, *10*(7), 1665.

Barel, O., and Yildiz, G. (2024). The effect of ultrasound pre-treatment on the color, antioxidant capacity and total phenolic content of freeze-dried banana slices, *Latin American Applied Research Journal,* 54(3): 329-335.

Baştürk, A., Ceylan, M. M., Alaca, K., and Yıldız, G. (2021). The Role of Selected Bioactive Compounds and Micronutrients with Immune-enhancing Activity on the Prevention and Mitigation of SARS-CoV-2. *European Journal of Nutrition & Food Safety*, 13(12): 25-59.

Berk, Ş.E., Demircan, H., Yıldız, G., Yıldız, G., Taşkın, O., and Oral, R. A. (2023). Microwave-Hot Air Hybrid Drying; The Impact of Pre-treatment Solution on Some Quality Characteristics of White Cherry Fruit. *EJONS International Journal on Mathematic, Engineering and Natural Sciences, 7*(2): 155–165.

Bhardwaj, R. L., Parashar, A., Parewa, H. P., & Vyas, L. (2024). An Alarming Decline in the Nutritional Quality of Foods: The Biggest Challenge for Future Generations' Health. *Foods*, *13*(6), 877.

Bhatta, S., Stevanovic Janezic, T., & Ratti, C. (2020). Freeze-Drying of Plant-Based Foods. *Foods*, *9*(1), 87.

Bolchini, S., Nardin, T., Morozova, K., Scampicchio, M., & Larcher, R. (2025). Antioxidant Maillard Reaction Products from Milk Whey: A Food By-Product Valorisation. *Foods*, *14*(3), 450.

Bourdoux, S., Li, D., Rajkovic, A., Devlieghere, F. and Uyttendaele, M. (2016), Performance of Drying Technologies to Ensure Microbial Safety of Dried Fruits and Vegetables. *Comprehensive Reviews in Food Science and Food Safety,* 15: 1056-1066.

Calín-Sánchez Á, Lipan L, Cano-Lamadrid M, Kharaghani A, Masztalerz K, Carbonell-Barrachina Á. A., Figiel A. Comparison of Traditional and Novel Drying Techniques and Its Effect on Quality of Fruits, Vegetables and Aromatic Herbs. *Foods.* 2020 Sep 9;9(9):1261.

Cemal, D., and Yildiz, G. (2022). The effect of drying temperature, cutting types, storage degree and period on physical, chemical and physicochemical properties of red capia pepper, *Latin American Applied Research Journal,* 52 (1):67-72.

Chemat, A., Song, M., Li, Y., & Fabiano-Tixier, A.-S. (2023). Shade of Innovative Food Processing Techniques: Potential Inducing Factors of Lipid Oxidation. *Molecules*, *28*(24), 8138.

Ciupei, D., Colişar, A., Leopold, L., Stănilă, A., & Diaconeasa, Z. M. (2024). Polyphenols: From Classification to Therapeutic Potential and Bioavailability. *Foods*, *13*(24), 4131.

Coelho, S. C., Estevinho, B. N., & Rocha, F. (2022). Recent Advances in Water-Soluble Vitamins Delivery Systems Prepared by Mechanical Processes (Electrospinning and Spray-Drying Techniques) for Food and Nutraceuticals Applications—A Review. *Foods*, *11*(9), 1271.

Corigliano, O., Morrone, P., & Algieri, A. (2025). Navigating the Challenges of Sustainability in the Food Processing Chain: Insights into Energy Interventions to Reduce Footprint. *Energies*, *18*(2), 296.

Coşkun, N., Sarıtaş, S., Jaouhari, Y., Bordiga, M., & Karav, S. (2024). The Impact of Freeze Drying on Bioactivity and Physical Properties of Food Products. *Applied Sciences*, *14*(20), 9183.

De Rossi, L., Rocchetti, G., Lucini, L., & Rebecchi, A. (2025). Antimicrobial Potential of Polyphenols: Mechanisms of Action and Microbial Responses—A Narrative Review. *Antioxidants*, *14*(2), 200.

Demircan, B., & Velioglu, Y. S. (2024). Control of Browning, Enzyme Activity, and Quality in Stored Fresh-cut Fruit Salads through Chitosan Coating Enriched with Bergamot Juice Powder. *Foods*, *13*(1), 147.

Divya, Joshi, S., Appukuttan, J., Chandrapala, J., & Majzoobi, M. (2025). Impact of Conventional and Advanced Techniques on Stability of Natural Food Colourants. *Foods*, *14*(18), 3187.

Dragoev, S. G. (2024). Lipid Peroxidation in Muscle Foods: Impact on Quality, Safety and Human Health. *Foods*, *13*(5), 797.

Duda-Seiman, C., Mititelu-Tartau, L., Biriescu, S., Almăşan, A. L., Bitu, B. O., Bucur, A. I., Luca, A., Hoinoiu, B., & Hoinoiu, T. (2025). A Comprehensive Study on the Nutritional Profile and Shelf Life of a Custom-Formulated Protein Bar Versus a Market-Standard Product. *Foods*, *14*(12), 2141.

El Hosry, L., Elias, V., Chamoun, V., Halawi, M., Cayot, P., Nehme, A., & Bou-Maroun, E. (2025). Maillard Reaction: Mechanism, Influencing Parameters, Advantages, Disadvantages, and Food Industrial Applications: A Review. *Foods*, *14*(11), 1881.

ElGamal, R., Song, C., Rayan, A. M., Liu, C., Al-Rejaie, S., & ElMasry, G. (2023). Thermal Degradation of Bioactive Compounds during Drying Process of Horticultural and Agronomic Products: A Comprehensive Overview. *Agronomy*, *13*(6), 1580.

Enaru, B., Drețcanu, G., Pop, T. D., Stănilă, A., & Diaconeasa, Z. (2021). Anthocyanins: Factors Affecting Their Stability and Degradation. *Antioxidants*, *10*(12), 1967.

Esteghlal, S., Gahruie, H. H., Niakousari, M., Barba, F. J., Bekhit, A. E.-D., Mallikarjunan, K., & Roohinejad, S. (2019). Bridging the Knowledge Gap for the Impact of Non-Thermal Processing on Proteins and Amino Acids. *Foods*, *8*(7), 262.

Fanzo J., Rudie C., Sigman I., Grinspoon S., Benton T. G., Brown M. E., Covic N., Fitch K., Golden C. D., Grace D., Hivert M. F., Huybers P., Jaacks L. M., Masters W. A., Nisbett N., Richardson R. A., Singleton C. R., Webb P., Willett W. C. Sustainable food systems and nutrition in the 21st century: a report from the 22nd annual Harvard Nutrition Obesity Symposium. *AmericanJournal of Clinical Nutrition* 2022 Jan 11;115(1):18-33.

Fekete, M., Lehoczki, A., Kryczyk-Poprawa, A., Zábó, V., Varga, J. T., Bálint, M., Fazekas-Pongor, V., Csípő, T., Rząsa-Duran, E., & Varga, P. (2025). Functional Foods in Modern Nutrition Science: Mechanisms, Evidence, and Public Health Implications. *Nutrients*, *17*(13), 2153.

Feng, B., An, D., Cheng, H., Zhang, X., & Zhao, J. (2025). Leaching Behavior and Mechanisms of Li, Rb, K, Sr, and Mg in Clay-Type Lithium Ore via a Roasting–Water Leaching Process. *Minerals*, *15*(9), 944.

Fernandes, L., & Tavares, P. B. (2024). A Review on Solar Drying Devices: Heat Transfer, Air Movement and Type of Chambers. *Solar*, *4*(1), 15-42.

Feszterová, M., Kowalska, M., & Mišiaková, M. (2023). Stability of Vitamin C Content in Plant and Vegetable Juices under Different Storing Conditions. *Applied Sciences*, *13*(19), 10640.

Gao, J. R., Li, M. Y., Cheng, Z. Y., Liu, X. Y., Yang, H., Li, M. T., He, R. Y., Zhang, Q., & Yang, X. H. (2024). Effects of Different Drying Methods on Drying Characteristics and Quality of Small White Apricot (*Prunus armeniaca* L.). *Agriculture*, *14*(10), 1716.

Generalić Mekinić, I., & Šimat, V. (2025). Bioactive Compounds in Foods: New and Novel Sources, Characterization, Strategies, and Applications. *Foods*, *14*(9), 1617.

Gómez, I., Janardhanan, R., Ibañez, F. C., & Beriain, M. J. (2020). The Effects of Processing and Preservation Technologies on Meat Quality: Sensory and Nutritional Aspects. *Foods*, *9*(10), 1416.

Grassino, A. N., Karlović, S., Šošo, L., Dujmić, F., Sabolović, M. B., Marelja, M., & Brnčić, M. (2024). Influence of Different Drying Processes on the Chemical and Texture Profile of *Cucurbita maxima* Pulp. *Foods*, *13*(4), 520.

Hajzer, Z. E., Alibrahem, W., Kharrat Helu, N., Oláh, C., & Prokisch, J. (2025). Functional Foods in Clinical Trials and Future Research Directions. *Foods*, *14*(15), 2675.

Hu, X., Zhang, B., Du, X., Zhang, H., Zhu, T., Zhang, S., Yang, X., Zhang, Z., Yang, T., Wang, X., & Wu, C. (2025). Recent Advances and Future Perspectives on Heat and Mass Transfer Mechanisms Enhanced by Preformed Porous Media in Vacuum Freeze-Drying of Agricultural and Food Products. *Foods*, *14*(17), 2966.

Izli, G., Izli, N., Taskin, O., and Yildiz, G. (2018). Convective drying of kumquat slices: Comparison of different drying temperatures on drying kinetics, colour, total phenolic content and antioxidant capacity. *Latin American Applied Research Journal,* 48:37-42.

İzli, G., and Yildiz, G. (2021). Evaluation of the high intensity ultrasound pre-treatment effects on the physical properties and bioactive compounds of convective dried quince samples, *International Journal of Fruit Science*, 21(1): 645-656.

İzli, G., Yildiz, G., and Berk, S. E. (2022). Quality retention in pumpkin powder dried by combined microwave-convective drying, *Journal of Food Science and Technology,* 59 (4): 1558-1569.

Jia, Z., Zhou, J., Wang, W., Liu, D., Zheng, X., Hu, M., Jiang, Y., & Lv, R. (2024). Optimization of Vacuum Freeze-Drying Process and Quality Evaluation of *Stropharia rugosoannulata*. *Applied Sciences*, *14*(22), 10158.

Kaźmierczak-Barańska, J., Halczuk, K., & Karwowski, B. T. (2025). Thiamine (Vitamin B1)—An Essential Health Regulator. *Nutrients*, *17*(13), 2206.

Kocira, A., Kozłowicz, K., Panasiewicz, K., Staniak, M., Szpunar-Krok, E., & Hortyńska, P. (2021). Polysaccharides as Edible Films and Coatings: Characteristics and Influence on Fruit and Vegetable Quality—A Review. *Agronomy*, *11*(5), 813.

Kulczyński, B., Suliburska, J., Gramza-Michałowska, A., Sidor, A., Kowalczewski, P. Ł., & Brzozowska, A. (2025). The Effect of Osmotic Dehydration Conditions on the Magnesium Content in Beetroot (*Beta vulgaris* L.). *Molecules*, *30*(14), 3051.

Ladika, G., Tsiaka, T., Stavropoulou, N. A., Strati, I. F., & Sinanoglou, V. J. (2024). Enhancing the Nutritional Value and Preservation Quality of Strawberries through an Optimized Osmotic Dehydration Process. *Applied Sciences*, *14*(20), 9211.

Li, Q., Li, L., Lai, P., Wei, Y., Lai, C., Liu, Y., Yang, M., Zhou, S., Chen, J., & Sun, J. (2025). Effects of Different Drying Methods on Physicochemical Properties and Nutritional Quality of Abalone Bioactive Peptides. *Molecules*, *30*(7), 1516.

Lisboa, H. M., Pasquali, M. B., dos Anjos, A. I., Sarinho, A. M., de Melo, E. D., Andrade, R., Batista, L., Lima, J., Diniz, Y., & Barros, A. (2024). Innovative and Sustainable Food Preservation Techniques: Enhancing Food Quality, Safety, and Environmental Sustainability. *Sustainability*, *16*(18), 8223.

Mahfoud, F., Frem, J., Assaf, J. C., Maache-Rezzoug, Z., Rezzoug, S.-A., Elias, R., Debs, E., & Louka, N. (2025). Evolution of Dried Food Texturization: A Critical Review of Technologies and Their Impact on Organoleptic and Nutritional Properties. *Processes*, *13*(3), 799.

Mansour, N. E., Villagran, E., Rodriguez, J., Akrami, M., Flores-Velazquez, J., Metwally, K. A., Alhumedi, M., Ahmed, A. F., & Elshawadfy Elwakeel, A. (2025). Effect of Drying Conditions on Kinetics, Modeling, and Thermodynamic Behavior of Marjoram Leaves in an IoT-Controlled Vacuum Dryer. *Sustainability*, *17*(13), 5980.

Mari, A., Parisouli, D. N., & Krokida, M. (2024). Exploring Osmotic Dehydration for Food Preservation: Methods, Modelling, and Modern Applications. *Foods*, *13*(17), 2783.

Mari, A., Manta, E., & Krokida, M. (2025). Innovative Microalgae-Based Edible Coatings with Encapsulated Bioactives: Enhancing Fresh Raspberry Shelf Life and Quality. *Processes*, *13*(4), 1193.

Moon, K. M., Kwon, E.-B., Lee, B., & Kim, C. Y. (2020). Recent Trends in Controlling the Enzymatic Browning of Fruit and Vegetable Products. *Molecules*, *25*(12), 2754.

Moutia, I., Lakatos, E., & Kovács, A. J. (2024). Impact of Dehydration Techniques on the Nutritional and Microbial Profiles of Dried Mushrooms. *Foods*, *13*(20), 3245.

Nainggolan, E. A., Banout, J., & Urbanova, K. (2024). Recent Trends in the Pre-Drying, Drying, and Post-Drying Processes for Cassava Tuber: A Review. *Foods*, *13*(11), 1778.

Neji, C., Semwal, J., Kamani, M. H., Máthé, E., & Sipos, P. (2022). Legume Protein Extracts: The Relevance of Physical Processing in the Context of Structural, Techno-Functional and Nutritional Aspects of Food Development. *Processes*, *10*(12), 2586.

Nielsen, S. D., Knudsen, L. J., Bækgaard, L. T., Rauh, V., & Larsen, L. B. (2022). Influence of Lactose on the Maillard Reaction and Dehydroalanine-Mediated Protein Cross-Linking in Casein and Whey. *Foods*, *11*(7), 897.

Nowacka, M., Dadan, M., & Tylewicz, U. (2023). Drying Technologies in Food Processing. *Applied Sciences*, *13*(19), 10597.

Nwankwo, C. S., Okpomor, E. O., Dibagar, N., Wodecki, M., Zwierz, W., & Figiel, A. (2023). Recent Developments in the Hybridization of the Freeze-Drying Technique in Food Dehydration: A Review on Chemical and Sensory Qualities. *Foods*, *12*(18), 3437.

Paparella, R., Panvino, F., Leonardi, L., Pucarelli, I., Menghi, M., Micangeli, G., Tarani, F., Niceta, M., Rasio, D., Pancheva, R., Fiore, M., & Tarani, L. (2025). Water-Soluble Vitamins: Hypo- and Hypervitaminosis in Pediatric Population. *Pharmaceutics*, *17*(1), 118.

Petković, M., Lukyanov, A., Đurović, I., & Miletić, N. (2025). A Novel Method for Analyzing the Kinetics of Convective/IR Bread Drying (CIRD) with Sensor Technology. *Applied Sciences*, *15*(9), 4964.

Postružnik, V., Stajčić, S., Borjan, D., Ćetković, G., Knez, Ž., Knez Marevci, M., & Vulić, J. (2024). Impact of Storage Conditions on Stability of Bioactive Compounds and Bioactivity of Beetroot Extract and Encapsulates. *Processes*, *12*(7), 1345.

Prosapio, V., & Lopez-Quiroga, E. (2020). Freeze-Drying Technology in Foods. *Foods*, *9*(7), 920.

Puente-Díaz, L., Spolmann, O., Nocetti, D., Zura-Bravo, L., & Lemus-Mondaca, R. (2020). Effects of Infrared-Assisted Refractance Window™ Drying on the Drying Kinetics, Microstructure, and Color of *Physalis* Fruit Purée. *Foods*, *9*(3), 343.

Radojčin, M., Pavkov, I., Bursać Kovačević, D., Putnik, P., Wiktor, A., Stamenković, Z., Kešelj, K., & Gere, A. (2021). Effect of Selected Drying Methods and Emerging Drying Intensification Technologies on the Quality of Dried Fruit: A Review. *Processes*, *9*(1), 132.

Qi, Y., Wang, W., Yang, T., Ding, W., & Xu, B. (2025). Maillard Reaction in Flour Product Processing: Mechanism, Impact on Quality, and Mitigation Strategies of Harmful Products. *Foods*, *14*(15), 2721.

Rana, M. S., Rahman, A. N. M. A., Ahmed, R., Hossain, M. P., Shadman, M. S., Majumdar, P. K., Islam, K. S., & Colton, J. (2024). Design, Fabrication, and Performance Evaluation of a Food Solar Dryer. *AgriEngineering*, *6*(4), 4506-4523.

Razzaque, M. S., & Wimalawansa, S. J. (2025). Minerals and Human Health: From Deficiency to Toxicity. *Nutrients*, *17*(3), 454.

Roohinejad, S., Koubaa, M., & Gharibzahedi, S. M. T. (2025). Drying Technologies for *Stevia rebaudiana* Bertoni: Advances, Challenges, and Impacts on Bioactivity for Food Applications—A Review. *Foods*, *14*(16), 2801.

Sablani, S. S. (2006). Drying of Fruits and Vegetables: Retention of Nutritional/Functional Quality. *Drying Technology*, 24(2), 123–135.

Santander, M., Chica, V., Correa, H. A. M., Rodríguez, J., Villagran, E., Vaillant, F., & Escobar, S. (2025). Unravelling Cocoa Drying Technology: A Comprehensive Review of the Influence on Flavor Formation and Quality. *Foods*, *14*(5), 721.

Santos, A. A. d. L., Leal, G. F., Marques, M. R., Reis, L. C. C., Junqueira, J. R. d. J., Macedo, L. L., & Corrêa, J. L. G. (2025). Emerging Drying Technologies and Their Impact on Bioactive Compounds: A Systematic and Bibliometric Review. *Applied Sciences*, *15*(12), 6653.

Schefer, S., Oest, M., & Rohn, S. (2021). Interactions between Phenolic Acids, Proteins, and Carbohydrates—Influence on Dough and Bread Properties. *Foods*, *10*(11), 2798.

Shahidi, F., & Hossain, A. (2022). Role of Lipids in Food Flavor Generation. *Molecules*, *27*(15), 5014.

Sławińska, N., & Olas, B. (2023). Selected Seeds as Sources of Bioactive Compounds with Diverse Biological Activities. *Nutrients*, *15*(1), 187.

Versino, F., Ortega, F., Monroy, Y., Rivero, S., López, O. V., & García, M. A. (2023). Sustainable and Bio-Based Food Packaging: A Review on Past and Current Design Innovations. *Foods*, *12*(5), 1057.

Wijegunawardhana, D., Wijesekara, I., Liyanage, R., Truong, T., Silva, M., & Chandrapala, J. (2024). Process-Induced Molecular-Level Protein–Carbohydrate–Polyphenol Interactions in Milk–Tea Blends: A Review. *Foods*, *13*(16), 2489.

Xie, M., Chen, Y., Sun, Y., Gao, Y., Wu, Z., Wu, R., Li, R., Hong, S., Wang, M., Zou, Y., Zhang, H., & Xiong, Y. (2023). Effect of Drying Kinetics, Volatile Components, Flavor Changes and Final Quality Attributes of *Moslae herba* during the Hot Air Thin-Layer Drying Process. *Molecules*, *28*(9), 3898.

Yang, Q., Yi, X., Xiao, H., Wang, X., Liu, L., Tang, Z., Hu, C., & Li, X. (2024). Effects of Different Drying Methods on Drying Characteristics, Microstructure, Quality, and Energy Consumption of Apricot Slices. *Foods*, *13*(9), 1295.

Yavruturk, M., Celik, P., Kazanci, H., Demiragac, R., and Yildiz, G. (2024). Potential uses of dried apricot pulp in obtaining functional products, *Latin American Applied Research Journal,* 54(4): 525–530.

Yıldız, G., İzli, G., Çavuş, M., and Ceylan, M.M. (2021). The Effect of Ultrasound Pre-treatment on the Quality Characteristics of Dried Iğdır Apricot. *Journal of the Institute of Science and Technology,* 11(1):303-313.

Yildiz, G., Ding, J., Gaur, S., Andrade, J., Engeseth, N. J., and Feng, H. (2018). Microencapsulation of docosahexaenoic acid (DHA) with four wall materials including pea protein-modified starch complex. *International Journal of Biological Macromolecules,* 114: 935–941.

Yildiz, G., and Izli, G. (2019a). The effect of ultrasound pretreatment on quality attributes of freeze-dried quince slices: Physical properties and bioactive compounds. *Journal of Food Process Engineering,* 42 (5): e13223.

Yildiz, G., and Izli, G. (2019b). Influence of microwave and microwave-convective drying on the drying kinetics and quality characteristics of pomelo. *Journal of Food Processing and Preservation,* 43 (6): e13812.

Yildiz, G. (2021). The Effect of Novel Microwave-Convective Drying on the Functional Properties of Dried-Pears (*Pyrus communis*), *Current Microwave Chemistry,* 8(1):22-26.

Yildiz, G. (2022). Color, microstructure, physicochemical, textural and sensory properties with the retention of secondary metabolites in convective-, microwave- and freeze-dried carrot (Daucus carota) slices, *British Food Journal,* 124 (11): 3922-3935.

Yildiz, G., and Bayat, G. (2022). The effect of different drying methods on the production and quality characteristics of home-made kurut, a traditional dried dairy product, *Latin American Applied Research Journal,* 52(4):365-369,

Yildiz, G., Gao, Y., Ding, J., Zhu, S., Chen, G., and Feng, H. (2025). Enhancing physicochemical, bioactive, and nutritional properties of sweet potatoes: Ultrasonic contact drying with slot jet nozzles compared to hot-air drying and freeze drying, *Ultrasonics Sonochemistry, 112*: 107216.

Zhang, X., & Liu, S.-Q. (2024). Effects of Reducing Sugars on Colour, Amino Acids, and Volatile Flavour Compounds in Thermally Treated Minced Chicken Carcass Hydrolysate. *Foods*, *13*(7), 991.

Zhang, X., Yin, Z., Xiang, S., Yan, H., & Tian, H. (2024). Degradation of Polymer Materials in the Environment and Its Impact on the Health of Experimental Animals: A Review. *Polymers*, *16*(19), 2807.

Zhao, J., Qin, X., Liu, Y., He, Q., Qin, J., Shen, F., & Wu, Z. (2025). Comparative Evaluation of Spray-Drying Versus Freeze-Drying Techniques on the Encapsulation Efficiency and Biofunctional Performance of Chenpi Extract Microcapsules. *Foods*, *14*(10), 1825.

Zhao, C., He, Z., Song, X., Zhang, X., Xiao, Y., Yu, J., Yang, M., & Tang, Z. (2025). Evaluation of Different Drying Methods on the Quality Parameters of *Acanthopanax senticosus* Fruits. *Foods*, *14*(7), 1100.

List of Contributors

Arzu Imece
Igdir University, Turkiye

Ayse Betul Demirbas Bilen
Istanbul Atlas University, Turkiye

Ayse Nur Elmaskaya
Igdir University, Turkiye

Bedriye Ural
University of Health Sciences, Turkiye

Derya Dikmen
Hacettepe University, Turkiye

Dilara Sokmen
Igdir University, Turkiye

Elif Feyza Topdas
Ataturk University, Turkiye

Emel Ucarer
Igdir University, Turkiye

Gulcin Yildiz
Igdir University, Iğdır, Turkiye

Gulen Ecem Kalkan
Istanbul Atlas University, Turkiye

Gulperi Demir Hakli
Selcuk University, Turkiye

Hakan Toguc
Inonu University, Turkiye

Memnune Sengul
Ataturk University, Turkiye

Merve Aydin Mende
Hacettepe University, Turkiye

Rabia Kaya Celik
Igdir University, Turkiye

Seda Ufuk
Ataturk University, Turkiye

Tuba Erkaya Kotan
Ataturk University, Turkiye

Umut Uyar
Igdir University, Turkiye

About the Editors

Dr. Gulcin Yildiz
Iğdır University, Turkiye

Dr. Gulcin Yildiz is an associate professor at Iğdır University with a Ph.D. and M.S. in food science from the University of Illinois at Urbana-Champaign, USA. She is currently conducting postdoctoral research in the United States. Her expertise spans functional foods, non thermal food processing (e.g., ultrasound, high-pressure), plant-based proteins, drying technologies, and food sustainability. Dr. Yildiz has published over 60 peer-reviewed articles, edited international books, and contributed extensively to book chapters on bioactive compounds and novel processing methods. She has received multiple TÜBİTAK International Publication Incentive Awards, a Best Researcher Award (USA, 2025), and other honors. She serves as associate editor and reviewer for leading journals and actively participates in European COST Actions related to fermented foods and green chemical engineering. Dr. Yildiz mentors graduate students and leads national and international projects funded by TÜBİTAK and USDA.

Dr. Arzu İmece
Iğdır University, Turkiye

Dr. Arzu İmece is an assistant professor in Gastronomy and Culinary Arts at Iğdır University. She earned her Ph.D. (2022) and M.S. (2016) in food engineering from Atatürk University, focusing on bioactive compounds affecting food quality and health. Her research includes food chemistry, processing technologies, bioactives, and public health, with publications on bee products, hydrogen-rich water, and functional foods such as black garlic and cherry. Dr. İmece teaches courses like menu planning and food processing, mentors students, and contributes to TÜBİTAK-funded multidisciplinary projects.

Index

A

B

C

D

E

Z

More books by Gulcin Yildiz can be found at:

https://novapublishers.com/writer/gulcin-yildiz/

More books by Arzu Imece can be found at:

https://novapublishers.com/writer/arzu-imece/